THE

INTERNATIONAL SERIES

OF

MONOGRAPHS ON PHYSICS

# PARTICLE-INTERACTION PHYSICS AT HIGH ENERGIES

BY

S. J. LINDENBAUM

SENIOR PHYSICIST, BROOKHAVEN NATIONAL LABORATORY, UPTON, NEW YORK

*and*

MARK W. ZEMANSKY PROFESSOR OF PHYSICS, THE CITY COLLEGE, THE CITY UNIVERSITY OF NEW YORK



CLARENDON PRESS · OXFORD

1973

*Oxford University Press, Ely House, London W. 1*

GLASGOW NEW YORK TORONTO MELBOURNE WELLINGTON
CAPE TOWN IBADAN NAIROBI DAR ES SALAAM LUSAKA ADDIS ABABA
DELHI BOMBAY CALCUTTA MADRAS KARACHI LAHORE DACCA
KUALA LUMPUR SINGAPORE HONG KONG TOKYO





FILMSET AT THE UNIVERSITIES PRESS, BELFAST.
PRINTED BY PHOTOLITHOGRAPHY AND BOUND IN
GREAT BRITAIN AT THE PITMAN PRESS, BATH.

# PREFACE

This book had its origins in a suggestion by Professor Denys Wilkinson in 1958 that the author write a short research monograph on pion physics. At that time, it appeared to us that pion physics had reached a relatively mature state and could be treated as a separate subject. Of course there were the unsolved, and not clearly related, phenomena of considerably smaller cross-section connected with curious or strange particles. As the book on pion physics progressed, the rapid advances made both experimentally and theoretically in strange particle research made it clear that pion physics was 'just the tip of the iceberg', and that in order to view the field properly a more complete and integrated work was required. Thus, several years after the work was begun, we agreed that a larger monograph was in order.

One of the author's major objectives was to treat the important aspects of the subject in proper historical perspective so that the reader could develop a good overall view of the field as well as a knowledge of its present state. Another major objective was to emphasize the integration of the development and status of both the experimental and theoretical aspects of the field.

Obviously, a work of this scope, which has extended over more than a decade, cannot be continuously updated in all aspects and the references cited give the reader a good indication of the status of the various sections. However, as the reader will note, the author has, in a number of instances which he considers particularly important, made a greater effort to update the material.

Although primarily developed as a research monograph, the author has, during the academic year 1970–1971, used the bulk of the material in this book as a basis for teaching graduate (second- and third-year level) courses in particle physics and advanced particle physics, in the City College of City University of New York. Encouraged by the students' reaction to the material, the author believes that this book has some merit as a graduate text or as one of several references for graduate courses in particle physics.

Many references are to other relevant books and review articles which cover various portions of the subject material, but obviously in a work of this size some references have been omitted. Similarly, it would have been impractical to include every research reference although the author has made an effort to make them as complete as possible.

The subject matter of the book ranges over so wide a field that it is not necessarily desirable or convenient to maintain a unique notation or set of units throughout. Rather, an attempt is made to facilitate understanding and use of the bibliography. To some extent, variations in notations are employed and this is made clear by the text and/or references cited. It is quite usual in the literature to set $\hbar = c = 1$. This leads to so-called 'natural units'. One can still fix one more quantity—either a mass, length, or time. A common set of natural units also sets the mass of the pion equal to 1. We will often make use of natural units when this simplifies the equations and follows customary usage in this field. In the earlier chapters, in particular, we shall generally retain the $\hbar$ and the $c$ (i.e. use classical cgs units). In radiation theory, it is common to employ Gaussian units which are unrationalized classical cgs units in which electric quantities are measured in esu and magnetic quantities in emu. One should note that in Gaussian units the electromagnetic fine structure constant is

$$\alpha = \frac{e^2}{\hbar c} = \tfrac{1}{137}.$$

The corresponding unrationalized coupling constants in meson theory are $g^2$ for direct coupling and $f^2$ for gradient coupling, where $g$ and $f$ are analogous to the mesonic charge. The exact definitions are given in the text. For convenience, we also make use of the rationalized dimensionless coupling constants $G^2$ and $F^2$ which are defined as

$$G^2 = 4\pi g^2 \quad \text{and} \quad F^2 = 4\pi f^2.$$

Three-vectors, and the three-vector part of a four-vector are indicated by using boldface type for the letter. Four-vectors are indicated by a Greek index such as $P_\mu$ where it is understood that $\mu = 0, 1, 2, 3$ (often authors use 4 instead of 0 for the time component). At times the index is omitted when it is obviously understood. Alternatively, when one wishes to distinguish several different four-vectors of the same type (such as P1, P2, P3, P4) a number or symbol will be used to distinguish them and it is to be understood that each of these quantities is a four-vector.

References will be consecutively numbered and assembled at the end of each chapter. A copy of relevant excerpts from the Particle Data Group tables will be found in an appendix.

The author wishes to particularly thank Professor Denys Wilkinson and the Clarendon Press for their valuable cooperation and suggestions. Their patience in waiting a decade for the manuscript was greatly appreciated. The author also wishes to thank the various authors and publishers for allowing reprints of relevant figures and material.

My wife, Leda, has assisted, substantially, in editing, preparing, and organizing the references.

S. J. L.

*New York*
*January 1972*

# CONTENTS

# 1

# INTRODUCTION TO PION PHYSICS

## 1.1. Yukawa's nuclear force meson

THE first domain investigated in the history of modern high energy physics was the study of pion physics. Pion physics began in 1935 in a paper [1] published by Yukawa. He was concerned with the explanation of the greatest physical mystery of the time, namely, the observed strong short-range forces acting between nucleons. All of the previously investigated (electromagnetic and gravitational) forces† involved (or appeared to involve) long-range forces where

$$\text{force} \sim \frac{\text{constant}}{r^2}.$$

On the other hand, the forces between nucleons seemed to be well defined in range (the nuclear force range is approximately $1{\cdot}4 \times 10^{-13} A^{\frac{1}{3}}$ cm, where $A$ is the atomic number).

Yukawa brilliantly demonstrated that the reason the electromagnetic force was long-range was a consequence of the zero rest mass of the photon, which was the field 'quantum' responsible for the electromagnetic interaction, and that the range of a force was intimately related to the rest mass of the quanta of the field responsible for the interaction.

The equation he considered as the wave equation for a non-zero rest mass particle is just the relativistic Klein–Gordon equation. This equation can easily be obtained by the usual quantum mechanical substitution of operators for observables in the corresponding classical equation, using the well known prescription

$$p = \frac{\hbar}{\mathrm{i}} \nabla, \tag{1.1}$$

$$E = \mathrm{i}\hbar \frac{\partial}{\partial t}, \tag{1.2}$$

and allowing the operators to operate on the wave function or state vector. The relativistic classical equation is

$$E^2 = p^2c^2 + m_0^2c^4,$$

where $E$ is the total energy, $p$ is the momentum, $c$ is the velocity of light, and

† It is obvious that the concept of force is a classical one, and when referred to in quantum phenomena it is a qualitative term.

$m_0$ is the particle rest mass. With the substitutions (1.1) and (1.2) we obtain

$$\left(\nabla^2 - \frac{m_0^2 c^2}{\hbar^2} - \frac{1}{c^2}\frac{\partial^2}{\partial t^2}\right)\varphi = 0, \tag{1.3}$$

where $\varphi$ is the particle wave function and is here considered to be a one-component wave function. Since we require invariance under Lorentz transformations, $\varphi$ must be a scalar or a pseudoscalar (a scalar quantity which changes sign upon space inversion or time inversion). Therefore $\varphi$ is suitable for describing spin-zero particles which require only a one component wave function. The nucleon spin wave functions which describe spin $\frac{1}{2}$ particles must also satisfy the Klein–Gordon equation as well as the Dirac equation. This can be deduced by noting that the correspondence principle of quantum mechanics requires that, at high quantum numbers, the proper relativistic quantum mechanics satisfies the classical relativistic equation. However, the nucleon spinor wave function has four components—two for the two spin positions of the particle and two for the two spin positions of the anti-particle. The pion is now known to be a spin-zero particle of negative intrinsic parity, therefore it must be described by a pseudoscalar wave function. If the intrinsic parity was positive, instead, we would have a scalar wave function for the pion.

We shall first develop the static scalar meson theory, since this will most easily illustrate the general principles. The Klein–Gordon equation describes a free pion field without sources—compare it with the electromagnetic scalar potential equation without sources

$$\left(\nabla^2 - \frac{1}{c^2}\frac{\partial^2}{\partial t^2}\right)\varphi = 0. \tag{1.4}$$

It is obvious that setting $m_0 = 0$ in the Klein–Gordon equation would yield the electromagnetic wave equation. The photon has a zero rest mass, but we must be careful of the above analogy since the photon has spin one. This spin property is reflected in the fact that the photon is described by a vector potential $A$ in addition to the scalar potential, and there is a corresponding vector potential wave equation. However, since we know experimentally that the pion has spin zero, the Klein–Gordon equation is the correct relativistic wave equation to describe it.

If we regard an infinitely heavy nucleon† as the source of the pion field,

† Owing to the Heisenberg uncertainty principle, if the nucleon is not considered infinitely heavy compared to the pion it cannot be localized by a $\delta$-function at the origin. For the actual physical case, the point nucleon has to be replaced by a properly smeared out distribution, but in this case we cannot have a Lorentz invariant theory.

analogously to the fixed point charge sources of the photon field in electromagnetic theory, the Klein–Gordon equation for scalar mesons with an infinitely heavy nucleon source of pions at the point $r_1$, becomes

$$\left(\nabla^2 - \frac{m_0^2 c^2}{\hbar^2} - \frac{1}{c^2}\frac{\partial^2}{\partial t^2}\right)\varphi = -4\pi g_1\, \delta(\mathbf{r}-\mathbf{r}_1)\, dr, \tag{1.5}$$

where $g_1$ is the mesonic charge analogous to $e$ in the electromagnetic case.

Now let us look for the static solution. By setting $\frac{\partial}{\partial t} = 0$, we find

$$\varphi(\mathbf{r}, \mathbf{r}_1) = -\frac{g_1}{|\mathbf{r}-\mathbf{r}_1|}\exp\left(\frac{-|\mathbf{r}-\mathbf{r}_1|}{\hbar/m_\pi c}\right). \tag{1.6}$$

The potential of a second nucleon located at point $r$ is, again analogously to the electromagnetic case,

$$V = g_2\varphi(\mathbf{r}, \mathbf{r}_1) = \frac{-g_1 g_2}{|\mathbf{r}-\mathbf{r}_1|}\exp\left(\frac{-|\mathbf{r}-\mathbf{r}_1|}{\hbar/m_\pi c}\right). \tag{1.7}$$

Since $g_1 - g_2 = g$,

$$V = \frac{-g^2}{|\mathbf{r}-\mathbf{r}_1|}\exp\left(\frac{-|\mathbf{r}-\mathbf{r}_1|}{\hbar/m_\pi c}\right), \tag{1.8}$$

where $m_\pi$ is the pion rest mass.

Eqn (1.8) represents the Yukawa potential. We observe that, if $m \to 0$, this becomes of the same form as the ordinary Coulomb potential. However, for $m$ greater than zero, this potential is characterized by a nominal range hereby defined as $\hbar/m_\pi c$ (i.e. the Compton wavelength of the particle), such that the potential decreases exponentially with this range in addition to the $\frac{1}{r}$ Coulomb potential distance dependence. We should note, also, that this potential is attractive. This line of reasoning led Yukawa to postulate the existence of a new particle or meson (which is now referred to as a pion) which was strongly coupled to nucleons with a sufficiently large coupling constant to explain the strength of nuclear forces and of mass such that the observed range of nuclear force was explained.

Yukawa further hypothesized that the meson which was created then decayed by electron emission with a lifetime $\sim 10^{-8}$ s through weak $\beta$-decay interaction.

## 1.2. Effective range of nucleon–nucleon forces

A careful phenomenological analysis, both theoretically and experimentally, of the nucleon–nucleon forces at the low incident kinetic energies

($\lesssim$10 MeV) has been made within the framework of the so-called 'effective-range theory'. This analysis demonstrates that the low energy nucleon–nucleon interaction (i.e. where only $s$-waves† are important) can be characterized by two parameters—a scattering length and an effective range. The effective range has been determined for various potential-well shapes for both the nucleon–nucleon singlet (total spin = 0) and the triplet (total spin = 1) interaction. It has been found that in the nucleon–nucleon scattering, for incident kinetic energy less than 10 MeV, each of the two spin states (singlet or triplet)‡ can be well described (to within a few per cent, or better) by the so-called 'shape independent approximation',

$$k \cot \delta = \frac{-1}{a} + \tfrac{1}{2} r_0 k^2, \tag{1.9}$$

where $k$ is the local wavenumber $\left(\text{i.e. } k = \frac{1}{\lambda\!\!\!^{-}}\right)$ of the relative motion and $a$ is the Fermi scattering length, which is a constant and is determined by the relationship that the total scattering cross-section at zero kinetic energy is

$$\sigma(E = 0) = 4\pi a^2. \tag{1.10}$$

Denoting $\delta$ as the $s$-wave phase shift in radians, $\delta$ is related to the spherically symmetric (in the centre of mass system—from now on referred to as the c.m.s.) $s$-wave total scattering cross-section by

$$\sigma_{\text{total}} = 4\pi k^2 \sin^2\delta. \tag{1.11}$$

$r_0$ is defined in eqn (1.9) by

$$r_0 = \int_0^\infty (v_0^2 - u_0^2)\, \mathrm{d}r, \tag{1.12}$$

where $\dfrac{v_0}{r}$ is the wave function inside the nuclear-force well at zero incident kinetic energy, and $\dfrac{u_0}{r}$ is the asymptotic wave function, outside the range

† The effective-range analysis has been extended (Hulthen and Sugawara [2]) to include the small amounts of $d$-wave coupled to the $s$-wave in the n–p triplet spin state by the non-central tensor force. It is shown that a consistent effective-range theory can be developed even when this $d$-wave is taken into account. The inclusion of the $d$-wave is of course essential to an explanation of the deuteron quadruple moment.

‡ This statement is independent of the charge state of the individual nucleons provided charge independence is valid for nuclear interactions. This assumption agrees with experiments within their errors and allowing for the estimated electromagnetic effects which are considerably smaller than nuclear effects but obviously violate charge independence.

of nuclear force at zero incident kinetic energy. The relationship between the two scattering parameters and the binding energy of the deuteron can be demonstrated to be ([2]–[4])

$$\frac{\sqrt{(2\,\mu\,|B|)}}{\hbar} = \frac{1}{a} + \tfrac{1}{2}r_0\,\frac{(2\,|B|)}{\hbar^2}, \tag{1.13}$$

where $\mu$ is the reduced mass and $B$ is the deuteron binding energy.

Many potential shapes have been investigated but the most commonly used are

Yukawa well $$V(r) = \frac{-V_0 e^{-r/\beta}}{r/\beta}, \tag{1.14}$$

Gaussian well $$V(r) = -V_0 e^{-r^2/\beta^2}, \tag{1.15}$$

Exponential well $$V(r) = -V_0 e^{-r/\beta}, \tag{1.16}$$

Square well $$\left.\begin{array}{ll} V(r) = -V_0 & r < \beta \quad (1.17)\\ V(r) = 0 & r > \beta \quad (1.18)\end{array}\right\}.$$

Any of these potential shapes and others (provided the decrease of potential with $r$ is fast enough) can be used to determine the two parameters $a$ and $r_0$. Conversely, given an $a$ and $r_0$, determined from the experimental data, then a $V_0$ and $\beta$ can be found for any of the wells considered. Hence we can say that the low energy experimental data determine only the depth and range of the potential.

Since it has been found that the binding energy of the deuteron ($B = 2{\cdot}1$ MeV) is small, compared to the triplet well depth ($V \approx -(20\text{–}30$ MeV)), and the virtual singlet level also has an energy ($\sim 0{\cdot}5$ MeV) small compared to the singlet well depth, it is useful to consider the concept of intrinsic range ($b$) which is defined as the limit of the effective range when the well depth is reduced just to the point where a state occurs at zero binding energy.

Hence,

$$b = \lim_{B \to 0} r_0. \tag{1.19}$$

The scattering length ($|a|$) tends to infinity, and its sign is positive for $B > 0$ and negative for $B < 0$.

The usefulness of the intrinsic range concept is that it allows definite unique comparisons between the intrinsic ranges of the various well shapes and, provided $\frac{|B|}{|V|} \ll 1$, the effective ranges are close to the intrinsic ranges.†

† For a bound state (positive scattering length) such as the n–p triplet the effective range is smaller than the intrinsic range. For a virtual state, such as the n–p, p–p, or n–n singlet (negative scattering length), the effective range is larger than the intrinsic range.

Furthermore, for a square well the intrinsic range is, identically, the well range. For the wells we have considered, we find

| | |
|---|---|
| Yukawa well | $b = 2{\cdot}120\beta$, |
| Gaussian well | $b = 1{\cdot}435\beta$, |
| Exponential well | $b = 3{\cdot}54\beta$, |
| Square well | $b = \beta$, |

we also find that the longer the tail of the well the larger the ratio of the intrinsic range to the parameter $\beta$.

The triplet n–p system [2] has been found to have an intrinsic range for the Yukawa well of $(2{\cdot}92 \pm 0{\cdot}12) \times 10^{-13}$ cm in the triplet state and $(2{\cdot}06 \pm 0{\cdot}21) \times 10^{-13}$ cm in the singlet n–p state.

For the Yukawa well we can evaluate the corresponding pion mass in each case, and we obtain $(280 \pm 12)m_e$† for the pion mass in the triplet spin state which agrees well with the observed value of $m_{\pi^+} = 273{\cdot}02 \pm 0{\cdot}03$. However, we should not take this apparent quantitative agreement too seriously, since we have not taken into account the known tensor force which gives rise to about 4 per cent *d*-wave in the deuteron ground state. Also, we have not considered the simultaneous exchange of two pions, which has a characteristic range half that of the single pion exchange, neither has the possibility of other shorter characteristic ranges due, for example, to $\rho$, $\omega$, $\phi$, and K-mesons been considered. For the singlet spin state the corresponding pion mass is $(398 \pm 42)m_e$. Here again, we should not be too concerned with the disagreement with the empirical pion mass. Of course, tensor forces vanish in the singlet state so they cannot be responsible. However the two-pion exchange, other shorter characteristic ranges, and the probable existence of a hard core in singlet-even states (as indicated in high energy p–p scattering) could all explain the difference.

## REFERENCES

1. YUKAWA, H. (1935). *Proc. phys. math. Soc. Japan* **17**, 48.
2. HULTHEN, L. and SUGAWARA, M. (1957). *Handbuch der Physik XXXIX, The two nucleon problem.* Springer Verlag, Berlin. p. 52–54.
3. BLATT, J. M. and WEISSKOPF, V. F. (1952). *Theoretical nuclear physics.* John Wiley and Sons, New York (Fourth Printing, 1958).
4. DEBENEDETTI, S. (1964). *Nuclear interactions.* John Wiley and Sons, New York. (Second Corrected Printing (1967) can serve as a more recent review than [2] and [3].)

† $m_e$ = electron mass.

# 2

# INTRODUCTION TO MESON FIELD THEORY

## 2.1. Lorentz invariance

IN Chapter 1, we have considered a classical approach to a specific pion–nucleon static interaction. In order to treat the more general problems within the framework of pion field theory, we must consider the field-theoretic approach. Our general requirement will be consistency with the special theory of relativity, or Lorentz invariance, of the field equations.

As an excellent general further reference on the material treated in this chapter the reader is referred to Schweber, Bethe, and de Hoffman [1].

A Lorentz transformation can be conveniently considered in the following way. Assume a four-dimensional space–time coordinate system $x_0 = ict$, $x_1, x_2, x_3$, where the zero-th, or time, coordinate is imaginary. It can then be shown that a rotation in the $(x_1, ict)$ plane by an imaginary angle $(\theta)$ is equivalent to a Lorentz transformation from one Lorentz frame $(L)$ to another $(L')$ moving with uniform velocity $v$ along the $x_1$ axis of $L$.

The relationship between $v$ and $\theta$ is $v = c \tanh \theta$, and it is easily shown that it is equivalent to a Lorentz transformation, as usually stated,

$$\begin{aligned} x_1 &= \frac{x_1' + vt'}{\sqrt{[1-(v/c)^2]}} \\ x_2 &= x_2' \\ x_3 &= x_3' \\ t &= \left(\frac{vx_1'}{c^2} + t'\right)\bigg/\sqrt{(1-v^2/c^2)}. \end{aligned} \tag{2.1}$$

The square of the length of a four-vector is, with the above conventions,

$$l'^2 = x_0^2 + x_1^2 + x_2^2 + x_3^2 = -c^2t^2 + x_1^2 + x_2^2 + x_3^2, \tag{2.2}$$

a Lorentz invariant, but the square of the length of a three-vector $(x_1, x_2, x_3)$ is not. The i can be dropped (and we do so) from the zero-th, or time, component, provided we remember the square of the length of a four-vector is given by $-x_0^2 + x_1^2 + x_2^2 + x_3^2$.

The change in proper time $(d\tau)$, or time measured by an observer moving with an object, is related to the change in space–time coordinates in any other Lorentz frame by

$$c^2\, d\tau^2 = c^2\, dt^2 - (dx_1^2 + dx_2^2 + dx_3^2). \tag{2.3}$$

It will be convenient to change the sign in eqn (2.2) and adopt as convention for the length of the four-vector

$$l^2 = x_0^2-(x_1^2+x_2^2+x_3^2) = -(l')^2, \tag{2.4}$$

which is the negative of the previously stated square of the length $(l')^2$. For calculation purposes this is equivalent to defining the components of our four-vector as $x_0, \mathrm{i}x_1, \mathrm{i}x_2, \mathrm{i}x_3$.

Let us define the fundamental tensor $g_{\mu\nu}$, with $\mu$, $\nu$ ranging from 0 to 3, as a $4\times 4$ matrix whose elements are $-\delta_\mu^\nu$ for $\mu, \nu = 1, 2, 3$ where $\delta_\mu^\nu = 1$ for $\mu = \nu$, $\delta_\mu^\nu = 0$ for $\mu \neq \nu$, and whose elements are $\delta_\mu^\nu$ whenever either $\mu$ or $\nu$ is 0. Thus,

$$g_{\mu\nu} = \begin{bmatrix} 1 & 0 & 0 & 0 \\ 0 & -1 & 0 & 0 \\ 0 & 0 & -1 & 0 \\ 0 & 0 & 0 & -1 \end{bmatrix}. \tag{2.5}$$

Therefore, we can denote contravariant vectors (those that transform like coordinates $x^0, x^1, x^2, x^3$) by $x^\mu$ and covariant vectors (those that transform like the gradient) by $x_\mu$ and treat tensors similarly. The prescriptions for changing from contravariant to covariant vectors, or vice versa, are then

$$x_\mu = \sum_{\nu=0}^{3} g_{\mu\nu} x^\nu, \tag{2.6}$$

$$x^\mu = \sum_{\nu=0}^{3} g^{\mu\nu} x_\nu, \tag{2.7}$$

$$\delta_\sigma^\nu = \sum_\mu g^{\nu\mu} g_{\mu\sigma}. \tag{2.8}$$

We can see from the above that changing from a contravariant to a covariant vector leaves the sign of the time component unchanged but reverses the signs of the space components. The scalar (or dot) product of two four-vectors, which is Lorentz invariant, is given by

$$\mathrm{X}.\mathrm{Y} = \sum_{\mu=0}^{\mu=3} x^\mu y_\mu = x_0 y_0 - x_1 y_1 - x_2 y_2 - x_3 y_3 = x^0 y^0 - x^1 y^1 - x^2 y^2 - x^3 y^3. \tag{2.9}$$

The energy–momentum four-vector is given by $x^0 = \dfrac{E}{c}$, $x^1 = p^1$, $x^2 = p^2$, $x^3 = p^3$, where $\mathbf{p}$ is the three-dimensional momentum. The length of this four-vector is

$$l^2 = E^2/c^2 - \mathbf{p}^2 = m_0^2 c^2.$$

Another useful four-vector is the space–time gradient of some scalar function of $(x_0, x_1, x_2, x_3)$ $(\Box\varphi)$ defined by

$$(\Box\varphi)_0 = \frac{1}{c}\,\partial\varphi/\partial t, \tag{2.10}$$

$$(\Box\varphi)_{1,2,3} = \partial\varphi/\partial x_{1,2,3}. \tag{2.11}$$

The contracted second-order derivative (with a change of sign for convenience) is the Lorentz invariant D'Alembertian,

$$\Box^2\varphi = 1/c^2\,\partial^2\varphi/\partial t^2 - \frac{\partial^2\varphi}{\partial x_1^2} - \frac{\partial^2\varphi}{\partial x_2^2} - \frac{\partial^2\varphi}{\partial x_3^2}. \tag{2.12}$$

Let $\Box^2\varphi = 0$, where $\Box^2 = \frac{1}{c^2}\frac{\partial^2}{\partial t^2} - \nabla^2$. The above is the wave equation of electromagnetic theory which is, obviously, Lorentz invariant.

In terms of $\Box^2$, the Klein–Gordon equation can be written as

$$\left(\Box^2 + \frac{m_0^2 c^2}{\hbar^2}\right)\varphi = 0, \tag{2.13}$$

with the present definition of the length of a four-vector (i.e. eqn (2.4)). A vector is space-like if $l^2 < 0$, time-like if $l^2 > 0$, and a null vector when $l^2 = 0$. If the distance of separation of two points is space-like, they cannot be connected by a light signal and, hence, measurements at one cannot interfere with the other. If the separation of two points is time-like they can be connected by a light signal and measurements at point (1) can interfere with those at point (2) provided $t_1 < t_2$.

## 2.2. Lagrangian and Hamiltonian methods

Field theory generally uses the Lagrangian or Hamiltonian formalisms to determine the field equations. Let us consider a real, one-component field which depends only on the space–time variable $x_\mu$. Then one defines the Lagrangian density $\mathscr{L}\left(\varphi, \frac{\partial\varphi}{\partial x_\mu}\right)$ in terms of the so-called action integral,

$$I = \int \mathscr{L}\left(\varphi, \frac{\partial\varphi}{\partial x_\mu}\right) \mathrm{d}x_0, \mathrm{d}x_1, \mathrm{d}x_2, \mathrm{d}x_3 = \int \mathscr{L}\left(\varphi, \frac{\partial\varphi}{\partial x_\mu}\right) \mathrm{d}^4x. \tag{2.14}$$

The $\mathscr{L}\left(\varphi, \frac{\partial\varphi}{\partial x_\mu}\right)$ is then determined by requiring that $I$ be stationary (i.e. $\delta I = 0$), where $\varphi$ is arbitrarily varied subject to the condition

$$\delta\varphi(t_1, \mathbf{x}) = \delta\varphi(t_2, \mathbf{x}) = 0, \tag{2.15}$$

where $\mathbf{x}$ is the space-vector describing a point and $t_1$ and $t_2$ are the time limits chosen for the integral. The integration volume can of course be arbitrarily chosen. If we vary $I$, we obtain

$$\delta I = \int \mathrm{d}^4x \left( \delta\varphi(x) \frac{\partial \mathscr{L}}{\partial \varphi(x)} + \delta \frac{\partial \varphi}{\partial x^\mu} \frac{\partial \mathscr{L}}{\partial(\partial\varphi/\partial x^\mu)} \right). \tag{2.16}$$

Integration by parts of the second term gives

$$\delta I = \int\limits_{\text{4-volume}} \mathrm{d}^4x \left( \frac{\partial \mathscr{L}}{\partial \varphi} - \frac{\partial}{\partial x^\mu} \frac{\partial \mathscr{L}}{\partial \left[ \dfrac{\partial \varphi}{\partial x^\mu} \right]} \right) \delta\varphi + \int\limits_{\substack{\text{surface}\\ \text{of volume}}} \mathrm{d}^3x \, \delta\varphi \left( \frac{\partial \mathscr{L}}{\partial \left( \dfrac{\partial \varphi}{\partial x^\mu} \right)} \right) \eta_\mu(x). \tag{2.17}$$

$\eta_\mu(x)$ is the outward normal to the surface at the point given by $x$. Making the reasonable physical assumption that $\varphi \to 0$ fast enough as $x \to \infty$ and letting our spatial dimensions $\to \infty$, the surface integral becomes zero. Since for arbitrary variations the value of the volume integral must be zero, its integrand must be zero, and this gives the field equations

$$\frac{\partial \mathscr{L}}{\partial \varphi} - \frac{\partial}{\partial x^\mu} \frac{\partial \mathscr{L}}{\partial(\partial\varphi/\partial x^\mu)} = 0. \tag{2.18}$$

We can show that an arbitrary number of four-divergences can be added to the Lagrangian density without affecting the field equations. The non-uniqueness of the Lagrangian (for example the addition of a four-divergence) will not affect the field equations derived from the Lagrangian. However, in general we will find that the Hamiltonian and canonical formalism will be different. Nevertheless, the physical information contained in the theory will be the same, and all predictions of calculations will be the same. In order to have a relativistically invariant field theory which yields relativistically invariant field equations, we must require that the action $I$ is Lorentz invariant. Since the four-dimensional space–time volume is a pseudo-scalar, we must choose $\mathscr{L}$ as a scalar, then we obtain a positive definite energy density for fields which have integral spin.

A relativistically invariant Hamiltonian theory has been developed by Weiss [2] and has been used in modern field theory [3]–[5]. Let us, however, proceed with a less ambitious project, namely, to develop the Hamiltonian formalism in analogy to the usual treatment in the classical mechanics of particles

$$H = \int \mathscr{H} \, \mathrm{d}^3x, \tag{2.19}$$

where $H$ is the Hamiltonian function and $\mathscr{H}$ is the Hamiltonian density given by

$$\mathscr{H} = \pi\dot{\varphi} - \mathscr{L}, \tag{2.20}$$

where $\mathscr{L}$ is the Lagrangian density in the classical problem. Then, following the same procedure we use in the classical mechanical problem, we obtain the canonical equations

$$\dot{\varphi}(x) = \frac{\delta H}{\delta \pi(x)} = \frac{\partial \mathscr{H}}{\partial \pi(x)}, \tag{2.21}$$

$$\dot{\pi}(x) = \frac{-\delta H}{\delta \varphi(x)},$$

where $\pi_\mu = \dfrac{\partial \mathscr{L}}{\partial \dot{\varphi}_\sigma}$ represents the canonically conjugate momentum and $\delta/\delta\varphi$ means the functional derivative as discussed in Schiff [6].

The field quantization can be carried out by the usual quantum mechanical commutation rules

$$\begin{aligned} [\pi_\sigma(\mathbf{x}, t), \varphi_{\sigma'}(\mathbf{x}', t)] &= -\mathrm{i}\hbar\, \delta_{\sigma\sigma'}(\mathbf{x}-\mathbf{x}'), \\ [\pi_\sigma(\mathbf{x}, t), \pi_{\sigma'}(\mathbf{x}', t)] &= [\varphi_\sigma(\mathbf{x}, t)\varphi_{\sigma'}(\mathbf{x}', t)] = 0. \end{aligned} \tag{2.22}$$

The particles which we have treated have Bose–Einstein statistics. If the particles have Fermi–Dirac statistics, then there are appropriate anti-commutation relations to replace the commutation relations.

Before we can proceed with the development of the interaction of the pion field with the nucleon fields, we must consider the Dirac equation which the nucleons with spin $\frac{1}{2}$ obey, namely,

$$\left(\boldsymbol{\alpha}.\nabla + \mathrm{i}\frac{\beta mc}{\hbar}\right)\Psi + \frac{1}{c}\frac{\partial \Psi}{\partial t} = 0, \tag{2.23}$$

where $\alpha$ and $\beta$ are the Dirac $4 \times 4$ matrices and $\Psi$ is the 4-component column matrix or nucleon spinor (2 components for the particle spins of $+\frac{1}{2}$ and $-\frac{1}{2}$, and two components for the anti-particle spins). $\alpha$ and $\beta$ must be Hermitian in order that the Hamiltonian be Hermitian, and they must satisfy the following conditions

$$\begin{aligned} \alpha^k\alpha^{k'} + \alpha^{k'}\alpha^k &= 2\delta^{kk'}, \\ \alpha^k\beta &= -\beta\alpha^k, \\ \beta^2 = (\alpha^k)^2 &= 1, \quad \text{where } k, k' = 1, 2, 3. \end{aligned} \tag{2.24}$$

Typical examples of a free-field Lagrangian density are

(a) For $\varphi$ (scalar or pseudoscalar)—for example a one-component pion field

$$\mathscr{L} = -\tfrac{1}{2}\mu^2\varphi^2 + \frac{1}{2}\frac{\partial\varphi}{\partial x_\nu}\frac{\partial\varphi}{\partial x^\nu}. \tag{2.25}$$

(b) For $\Psi$ (spinor)—for example a nucleon field

$$\mathscr{L} = -\tfrac{1}{2}\overline{\Psi}\left(-\mathrm{i}\frac{\gamma^\mu\partial}{\partial x^\mu}+m\right)\Psi-\frac{1}{2}\left(\mathrm{i}\frac{\partial\overline{\Psi}}{\partial x^\mu}\gamma^\mu+m\overline{\Psi}\right)\Psi. \tag{2.26}$$

The notation and terms used in eqn (2.26) will now be defined. Let us introduce the $\gamma$ matrices as follows

$$\begin{aligned}\gamma^0 &= \beta,\\ \gamma^k &= \beta\alpha^k, \qquad k = 1,2,3,\end{aligned} \tag{2.27}$$

or, equivalently,

$$\boldsymbol{\gamma} = \boldsymbol{\beta\alpha}.$$

With these definitions we can write, as a non-unique set of representations,

$$\boldsymbol{\gamma} = \begin{bmatrix} 0 & \boldsymbol{\sigma} \\ -\boldsymbol{\sigma} & 0 \end{bmatrix}, \tag{2.28}$$

where $\boldsymbol{\sigma}$ is represented by the three Pauli spin matrices

$$\begin{aligned}\sigma_1 &= \begin{bmatrix} 0 & 1 \\ 1 & 0 \end{bmatrix}, \qquad \sigma_2 = \begin{bmatrix} 0 & -\mathrm{i} \\ +\mathrm{i} & 0 \end{bmatrix},\\ \sigma_3 &= \begin{bmatrix} 1 & 0 \\ 0 & -1 \end{bmatrix}, \qquad \beta = \begin{bmatrix} \mathrm{I}_2 & \mathrm{O}_2 \\ \mathrm{O}_2 & -\mathrm{I}_2 \end{bmatrix},\end{aligned} \tag{2.29}$$

where $\mathrm{I}_2 = \begin{bmatrix} 1 & 0 \\ 0 & 1 \end{bmatrix}$, and $\mathrm{O}_2 = \begin{bmatrix} 0 & 0 \\ 0 & 0 \end{bmatrix}$,

$$\boldsymbol{\alpha} = \begin{bmatrix} \mathrm{O}_2 & \boldsymbol{\sigma} \\ \boldsymbol{\sigma} & \mathrm{O}_2 \end{bmatrix}.$$

The $\gamma$ matrices satisfy the commutation rules

$$\begin{aligned}\gamma^\mu\gamma^\nu+\gamma^\nu\gamma^\mu &= 2g^{\mu\nu}I,\\ \gamma_\mu\gamma_\nu+\gamma_\nu\gamma_\mu &= 2g_{\mu\nu}I,\end{aligned} \tag{2.30}$$

where we raise or lower indices on $\gamma$ as we do for a four-vector. In terms of the $\gamma$ matrices we can now write the Dirac equation in the symmetrical space–time covariant form

$$-\sum_{\mu=0}^{\mu=3}\mathrm{i}\gamma^\mu\frac{\partial\Psi}{\partial x^\mu}+\frac{mc}{\hbar}\Psi = 0. \tag{2.31}$$

We define $\overline{\Psi} = \Psi^+\beta = \Psi^+\gamma^0$, where the symbol ($^+$) denotes the Hermitian adjoint (or conjugate) (i.e. $M^{+}_{ln} = M^{*}_{nl}$, where * means the complex conjugate of each element is substituted for the element). Then we can show that, under a Lorentz transformation, $\overline{\Psi}\Psi$ is a scalar covariant, except for time reversals for which it is a pseudoscalar.

In constructing a relativistically invariant interaction term between the nucleon spinor field and the pion field to insert into the Lagrangian or Hamiltonian, we shall consider the insertion of an arbitrary $4\times 4$ matrix between $\overline{\Psi}$ and $\Psi$ and the multiplication of the resultant product by a suitable function of the pion field, or its derivatives, to form a relativistically invariant interaction term which is Hermitian. An even number of spinor functions is required in order to obtain physically meaningful results† and, therefore, a bilinear term in the nucleon wave function is the lowest order we can consider. We can use the four $\gamma$ matrices to form new matrices by taking all possible products of them. Consider the products of two or more $\gamma$ matrices. Due to the anti-commutation rules, a change in order can only change the sign. Since $(\gamma^{\mu})^2 = \pm I$, only products with all $\gamma$'s different are linearly independent and, therefore, there are obviously a combined total of sixteen of these linearly independent products of the $\gamma$-matrices. Since an arbitrary $4\times 4$ matrix has sixteen elements, we can express an arbitrary $4\times 4$ matrix in terms of the sixteen linearly independent products of the matrices. One of these products which is of special importance is defined as

$$\gamma_5 = \gamma^0\gamma^1\gamma^2\gamma^3. \tag{2.32}$$

It should be noted that the index 5 on $\gamma_5$ cannot be raised or lowered. It follows that

$$\gamma_5\gamma_\mu + \gamma_\mu\gamma_5 = 0 \tag{2.33}$$

and

$$\gamma_5^2 = -I.$$

The Hermitian form of the others is $I^{\nu}$, $\gamma^{\mu}$, $\mathrm{i}\gamma^{\mu}\gamma^{\nu}$, $\mathrm{i}\gamma^{\mu}\gamma^{\nu}\gamma^{k}$. When interposed between $\overline{\Psi}$ and $\Psi$ the Lorentz transformation properties of these products are‡

$$\overline{\Psi}I\Psi \quad \text{(scalar)}, \tag{2.34a}$$

$$\overline{\Psi}\gamma^{\mu}\Psi \quad \text{(vector)}, \tag{2.34b}$$

$$\overline{\Psi}\mathrm{i}\gamma^{\mu}\gamma^{\nu}\Psi \quad \text{(tensor)}, \tag{2.34c}$$

$$\overline{\Psi}\mathrm{i}\gamma^{\mu}\gamma^{\nu}\gamma^{k}\Psi \quad \text{(pseudovector)}, \tag{2.34d}$$

$$\overline{\Psi}\gamma_5\Psi \quad \text{(pseudoscalar)}. \tag{2.34e}$$

† Otherwise, for example, a rotation of the coordinate system by $2\pi$ radians will lead to a sign change.

‡ We consider all transformations except time reversals.

With regard to Lorentz transformations, we should note that the so-called proper Lorentz transformations are built up by continuous variation from the unit transformation and are characterized by an invariant four-volume

$$d^4x = d^4x'.$$

In addition to the proper Lorentz transformation there are the improper Lorentz transformations, which can be constructed by using an appropriate proper Lorentz transformation and reflections of one or more of the space–time axes. Reflection of the time-axis $x_0' = -x_0$ is called time reversal. Reflection of all three space-axes, $x_1' = -x_1$, $x_2' = -x_2$, $x_3' = -x_3$, is called space inversion. Reflection of all four space–time axes is called space and time reversal or inversion.

We should note that the term pseudoscalar generally refers to a scalar quantity (i.e. scalar product of two four-vectors) which changes sign under space reflection or inversion. However, there can be scalar quantities which change sign (i.e. are pseudoscalar) under time inversion. The scalar product of two four-vectors (of the coordinate type) is a scalar (i.e. does not change sign under space inversion or time reversal). There is another class of vectors called pseudovectors which exhibit the characteristics of vectors in all respects, except that the scalar product of a pseudovector with a vector yields a pseudoscalar instead of a scalar.

The non-relativistic three-dimensional (polar) vectors which represent coordinates are the analogues of our vector while the three-dimensional cross product of two (polar) vectors are the analogues of our pseudovector (one example is the nucleon spin $\boldsymbol{\sigma}$). Space inversion changes the sign of the components of a three-dimensional vector but does not affect the sign of the components of a three dimensional pseudovector.

### 2.3. Interacting fields

Let us now consider some of the simplest interaction terms in the Lagrangian between the nucleon spinor field and various types of pion fields which are relativistically invariant. We restrict ourselves to linear terms in the pion field or the first derivatives, and consider local couplings only (i.e. all factors refer to the same space–time point). For the pion field we consider the scalar, or pseudoscalar (for spin 0), and the vector, or pseudovector (for spin 1). The coupling can be non-derivative, or so-called direct, with a dimensionless coupling constant denoted by $G$, or derivative coupling with a coupling constant denoted by $F/\mu$, where $\mu$ is the mass of the particle and is introduced to make the coupling constant dimensionless. The eight possibilities are given and justified below. Notice that the usual relativistic convention of summing over a repeated index is used. $\partial/\partial x_\nu$ is the four-component derivative (See Bethe and de Hoffman for the transformation properties of $\gamma$ matrices). For convenience, we shall generally be using natural units, with $\hbar = c = 1$.

(a) *Scalar field* (*scalar coupling*)

$$\mathrm{S(S)}, \qquad G(\overline{\Psi}\Psi)\varphi.$$

The product of $\varphi$ (scalar) and $\overline{\Psi}\Psi$ (scalar covariant) gives Lorentz invariance.

Pion spin $= 0$, parity is even.

(b) *Scalar field* (*vector coupling*)

$$\mathrm{S(V)}, \qquad \frac{F}{\mu}\overline{\Psi}\gamma_\nu\Psi\frac{\partial\varphi}{\partial x_\nu}.$$

The product of $\dfrac{\partial\varphi}{\partial x_\nu}$ (vector and $\overline{\Psi}\gamma_\nu\Psi$ (vector covariant) gives Lorentz invariance.

Pion spin $= 0$, parity is even.

(c) *Pseudoscalar field* (*pseudoscalar coupling*)

$$\mathrm{PS(PS)}, \qquad G\overline{\Psi}\gamma_5\Psi\varphi.$$

The product of $\varphi$ (pseudoscalar) and $\overline{\Psi}\gamma_5\Psi$ (pseudoscalar covariant) gives Lorentz invariance.

Pion spin $= 0$, parity is odd.

(d) *Pseudoscalar field* (*pseudovector coupling*)

$$\mathrm{PS(PV)}, \qquad \mathrm{i}\frac{F}{\mu}\overline{\Psi}\gamma_5\gamma_\nu\Psi\frac{\partial\varphi}{\partial x_\nu}.$$

The product of $\dfrac{\partial\varphi}{\partial x_\nu}$ (pseudovector) and $\overline{\Psi}\gamma_5\gamma_\nu\Psi$ (pseudovector covariant) gives Lorentz invariance.

Pion spin $= 0$, parity is odd.

(e) *Vector field* (*vector coupling*)†

$$\mathrm{V(V)}, \qquad G\overline{\Psi}\gamma^\mu\Psi\varphi_\mu,$$

† This is the type of interaction term found in electrodynamics resulting from coupling between charges and fields, namely,

$$A_\mu\overline{\Psi}\gamma^\mu\Psi.$$

The product of $\varphi_\mu$ (vector) and $\overline{\Psi}\gamma^\mu\Psi$ (vector covariant) gives Lorentz invariance.
Pion spin $= 1$, parity is even.

(f) *Vector field* (*tensor coupling*)

$$\text{V(T)}, \qquad \frac{F}{\mu}\overline{\Psi}\gamma_\mu\gamma_\nu\Psi\left(\frac{\partial\varphi_\mu}{\partial x_\nu}-\frac{\partial\varphi_\nu}{\partial x_\mu}\right).$$

The product of vector field $\left(\dfrac{\partial\varphi_\mu}{\partial x_\nu}\right)$ with the anti-symmetric tensor covariant $(\gamma_\mu\gamma_\nu)$ gives Lorentz invariance.
Pion spin $= 1$, parity is even.

(g) *Pseudovector field* (*pseudovector coupling*)

$$\text{PV(PV)}, \qquad \mathrm{i}G\overline{\Psi}\gamma_5\gamma^\mu\Psi\varphi_\mu.$$

The product of $\varphi_\mu$ (pseudovector) with $\overline{\Psi}\gamma_5\gamma^\mu\Psi$ (pseudovector covariant) gives Lorentz invariance.
Pion spin $= 1$, parity is odd.

(h) *Pseudovector field* (D tensor)

$$\text{PV(DT)}, \qquad \mathrm{i}\frac{F}{2\mu}(\overline{\Psi}\gamma_5\gamma_\mu\gamma_\nu\Psi)\left(\frac{\partial\varphi_\nu}{\partial x_\mu}-\frac{\partial\varphi_\mu}{\partial x_\nu}\right)$$

Pion spin $= 1$, parity is odd and the term is Lorentz invariant.

In the non-relativistic limit, the PS(PV) interaction can be shown to be equivalent to $\dfrac{F}{\mu}\,\Psi^+\boldsymbol{\sigma}\Psi\,.\,\nabla\varphi$. There is an equivalence theorem for the pseudo-scalar meson field which states that, if we consider first-order terms in the coupling constant (i.e. terms proportional to $G$ in the Hamiltonian or matrix element which represent terms proportional to $G^2$ in the cross-section) the pseudovector or pseudoscalar couplings give the same results [7] provided $F = \dfrac{\mu}{2M}G$.

The total Lagrangian which corresponds to two interacting fields is the sum of the two free-field Lagrangians, and an additional term due to the interaction. Let us consider a spinor (nucleon) field and a scalar boson field (meson) interacting in a manner described by a direct coupling term. The interaction Lagrangian has been previously given in (a) and is

$$\mathscr{L}_{\text{int}} = G(\overline{\Psi}\Psi)\varphi. \tag{2.35}$$

The total Lagrangian is, therefore,

$$\mathscr{L} = -\tfrac{1}{2}\overline{\Psi}\left(-\mathrm{i}\gamma^{\mu}\frac{\partial}{\partial x^{\mu}}+m\right)\Psi-\frac{1}{2}\left(\mathrm{i}\frac{\partial\overline{\Psi}\gamma^{\mu}}{\partial x^{\mu}}+m\overline{\Psi}\right)\Psi \qquad (2.36)$$

(nucleon field)

$$+\frac{1}{2}\frac{\partial\varphi}{\partial x_{\nu}}\frac{\partial\varphi}{\partial x^{\nu}}-\tfrac{1}{2}\mu^{2}\varphi^{2}$$

(meson field)

$$-G\overline{\Psi}\Psi\varphi.$$

(interaction term)

If we now vary $\mathscr{L}$ with respect to $\overline{\Psi}$, we obtain the field equation for the nucleon ($\Psi$) field, which is

$$\left(-\mathrm{i}\gamma^{\mu}\frac{\partial}{\partial x^{\mu}}+m\right)\Psi = -G\varphi\Psi. \qquad (2.37)$$

The right-hand side represents the interaction with the meson field. We obtain the $\varphi$ field equation by varying the total Lagrangian with respect to $\varphi$, which gives

$$(\Box^{2}+\mu^{2})\varphi = -G\overline{\Psi}\Psi. \qquad (2.38)$$

The right-hand side is the nucleon source term for the boson (meson) field. The above equation is equivalent to the Yukawa equation which we previously investigated.

In order to see this consider an infinitely heavy point nucleon. Then the source term can be replaced by $4\pi g'\delta(\mathbf{r})\,\mathrm{d}\tau$. (The change of units, from rationalized to unrationalized, by a factor of $4\pi$ is unimportant. Hence we have

$$\left(\nabla^{2}-\frac{m_{0}^{2}c^{2}}{\hbar^{2}}-\frac{1}{c^{2}}\frac{\partial}{\partial t^{2}}\right)\varphi = 4\pi g'\delta(\mathbf{r})\,\mathrm{d}\tau, \qquad (2.39)$$

which is the same as the Yukawa equation (eqn (1.5)).

To treat the above problem properly within the framework of quantum mechanics, we should quantize the field variables with the usual technique in the Hamiltonian method of substituting commutators for the Poisson brackets. This results in the field variables becoming non-commuting operators.

The general physical case of the interaction of a boson (meson) and a fermion (physical nucleon) field cannot be solved in closed form, even for the simple interactions which we have considered, but there are two limiting approximations for which approximate solutions may be obtained—the weak coupling approximation and the strong coupling approximation. These will be considered in the following sections.

### 2.4. Weak coupling approximation

When $G \ll 1$ we have the so-called weak coupling approximation, which allows perturbation theory methods similar to those used in quantum electrodynamics to be employed. Solutions are expanded in a series of powers of the coupling constant and, as the coupling constant is made smaller, the convergence of the series is increased so that only the first few terms would be important if the coupling constant were small enough. The approach is quite similar to that used quite successfully in quantum electrodynamics. However, whereas in quantum electrodynamics $\alpha = e^2/\hbar c \sim \frac{1}{137}$,† in the meson case experiments indicate that $G \gtrsim 10$ (and $g^2 \sim 10$). Therefore the method is of very limited value, since the series can diverge or converge only slowly. However, since in many cases a considerable amount of qualitative progress has been made using the methods of the weak coupling approximation, we will discuss this approach, briefly.

### 2.5. Feynman diagrams

A most useful form of the weak coupling theory is the formalism of Feynman and Dyson. The Feynman–Dyson perturbation expansion exhibits the covariant form of the theory at every step. The conserved quantities, at each step, are energy and momentum. The mass is not conserved but, of course, is an invariant quantity. The earlier perturbation theory conserved momentum and mass but not energy.

Feynman developed his perturbation treatment into a set of simple and complete rules for writing down matrix elements in terms of increasing order. He evolved the method in terms of diagrams ('Feynman diagrams') which give a powerful pictorial aid to grasping the physical meaning of the processes involved and make it easier to avoid mistakes and omissions, and last, but not least, lead to a combination of terms which reduced the over-all number of terms. Previously, the treatment of an interaction which proceeded through $n$ intermediate states required evaluating $2^n$ matrix-elements, whereas the Feynman method replaced this by one expression, generally less cumbersome than each of the previous multitudinous number of terms. Feynman treated both electrodynamics and meson interactions by these methods.

† The analogous quantity in meson theory would be $\dfrac{g^2}{\hbar c} = \dfrac{G^2}{4\pi\hbar c}$. In natural units these quantities become $g^2 = \dfrac{G^2}{4\pi} \sim 10$.

Let us now consider some typical 'Feynman diagrams' in say a spinor–spinor interaction via a boson field. One example of this would be a nucleon–nucleon interaction via interaction with the meson field. Consider a two-dimensional plane in which time is along the horizontal axis of this paper, and space is along the vertical axis as follows.†

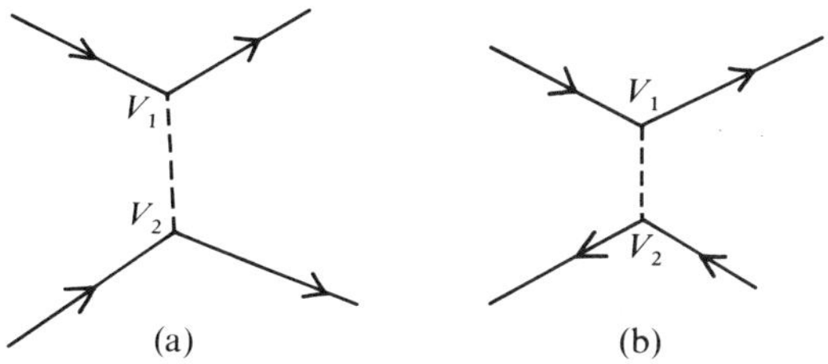


FIG. 2.1. (a)–(b).

A spinor particle line is shown diagramatically as progressing positively in time (indicated by arrow). A boson (meson) is shown by the dashed line. Therefore the diagram which represents nucleon–nucleon scattering in lowest order via a single meson exchange (i.e. to order $G^2$) is as shown in Fig. 2.1(a). Feynman's form of the perturbation theory is a non-time-ordered form. That is, in Fig. 2.1(a) it does not matter whether the vertex $V_1$ corresponds to an earlier time than vertex $V_2$ or vice versa.

'Feynman diagrams' may be topologically distorted so that the vertices are interchanged provided the external lines in the initial and final state are not changed and the diagrams are still topologically equivalent. Furthermore, Feynman's expression automatically contains the sum of the contribution of all diagrams obtained by all possible permutations of the vertices. This means that in Fig. 2.1(a), which represents a single exchange of a pion between two nucleons, it does not matter which nucleon emits the meson and which absorbs it. Feynman's expression automatically represents the sum of all relevant contributions to this second order process. Nucleon–anti-nucleon scattering via exchange of a meson is represented by Fig. 2.1(b). The arrow direction travelling backward in time on the nucleon line is equivalent to an anti-nucleon travelling forward in time.

Let us now consider the diagrams for the nucleon–anti-nucleon interaction in lowest order. In this case the Figs. 2.1(c) and 2.1(d) are appropriate.

† Feynman diagrams are often shown with the time-axis vertical and the space-axis horizontal, but the inversion we have employed is convenient for treating high energy reactions.

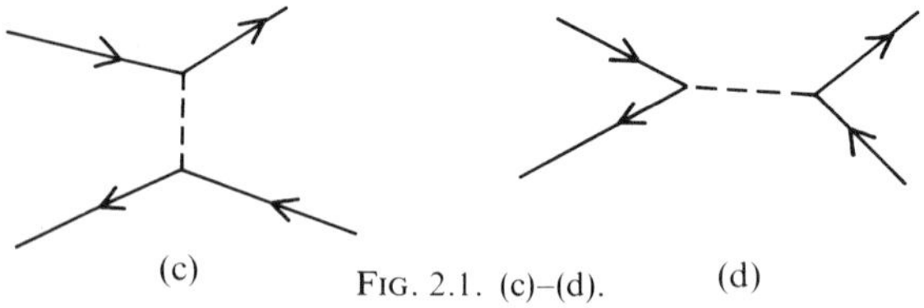


FIG. 2.1. (c)–(d).

Fig. 2.1(c) depicts a single meson exchange between nucleon and anti-nucleon, whereas Fig. 2.1(d) depicts the virtual annihilation of the nucleon–anti-nucleon pair with the emission of a virtual meson which then creates a nucleon–anti-nucleon pair in the final state.

The Feynman diagrams considered above represent the processes which involve two fermion lines in the initial and final states, plus an additional boson line proceeding from intermediate state vertices. Another common situation of interest are those processes which in the initial and final states involve one fermion and one boson line plus an additional fermion line

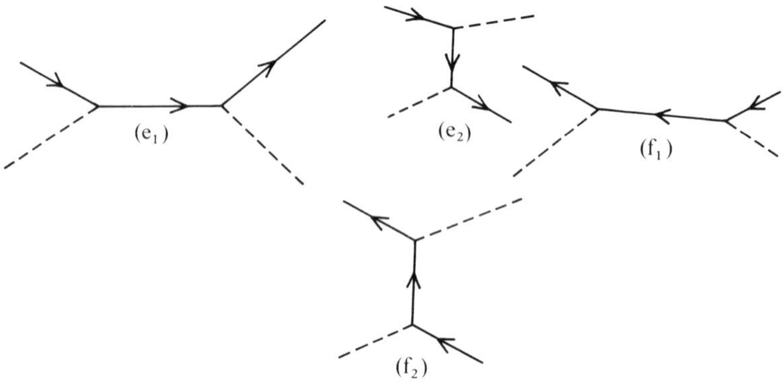


FIG. 2.1.(e) A meson is scattered by a nucleon. (e1) The initial meson is absorbed in initial state and then re-emitted. (e2) The final meson is emitted first and then the initial meson is absorbed. FIG. 2.1.(f) A meson is scattered by an anti-nucleon, (f1) The initial meson is absorbed by an anti-nucleon and then re-emitted. (f2) The final meson is emitted and then the initial meson is absorbed by an anti-nucleon.

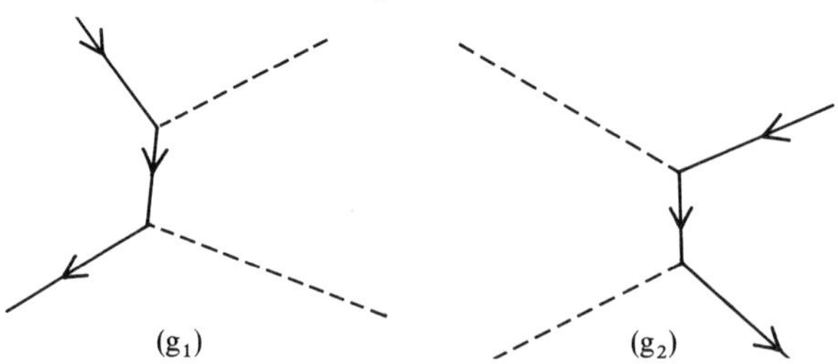


FIG. 2.1. (g1) Annihilation of a nucleon–anti-nucleon pair by double meson emission. (g2) Production of a nucleon–anti-nucleon pair by two mesons.

proceeding from the intermediate state vertices as shown in Figs. 2.1(e), 2.1(f), and 2.1(g).

## 2.6. *S*-matrix

The perturbation theory approach can be illustrated by use of the *S*-matrix approach introduced [8] by Heisenberg, Wheeler, and others. The *S*-matrix is a matrix with elements, between specified initial and final states, which are proportional to the amplitude for transition from initial to final states.

The observables in *S*-matrix theory are as follows.

(a) Energy and momentum of isolated non-interacting particles.

(b) In processes which involve collisions of particles, absorption, and emission of particles, the asymptotic solutions at large distances for the wave functions are considered as the only observables. This, of course, yields the transition rate or cross-section for such processes.

(c) In the case of stationary state closed systems the energy levels would be the observables.

Feynman's quantum electrodynamic treatment was an *S*-matrix approach which gave a detailed prescription for calculating the *S*-matrix elements. For a summary of rules for writing Feynman matrix elements in electrodynamics and meson theory and a more detailed discussion of some of these points, see Schweber, Bethe, and De Hoffman (Vol. 1, 218–254).

The *S*-matrix elements of *n*th order have a one–to–one correspondence to Feynman diagrams with *n* corners (identified). This characteristic allows us to evaluate a matrix element in the following simple fashion. First, we can draw all topologically different Feynman graphs of any order *n* for the interaction. Then, we can obtain the matrix elements corresponding to any *n*th order diagram by simply following the Feynman diagram rules.

The transition probability amplitude is the *S*-matrix element between the specified initial and final state.

$$T_{\mathrm{FI}} = (2\pi)^4\,\delta^4\Big(\sum_{p_{\mathrm{F}}} - \sum_{p_{\mathrm{I}}}\Big)\langle F|\,M\,|I\rangle. \tag{2.39}$$

The transition probability to a set of final states of density $\rho_{\mathrm{F}}$, per unit time and volume, is given by

$$(\mathrm{TP})_{\mathrm{FI}} = (2\pi)^4 \int \mathrm{d}\rho_{\mathrm{F}}\,\delta^4\Big(\sum_{p_{\mathrm{F}}} - \sum_{p_{\mathrm{I}}}\Big)|\langle F|\,M\,|I\rangle|^2 \tag{2.40}$$

The cross-section is then

$$\sigma_{\mathrm{FI}} = (\mathrm{TP})_{\mathrm{FI}} \Big/ \begin{matrix}\text{Incident particle flux}\\ \text{per unit area, per unit time.}\end{matrix} \tag{2.41}$$

## 2.7. Charge independence

For over three decades, the great mass of experimental data compiled on the forces between nucleons has been consistent (within reason) with two principles referred to as 'charge symmetry' and 'charge independence'. Charge symmetry states that the nuclear interaction (or forces) between two neutrons in a particular spin and angular momentum state is the same as between two protons (excluding Coulomb forces) which are in the same spin and angular momentum state. Charge independence states that the nuclear interaction (or forces) between any two nucleons in the same spin and angular momentum state is the same (i.e. N–N, P–P, and N–P forces are the same if the spin and angular momentum are the same). The charge independence principle was first proposed by Breit and collaborators, in 1936 [9], and includes charge symmetry, which is less restrictive. We should note that the Coulomb interactions (or forces) are manifestly charge dependent and violate both charge symmetry and charge independence. Therefore, these principles will be useful to the degree of approximation where nuclear forces are much larger than Coulomb forces.

## 2.8. Isotopic spin

The charge independence principle can most usefully be formulated in the language of isotopic spin first introduced by Cassen and Condon (1936) [10]. The two possible charge states of a nucleon are described by an isotopic spin $= \frac{1}{2}\boldsymbol{\tau}$, analogous to the ordinary nucleon spin $\frac{1}{2}\boldsymbol{\sigma}$, which is represented by the usual Pauli spin matrices in spin-space. Thus, $\boldsymbol{\tau}$ is represented by the same three matrices in charge space

$$\tau_x = \begin{bmatrix} 0 & 1 \\ 1 & 0 \end{bmatrix}, \qquad \tau_y = \begin{bmatrix} 0 & -\mathrm{i} \\ \mathrm{i} & 0 \end{bmatrix}, \qquad \tau_z = \begin{bmatrix} 1 & 0 \\ 0 & -1 \end{bmatrix}. \tag{2.42}$$

The proton and neutron wave functions in charge-space are represented by the two-element matrices

$$\eta(\mathrm{p}) = \begin{bmatrix} 1 \\ 0 \end{bmatrix}, \qquad \eta(\mathrm{n}) = \begin{bmatrix} 0 \\ 1 \end{bmatrix}. \tag{2.43}$$

By definition, the above $\tau$ matrices have identical algebraic properties to the $\sigma$ matrices. Therefore

$$\tau_x^2 = \tau_y^2 = \tau_z^2 = 1, \tag{2.44a}$$

$$\tau_x\tau_y = \mathrm{i}\tau_z, \; \tau_y\tau_z = \mathrm{i}\tau_x, \; \tau_z\tau_x = \mathrm{i}\tau_y. \tag{2.44b}$$

Also, $\tau^2$ and one of its components can be diagonalized simultaneously, in analogy with the ordinary spin. Conventionally, $\tau_z$ is taken to be the diagonalized component in order to satisfy charge conservation and the relation between $\tau_z$ and $q$. We can demonstrate that

$$\frac{\tau_z}{2}\,\eta(\mathrm{p}) = +\tfrac{1}{2}\eta(\mathrm{p}), \tag{2.45a}$$

$$\frac{\tau_2}{2}\,\eta(\mathrm{n}) = -\tfrac{1}{2}\eta(\mathrm{n}). \tag{2.45b}$$

Therefore if the eigenvalues of $\tau_z$ operating on the nucleon charge wave function are denoted by $2t_z$, $t_z = +\frac{1}{2}$ for proton and $t_z = -\frac{1}{2}$ for neutron. The charge of a nucleon ($q$) in units of the electron charge ($e$) is then given by

$$q = t_z + \tfrac{1}{2} = \begin{pmatrix} 1 \text{ for proton} \\ 0 \text{ for neutron} \end{pmatrix}. \tag{2.46}$$

The nucleon is then said to have an isotopic spin of $t = \frac{1}{2}$, although the eigenvalue of $t^2$ is obviously $t(t+1) = \frac{3}{4}$. Using the language of isotopic spin, we can write a generalized wave function in which neutrons and protons are distinguished by the isotopic spin coordinate $\tau_z = \pm 1$. Then it can be proved that the Pauli exclusion principle is replaced by the so-called generalized Pauli exclusion principle, which states that the total wave function representing a system of nucleons must be anti-symmetric when all the coordinates (i.e. space, spin, and isotopic spin) of two nucleons are exchanged.

Consider the operators

$$\tau_+ = \tfrac{1}{2}(\tau_x + \mathrm{i}\tau_y) = \begin{bmatrix} 0 & 1 \\ 0 & 0 \end{bmatrix}, \tag{2.47a}$$

$$\tau_- = \tfrac{1}{2}(\tau_x - \mathrm{i}\tau_y) = \begin{bmatrix} 0 & 0 \\ 1 & 0 \end{bmatrix}. \tag{2.47b}$$

We can easily verify that

$$\tau_+\eta(\mathrm{n}) = \begin{bmatrix} 0 & 1 \\ 0 & 0 \end{bmatrix}\begin{bmatrix} 0 \\ 1 \end{bmatrix} = \begin{bmatrix} 1 \\ 0 \end{bmatrix} = \eta(\mathrm{p}); \tag{2.48a}$$

$$\tau_-\eta(\mathrm{p}) = \begin{bmatrix} 0 & 0 \\ 1 & 0 \end{bmatrix}\begin{bmatrix} 1 \\ 0 \end{bmatrix} = \begin{bmatrix} 0 \\ 1 \end{bmatrix} = \eta(\mathrm{n}). \tag{2.48b}$$

In a system containing $A$ nucleons (such as a nucleus) let us define a total isotopic spin† operator $\mathbf{T}$, in analogy to the total nucleon spin operator $\mathbf{S}$,

$$\mathbf{T} = \frac{1}{2}\sum_{j=1}^{A} \boldsymbol{\tau}_j. \tag{2.49}$$

† It has become quite common also to designate isotopic spin by $I$.

The addition implied in the last equation must be performed according to the usual quantum mechanical laws for vector addition of angular momenta. Hence the eigenvalues of $(\mathbf{T})^2$ are $T(T+1)$. The $T$ value is called the total isotopic spin. The total electric charge of a system of nucleons, which we denote by $Z$, is given by

$$Z = \tfrac{1}{2}A + T_z, \tag{2.50}$$

where $T_z$ is the eigenvalue of $\mathbf{T}_z$. From charge conservation, it is obvious that $T_z$ must be conserved in all interactions. The most general and useful form of the charge independence hypothesis is that the value of $T$ is conserved in all nuclear interactions and, therefore, $T$ is a good quantum number. This form of the charge independence hypothesis is easily generalized to include pion–nucleon interactions as well as interactions involving so-called 'strange particles'. If $T$ is conserved $(\mathbf{T})^2$ must commute with the Hamiltonian of the system and the Hamiltonian of the system must be invariant with respect to rotations in charge space, therefore, the properties of a nuclear state must depend only on $T$ and not on $T_z$.

To show that the conservation of isotopic spin is equivalent to the older definition of charge independence and also to indicate the use of isotopic spin, let us now consider a two-nucleon system which has a space–spin wave function denoted by $\Psi_{L,S}$ (where $L$ is the orbital angular momentum, and $S$ is the spin) and a charge-space wave function denoted by $\chi_{T,T_z}$. Obviously, for two nucleons each of isotopic spin $\frac{1}{2}$, the resultant charge-space wave function can have a total isotopic spin $T = 0$ or $1$ and with a $T_z = 0$ or $1, 0, -1$, respectively. There are four charge-space wave functions. In order to satisfy the generalized Pauli principle, the charge wave function must be either symmetric or anti-symmetric with respect to exchange of the two nucleons. Hence the four unique possibilities are

$$\chi_{1,1} = \eta_1(\mathrm{p})\eta_2(\mathrm{p}) \quad \text{for} \quad \mathrm{p+p}, \tag{2.51a}$$

$$\chi_{1,-1} = \eta_1(\mathrm{n})\eta_2(\mathrm{n}) \quad \text{for} \quad \mathrm{n+n}, \tag{2.51b}$$

$$\chi_{1,0} = \frac{1}{\sqrt{2}}\big(\eta_1(\mathrm{p})\eta_2(\mathrm{n}) + \eta_1(\mathrm{n})\eta_2(\mathrm{p})\big) \quad \text{for} \quad \mathrm{n+p}, \tag{2.51c}$$

$$\chi_{0,0} = \frac{1}{\sqrt{2}}\big(\eta_1(\mathrm{p})\eta_2(\mathrm{n}) - \eta_1(\mathrm{n})\eta_2(\mathrm{p})\big) \quad \text{for} \quad \mathrm{n+p}. \tag{2.51d}$$

We should note that the subscript on the nucleon charge wave function, $\eta$, distinguishes the first and second nucleon.

The $T = 1$ isotopic spin state is a charge triplet with a symmetric charge space wave function, and the $T = 0$ isotopic spin state is a charge singlet state with an anti-symmetric charge-space wave function. Since the symmetry of a two-nucleon isotopic spin wave function depends only on $T$ and not on $T_z$, it is obvious that for a particular space–spin wave function symmetry, for example space–spin anti-symmetric, any one of the three $T = 1$ symmetric charge-space wave functions can be combined with the same anti-symmetric space–spin state and satisfy the generalized Pauli principle.

Now let us recall the original charge independence hypothesis that if the space–spin states are the same for two nucleons then the nuclear forces, or interactions, are independent of the charge state. This then requires that the nuclear forces in the isotopic spin state $T = 1$ are independent of the value of $T_z$, provided that the space–spin wave function is the same. In other words, for the same space–spin state and for the isotopic spin state $T = 1$, the p+p, n+n, and n+p nuclear forces are the same. It is obvious that the $T = 0$ charge wave function, which is anti-symmetric, must be combined with symmetric space–spin wave functions to have the product wave function anti-symmetric. Therefore charge independence does not in any way relate the $T = 0$ to the $T = 1$ nucleon–nucleon interaction. However, charge independence does say that the nucleon–nucleon interaction in a particular $T$ state is independent of $T_z$ (i.e. p+p, n+n, and n+p systems in the $T = 1$ state all have the same nuclear interactions). We have, previously, demonstrated that if $T$ as well as $T_z$ is conserved and is a good quantum number then the products of a nuclear state must depend only on $T$ and not on $T_z$, and from the foregoing considerations it is clear that this is just what one deduces for the two nucleon system from the original definition of charge independence.

The foregoing arguments can easily be generalized to a many nucleon system for which the charge independence can be stated as

$T$ is conserved by nuclear interactions and is a good quantum number and, therefore, although $T_z$ is also conserved due to charge conservation, the properties of a nuclear state depend only on $T$ and are independent of $T_z$.

For a system of $A$ nucleons, the possible $T$ values will obviously be integral for $A$ even and half integral for $A$ odd. The possible range of $T$ values will be $A/2$, $A/2\text{-}1$, $A/2\text{-}2$, ..., 0 for $A$ even and $A/2$, $A/2\text{-}1$, $A/2\text{-}2$, ..., $\frac{1}{2}$ for $A$ odd.

Since the Hamiltonian must be a scalar, the only possible scalar that can be formed in charge-space for two nucleons is $\boldsymbol{\tau}_1 . \boldsymbol{\tau}_2$, where the subscripts 1 and 2 identify the individual nucleons. Therefore, the Hamiltonian or potential between two nucleons is of the form

$$V = V_1 + V_2 \boldsymbol{\tau}_1 . \boldsymbol{\tau}_2,$$

with $V_1$ and $V_2$ functions of the distance between the nucleons, their spins, etc.

As a result of an extensive investigation of pion production in nucleon–nucleon interactions and a study of pion–nucleon interactions extending over nearly two decades, it has been well established that there are three charge-states of the pion (positive ($\pi^+$), neutral ($\pi^0$), and negative ($\pi^-$)) which appear to differ only in the charge. Furthermore, it is found that they can be described well by an isotopic spin $T = 1$ triplet state. Once the existence of singly-charged pions had been discovered, it was recognized by Fröhlich, Heitler, and Kemmer [11] that a neutral pion which forms an isotopic spin triplet with the charged pions should exist if the exchange of pions were to be considered the main vehicle of the nuclear interaction and, at the same time, charge independence of nuclear forces was to be valid. A loose illustration of this is the following. Consider the three possible pairs of two nucleons p+p, n+n, n+p, when in the same space–spin state. Then if there were only charged pions, a single pion exchange could occur only in the n+p case, whereas at least a double pion exchange would be required in the p+p and the n+n cases. It can be shown that this differentiation would violate the requirement of charge independence. Fröhlich *et al.* considered only single pion exchanges which could be correct if the pion–nucleon coupling constant was weak enough. Kemmer then [12] developed a symmetrical theory of pion–nucleon interactions which involved $\pi^+$, $\pi^0$, and $\pi^-$ coupled to nucleons in a manner selected to preserve charge independence.

Let us now consider the general charge independent interaction of a pion and a nucleon in which no new particles are produced and none disappear (i.e. elastic scattering and charge exchange).

Since the pion has isotopic spin $T = 1$ and the nucleon has isotopic spin $T = \frac{1}{2}$, the pion–nucleon system will have isotopic spin $T = \frac{3}{2}$ or $T = \frac{1}{2}$. Therefore, the pion–nucleon system can be described by two independent isotopic spin amplitudes $a_{\frac{3}{2}}$ and $a_{\frac{1}{2}}$. In the $T = \frac{3}{2}$ state, the $T_z$ values can be $+\frac{3}{2}$, $+\frac{1}{2}$, $-\frac{1}{2}$, and $-\frac{3}{2}$, but the isotopic spin amplitude is the same for all the substates (i.e. $a_{\frac{3}{2}} = a_{\frac{3}{2},\frac{3}{2}} = a_{\frac{3}{2},\frac{1}{2}} = a_{\frac{3}{2},-\frac{1}{2}} = a_{\frac{3}{2},-\frac{3}{2}}$), where the first index $= T$ and the second index $= T_z$. In the $T = \frac{1}{2}$ state, the $T_z$ values can be $\frac{1}{2}$, $-\frac{1}{2}$ and we have, similarly, $a_{\frac{1}{2}} = a_{\frac{1}{2},\frac{1}{2}} = a_{\frac{1}{2},-\frac{1}{2}}$. Now let us write out the possible charge states that contribute to each of the six amplitudes

| AMPLITUDE | CHARGE STATE |
|---|---|
| (1) $a_{\frac{3}{2},\frac{3}{2}}$ | $\pi^+ + \mathrm{p}$ |
| (2) $a_{\frac{3}{2},\frac{1}{2}}$ | $\begin{cases}\pi^+ + \mathrm{n}\\ \pi^0 + \mathrm{p}\end{cases}$ |
| (3) $a_{\frac{3}{2},-\frac{1}{2}}$ | $\begin{cases}\pi^- + \mathrm{p}\\ \pi^0 + \mathrm{n}\end{cases}$ |
| (4) $a_{\frac{3}{2},-\frac{3}{2}}$ | $\pi^- + \mathrm{n}$ |
| (5) $a_{\frac{1}{2},\frac{1}{2}}$ | $\begin{cases}\pi^0 + \mathrm{p}\\ \pi^+ + \mathrm{n}\end{cases}$ |
| (6) $a_{\frac{1}{2},-\frac{1}{2}}$ | $\begin{cases}\pi^- + \mathrm{p}\\ \pi^0 + \mathrm{n}\end{cases}$ |

Since both $T$ and $T_z$ must be conserved and there are no doubly charged or negatively charged protons, it is obvious that only direct scattering (i.e. no change of charge of pion or nucleon) can occur in the $a_{\frac{3}{2},\frac{3}{2}}(\pi^+ + \mathrm{p})$ state and the $a_{\frac{3}{2},-\frac{3}{2}}(\pi^- + \mathrm{n})$ state which are both pure $T = \frac{3}{2}$ states, and the pion–nucleon interaction in these states depends only on $a_{\frac{3}{2}}$. However, in $a_{\frac{3}{2},\frac{1}{2}}$, $a_{\frac{3}{2},-\frac{1}{2}}$, $a_{\frac{1}{2},\frac{1}{2}}$, and $a_{\frac{1}{2},-\frac{1}{2}}$ states, since there are always two charge possibilities, charge exchange as well as direct scattering can occur between initial and final states. Furthermore, the physical states $\pi^+ + \mathrm{n}$, $\pi^0 + \mathrm{p}$, $\pi^- + \mathrm{p}$, and $\pi^0 + \mathrm{n}$ involve mixtures of $T = \frac{3}{2}$ and $T = \frac{1}{2}$ states and, therefore, the interaction in these states depends on both $a_{\frac{3}{2}}$ and $a_{\frac{1}{2}}$. The appropriate decomposition of a particular physical pion–nucleon state such as $\pi^- + \mathrm{p}$ into a superposition of isotopic spin states is performed by treating $T$ as an angular momentum according to the method for decomposition of angular momentum eigenstates using Clebsch–Gordan coefficients. We shall illustrate this method in more detail in Chapter 4.

### 2.9. Summary of conservation laws

For our future use, it may be desirable at this time to summarize the most basic conservation laws generally accepted as valid. Since the conservation laws which apply depend on the nature of the interaction, we shall distinguish the following types.

I. Strong interactions,
   (a) Nuclear interactions (N) with a coupling strength of $\dfrac{g^2}{\hbar c} \gtrsim 1$.
   (b) Electromagnetic interactions (E.M.) with a coupling strength $\dfrac{e^2}{\hbar c} \sim \dfrac{1}{137}$.

II. Weak interactions (W) with a coupling strength† $\lesssim 10^{-11}$.

The classes of interactions for which a conservation law or invariance of a symmetry is assumed valid will be indicated in brackets

1. Lorentz invariance.
   A. Proper Lorentz transformations (all interactions),
      (a) translational invariance results in a conservation of energy and momentum.
      (b) rotational invariance results in a conservation of angular momentum.
   B. Improper Lorentz transformations,
      (a) conservation of parity (invariance under inversion of space coordinates) (N and E.M.),
      (b) time reversal invariance (N and E.M.).
2. Charge conjugation invariance (N and E.M.).
3. Conservation of charge (gauge invariance) (all interactions).
4. Conservation of baryon numbers (all interactions).
5. Conservation of strangeness (N and E.M.).
6. TCP invariance (all interactions).
7. Conservation of leptons (all interactions i.e. E.M. and W) (see Chapter 12).
8. CP (or T) invariance—CP invariance has been observed to be violated by a small amount ($\sim 10^{-3}$ level) in weak interactions (so far only in kaon decays). So far there have been no observed violations in strong interactions or in E.M. interactions. The origin of the violation is still obscure and is discussed in Chapter 12, §12.15.
9. Isotopic spin (SU(2)) invariance (N).
10. SU(3) and SU(6) symmetry groups—This subject is discussed in Chapter 10. These symmetries are assumed to be valid for so-called very strong interactions and are violated by medium strong, electromagnetic, and weak interactions.

† In the case of the weak interactions the coupling constant squared is not dimensionless but has the dimensions of mass$^{-2}$. Thus, it is appropriate to form the product $\frac{(Gm^2)^2}{4\pi\hbar c}$. Typically, $m$ is taken equal to the nucleon mass. In that case, $\frac{(Gm_N^2)^2}{4\pi\hbar c} \lesssim 10^{11}$. This is the characteristic coupling strength for weak interaction decays. In the case of weak interaction scattering processes, the appropriate coupling strength is about five orders of magnitude larger $\left(\text{i.e.}\frac{(Gm_N^2)}{4\pi\hbar c} \sim 10^{-6}\right)$. The beta, or leptonic, decay of nucleons or mesons or decays of strange particles involving strangeness violation are weak interactions. The coupling constant that is appropriate is $\lesssim 10^{-11}$.

11. G. parity conjugation invariance (N).

At this point we shall temporarily terminate the general theoretical introduction† and add to it when appropriate subject matter is developed.

## REFERENCES


1*a*. SCHWEBER, S., BETHE, H., and de HOFFMAN, F. (1955). *Mesons and fields*, Vol. 1 Row, Peterson and Co. Evanston, Illinois (U.S.).

1*b*. BETHE, H., and de HOFFMAN, F. (1955). *Mesons and fields*, Vol. 2. Row, Peterson and Co.

2. WEISS, P. (1936). *Proc. R. Soc. London*. **A156,** 192; (1938). **A169,** 102; (1938). **A169,** 119.
3. TOMANAGA, S. (1946). *Prog. theor. Phys., Osaka*. **1,** 27.
4. DIRAC, P. A. M. (1945). *Phys. Rev.* **73,** 1092.
5. SCHWINGER, J. (1948). *Phys. Rev.* **74,** 1439.
6. SCHIFF, L. I. (1949). *Quantum mechanics*. McGraw-Hill, New York.
7. CASE, K. M. (1949). *Phys. Rev.* **76,** 14; MOLDAUER, P. and CASE, K. M. (1953). *Phys. Rev.* **A91,** 459. (More generalized equivalence theorems.)
8. WENZEL, G. (1947). *Rev. mod. Phys.* **19,** 1.
9. BREIT, G., CONDON, E. U., and PRESENT, R. D. (1936). *Phys. Rev.* **50,** 825; BRETT, G. and FETNBERG, E. (1936) *Phys. Rev.* **50,** 850; BREIT, G. and STEHN, J. R. (1937). *Phys. Rev.* **52,** 396.
10. CASSEN. B., and CONDON, E. U. (1936). *Phys. Rev.* **50,** 846.
11. FRÖHLICH, H., HEITLER, W., and KEMMER, N. (1938). *Proc. R. Soc.* **A166,** 154.
12. KEMMER, N. (1938). *Proc. Camb. phil. Soc. math. phys. Sci.* **34,** 354.
13. HAMILTON, J. (1959). *Theory of elementary particles*. Clarendon Press, Oxford.


† The reader is referred to Hamilton's book [13] for a general theoretical treatment.

# 3

# PROPERTIES OF THE PION

## 3.1. Background

IN addition to predicting the mass of the pion ($\sim$200–300 $m_e$) on the basis of the nuclear force range, Yukawa also realized that the pion could decay via electron emission and, in fact, postulated this mechanism to explain the nuclear $\beta$-decay. This fixed the estimated lifetime of the pion electron decay at $\sim 10^{-8}$ s. The third general characteristic of the pion to emerge from Yukawa's work was the coupling constant which, when related to the observed strength of nuclear forces, was estimated to be $g^2/\hbar c \sim 1$.

## 3.2. Discovery of mesons

In the next few years following Yukawa's work, cosmic ray investigations at mountain altitudes indicated that a sizeable flux of penetrating particles of mass $\sim$200 $m_e$ existed, and that these particles were unstable and decayed [1]–[3]. This was deduced from observations that the absorption was greater in a given mass of air than in an equivalent mass of solid absorber.

The electronic decay of one of these intermediate mass particles was finally observed in a cloud chamber (1940) [4] with magnetic field, and the lifetime of the particles was determined thereafter by counter techniques as $\approx 2{\cdot}2 \times 10^{-6}$ s. Although this lifetime was about two orders of magnitude longer than Yukawa's prediction, there was a general tendency, at that time, to associate these particles with the Yukawa nuclear meson. However, counter experiments in 1945–47 [5] demonstrated that the negative mesons in cosmic rays were not absorbed as expected when stopped in matter and attracted by positive nucleons. Therefore they exhibited a very weak interaction with nuclear matter, implying a coupling constant so small that the observed nuclear force strength could not be explained.

On the basis of this anomaly and other cosmic ray work, the two meson hypotheses were suggested by Marshak and Bethe (1947) [6], who assumed that the Yukawa nuclear force meson was the parent particle which decayed to the observed cosmic ray meson. In the same year, Lattes *et al.* [7] scanned photographic emulsions exposed to cosmic rays at high altitudes. They found events which were clearly interpretable as a coming to rest of a positive heavy meson, denoted as $\pi$ ($m \sim 300\ m_e$), followed by emission of a lighter meson, denoted as $\mu$ ($m \sim 200\ m_e$), of constant range or energy, which meant a two-body decay later identified as $\pi^+ \to \mu^+ + \nu$, where $\nu$ is a neutrino. Lattes *et al.* found events clearly interpretable as the capture of $\pi^-$ mesons at rest by emulsion nuclei, with the considerable disruption of the nucleus evidenced by emission of several charged particles. This meant that negative pions had the strong nuclear interaction required for the Yukawa meson. It was clear that the positive pions could not be captured at rest by a nucleus owing to the Coulomb barrier and, therefore, decayed to a muon and neutrino.

Using photographic emulsion sensitive to minimum ionization, the positive muons were shown to decay to an electron of continuous energy spectrum which, on average, had one third of the available energy. This led to the conclusion that the decay process involved two neutral particles which it has since been concluded are neutrinos. Soon after, a cosmic ray experiment measured the lifetime of the pion as $\sim 10^{-8}$ s, which was the order of magnitude predicted by Yukawa.

Hence it was finally clear that the pion had the three essential properties of the Yukawa meson, namely, the right order of magnitude of mass to explain the observed range of nuclear forces, the right order of magnitude of coupling constant to nucleons to explain the strength of nuclear forces, and the right order of magnitude of decay lifetime.

In the next few years, the confirmation of the existence of positive and negative pions and a determination of most of their properties as well as the discovery of a neutral pion resulted [8] from intensive pion investigations at the large synchro-cyclotrons (energy $\sim$ 200–400 MeV), electron synchrotrons, and linear accelerators.

### 3.3. Pion charges

Many experiments at the various particle accelerators have demonstrated that pions have one electronic unit (positive or negative) or a neutral charge. Direct measurements of the positive and negative pion charge are continually made in experiments where the ionization density (as a function of momentum) is measured or when, given a knowledge of the pion mass, the relation between momentum and velocity is measured. These experiments fix the pion charge to values close to one electronic unit. However, the proof that the pion charge is exactly one electronic unit is dependent upon the validity of the law of charge conservation in observed reactions of the following types (for which the known charge of other particles is written under the particle symbol and the required charge deduced for the pion, in order to satisfy the law of charge conservation, is written under the pion symbol)

$$\begin{array}{ccccccc} \pi^+ & \rightarrow & \mu^+ & +\nu & \rightarrow & e^+ & +3\nu \\ +1 & & +1 & 0 & & +1 & 0, \end{array}$$

mesonic disintegration of the deuterons

$$\begin{array}{cccc} \pi^+ + \mathrm{D} & \rightarrow & \mathrm{p} + \mathrm{p} \\ +1 \quad +1 & & +1 \quad +1, \end{array}$$

capture of a negative pion in hydrogen

$$\begin{array}{ccc} \pi^- + \mathrm{p} & \rightarrow & \mathrm{n} + \gamma \\ -1 \quad +1 & & 0 \quad 0, \end{array}$$

$$\begin{array}{ccc} \pi^- + \mathrm{p} & \rightarrow & \mathrm{n} + \pi^0 \\ -1 \quad +1 & & 0 \quad 0. \end{array}$$

Many experiments [8]–[10] have demonstrated that neutral pions exist with a mass a few MeV smaller than charged pions, and decay via the electromagnetic interaction in $\sim 10^{-16}$ s, predominantly to two $\gamma$-rays (i.e. $\pi^0 \rightarrow 2\gamma$). An alternative, but rare, mode of decay is $\pi^0 \rightarrow e^+ + e^- + \gamma$, which has been found experimentally to occur about 0·7 per cent of the time [11]. This process can be considered as an internal pair production by one of the two decay $\gamma$-rays of the $\pi^0$. Dalitz [12] has shown that the relative probability of this process can be calculated simply from our knowledge of quantum electrodynamics and does not depend on the details of the $\pi^0$ decay mechanism. We would expect that the order of magnitude of the relative probability would be of the order $\alpha \approx \frac{1}{137}$. It is estimated that the process should occur 0·7 per cent of the time, in excellent agreement with the experimental results. The internal pair production by both $\gamma$-rays is also possible, we then have the process $\pi^0 \rightarrow e^+ + e^- + e^+ + e^-$. We would obviously expect this process to be of the order of magnitude $\alpha^2$. The Dalitz estimates and the experimental results agree.

The fact that the $\pi^0$ decays electromagnetically in $\sim 10^{-16}$ s via 2 $\gamma$-rays for 99·3 per cent of the time means that the mean distance of travel before decay is of order $3 \times 10^{-6}$ cm.† Therefore all experiments which involve detection of the $\pi^0$ are indirect in the sense that only the electromagnetic decay products can be observed.

### 3.4. Charged pion mass

The first precision value of the positive pion mass was obtained from a study of the $\pi^+ \rightarrow \mu^+ + \nu$ decay, when the $\pi^+$ is at rest. Magnetic field deflection was used [13] to determine the value of the $\mu^+$ momentum ($P_{\mu^+}$)

† This is true for the Lorentz factor $\beta\gamma = \dfrac{\beta}{\sqrt{(1-\beta^2)}} \sim 1$. Obviously, this distance must be multiplied by $\gamma$ for high energy $\pi^0$s.

and, thereby, assuming that the neutrino has zero rest mass and using the best experimental values for the $\mu^+$ mass, the $\pi^+$ mass could be determined from conservation of energy and momentum, since

$$m_{\pi^+} = P_{\mu^+} + \sqrt{(P_{\mu^+}^2 + m_\mu^2)} = 139{\cdot}58 \pm 0{\cdot}05 \text{ MeV}.$$

The mass ratio $m_{\pi^-}/m_{\pi^+}$ was measured, also, and was found [13] to be $0{\cdot}998 \pm 0{\cdot}002$. Later experiments by this group and others, using counters, gave the best weighted value of the charged (±) pion mass (given in the Particle Data Group tables [14] as $134{\cdot}578 \pm 0{\cdot}013$). The various sources of data and methods used are referred to in the data card listing.

### 3.5. Neutral pion mass and lifetime

The short lifetime of the neutral pion means that only indirect means can be used effectively to measure accurately its mass and lifetime. The first reaction used was

$$\pi^- + \mathrm{p} \rightarrow \mathrm{n} + \pi^0 \rightarrow \mathrm{n} + 2\gamma.$$

The negative pions were allowed to come to rest in hydrogen, and since the reaction was observed it was clear that the $\pi^0$ was lighter. The negative pions are captured in an $s$-orbit around the protons in $\sim 10^{-12}$ s [15]. Owing to the kinetic energy which the lighter $\pi^0$ carries, the two photons have different energies and the observed width of the spectrum, and the minimum opening angle for the symmetrical two photon decay, allows us to determine the difference between the $m_{\pi^-}$ and $m_{\pi^0}$. From these minimum angle measurements we obtain [9]

$$m_{\pi^-} - m_{\pi^0} = 4{\cdot}5 \pm 0{\cdot}3 \text{ MeV}.$$

Modern fast electronic techniques have allowed a much more accurate measurement [10] of $m_{\pi^-} - m_{\pi^0}$. The main features of the experimental arrangement are sketched in Fig. 3.1. Basically, the method consists of accurately measuring the time of flight of the slow recoiling neutron accompanying the $\pi^0$. The result obtained was

$$m_{\pi^-} - m_{\pi^0} = 4{\cdot}59 \pm 0{\cdot}01 \text{ MeV},$$
$$m_{\pi^0} = 134{\cdot}99 \pm 0{\cdot}05 \text{ MeV}.$$

The reaction $\pi^- + \mathrm{p} \rightarrow \mathrm{n} + \gamma$ was observed, also. It served as a check on the apparatus and also gave an independent measurement of $m_{\pi^-} = 139 \pm 0{\cdot}6$ MeV which agrees, within the considerably larger error, with the more accurate measurement described earlier. In view of the $s$-wave character of the initial state and the zero spin of pions, the high cross-section observed

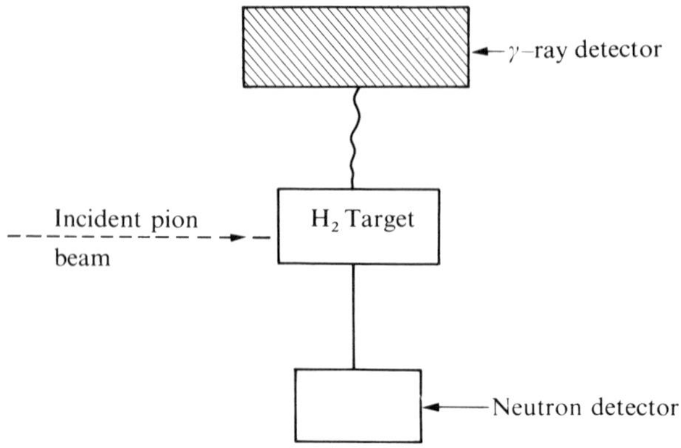


FIG. 3.1.

for zero-energy charge exchange strongly supports the assignment to the $\pi^0$ of the same parity (odd) as that of $\pi^-$. A series of counter results of similar, or higher, accuracy have been reported (see [14] for complete references to the various papers).

The present best values are

$$m_{\pi^\pm} - m_{\pi^0} = 4{\cdot}6041 \pm 0{\cdot}0037,$$
$$m_{\pi^0} = 134{\cdot}975 \pm 0{\cdot}015.$$

The $\pi^0$ lifetime has been measured both by direct observation, using emulsions, and by indirect observation, using counter techniques. An example of the former method was an experiment by Glaser, Seeman, and Stiller [16*a*]. In their experiment they observed $K^+$ mesons, which when brought to rest in photographic emulsion decay by emitting a $\pi^+$ and a $\pi^0$ in opposite directions. The $\pi^0$ travels until it decays electromagnetically, and they only selected the decays which involve one $\gamma$ and an $e^+ + e^-$ pair, the latter having visible tracks. Thus, they can locate the $\pi^0$ decay point and, knowing the end of the $K^+$ track and the velocity of the $\pi^0$, they could determine the lifetime to be $\approx(1{\cdot}9 \pm 0{\cdot}5) \times 10^{-16}$ s. The accurate measurements have been obtained using counter techniques [16] which measure the lifetime indirectly (see [14]). The weighted average [14] is $(1{\cdot}12 \pm 0{\cdot}20) \times 10^{-16}$ s.

### 3.6. The spin of the pion

The spin of the positive pion has been determined using the principle of detailed balance applied to the reaction $\pi^+ + D \rightarrow p + p$, and its inverse reaction $p + p \rightarrow D + \pi^+$. Although historically this principle has been used, the more fundamental, although almost equivalent approach in cases involving spin, is the reciprocity theorem of nuclear physics as stated by Blatt and Weisskopf ([3], Chapter 1), which is based on the assumption that time reversal invariance holds. The theorem states that the probability for a transition proceeding one way in time is equal to the probability for the same transition but with the sense of time reversed, and with the spin directions of both the emerging particles from the scattering centre reversed. If $S_{\alpha\beta}$ is the scattering matrix, or the transition matrix, for the process $\alpha \to \beta$, then the reciprocity theorem gives $S_{\alpha\beta} = S_{-\beta-\alpha}$, where we define the 'time reversed channel' $-\alpha$, as identical to $\alpha$ in all respects except that the spin directions of both emerging particles from the scattering centre have been reversed.

The detailed balance principle states $S_{\alpha\beta} = S_{\beta\alpha}$ and is usually derived from perturbation theory arguments or statistical consideration of particles in a box, thus it is not as general as the reciprocity theorem. For spinless particles, the two statements are identical and, in most cases where one averages over spin directions, the distinction is removed.

Marshak (1951) and Cheston first suggested [17] the application of detailed balance to the processes $\pi^+ + \mathrm{D} \to \mathrm{p}+\mathrm{p}$ for determination of the pion spin. Since the $S$-matrix element is the same for the original and time reversed reaction we can easily demonstrate the following relation, for the same c.m.s. energy in the two cases at the same c.m.s. angle ($\theta$),

$$\frac{\mathrm{d}\sigma}{\mathrm{d}\Omega(\theta)}(\mathrm{p}+\mathrm{p} \to \mathrm{D}+\pi^+)p^2(2s_\mathrm{p}+1)(2s_\mathrm{p}+1)$$

$$= \frac{\mathrm{d}\sigma}{\mathrm{d}\Omega(\theta)}(\pi^+ + \mathrm{D} \to \mathrm{p}+\mathrm{p})q^2(2s_\pi+1)(2S_\mathrm{D}+1),$$

where $p$ and $q$ are the proton and pion c.m.s. momenta, respectively, $s_\mathrm{p}$ is the proton spin ($=\frac{1}{2}$), $S_\mathrm{D}$ is the deuteron spin ($= 1$), and $s_\pi$ is the pion spin, which is to be determined. Putting in the known spin values, we obtain

$$\frac{\mathrm{d}\sigma}{\mathrm{d}\Omega(\theta)}(\mathrm{p}+\mathrm{p} \to \mathrm{D}+\pi^+) = \frac{\mathrm{d}\sigma}{\mathrm{d}\Omega(\theta)}(\pi^+ + \mathrm{D} \to \mathrm{p}+\mathrm{p}) \times \frac{q^2}{p^2} \times \tfrac{3}{4} \times (2s_\pi+1).$$

This expression is, of course, only applicable if the protons are unpolarized. Therefore it is clear that if both of the reactions $\mathrm{p}+\mathrm{p} \rightleftarrows \pi^+ + \mathrm{D}$ are measured experimentally and compared, the pion spin $s_\pi$ can be determined. $s_\pi$ must be an integer, since the pion is a boson and bosons can be shown to have integral spins. Therefore, in particular, if $s_\pi = 0$ or $s_\pi = 1$ there will be a factor of 3 difference in the right-hand side of the equation. Cartwright *et al.* [18] (1953) measured the differential cross-section for $\mathrm{p}+\mathrm{p} \to \mathrm{D}+\pi^+$

at the Berkeley cyclotron. The incident proton kinetic energy was 341 MeV corresponding to a c.m.s. pion kinetic energy of 21·4 MeV.

Durban *et al.* [19*a*] (1951) determined the differential and total cross-sections of $\pi^+ + D \rightarrow p + p$ at a c.m.s. pion energy of 25 MeV. Clark *et al.* determined the total cross-section only for $\pi^+ + D \rightarrow p + p$, at a c.m.s. pion energy of ~23 MeV. Cartwright *et al.* [18] then calculated the expected results using detailed balance for the inverse reaction observed by Durbin *et al.*, and found that $s_\pi = 0$ fit very well, but $s_\pi = 1$ gave a result differing by more than three error deviations on the differential cross-sections and more than four deviations on the total cross-section comparison. The total cross-section work of Clark *et al.* [19*b*], done at about the same time, was also strongly consistent with a spin of 0.

We might hope that the negative pion spin could be determined by an analogous method. The appropriate reaction would be

$$\pi^- + D \rightleftarrows n + n.$$

Unfortunately, a pure n+n interaction is not obtainable experimentally. The best one could do would be to study n+D and consider only those events in which the proton is only a spectator in the sense of the impulse approximation. This, unfortunately, would not be a clean experiment, and the interpretation of the results would be open to question. However, it should be remarked that either charge symmetry or charge independence of pion nucleon forces strongly imply that the spin of the $\pi^-$ must be the same as that of the $\pi^+$.

### 3.7. Spin of the neutral pion

Experiments have shown that the neutral pion readily decays into two photons ($\pi^0 \rightarrow 2\gamma$) with a lifetime which is expected for this electromagnetic decay. Some time ago, Yang [20] showed that the $2\gamma$ decay is strictly forbidden for odd spin particles and, hence, the $\pi^0$ must have even spin. The validity of charge independence, which is observed, depends upon the $\pi^0$ spin being zero. The charge exchange scattering $\pi^- + p \rightarrow \pi^0 + n$ would not take place at rest (i.e. near zero energy) with the observed large cross-section if the spin of the $\pi^0$ were two, thus the $\pi^0$ spin must be zero.

The parity of the $\pi^0$ can be determined, in principle, by measuring the plane of polarization of the two photons. If one analyses, in terms of linearly polarized photons for a pseudoscalar $\pi^0$ (odd parity), the two photons should have planes of polarization which are orthogonal. For a scalar $\pi^0$ the planes of polarization for the two photons should be parallel. The experimental determination of the plane of polarization has not been practicable since the accuracy attainable in measurements of the plane of polarization of photons at high energies is poor. However, when the two $\gamma$'s both convert to Dalitz $e^+ + e^-$ pairs, the correlation between the planes of the pairs reflects the polarization correlation between the photons. The experimentally determined correlation [21*a*] (Plano 1959) is clearly that expected [21*b*] for an odd parity $\pi^0$.

We should remember that the observed charge independence strongly requires an odd parity $\pi^0$ also and, therefore, it would be almost impossible to reconcile the experimental data with an even parity $\pi^0$.

### 3.8. Parity of charged mesons

The determination as to whether the pion has a scalar or pseudoscalar wave function (odd or even intrinsic parity) can be made from a consideration of the absorption of stopped $\pi^-$ mesons in deuterium. The following reactions take place,

$$\begin{aligned}
&\text{(a)}\ \pi^- + \mathrm{D} \to 2\mathrm{n},\\
&\text{(b)}\ \pi^- + \mathrm{D} \to 2\mathrm{n} + \pi^0 \to 2\mathrm{n} + 2\gamma,\\
&\text{(c)}\ \pi^- + \mathrm{D} \to 2\mathrm{n} + \gamma.
\end{aligned}$$

Panofsky and co-workers [22] (1951) did the original work using $\pi^-$ meson beams at the Berkeley cyclotron, and directly observed reaction (b) by detecting the $\gamma$-rays. The slowing down and capture process was carefully analysed theoretically by Wightman [23]. He estimated that in liquid deuterium a negative pion that has been slowed down to 200 keV by the collision loss mechanism is slowed down further and captured in an outer orbit about the deuteron, in about $2 \times 10^{-12}$ s. The $\pi^-$ then finally cascades down to the $k$-shell of the mesonic-deuteron atom in a comparably short time and is captured from the $l = 0$, $k$-shell orbit. The sum of the capture rates of competing processes for the capture of the $\pi^-$ from orbits of $l \neq 0$ are negligible compared to capture from $l = 0$ orbit transition rate. This was demonstrated by Brueckner *et al.* [24] (1951) for the reaction (a) (in which we will be most interested) by considering the most important of the $l \neq 0$ capture orbits, namely, $l = 1$. These authors used an analysis of the inverse reaction $\mathrm{n} + \mathrm{n} \to \pi^- + \mathrm{D}$ to demonstrate that the capture rate from the $l = 1$ state is negligible. Since the characteristic slow-down time is very short compared to the $\pi^- \to \mu^- + \nu$ decay lifetime, this decay process occurs to a negligible degree before capture. Therefore, in the Panofsky experiments the pion is captured by the deuteron from an $S$-state, and the negative pion rest mass energy, minus the deuteron binding energy, is the available kinetic energy in each final state. Hence, in reaction (c) the $\gamma$-ray energy would be expected to have a phase-space distribution which would correspond to $2n+\gamma$, with the expected modifications due to the neutron–neutron interaction in the final state, which is strongest at low relative neutron–neutron energies. Therefore, the $\gamma$-ray energy would tend toward the pion rest mass energy in reaction (c). On the other hand, in reaction (b), since the $\pi^0$ decays into 2 $\gamma$-rays, the average $\gamma$-ray energy would tend to be somewhat less than $\frac{1}{2}$ the pion rest mass. Panofsky and co-workers found evidence for reaction (c) but no evidence for reaction (b). That reaction (b) does not occur was subsequently verified by Steinberger and Chinowsky [25]. They had previously studied the capture of negative pions in hydrogen in which case practically every pion (to an accuracy of 1 part in $\sim 10^4$) gives rise to one of the two reactions

$$\text{(d)}\ \pi^- + p \to \pi^0 + n \to 2\gamma + n,$$
$$\text{(e)}\ \pi^- + p \to n + \gamma.$$

The $\gamma$-rays from the two reactions can be easily separated since those from (e) are monochromatic and occur at the upper energy cut-off of the $\gamma$-ray spectrum due to (d), and the relative frequency of (d) to (e) was determined. Then by comparing the number of $\gamma$-rays produced in a hydrogen and a deuterium target by the same number of stopping negative pions and making the plausible assumption that the capture rate in deuterium is the same as the capture rate in hydrogen,† these authors were able to determine the total number of capture interactions which gave rise to the process (a) ($\pi^- + D \to 2n$), since, for the same number of stopping pions in hydrogen or in deuterium,

$$\underset{\text{deduced}}{\underset{\downarrow}{\text{No. of (a)}}} + \underset{\text{measured}}{\underset{\downarrow}{\text{No. (b)}}} + \underset{\text{measured}}{\underset{\downarrow}{\text{No. (c)}}} = \underset{\text{measured}}{\underset{\downarrow}{\text{No. (d)}}} + \underset{\text{measured}}{\underset{\downarrow}{\text{No. (e)}}}.$$

They found $\dfrac{\text{No. (a)}}{\text{No. (c)}} = (2{\cdot}4 \pm 0{\cdot}5):1.$

Subsequently, Chinowsky and Steinberger observed [25] both neutrons in coincidence in process (a) and directly demonstrated the existence of this process in a sizeable fraction of the captures, and approximately estimated the ratio $\dfrac{\text{No. (a)}}{\text{No. (c)}} \sim 1{\cdot}5 \pm 0{\cdot}8.$

† Since the initial capture occurs in the outer atomic orbits, it is only determined by the Coulomb field in the region of the outer orbits and should, therefore, be almost identical for hydrogen or deuterium.

Now that we have demonstrated the occurrence of the capture process

$$\underset{\substack{\downarrow \\ \text{at rest}}}{\pi^-} + \mathrm{D} \rightarrow 2\mathrm{n}$$

with a reasonable branching ratio and have previously concluded that the $\pi^-$ is captured in an $S$ state, we can [26] apply conservation of parity, conservation of angular momentum, and the requirements of the Pauli exclusion principle to this process and, thereby, deduce the $\pi^-$ parity.

The initial $\pi^- + \mathrm{D}$ state has an angular momentum of 1, since the deuteron angular momentum is 1 and the pion is a spin zero particle in an $S(l = 0)$ state. Therefore the final $2n$ state must also have an angular momentum of 1, due to conservation of angular momentum. Therefore let us now consider all possible final state wave functions which have an angular momentum of 1 and also satisfy the Pauli exclusion principle, which requires that the $2n$ state be anti-symmetric under exchange of both spin and space coordinates. Hence, if the orbital angular momentum is even (even parity), the $2n$ spin function must be anti-symmetric under spin exchange ($S = 0$), whereas if $l$ is odd, the spin wave function must be symmetric ($S = 1$) under spin exchange. Since $S$ can only be 0 or 1, $l$ must be 0, 1, or 2 to form a $J = 1$ state since $\mathbf{J} = \mathbf{L}+\mathbf{S}$. With $l = 0$ or $l = 2$, the Pauli principle requires $S = 0$. Hence we cannot construct a $J = 1$ final state. Therefore the only possibility is $l = 1$ which then requires $S = 1$ with which we do indeed have a $J = 1$ final state ($L = J = S = 1$) which satisfies the Pauli principle. But now the final 2n state has odd parity ($L = 1$). Therefore, by parity conservation the initial $L = 0$ state of the $\pi^- + \mathrm{D}$ system must have odd parity. The deuteron, as is well known, has even parity since it is a mixture of even angular momentum states ($S$, and a few per cent of $D$). Therefore the parity of the initial system state will be even or odd depending upon whether the intrinsic pion parity is even or odd, respectively. We have concluded, already, that the final state and, therefore, the initial state must have odd parity by the law of parity conservation. Hence we have proved that the negative pion has odd intrinsic parity.

We should note, at this point, that the concept of intrinsic parity of a particle which can be emitted or absorbed by another particle (in the present case, the $\pi^-$ is absorbed by a nucleon) is a relative one, which is defined by the change of the parity of the original system (nucleon) upon absorbing the particle. Wick, Wightman, and Wigner (1952) [27] have discussed this concept and its limitations in more detail.

As previously mentioned, the reaction for stopped negative pions ($\pi^- + \mathrm{D} \rightarrow 2\mathrm{n} + \pi^0$) is not observed. This is not too surprising since the initial state has $J = 1$ and, therefore, the final state must have $J = 1$, which requires $l = 1$ in the final state. However, only 2·3 MeV is available, which would tend to make the reaction unlikely.

The parity of the $\pi^+$ cannot be as directly demonstrated, but the basic concepts of charge symmetry and charge independence depend on the $\pi^+$ having the same parity as the $\pi^-$. Incidentally, charge independence requires the same parity for the $\pi^0$. Since many tests of charge symmetry and charge independence exist, it is extremely probable that the $\pi^+$ has odd intrinsic parity. Furthermore, as we shall see later, there is reasonable evidence for a gradient coupling between the nucleon and the pion in the non-relativistic approximation. In the simple case of a direct pseudoscalar pion–nucleon coupling (which reduces to gradient coupling for non-relativistic energy according to the equivalence theorems) a spin-zero pion would require odd pion parity. There are many other cases of consistency with the assumption that the positive pion has odd parity. However, a conclusive direct experimental verification of the odd parity for the $\pi^+$, using the same method as for the $\pi^-$, has not been feasible since one would have to study the reaction $\pi^+ + D \rightarrow p + p$. Unfortunately, if one used stopping $\pi^+$, the repulsive Coulomb barrier would prevent absorption of the $\pi^+$ by deuterium and the $\pi^+$ would decay to $\mu^+$. We could, of course, study this reaction at very low energies (up to a few MeV) but the capture rate would be lower, and we would not be completely sure that the initial state is an $S$-state.

### 3.9. Lifetime and decay modes of the pion

The decay of the positive pion into a positive muon of unique range was first observed in the Lattes, Occhialini, and Powell experiments, which resulted in the discovery of the pion. The known decay processes of a positive pion at rest and the relative branching ratios are [14, 28]

| | |
|---|---|
| (a) $\pi^+ \rightarrow \mu^+ + \nu_\mu$ | 1 |
| (b) $\pi^+ \rightarrow e^+ + \nu_e$ | $(1{\cdot}24 \pm 0{\cdot}03) \times 10^{-4}$ |
| (c) $\pi^+ \rightarrow \mu^+ + \nu + \gamma$ | $(1{\cdot}24 \pm 0{\cdot}25) \times 10^{-4}$ |
| (d) $\pi^+ \rightarrow \pi^0 + e^+ + \nu$ | $(1{\cdot}02 \pm 0{\cdot}07) \times 10^{-8}$ |
| (e) $\pi^+ \rightarrow e^+ + \nu + \gamma$ | $(3{\cdot}0 \pm 0{\cdot}5) \times 10^{-8}$. |

It is clear that monoenergetic muon decay occurs with a frequency of about four orders of magnitude greater than the next most important mode.

In the monochromatic muon decay reaction (a) the conservation of energy, momentum, and charge requires that a second neutral particle of low rest mass ($<$ few $m_e$) be emitted. $\gamma$-ray emission is ruled out since electron showers or pairs are not observed when positive pions decay. Furthermore, no strong nuclearly interacting particle is observed. The lifetime of the $\pi^+$ meson can be studied by electronic measurement of the decay distribution of the stopped $\pi^+$. The lifetime of the $\pi^+$ decay is given in [14] as having a mean life of $(2{\cdot}603 \pm 0{\cdot}006) \times 10^{-8}$ s. As we shall see later, this lifetime is characteristic of what we would expect for a universal Fermi weak interaction. Therefore the neutrino is the only known weak interaction particle of zero or very near zero rest mass, and it is reasonable to assume it is the partner of the muon. This also satisfies the requirement of conservation of statistics, since a boson (the pion) must transform into an even number of fermions to conserve statistics. Direct demonstration that the neutral partner of the muon in pion decay is a particle which interacts with nuclear matter as a neutrino would be expected to interact has recently been accomplished [29] in an experiment at the Brookhaven Alternating Gradient Synchrotron. This experiment also demonstrated that it is highly probable that the neutrino associated with a muon is different from the neutrino associated with an electron in $\beta$-decay. Since, as previously noted, the $\pi^-$ is captured at rest, its weak decay characteristics can only be studied in flight.

In recent tests of CPT invariance, the difference of the lifetimes of $\pi^+$ and $\pi^-$ have been measured to good precision by observing the decays of the same momentum pions in flight. The result [14] is

$$\frac{(\tau^+ - \tau^-)}{\bar{\tau}} = (0{\cdot}053 \pm 0{\cdot}068) \text{ per cent.}$$

Thus it has been found that the $\pi^+$ and $\pi^-$ have the same lifetime within a small error strongly supporting CPT conservation. As previously discussed, the neutral pion decays into two $\gamma$-rays $\pi^0 \to 2\gamma$ with a lifetime of $\tau_{\pi^0} = (0{\cdot}89 \pm 0{\cdot}18) \times 10^{-16}$ s.

## REFERENCES


1. Anderson, C. D., and Neddermeyer, S. H. (1937). *Phys. Rev.* **51,** 884; (1938). **54,** 88.
2. Street, C., and Stevenson, C. (1937). *Phys Rev.* **51,** 1005.
3. Auger, P., Ehrenfest, P., Fereonand, A., and Fournier, M. (1937). *C. r. hebd. Séanc. Acad. Sci., Paris.* **204,** 25.
4. Williams, E. J., and Roberts, G. E. (1940). *Nature,* **145,** 102.
5. Conversi, M., Pancini, E., and Piccioni, O. (1947). *Phys. Rev.* **71,** 209L.
6. Marshak, R. E. and Bethe, H. A. (1947). *Phys. Rev.* **72,** 506.
7. Lattes, C. M. G., Muirhead, H., Occhialini, G. P. S., and Powell, C. F. (1947). *Nature.* **159,** 694; Lattes, C. M. G., Occhialini, G. P. S., and Powell, C. F. (1947). *Nature.* **160,** 453, 484.

8. BORKLUND, R., CRANDALL, W. E., MOYER, B. J., and YORK, H. F. (1950). *Phys. Rev.* **77,** 213.
9. CHINOWSKY, W. and STEINBERGER, J. (1954). *J. Phys. Rev.* **93,** 586.
10. HADDOCK, R. P., ABASHIAN, A., CROWE, K. M., and CZIRR, J. B. (1959). *Phys. Rev. Lett.* **3,** 478.
11. LINDENFELD, P., SACHS, A., and STEINBERGER, J. (1953). *Phys. Rev.* **89,** 531.
12. DALITZ, R. H. (1951). *Proc. phys. Soc., London* A**64,** 667.
13. BARKAS, W. H., BIRNBAUM, W., and SMITH, F. M. (1952). *Phys. Rev.* **101,** 778.
14. PARTICLE DATA GROUP. (1970). Review of particle properties. *Rev. mod. Phys.* **42,** 87.
15. FIELDS, T. H., YODH, G. B., DERRICK, M., and FETKOVICH, J. G. (1960). *Phys. Rev. Lett.* **5,** 69; LEON, M. and BETHE, H. A. (1962). *Phys. Rev.* **127,** 636.
16*a*. GLASSER, R. G., SEEMAN, N., and STILLER, B. (1961). *Phys. Rev.* **123,** 1014.
16*b*. VON DARDEL, G., DEKKERS, D., MERMOD, R., VAN PUTTEN, J., VIVARGENT, M., WEBER, G., and WINTER, K. (1963). *Phys. Lett.* **4,** 51.
17. MARSHAK, R. E. (1951). *Phys. Rev.* **82,** 313; CHESTON, W. B. (1951). *Phys. Rev.* **83,** 1118.
CARTWRIGHT, W. F., RICHMAN, C., WHITEHEAD, M. N., and WILCOX, H. A. (1953). *Phys. Rev.* **91,** 677.
19*a*. DURBIN, R., LOAR, H., and STEINBERGER, J. (1951). *Phys. Rev.* **83,** 646.
19*b*. CLARK, D. L., ROBERTS, A., and WILSON, R. (1951). *Phys. Rev.* **83,** 649L; (1952). *Phys. Rev.* **85,** 523.
20. YANG, C. N., (1950). *Phys. Rev.* **77,** 242, 722.
21*a*. PLANO, R., PRODELL, A., SAMIOS, N., SCHWARTZ, M., and STEINBERGER, J. (1959). *Phys. Rev. Lett.* **3,** 525.
21*b*. KROLL, N., and WADA, W. (1955). *Phys. Rev.* **98,** 1355.
22. PANOFSKY, W. K. H., AAMODT, R. L., and HADLEY, J. (1951). *Phys. Rev.* **81,** 565.
23. WIGHTMAN, A. S. (1950). *Phys. Rev.* **77,** 521.
24. BRUECKNER, K. A., SERBER, R., and WATSON, K. (1951). *Phys. Rev.* **81,** 575.
25. STEINBERGER, J. K. and CHINOWSKY, W. (1954). *Phys. Rev.* A**95,** 623; (1954). *Phys. Rev.* **95,** 1561; (1955). *Phys. Rev.* **100,** 1476.
26. FERETTI, B. (1947). *Proc. phys. Soc. Lond.* **1,** 75.
27. WICK, G. C., WIGHTMAN, A. S., and WIGNER, E. P. (1952). *Phys. Rev.* **88,** 101.
28. ASHKIN, J., FAZZINI, T., FIDECARO, G., GOLDSCHMIDT-CLERMONT, Y., LIPMAN, N. H., MERRISON, A. W., and PAUL, H. (1960). *Nuovo Cim.* **16,** 490.
29. DANBY, G., GAILLARD, J., GOULIANOS, K., LEDERMAN, L., MISTRY, N., SCHWARZ, M., and STEINBERGER, J. (1962). *Phys. Rev. Lett.* **9,** 36. (A similar, higher statistics, experiment was also performed later at CERN by BERNARDINI, G., *et al.* (1964). *Phys. Lett.* **13,** 86.)

# 4

# PION–NUCLEON SCATTERING

## 4.1. Introduction

IN Chapter 2 (§ 2.3), we considered the various simple Lorentz covariant forms of the pion–nucleon interaction. In Chapter 3, we demonstrated that the pion has spin zero and odd intrinsic parity and, therefore, should be represented by a pseudoscalar wave function. Therefore there are only two remaining possible forms of the interactions (see § 2.3) considered for the pion–nucleon case—a pseudoscalar pion field with pseudoscalar coupling $(G\bar{\Psi}\gamma_5\Psi\varphi)$ or a pseudoscalar pion field with pseudovector coupling $\left(\mathrm{i}\frac{F}{\mu}\bar{\Psi}\gamma_5\gamma_\nu\Psi\frac{\partial\phi}{\partial x_\nu}\right)$. The latter, PS(PV), gives a better direct explanation of the low energy pion–nucleon interaction but is not renormalizable like the PS interaction. Fortunately, it has been shown that at low energies the two forms of the interaction are in many cases equivalent.† That is to say that as long as the nucleon recoil energy is small the PS interaction has as its dominant term the pseudovector interaction. For convenience, we are using units with $\hbar = c = 1$. The charge independent PS interaction term in the Hamiltonian is (Kemmer 1938)

$$H_{\text{int}_{\text{PS}}} = \mathrm{i}G\int \mathrm{d}^3x\bar{\Psi}(x,t)\gamma_5[\boldsymbol{\tau}.\boldsymbol{\varphi}(x,t)]\Psi(x,t).$$

$G$ is the (pseudoscalar) coupling constant, $\bar{\Psi}$ and $\Psi$ are the nucleon field operators, $\frac{1}{2}\tau$ is the nucleon isospin, and $\boldsymbol{\varphi}$ (the pion amplitude) is an isotopic spin vector. The quantity $\boldsymbol{\tau}.\boldsymbol{\varphi}$ is invariant with respect to rotation in charge-space and, therefore, assures isotopic spin conservation and hence charge independence. At low energies compared to the creation of a nucleon–antinucleon pair ($\lesssim$300 MeV), we have for pseudovector coupling‡

$$H_{\text{PV}} = \frac{f}{\mu}\int \mathrm{d}^3x\bar{\Psi}(\boldsymbol{\sigma}'.\boldsymbol{\nabla})(\boldsymbol{\tau}.\boldsymbol{\varphi})\Psi. \tag{4.1}$$

Let us now consider the virtual process of emission and absorption of a pion by a nucleon. It is to be expected that this process will give the largest range force in the pion–nucleon and the nucleon–nucleon interactions and, therefore, we should expect to have a high cross-section for single pion

† The equivalence theorem does not hold, for example, in weak coupling theory $\pi^- + \mathrm{n}$ scattering, since the intermediate state does not correspond to a free proton.

‡ $\boldsymbol{\nabla}$ does not act on $\tau$.

exchange. The pion pair exchange or nucleon–anti-nucleon pair exchange will be expected to have, respectively, smaller cross-section effects. When a nucleon at rest emits a low energy pion, the nucleon recoil energy is small and the nucleon to a high degree of approximation will remain in an $s$-state. Since the pion has odd intrinsic parity, it must have $l = 1, 3, 5, \ldots$ in order to conserve parity. In order to conserve angular momentum $l = 1$ is the only possibility, since this can be allowed when accompanied by nuclear spin flip, but no higher $l$ can be accommodated. The virtual pion emission process and its inverse (absorption of a low energy pion by a nucleon) are illustrated in Fig. 4.1.

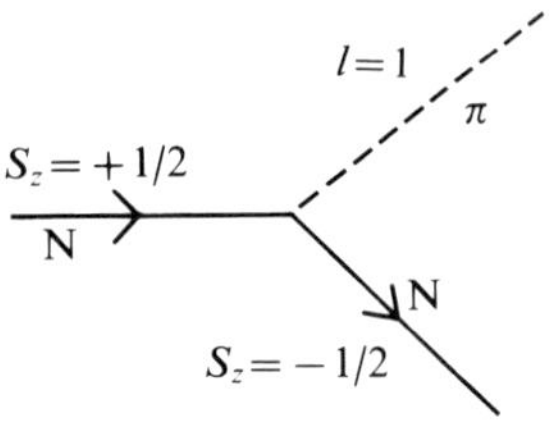


FIG. 4.1.(a) Virtual pion emission.

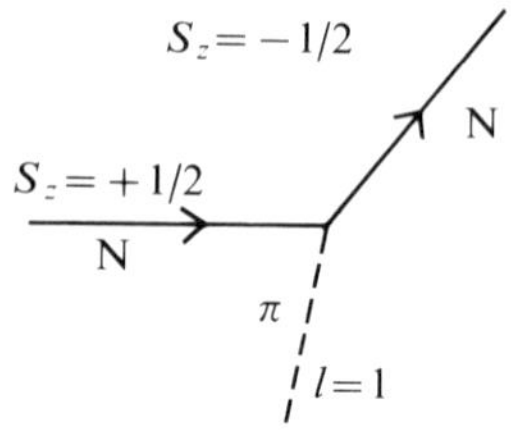


FIG. 4.1.(b) Virtual pion absorption.

Conservation of angular momentum requires that the $\pi$–N state have $J = \frac{1}{2}$, $l = 1$. If we further require charge independence, the isotopic spin state must be $T = \frac{1}{2}$: The $\boldsymbol{\sigma}.\boldsymbol{\varphi}$ in the PV interaction has just these characteristics, and strongly favours $p$-wave pions,† and provides the nucleon spin flip required for this simple virtual picture of the absorption and emission of a single pion.

Let us now look at physical $\pi$–p scattering at low energies selecting those diagrams which involve the emission and absorption of single pions.

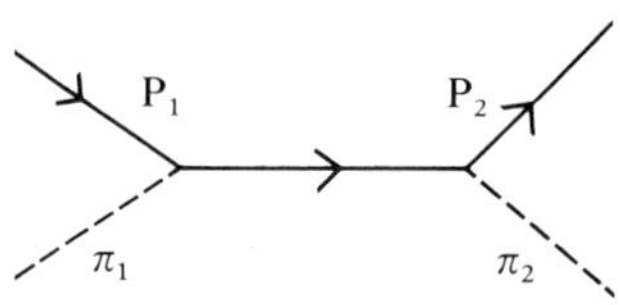


FIG. 4.1.(c) Pion absorption followed by pion emission. This is referred to as a direct Born term.

Figs. 4.1(c) and 4.1(d) have already been considered (§ 2.5). Fig. 4.1(c), absorption followed by emission, is referred to as a direct Born term when a lowest order relativistic perturbation theory calculation is made. In order to conserve parity, angular momentum, and isospin, it is required that the initial state from which this absorption occurs must have $J = \frac{1}{2}$, $l = 1$, $T = \frac{1}{2}$. Thus, this process cannot occur from $s$-wave initial states. Fig. 4.1(d),

† Owing to nucleon recoil and other relativistic effects $S$ and higher angular momentum states also contribute to a much smaller extent.

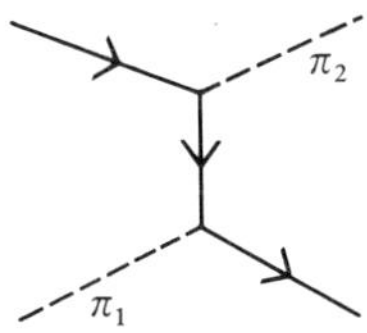


FIG. 4.1.(d) Pion emission followed by pion absorption. This is referred to as a nucleon exchange.

emission of a pion by the nucleon followed by absorption of a pion by the nucleon, is referred to as nucleon exchange since (as we can see from the diagram) the nucleon line crosses over to what was originally the pion line. Nucleon exchange should be expected to occur most strongly in the $J = \frac{3}{2}$, $l = 1$, $T = \frac{3}{2}$ state, since the Clebsch–Gordan coefficients are most favourable in the case where the angular momenta and isotopic spin vectors of both pion and nucleon are aligned. Calculations show that the nucleon exchange for the $T = J = \frac{3}{2}$ state is reduced by a factor of two when either $J$ or $T = \frac{1}{2}$, and about a factor of 4 when $J = T = \frac{1}{2}$.

In $s$-wave pion–nucleon states there can obviously be no contribution from the nucleon exchange process considered above. From the nature of the pseudoscalar interaction, it is expected that the most important $s$-wave contribution will come from Fig. 4.1(e).

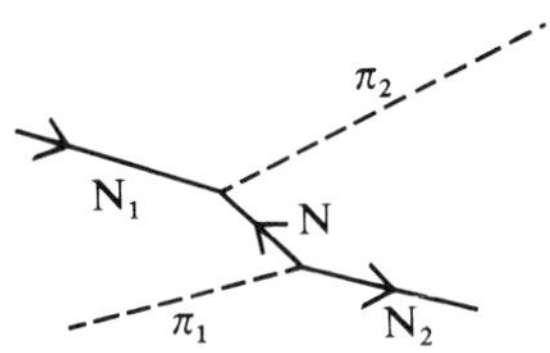


FIG. 4.1.(e)

Thus in the $p$-wave case we are exchanging single pions, which are primarily responsible for the nuclear force range and, therefore, the range of the $p$-wave interaction is expected to be $\sim 1{\cdot}4 \times 10^{-13}$ cm. In the $s$-wave case (see Fig. 4.1(e)) we are exchanging a nucleon–anti-nucleon pair, and its much greater mass causes us to expect a much shorter range, say $\sim 10^{-14}$ cm. Therefore, from the above consideration, if we perform a phase shift analysis of the low energy $\pi$–p scattering experiments, we should expect to find a small $s$-wave important near zero pion energy, with a rapidly increasing $p$-wave contribution, and little $d$- and higher angular momentum waves up to several hundred MeV. The strongest interactions are to be expected in the $T = J = \frac{3}{2}$ state. In fact, according to the strong coupling theory, a nucleon isobar or resonance in the $T = J = \frac{3}{2}$ state could be expected to dominate the low energy $\pi$–p interaction. Experimentally, this has turned out to be the case, although presenting these theoretical arguments before the experiments is an excellent example of 'putting the cart before the horse', which the author hopes will be excused in the interests of effective presentation.

### 4.2. Strong coupling meson theory

The strong coupling meson theory had been developed by a number of workers, and an account of its basic calculations and results was summarized by Pauli [1] (1946) and reviewed by Wenzel [2] (1947). At the time there were no satisfactory methods for treating strong coupling mesons via a quantized field theory, since neither higher order terms in the coupling constant nor self-energy problems could be handled. It was customary to argue that, in the strong coupling limit, one would expect many virtual mesons to exist around the nucleon and, therefore, a classical treatment of the meson field could be considered a reasonable approximation.

At present, in spite of the fact that self-energy problems can be treated by renormalization, we still have not developed a good consistent treatment for the strong coupling problem. In the classical strong coupling theory calculations, the nucleons were treated non-relativistically, and a finite size and definite position was assigned to the nucleons. The mesons were treated as a classical field.

The nuclear force characteristics favoured the symmetrical pseudoscalar theory. Pseudovector coupling was chosen since pseudoscalar coupling involves the treatment of nucleon pairs which cannot be done in a non-relativistic treatment. On the other hand, the divergence difficulty in pseudovector coupling calculations was obviated because the introduction of a finite nucleon radius led to a cut-off of the divergences at the higher momenta. The meson–nucleon interaction term in the Hamiltonian was assumed to be

$$\mathrm{H}_{\mathrm{int}} = -(F(\mu) \int V(\mathbf{r})\boldsymbol{\sigma}.\nabla\varphi \, \mathrm{d}^3r, \tag{4.2}$$

where $\int V(\mathbf{r})\,\mathrm{d}^3r = 1$. $V(\mathbf{r})$ represents a 'shape function' of the nucleon.

Let us now determine the form factor by taking the Fourier transform of $V(\mathbf{r})$,

$$u(\mathbf{k}) = \int V(\mathbf{r})\mathrm{e}^{-\mathrm{i}\mathbf{k}.\mathbf{r}}\,\mathrm{d}^3r. \tag{4.3}$$

For $\mathbf{k} = 0$,

$$u(\mathbf{k}) = \int V(\mathbf{r})\,\mathrm{d}^3r = 1. \tag{4.4}$$

We then define the effective radius of the nucleon in terms of the form factor as follows

$$r_0 = \frac{\pi}{2\int |u(\mathbf{k})|^2\,\mathrm{d}^3k}, \tag{4.5}$$

It is obvious that for meson wavelengths, $\dfrac{1}{k} \ll a$. The form factor greatly reduces their contributions. Hence in what follows the approximation is made that the meson wavelength is large enough, so that $|u(\mathbf{k})| \approx 1$.

The most famous, and the most relevant, prediction of the old strong coupling theory calculations is that the nucleon, owing to its interaction with the meson field, is accompanied by a series of 'isobar states' or excited states of angular momentum $J = \frac{1}{2}, \frac{3}{2}, \frac{5}{2}, \frac{7}{2}, \ldots$, the first of which is the nucleon ground state. The excitation energy is given by

$$E = 2\omega_0(J+\tfrac{3}{2})(J-\tfrac{1}{2}), \tag{4.6}$$

where the constant $\omega_0$, which controls the spacing, is given by

$$\omega_0 = 6\pi r_0\left(\frac{\mu}{F}\right)^2 = 6\pi r_0\left(\frac{2M}{G}\right)^2 = \frac{3}{2}\left\{r_0\left(\frac{\mu}{f}\right)^2\right\} = \frac{6r_0M^2}{g^2}. \tag{4.7}$$

For the charge independent case the theory predicts that the lowest lying states would have $T = J = \frac{3}{2}, \frac{5}{2}, \frac{7}{2}, \ldots$, etc. The states which have $T \neq J$ would have an excitation energy $\sim f^2/r_0$, which in the strong coupling limit would correspond to much higher energies. For intermediate strength couplings (which appear to be the case) the $T \neq J$ states excitation energy would decrease, while the $T = J$ states excitation energy would increase, and they could become comparable.†

The nature of the $T = J = \frac{3}{2}$ state is that of an isobar state consisting of a pion (in a $p$ state) and the nucleon. The $T = J = \frac{5}{2}$ state consists of two pions and the nucleon. The $T = J = \frac{7}{2}$ state contains three pions and the nucleon, and so on.

The meson scattering cross-section was also derived in these calculations. The original meson field of the nucleon, which contains many mesons, was considered to be slightly perturbed by the incident meson which penetrated inside the physical nucleon. The main vehicle for the interaction involves the gradient coupling of the meson field with the nucleon spin, which leads to a self-energy of the nucleon spin and an associated spin inertia proportional to $f^2/r_0$. Pauli developed general formulae which reduce as follows. In the case when $\dfrac{1}{r_0} \gg \mu$ in the strong coupling case, where $\dfrac{1}{r_0}$ is of the order of the

† i.e. the ratio

$$\frac{\omega_0}{f^2/r_0} = \frac{T = J \text{ states spacing}}{T \neq J \text{ states spacing}} = \frac{\frac{3}{2}r_0(\mu/f)^2}{f^2/r_0} \approx \frac{\frac{3}{2}\mu^2r_0^2}{f^4} = \frac{\frac{3}{2}\times\frac{1}{50}}{0{\cdot}01} \approx 3.$$

nucleon mass, and at low energies where $\omega \ll \frac{1}{r_0}$, $u(\mathbf{k}) \approx 1$, we obtain

$$\sigma_{\text{total}} = \frac{32\pi}{9}\left(\frac{f^2}{\mu}\right)^2 \frac{k^4}{\omega^2\mu^2}\left[\frac{1+\left(\dfrac{\omega}{\omega_0}\right)^2}{\{1-(\omega/\omega_0)^2\}^2+\left(\dfrac{2r_0k^3}{\omega_0\omega}\right)^2}\right]. \tag{4.8}$$

The most relevant feature for present day considerations is the resonance denominator in the last factor of eqn (4.8), namely,

$$\frac{1+\left(\dfrac{\omega}{\omega_0}\right)^2}{\{1-(\omega/\omega_0)^2\}^2+\left(\dfrac{2r_0k^3}{\omega_0\omega}\right)^2}.$$

When $\omega = \omega_0$, there is a resonance. If $\omega_0 > \mu$, there are no bound state isobars. If $\omega_0 < \mu$, we would expect bound state isobars. As we shall discuss later, the first isobar (i.e. the $T = J = \frac{3}{2}$ isobar) is observed to have a resonance kinetic energy of about 190 MeV in the laboratory system. From this we obtain $\omega_0 \approx 2{\cdot}3\ \mu$. Since it is reasonable to assume $r_0M \approx 1$ (i.e. that the radius of the nucleon is approximately equal to its Compton wavelength) we obtain, from eqn (4.7),

$$\omega_0 = 2{\cdot}3\ \mu = \frac{3}{2}\frac{6r_0M^2}{g^2} = \frac{3}{2}\frac{r_0\mu^2}{f^2}, \tag{4.9}$$

or, setting $r_0M = 1$,

$$g^2 = \frac{6M}{2{\cdot}3\ \mu} \approx 17, \tag{4.10a}$$

$$f^2 = \frac{3}{2}\frac{\mu\times\dfrac{1}{6{\cdot}7}}{2{\cdot}3\ \mu} = \frac{3}{2\times2{\cdot}3\times6{\cdot}7} \approx 0{\cdot}1. \tag{4.10b}$$

These coupling constant values are close to the characteristic accepted values of $g^2$, which are typically $\sim$10–15 and $f^2$, which are typically $\approx$0·08–0·1. Of course, the fact that we obtain $f^2 \approx 0{\cdot}1$ means that at energies up to the resonance there are typically one to at times a few mesons present in the nucleon field and, hence, the classical treatment, without quantization, is unjustified. The lack of renormalization, the lack of the possibility of treatment of $s$-wave scattering, and the necessity of introducing an extended source are all limitations of this simple treatment.

It appears that the pion–nucleon coupling is strong enough to form nucleon isobars and, hence, we have some of the general features of the strong coupling limit. However, especially at low energies (i.e. $\omega \ll M$), where the pseudovector coupling reduces the interaction, the strong coupling limit assumption that on the average there are many mesons in the field is unjustified. Nevertheless, it is comforting that the most important features of low energy $\pi$–p scattering can be understood in the above terms.

### 4.3. Weak coupling theory

Since the experimental facts, such as the existence of nucleon isobars (especially the $T = J = \frac{3}{2}$), contradict the basic assumption of weak coupling theory, we shall treat this subject only briefly.

In analogy to electromagnetic theory, weak coupling theory is most useful when $g^2$ and $f^2$ are small enough for only the lowest order terms in the perturbation theory treatment to contribute appreciably. This is, obviously, not the case for the pion–nucleon interaction. Nevertheless, it is instructive to consider the weak coupling theory prediction for $\pi$–N scattering. A detailed presentation of weak coupling theory calculation is given in Marshak's book [3]. First let us consider the pseudoscalar pion with pseudoscalar coupling

$$H_{\text{int}} \sim G\overline{\Psi}\gamma_5\Psi\varphi. \tag{4.11}$$

The $\gamma_5$ operator transforms the nucleon from positive to negative energy states analogous to the scattering of radiation by electrons, where the operator $\gamma_\mu$ has large matrix elements between positive and negative energy states.

The Thomson cross-section formula for low-energy $\gamma$ scattering, by electrons is,

$$\sigma_{\text{Thomson}} = \tfrac{2}{3}\{4\pi\alpha^2(\hbar/mc)^2\},$$

where $\alpha = e^2/\hbar c$ is the dimensionless coupling constant. Similarly, for pseudoscalar mesons with pseudoscalar coupling, we obtain

$$\sigma_{\pi-\text{N}}[\text{PS}] \sim 4\pi g^4(\hbar/Mc)^2.$$

This approximately constant cross-section is in complete disagreement with the low energy data since, using $g^2 \approx 15$, $\sigma[\text{PS}] \approx 750$ mb. Of course the small $s$-wave phase shifts can be explained by a mechanism of this type, but very heavy use of pair suppression arguments is required to get quantitative agreement.

For pseudovector coupling the matrix elements $M_{\text{if}}$ and $M_{\text{fi}}$ are proportional to

$$\frac{f}{\mu}\frac{\boldsymbol{\sigma}.\mathbf{k}}{\sqrt{(\omega(k))}}. \tag{4.12}$$

Hence,

$$\sigma[\text{PV}] \sim \left(\frac{k^2}{\omega}\right)^2\left(\frac{f}{\mu}\right)^4 \sim \left(\frac{k^2}{\omega\mu}\right)^2\left(\frac{\hbar}{\mu c}\right)^2 f^4. \tag{4.13}$$

The result of a perturbation theory calculation is

$$\sigma = \frac{32\pi}{9}\left(\frac{f^2}{\mu}\right)^2 \frac{k^4}{\omega^2\mu^2}. \tag{4.14}$$

The $k^4$ factor is introduced by the pseudovector coupling. With the exception of the last resonance factor, contained in the bracket of eqn (4.8), this is the same as the strong coupling theory formula previously developed. Therefore at low pion energy ($\omega \ll \mu$) it gives the same prediction, but its continuous and rapid increase with increasing $k$ disagrees with experiment at the higher energies.

As was previously mentioned, at very low energies even the strong coupling theory corresponds to an effective weak coupling interaction which qualitatively explains the similarity in the result.

### 4.4. Early measurements of low energy ($\lesssim$200 MeV) pion–nucleon scattering

The first experimental picture of the nature of the pion–nucleon interaction began to emerge as a result of the study of $\pi^{\pm}$–p total cross-sections with the F.M. cyclotron at Chicago.

Anderson, Fermi *et al.* measured the total hydrogen cross-section for positive and negative pions from low energy to 135 and 200 MeV, respectively [4]. They determined the total cross-sections using the conventional counter telescope transmission measurement method. Typical accuracies were 10–20 per cent. They observed the sharp rise of the cross-sections with energy from very low values at low energy, which is illustrated in Fig. 4.2(a) and which strongly suggested that gradient coupling with a *p*-wave was predominantly responsible for the interaction.† The data shown in Fig. 4.2(a) generally satisfied the predictions [3] of PS(PV) theory for the low energy behaviour $\sigma \propto k^4$, with a value of $f^4 \sim 0{\cdot}1$. We now know, experimentally, that $f^4 \sim 0{\cdot}01$ is more appropriate. The continuous increase in cross-section with increasing momentum in the weak coupling calculation is not physically

† The high ($\sim$ 500–750 mb) constant cross-section predicted by the PS(PS) interaction was grossly contradicted by the experiments.

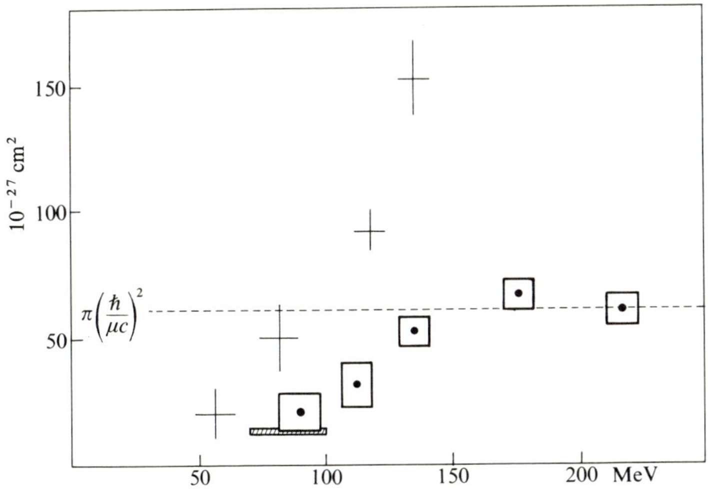


FIG. 4.2.(a) Total cross-sections of negative pions in hydrogen (the sides of the rectangle represent the error), and positive pions in hydrogen (the arms of the cross represent the error). The cross-hatched rectangle is the Columbia result. The black square is the Brookhaven result and does not include the charge-exchange contribution. (From Anderson, Fermi, Long, and Nagle. (1952). *Phys. Rev.* **85,** 936.)

significant. If an appropriate strong coupling treatment is made, an isobar which corresponds to a definite angular momentum state, like the $T = J = \frac{3}{2}$ isobar, would exhibit a resonance at the resonant energy of the isobar. We have already found an indication of this in eqn (4.8). If an appropriate phase shift analysis were made, the maximum cross-section would be limited to less than the total cross-section contained in the incident channel wave—the maximum permissible total cross-section being attained at the resonance energy. For example for $T = J = \frac{3}{2}$, the maximum cross-section would be $8\pi\lambda\!\!\!^{-2}$.† These early measurements were explained by Brueckner [5] in a theoretical treatment based on the strong coupling meson theory in which he took the above effect into account. He calculated the scattering to be expected from the existence of the lowest lying nucleon isobar of $T = J = \frac{3}{2}$ as predicted for the pseudoscalar meson field in the strong coupling theory.

The resonance level parameters for a resonance in the state $T = J = \frac{3}{2}$ were assigned phenomenologically, using the methods of Wigner and Eisenbud and imposing charge independence approximately to fit the total cross-section data (Fig. 4.2(b)).

† $\lambda\!\!\!^{-} = \dfrac{\lambda_{\text{cms}}}{2\pi} = \dfrac{1}{k}$.

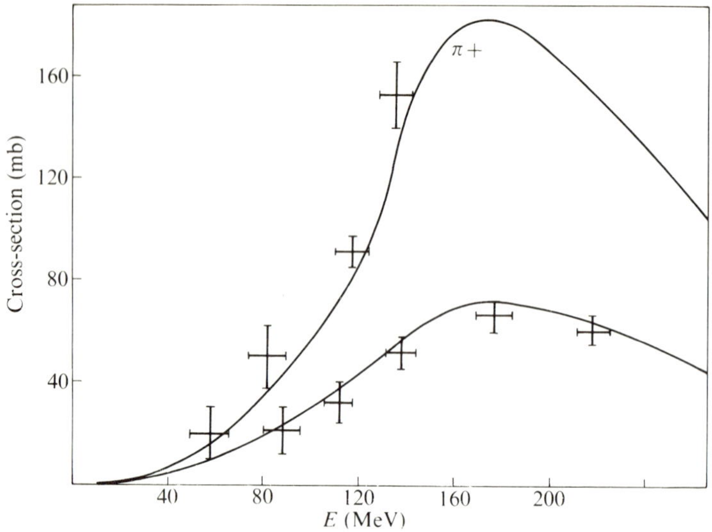


FIG. 4.2.(b) Total cross-sections for scattering of $\pi^+$ and $\pi^-$ mesons in hydrogen (including charge-exchange). (From Brueckner. (1952). *Phys. Rev.* **86,** 106.)

The result was of the form

$$\sigma_{\text{total}} \approx 8\pi\lambda\!\!\!^{-2}\frac{\rho^2}{\rho^2+\epsilon^2}+\text{smaller terms}, \qquad (4.15)$$

where

$$\rho = \left(\frac{(a/\lambda\!\!\!^{-})^3}{1+(a/\lambda\!\!\!^{-})^2}\right)b$$

and $\epsilon = (\omega_r - \omega)$, where $\omega$ is the total pion c.m.s. energy and $\omega_r$ is its value at the resonant energy; $a$ is the channel radius and is of order $\hbar/\mu c$; $b$ is a constant related to the width of the resonance. This form insured that $\sigma_{\text{total}} \leq 8\pi\lambda\!\!\!^{-2}$ with the equality† applying at the resonance. The $T = J = \frac{3}{2}$ phase shift would go through 90° at the resonance. Subsequent, more complete total cross-section data and extensive negative pion on hydrogen differential data [6] were phase shift analysed by Fermi *et al.* [7] assuming *s*- and *p*-waves only, in a best-fit computer programme. At that time, the $\pi^+$–p total cross-sections were still increasing sharply with energy to 135 MeV (the highest energy then available at Chicago). The $\pi^-$–p total cross-sections exhibited a sharp rise with increasing energy, and then a

† This presupposes setting the 'smaller terms' equal to zero, which is appropriate if a resonance in the $T = J = \frac{3}{2}$ state is the only interaction.

broad plateau to 220 MeV. These authors [7] never found a resonant solution, and their conclusion was that the results could be best explained by a large $T = J = \frac{3}{2}$ phase shift which rose to about 50°, and then fell again at the higher energies (i.e. did not go through resonance), accompanied by a rather large *s*-wave phase shift toward the high energy end. This implied that the earlier fit to the data by the Brueckner formula was an accident. Further support for this type of behaviour was given by an intermediate coupling extended source theory by Chew [8] which was able to fit the then-known total cross-section variation with energy reasonably well, without going through a resonance.

### 4.5. Early intermediate energy (~150–750 MeV) $\pi^{\pm}$–p total cross-section measurements

Early in 1953, the higher energy beams which became available at the Brookhaven Cosmotron were used to determine the behaviour of $\sigma(\pi^{\pm}\text{–p})$ from 150 to 750 MeV. A prominent peak at about 180–200 MeV and a rapid decrease at higher energies was observed [9]. For positive pions, particularly, the decrease after the peak was considerably faster than $\frac{1}{\lambda\!\!\!^{-2}}$. This behaviour was very suggestive of a resonance, and it was shown, by Lindenbaum and Yuan, that the Brueckner resonance formula fits the data as illustrated in Fig. 4.3.

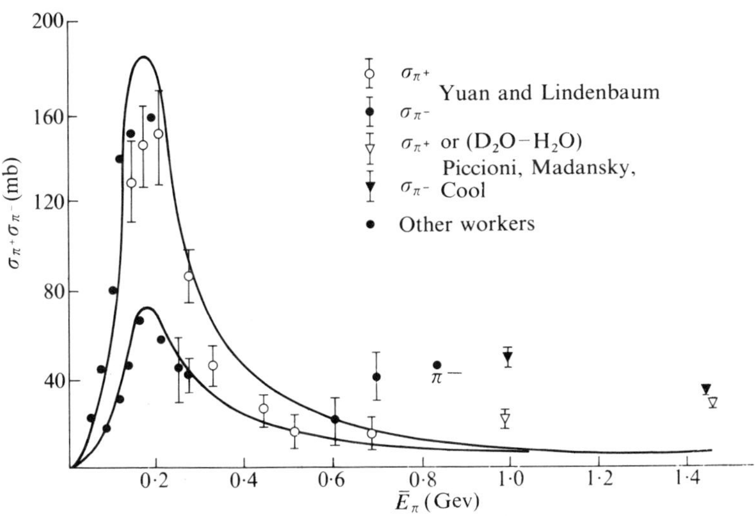


FIG. 4.3. (From Lindenbaum and Yuan [9*b*].)

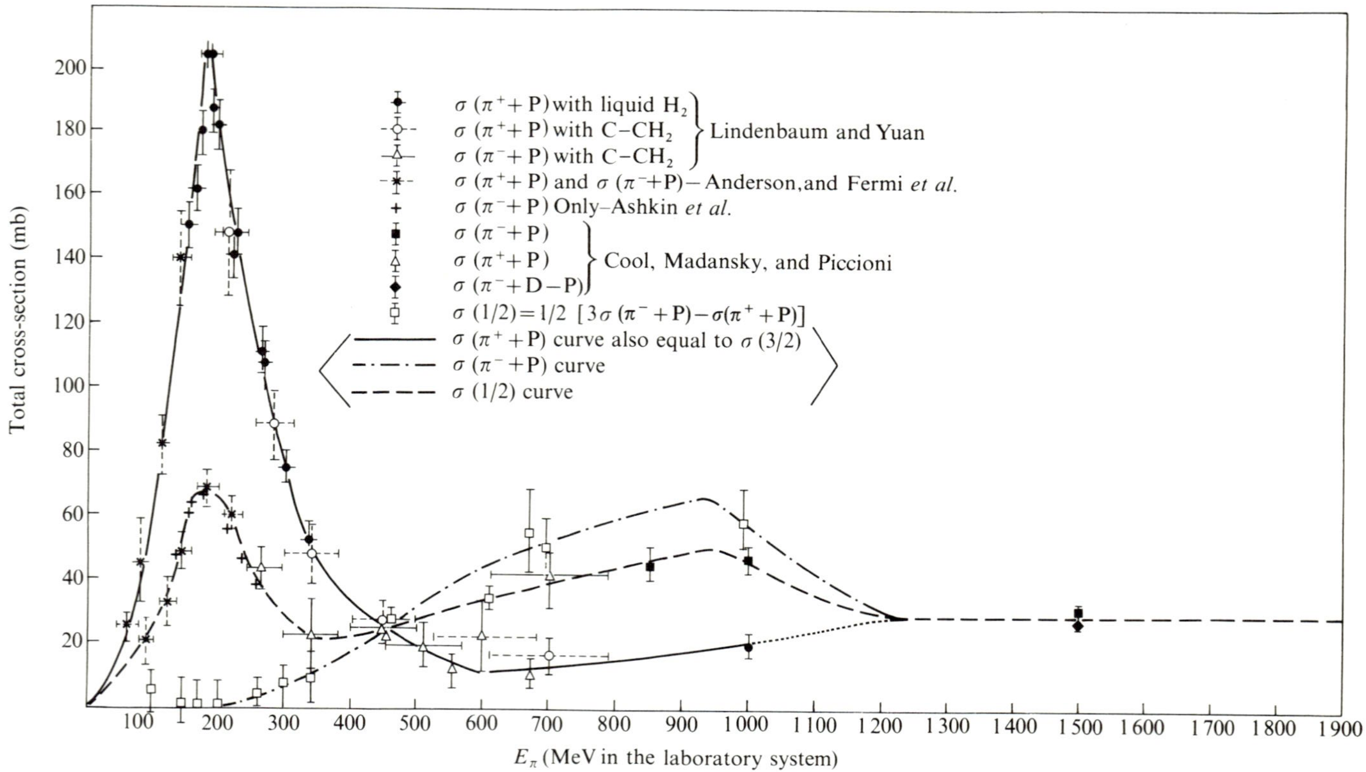


FIG. 4.4. The total cross-section of hydrogen for positive and negative pions from 0 to 1500 MeV. The curves also include some cross-sections (at the higher energy end) which are equal to deuterium–hydrogen differences. The deduced values for $\sigma(\frac{3}{2})$ or $\sigma(\frac{1}{2})$ are also shown. (From [10]. 1955.)

Although the absolute magnitude of the $\pi^{\pm}$–p raw peak cross-section was below $8\pi\lambda\!\!\!^{-2}$ by about 15 per cent, it was concluded that, within the resolution errors of the early measurements which used a $C$–$CH_2$ difference, a value of $8\pi\lambda\!\!\!^{-2}$ was attained. In addition, the data revealed that $\sigma(\pi^+\text{–p}) \approx 3\sigma(\pi^-\text{–p})$. If one assumes charge independence (i.e. conservation of isotopic spin) then for the $\pi$–p system there are two total cross-sections, one for the $T = \frac{3}{2}$ state and one for the $T = \frac{1}{2}$ state. As we shall see in § 4.6, the result is

$$\sigma(T = \tfrac{3}{2}) = \sigma(\pi^+\text{–p}) \tag{4.16}$$

$$\sigma(T = \tfrac{1}{2}) = \tfrac{1}{2}[3\sigma(\pi^-\text{–p}) - \sigma(\pi^+\text{–p})]. \tag{4.17}$$

Since we have just noted that $3\sigma(\pi^-\text{–p}) \approx \sigma(\pi^+\text{–p})$, it follows that $\sigma(T = \frac{1}{2}) \approx 0$, while $\sigma(T = \frac{3}{2})$ was behaving as if it was going through a resonance at ~180–200 MeV. Thus, we had a striking confirmation of the general features of the strong coupling theory predictions.

Subsequent higher precision measurements of these data were made by several groups [10]–[12] at low and higher energies. In mid-1955 the total cross-section situation was summarized [10] in Fig. 4.4. It was by now clearly demonstrated that the low energy ($(E_\pi)_{\text{lab}} \leq 300$ MeV) total cross-section behaviour was that expected for a resonance in the $T = J = \frac{3}{2}$ state, with little or no interaction cross-section in the $T = \frac{1}{2}$ state below the resonance energy, and an increasing interaction cross-section with increasing energy thereafter, which became larger than the $T = \frac{3}{2}$ cross-section at about 450 MeV. The $T = \frac{1}{2}$ cross-section data exhibited a broad peak in the neighbourhood of 700–1000 GeV, while the $T = \frac{3}{2}$ cross-section exhibited a broad minimum in this region.

Since it had also been shown [13] that the characteristics of the production of pions, in p–Be and p–p collisions, were those to be expected if pion production proceeded almost entirely through excitation of one, or both, nucleons to the $T = J = \frac{3}{2}$ isobaric state, it was by now quite strongly established by experimental evidence that this state dominated the low energy $\pi$–p interaction (see Chapter 6).

The mystery as to why the phase shift analysis solution of Fermi, Metropolis, and Alei did not show this $T = J = \frac{3}{2}$ resonance had been resolved, also, by de Hoffman, Bethe, and Metropolis. They showed [14] that we could obtain a solution with $\delta_{33}$ resonating which fitted the data about as well as the original Fermi–Metropolis solutions. They used arguments of continuity, together with physical reasoning and some qualitative results of the theory, to require that the program used in the Maniac computer looked for a $\delta_{33}$ resonant solution, and they then found one.

The original Fermi–Metropolis work required merely a best fit to the phase shift problem and, hence, did not obtain this resonant solution since it was a slightly (but not significantly) worse fit.

### 4.6. Phase-shift analysis

In an isolated system of two spinless particles, the total orbital angular momentum $L$ and its $z$-component $L_z$ are conserved.

The solutions of the angular part of the separated wave function (i.e. spherically symmetrical potential) are of the form

$$Y_{l,m}(\theta, \phi) = C_{l,m} P_l^m(\cos\theta) e^{im\varphi}. \tag{4.18}$$

The total wave function is of the form

$$\psi(r, \theta, \varphi) = R(r) Y(\theta, \varphi), \tag{4.19}$$

where

$$P_l^m(\cos\theta) = \frac{\left(\frac{m}{|m|}\right)^{|m|}}{2^l l!} (\sin\theta)^{|m|} \left(\frac{d}{d(\cos\theta)}\right)^{l+|m|} (\sin\theta)^{2l}.$$

$Y_{l,m}(\theta)$ are the normalized spherical harmonics and are given by

$$Y_{l,m}(\theta, \varphi) = \frac{(-1)^{l+|m|}}{2^l l!}\left(\frac{m}{|m|}\right)^{|m|}\left\{\frac{(2l+1)(l-|m|)!}{4\pi(l+|m|)!}\right\}^{\frac{1}{2}} \times$$
$$\times \left\{(\sin\theta)^{|m|}\left(\frac{d}{d(\cos\theta)}\right)^{l+|m|}(\sin\theta)^{2l}\right\} e^{im\varphi}, \tag{4.20}$$

where

$$\int_{-1}^{1}\int_{0}^{2\pi} Y_{lm}^* Y_{l'm'}\, d(\cos\theta)\, d\varphi = \delta_{ll'}\, \delta_{mm'} \tag{4.21}$$

and

$$Y_{l,-m}(\theta, \phi) = (-1)^m Y_{l,m}^*(\theta, \phi).$$

These spherical harmonics have the property that†

$$L^2 Y_{l,m}(\theta, \varphi) = l(l+1)\hbar^2 Y_{l,m}(\theta, \varphi), \tag{4.22}$$

$$L_z = i\hbar \frac{\partial}{\partial\varphi}, \tag{4.23}$$

$$L_z Y_{l,m}(\theta, \varphi) = m\hbar Y_{l,m}(\theta, \varphi). \tag{4.24}$$

Furthermore, it can easily be shown that $\psi(r, \theta, \phi)$ has even or odd parity

† We are using the Condon–Shortley phase convention.

according to whether $l$ is even or odd, respectively,

$$P\psi(r, \theta, \phi) = (-1)^l \psi(r, \theta, \phi). \tag{4.25}$$

Hence, a general solution for the angular part of the wave function for a particular value of $l$ is

$$\sum_{m=-l}^{m=+l} Y_{l,m}(\theta, \varphi). \tag{4.26}$$

If one particle has spin, the total angular momentum $\mathbf{J} = \mathbf{L}+\mathbf{S}$, and its $z$ component is conserved, also. Then the following operators commute with the Hamiltonian,

$$J^2, J_z, L^2, L_z, S^2. \tag{4.27}$$

If both particles have spin then there can be non-central tensor forces of the form

$$\frac{3}{r^2}(\boldsymbol{\sigma}_1 . \mathbf{r})(\boldsymbol{\sigma}_2 . \mathbf{r}) - \boldsymbol{\sigma}_1 . \boldsymbol{\sigma}_2,$$

which prevent $L^2$ and $S^2$ from commuting with the Hamiltonian. Hence only $J^2$ and $J_z$ commute with the Hamiltonian, and are conserved.

We can consider the general case where one, or both, particles has spin as follows. Let the first particle be described by the angular wave function $Y_{j,m}(1)$ and the second by the wave function $Y_{j',m'}(2)$. Then the product wave function is

$$Y_{j,m}(1) Y_{j',m'}(2). \tag{4.28}$$

$$J_z Y_{j,m}(1) Y_{j',m'}(2) = (J_{1z} + J_{2z}) Y_{j,m}(1) Y_{j',m'}(2) = (m+m') Y_{j,m}(1) Y_{j',m'}(2). \tag{4.29}$$

There exist linear combinations of $Y_{j,m}(1)Y_{j',m'}(2)$, which are simultaneous eigenfunctions of $J^2$ with eigenvalue $J(J+1)$, and also eigenfunctions of $J_z$ with eigenvalue $M$.

These eigenfunctions of the total angular momentum will be denoted by $\mathcal{Y}^{M}_{J,j,j'}$, where

$$\mathcal{Y}^{M}_{J,j,j'} = \sum_{m=-j}^{j} \sum_{m'=-j'}^{m'=j'} C_{j,j'}(J, M, m, m') Y_{j,m}(1) Y_{j',m'}(2). \tag{4.30}$$

The $C_{jj'}(J, M, m, m')$ are called the Clebsch-Gordan coefficients. $C_{jj'}$ vanish unless $m+m' = M$. Therefore, $m' = M-m$. Hence,

$$\mathcal{Y}^{M}_{J,j,j'} = \sum_{m=-j}^{j} C_{j,j'}(J, M, m, M-m) Y_{j,m}(1) Y_{j',M-m}(2). \tag{4.31}$$

Conversely,

$$Y_{j,m}(1) Y_{j',m'}(2) = \sum_{|J|=|j-j'|}^{j+j'} \sum_{M=-J}^{M=+J} C_{j,j'}(J, M, m, m') \mathcal{Y}^{M}_{J,j,j'}, \tag{4.32}$$

but, since $M = m+m'$, it follows that

$$Y_{j,m}(1)Y_{j',m'}(2) = \sum_{J=|j-j'|}^{j+j'} C_{j,j'}(J, M = m+m', m, m')\mathscr{Y}_{J,j,j'}^{M=m+m'}. \quad (4.33)$$

Certain orthogonality relations can be derived, since the functions $Y_{j,m}$ and also the functions $\mathscr{Y}_{j,j'}^{M}$ form orthogonal sets, the Clebsch–Gordan coefficients are matrix elements of a unitary transformation and are real numbers. Thus the transformation is an orthogonal transformation. We, therefore, have the following properties

$$\sum_{m=-j}^{j} \sum_{m'=-j'}^{j'} C_{j,j'}(J, M, m, m')C_{jj'}(J', M', m, m') = \delta_{JJ'}\,\delta_{MM'}, \quad (4.34)$$

$$\sum_{J=|j-j'|}^{j+j'} \sum_{M=-J}^{J} C_{j,j'}(J, M, m, m')C_{j,j'}(J, M, m'', m''') = \delta_{mm''}\,\delta_{m'm'''}, \quad (4.35)$$

$$\sum_{M=-J}^{J} \sum_{m''=-j''}^{j''} C_{j,j''}(J, M, m, m'')C_{j',j''}(J, M, m', m'') = \frac{2J+1}{2j+1}\,\delta_{jj'}\,\delta_{mm'}. \quad (4.36)$$

Wigner and Condon and Shortley derived the general formulae for the Clebsch–Gordan coefficients. Condon and Shortley [15] have tabulated $j' = \frac{1}{2}$, 1, $\frac{3}{2}$, 2 such that these tables can (by utilizing various symmetry relations) be used when any one of $j$, $j'$, or $J$, is $\frac{1}{2}$, 1, $\frac{3}{2}$, or 2. Our treatment has been similar to that of Blatt and Weiskopf [16] to which the reader is referred for more detail. The symbol $C_{j,j'}(J, M, m, m')$ can be related to Condon and Shortley's symbol $(jj'mm' \mid jj'JM)$.

The Clebsch–Gordan coefficients are directly useable in determining isotopic spin functions, since these have properties identical to angular momentum. The Clebsch–Gordan coefficients for $j' = \frac{1}{2}$ and $j' = 1$ are tabulated in Tables 4.1 and 4.2.

Let us consider pion scattering in c.m.s. The incident pion momentum along the $z$ direction is denoted by $\mathbf{k}$. Hence the incident plane pion wave is represented by $e^{ikz}$. As a result of a scattering, the total wave function becomes

$$\psi = e^{ikz} + e^{ikr}\frac{f(\theta)}{r}, \quad (4.37)$$

where the second term represents the small additional spherical scattered wave. We can expand the incident wave into a combination of an incoming spherical wave and an outgoing spherical wave as follows

$$e^{ikz} = e^{ikr\cos\theta} = \sum_{l=0}^{\infty} A_l(r)Y_{l,o}(\theta), \quad (4.38a)$$

TABLE 4.1

*The Clebsch–Gordan coefficients from Condon and Shortley, and Blatt and Weisskopf* [16]

*Clebsch–Gordan coefficients* $C_{jj'}(J, M, m, m')$ *for* $j' = \frac{1}{2}$

| | $m' = \frac{1}{2}$ | $m' = -\frac{1}{2}$ |
|---|---|---|
| $J = j+\frac{1}{2}$ | $\left[\frac{j+M+\frac{1}{2}}{2j+1}\right]^{\frac{1}{2}}$ | $\left[\frac{j-M+\frac{1}{2}}{2j+1}\right]^{\frac{1}{2}}$ |
| $J = j-\frac{1}{2}$ | $-\left[\frac{j-M+\frac{1}{2}}{2j+1}\right]^{\frac{1}{2}}$ | $\left[\frac{j+M+\frac{1}{2}}{2j+1}\right]^{\frac{1}{2}}$ |

where

$$A_l(r) = \mathrm{i}^l\sqrt{[4\pi(2l+1)]}[j_l(kr)] = \mathrm{i}^l\sqrt{[4\pi(2l+1)]}[\sqrt{(\pi/2kr)}]J_{l+\frac{1}{2}}(kr). \quad (4.38\text{b})$$

$J_n(z)$ is a Bessel function of the first kind of order $n$ and $j_l(z)$ is a spherical Bessel function of order $l$. The following asymptotic expressions can be derived

$$j_l(kr) \cong \frac{(kr)^l}{1\times 3\times 5\ldots(2l+1)}\,\mathrm{e}^{ikr\cos\theta}, \qquad \text{for } kr \ll l. \quad (4.39\text{a})$$

$$j_l(kr) \cong \frac{\sin(kr-\frac{1}{2}l\pi)}{kr} = -\frac{1}{2\mathrm{i}kr}[\mathrm{e}^{-\mathrm{i}(kr-l\pi/2)} - \mathrm{e}^{+\mathrm{i}(kr-l\pi/2)}], \quad (4.39\text{b})$$

for $kr \gg l$.

TABLE 4.2 (from [16])

*Clebsch–Gordan coefficients* $C_{jj'}(J, M, m, m')$ *for* $j' = 1$

| | $m' = 1$ | $m' = 0$ | $m' = -1$ |
|---|---|---|---|
| $J = j+1$ | $\left[\frac{(j+M)(j+M+1)}{(2j+1)(2j+2)}\right]^{\frac{1}{2}}$ | $\left[\frac{(j-M+1)(j+M+1)}{(2j+1)(j+1)}\right]^{\frac{1}{2}}$ | $\left[\frac{(j-M)(j-M+1)}{(2j+1)(2j+2)}\right]^{\frac{1}{2}}$ |
| $J = j$ | $-\left[\frac{(j+M)(j-M+1)}{2j(j+1)}\right]^{\frac{1}{2}}$ | $\frac{M}{[j(j+1)]^{\frac{1}{2}}}$ | $\left[\frac{(j-M)(j+M+1)}{2j(j+1)}\right]^{\frac{1}{2}}$ |
| $J = j-1$ | $\left[\frac{(j-M)(j-M+1)}{2j(2j+1)}\right]^{\frac{1}{2}}$ | $-\left[\frac{(j-M)(j+M)}{j(2j+1)}\right]^{\frac{1}{2}}$ | $\left[\frac{(j+M+1)(j+M)}{2j(2j+1)}\right]^{\frac{1}{2}}$ |

Thus,

$$e^{ikz} \cong \frac{\pi^{\frac{1}{2}}}{kr}\sum_{l=0}^{\infty}[\surd(2l+1)]i^{l+1}[e^{-i(kr-l\pi/2)}-e^{+i(kr-l\pi/2)}]Y_{l,0}, \quad (4.40)$$

for $kr \gg l$.

We can also expand

$$f(\theta) = \sum_{l} F_l Y_{l,0}(\theta). \quad (4.41)$$

Hence combining the above equations we can write

$$\psi_{\text{final}} \approx \frac{\pi^{\frac{1}{2}}}{kr}\sum_{l=0}^{\infty}\surd(2l+1)i^{l+1}[e^{-i(kr-l\pi/2)}-\eta_l e^{i(kr-l\pi/2)}]Y_{l,0}, \quad (4.42)$$

where the complex constant $\eta$ represents the modification to the outgoing spherical wave component of the incident wave, but

$$\psi_{sc} = \psi(r)-e^{ikz} = \frac{\pi^{\frac{1}{2}}}{kr}\sum_{l=0}^{\infty}\surd(2l+1)i^{l+1}(1-\eta_l)e^{i(kr-l\pi/2)}Y_{l,0}. \quad (4.43)$$

The number of scattered particles per second ($N_{sc}$) is obtained by calculating the flux due to $\psi_{sc}$ through a sphere surrounding the scattering centre. The radius $r_0$ of the sphere is chosen such that $r_0 \to \infty$.

$$N_{sc} = \frac{\hbar}{2iM}\int\left(\frac{\partial\psi_{sc}}{\partial r}\psi_{sc}^{*}-\frac{\partial\psi_{sc}^{*}}{\partial r}\psi_{sc}\right)r_0^2 \sin\theta \, d\theta \, d\varphi. \quad (4.44)$$

Thus,

$$N_{sc} = \frac{\pi v}{k^2}\sum_{l=0}^{\infty}(2l+1)\,|1-\eta_l|^2, \quad (4.45)$$

where $v$ is the c.m.s. velocity. $\sigma_{sc} = N_{sc}/v$, thus

$$\sigma_{sc} = \pi\lambda\!\!\!^{-2}\sum_{l=0}^{\infty}(2l+1)\,|1-\eta_l|^2.$$

From eqn (4.37), and the definition of $\psi_{sc}$, we have

$$f(\theta) = \frac{\pi^{\frac{1}{2}}}{k}\sum_{l=0}^{\infty}(1-\eta_l)\surd(2l+1)\times i^{l+1}e^{-il\pi/2}Y_{l,0}, \quad (4.46a)$$

$$f(\theta) = (i\pi^{\frac{1}{2}}/k)\sum_{l=0}^{\infty}(1-\eta_l)\surd(2l+1)Y_{l,0}. \quad (4.46b)$$

The scattering process can be conveniently represented in terms of phase shifts ($\delta_l$) which are angles which represent the change in phase of the scattered wave with respect to the incoming wave by an angle $\delta_l$. It can easily be shown that, in terms of those phase shifts,

$$\eta_l = e^{2i\delta_l}, \tag{4.47a}$$

$$f(\theta) = \frac{1}{2ik}\sum_l (2l+1)(e^{2i\delta_l}-1)P_l(\cos\theta). \tag{4.47b}$$

$\delta_l$ is real when there is no absorption, and must be complex with a positive imaginary part when there is absorption.

The partial wave scattering amplitude ($a_l$) is a useful quantity which describes the effect of the scattering in a particular $l$-state. $a_l$, when written in units of $\lambda\!\!\!^-$ becomes, in terms of $\delta$,

$$a_l = \frac{1}{2i}(e^{2i\delta_l}-1) = \frac{1}{2i}e^{i\delta_l}(e^{i\delta_l}-e^{-i\delta_l})$$
$$= \frac{1}{2i}e^{i\delta_l}(2i\sin\delta_l) = e^{i\delta_l}\sin\delta_l. \tag{4.48a}$$

From our definition,

$$f(\theta) = \lambda\!\!\!^- \sum_l (2l+1)a_l P_l(\cos\theta). \tag{4.48b}$$

By definition,

$$\psi_{sc} = \psi(r) - e^{ilk\cos\theta} = \frac{e^{ikr}f(\theta)}{r},$$

$$dN_{sc} = \frac{\hbar}{2iM}\left(\frac{\partial\psi}{\partial r}\psi_{sc}^* - \frac{\partial\psi_{sc}^*}{\partial r}\psi_{sc}\right)r^2\sin\theta\,d\theta\,d\varphi,$$

$$dN_{sc} = \frac{\hbar k}{M}|f(\theta)|^2\,d\Omega,$$

where $d\Omega$ is equal to a differential solid angle and

$$d\sigma = \frac{1}{v}dN_{sc} = |f(\theta)|^2\,d\Omega.$$

Thus we obtain

$$\frac{d\sigma}{d\Omega} = |f(\theta)|^2 = \lambda\!\!\!^{-2}\left|\sum_l (2l+1)\exp(i\,\delta_l)\sin\delta_l P_l(\cos\theta)\right|^2, \tag{4.49}$$

where $\Omega$ is the solid angle in steradians.

$$\sigma_{\text{total}} = \int_{\theta=0}^{\theta=\pi} \frac{d\sigma}{d\Omega}\,d\Omega = 4\pi\lambda\!\!\!^{-2}\sum_l (2l+1)\sin^2\delta_l. \tag{4.50}$$

In the above treatment we have ignored the fact that, in $\pi$–p scattering, the proton has a spin. To take account of the fact that the proton spin can be flipped in a collision, we should have used the following expanded wave function in our treatment

$$\psi \approx e^{ikz}\chi_{m_s} + (f(\theta)\chi_{m_s} + g(\theta,\varphi)\chi_{-m_s})\frac{e^{ikr}}{r}, \tag{4.51}$$

where $f(\theta)$ is now the amplitude for the ordinary (non-spin flip) scattering, and $g(\theta, \varphi)$ is the amplitude for the spin flip scattering.

We can expand as follows

$$g(\theta,\varphi) = \sum_l G_l Y_l^{2m_s}(\theta,\varphi) \tag{4.52}$$

and, as before,

$$f(\theta) = \sum F_l Y_l^0(\theta). \tag{4.53}$$

We obtain

$$F_l = \frac{1}{k}\sqrt{\frac{4\pi}{2l+1}}\,[(l+1)a_{l,l+\frac{1}{2}} + l a_{l,l-\frac{1}{2}}], \tag{4.54}$$

$$G_l = \frac{1}{k}\sqrt{\frac{4\pi}{2l+1}}\,\sqrt{[l(l+1)]}[a_{l,l+\frac{1}{2}} - a_{l,l-\frac{1}{2}}]. \tag{4.55}$$

If the initial proton target is unpolarized and we sum over the recoil spin proton direction, the differential cross-section becomes

$$\frac{d\sigma}{d\Omega} = |f(\theta)|^2 + |g(\theta,\varphi)|^2 = \left|\sum_l F_l Y_l^0\right|^2 + \left|\sum_l G_l Y_l^{2m_s}\right|^2. \tag{4.56}$$

For pion energies (c.m.s.) $\sim \mu$, $pr_0 \sim 1$ hence, including $s$- and $p$-waves in the phase shift analysis should allow for the most important effects. This seems to be the case experimentally and was noticed in the early analyses.

With spin taken into account, the generalized partial wave scattering amplitude is defined as

$$a_{l,j} = \frac{1}{2\mathrm{i}}\,(\mathrm{e}^{2\mathrm{i}\delta_{l,j}}-1) = \mathrm{e}^{\mathrm{i}\delta_{l,j}}\sin\delta_{l,j}. \tag{4.57}$$

With only *s*- and *p*-waves, we need to calculate only the following

$$F_0 = \frac{\sqrt{(4\pi)}}{k}\,a_{0,\frac{1}{2}}, \tag{4.58}$$

$$F_1 = \frac{1}{k}\sqrt{(4\pi/3)}(a_{1,\frac{3}{2}}+a_{1,\frac{1}{2}}), \tag{4.59}$$

$$G_0 = 0, \tag{4.60}$$

$$G_1 = \frac{1}{k}\sqrt{(8\pi/3)}(a_{1,\frac{3}{2}}-a_{1,\frac{1}{2}}). \tag{4.61}$$

Hence,

$$\frac{\mathrm{d}\sigma}{\mathrm{d}\Omega} = \frac{1}{k^2}\,(|a_{0,\frac{1}{2}}+(2a_{1,\frac{3}{2}}+a_{1,\frac{1}{2}})\cos\theta|^2+|a_{1,\frac{3}{2}}-a_{1,\frac{1}{2}}|^2\sin^2\theta). \tag{4.62}$$

The formulae we have just derived apply for each charge state of the $\pi$–N system.

Since as we have noticed previously that owing to isotopic spin conservation there are only two independent amplitudes, then we need only double our three parameters to describe the $\pi$–N system and obtain a total of six independent amplitudes, which are sufficient to describe both the $\pi$–p and $\pi$–n systems. If isotopic spin were not conserved, we would need nine independent amplitudes to describe the $\pi$–p system with *s*- and *p*-waves only, and if charge symmetry were not valid, we would need another nine for the $\pi$–n system. Let us now write out these six amplitudes but change the notation, as follows.

For $\pi^+$–p $\to$ $\pi^+$–p,

$$a_{l,j}(\pi^+ \to \pi^+) = a_{l,j}(\tfrac{3}{2}), \tag{4.63}$$

where the number in parenthesis is the isotopic spin.

For $\pi^-$–p $\to$ $\pi^-$–p,

$$a_{l,j}(\pi^- \to \pi^-) = \tfrac{1}{3}[a_{l,j}(\tfrac{3}{2})+2a_{l,j}(\tfrac{1}{2})]. \tag{4.64}$$

For $\pi^-$–p $\to$ $\pi^0$–n,

$$a_{l,j}(\pi^- \to \pi^0) = \frac{\sqrt{2}}{3}\,[a_{l,j}(\tfrac{3}{2})-a_{l,j}(\tfrac{1}{2})]. \tag{4.65}$$

We should notice that since all spin effects were taken into account, the generality of the treatment of the phase shift analysis makes these formulae completely relativistic provided the relativistic value for $k$, or $\lambda\!\!\!^{-}$, is inserted. In fact, the validity of the phase shift analysis depends only on the conditions of the covariance and unitarity of the $S$-matrix.

### 4.7. Proton polarization in π–p scattering

In eqns (4.51–57) we have considered and explicitly taken into account the proton spin. This treatment can be extended to answer the question as to whether there is any resulting polarization of the recoiling proton after the scattering.

For convenience, we have in our previous treatment chosen the axis of quantization of the $z$-component of orbital angular momenta or spins along the $z$-axis, which is the direction of the incident pion, as depicted in the following diagram.

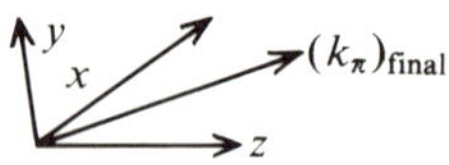


The total scattering amplitude then contains two terms (see eqn (4.51)) $f(\theta)\chi_{m_s}$ and $g(\theta, \varphi)\chi_{-m_s}$, where $f(\theta)$ is the amplitude for ordinary (non-spin flip) scattering, $g(\theta, \varphi)$ is the amplitude for spin flip scattering, and $m_s$ is the value of the $z$-component of the spin.

We can show, from the requirements of parity and angular momentum conservation in strong interactions, that only the component of proton spin perpendicular to the scattering ($xz$) plane can be polarized.

The independent vectors we have at our disposal are $\mathbf{k}_i$, $\mathbf{k}_f$, and $\boldsymbol{\sigma}$. The independent rotationally invariant quantities that we can form from these vectors, that also conserve parity, are

(a) $\mathbf{k}_i . \mathbf{k}_f$,
(b) $\boldsymbol{\sigma}.(\mathbf{k}_i \times \mathbf{k}_f)$.

The non-spin flip amplitude can be a function only of (a). The spin flip amplitude can include (b). However, the expectation value of (b) can only be non-zero for the component of $\boldsymbol{\sigma}$ which is parallel to the normal to the scattering ($xz$) plane. Thus, the polarization must be along the normal to the scattering plane. We lose no generality, and certainly simplify the treatment of determining the polarization, if we always choose the scattering plane as the $xz$ plane, for which case $\phi = 0$. Since we know that any resulting proton spin polarization will be in the $y$-direction, we wish to express the eigenstates of the spin wave function along the $y$-direction in terms of the eigenstates of the spin wave function along the $z$-direction. Using the usual relationships,

$$\xi^{(y)}_{\pm\frac{1}{2}} = \frac{1}{\sqrt{2}}(\xi^{(z)}_{\frac{1}{2}} \pm \mathrm{i}\xi^{(z)}_{-\frac{1}{2}}),$$

where in our treatment so far we have left out the $(z)$ superscript as it was understood from our defining statements. Clearly

$$\sigma_y \xi^{(y)}_{\pm\frac{1}{2}} = \pm\xi^{(y)}_{\pm\frac{1}{2}},$$
$$\sigma_z \xi^{(z)}_{\pm\frac{1}{2}} = \pm\xi^{(z)}_{\pm\frac{1}{2}}.$$

We will wish to reformulate our previous treatment expressing the results in terms of $\xi^{(y)}_{\pm\frac{1}{2}}$.

Let us now, in reformulating, consider separately the two cases where the proton spin (for an unpolarized target) has $z$ eigenvalues $\pm\frac{1}{2}$ (i.e. is polarized along the positive $z$-direction, or negative $z$-direction, respectively). The scattering amplitude for the two cases when expressed in terms of $\xi^{(y)}_{\pm\frac{1}{2}}$ becomes

(a) $m_s = m_z = +\frac{1}{2}$,

$$f(\theta)\xi_{\frac{1}{2}} + g(\theta, 0)\xi_{-\frac{1}{2}} = [f(\theta) - \mathrm{i}g(\theta, 0)]\frac{\xi^{(y)}_{+\frac{1}{2}}}{\sqrt{2}} + [f(\theta) + \mathrm{i}g(\theta, 0)]\frac{\xi^{(y)}_{-\frac{1}{2}}}{\sqrt{2}},$$

(b) $m_s = m_z = -\frac{1}{2}$,

$$f(\theta)\xi_{-\frac{1}{2}} - g(\theta, 0)\xi_{+\frac{1}{2}} = [f(\theta) - \mathrm{i}g(\theta, 0)]\frac{\xi^{(y)}_{+\frac{1}{2}}}{\mathrm{i}\sqrt{2}} - [f(\theta) + \mathrm{i}g(\theta, 0)]\frac{\xi^{(y)}_{-\frac{1}{2}}}{\mathrm{i}\sqrt{2}}.$$

In this latter equation the simplification introduced by always defining $\phi = 0$ (i.e. $xz$ as the scattering plane) allowed us to avoid using different functions for $g$ in the two equations which would be necessary in the general case. Since the $\exp(\mathrm{i}m\phi)$ factor in the spherical harmonics has the opposite phase, for the two cases when $\phi \neq 0$.

Thus, since the differential scattering cross-section is the square of the scattering amplitude, the results for the two cases $\xi^{(y)}_{\pm\frac{1}{2}}$ are

$$\frac{\mathrm{d}\sigma}{\mathrm{d}\Omega}(\xi^{(y)}_{+\frac{1}{2}}) = \tfrac{1}{2}\,|f(\theta) - \mathrm{i}g(\theta, 0)|^2,$$

$$\frac{\mathrm{d}\sigma}{\mathrm{d}\Omega}(\xi^{(y)}_{-\frac{1}{2}}) = \tfrac{1}{2}\,|f(\theta) + \mathrm{i}g(\theta, 0)|^2.$$

The polarization is defined as

$$P = \frac{\dfrac{d\sigma}{d\Omega}(\xi^{(y)}_{+\frac{1}{2}}) - \dfrac{d\sigma}{d\Omega}(\xi^{(y)}_{-\frac{1}{2}})}{\dfrac{d\sigma}{d\Omega}(\xi^{(y)}_{+\frac{1}{2}}) + \dfrac{d\sigma}{d\Omega}(\xi^{(y)}_{-\frac{1}{2}})} = \frac{\dfrac{d\sigma}{d\Omega}(\xi^{(y)}_{+\frac{1}{2}}) - \dfrac{d\sigma}{d\Omega}(\xi^{(y)}_{-\frac{1}{2}})}{\left(\dfrac{d\sigma}{d\Omega}\right)_{\text{unpolarized}}},$$

where $\left(\dfrac{d\sigma}{d\Omega}\right)_{\text{unpolarized}}$ represents the case of an unpolarized proton target which has equal numbers of $m_s = +\frac{1}{2}$, and $m_s = -\frac{1}{2}$, protons,

$$P = \frac{\frac{1}{2}\{f(\theta) - ig(\theta, 0)\}\{f^*(\theta) + ig^*(\theta, 0)\} - \frac{1}{2}\{f(\theta) + ig(\theta, 0)\}\{f^*(\theta) - ig^*(\theta, 0)\}}{\frac{1}{2}(f(\theta) - ig(\theta, 0))(f^*(\theta) + ig^*(\theta, 0)) + \frac{1}{2}(f(\theta) + ig(\theta, 0))(f^*(\theta) - ig^*(\theta, 0))}.$$

To simplify evaluating we refer to $f$ and $g$ only. Thus,

$$P = \frac{\frac{1}{2}[fig^* - igf^* + fig^* - f^*ig]}{[|f|^2 + |g|^2]}.$$

$$P = \frac{ifg^* - if^*g}{|f|^2 + |g|^2} = \frac{-i[f^*g - fg^*]}{|f|^2 + |g|^2} = \frac{-i[+2i\,\mathrm{Im}(f^*g)]}{|f|^2 + |g|^2} = \frac{2\,\mathrm{Im}(f^*g)}{|f|^2 + |g|^2},$$

and finally, re-introducing our full notation† and explicitly showing the direction of $P$, we obtain

$$\mathbf{P} = \frac{2\,\mathrm{Im}(f^*(\theta)g(\theta))\hat{\mathbf{n}}}{|f(\theta)|^2 + |g(\theta)|^2},$$

where

$$\hat{\mathbf{n}} = \frac{\mathbf{k}_i \times \mathbf{k}_f}{|k_i|\,|k_f|}$$

is the normal to the scattering plane. If the target were polarized, say along the $y$-direction with a value $P_t$, then it can be shown that, if the polarization is such that $P \neq 0$, there would be a left–right 'asymmetry' in the scattering proportional to the product $PP_t$. Targets of various types containing polarized protons as part of a more complex structure have been utilized for some time now. Single scattering of pions from such a target can be used to determine the parameter $P$. Another way of determining the parameter $P$ is to scatter the recoil protons from an unpolarized target a second time from an 'analyser' material which has a known polarization parameter and,

† $\phi$ is by choice always zero, so we omit it.

thus, then determine the parameter $P$ from the observed left–right asymmetry of the second scattering.

### 4.8. Coulomb scattering

In the above treatment we have obviously neglected the Coulomb interaction.

The Coulomb scattering amplitude can be well represented to order $\alpha = \frac{e^2}{\hbar c} \approx \frac{1}{137}$ by the amplitude that corresponds to Rutherford scattering, and which can be obtained in Born approximation as a Fourier transform of the Coulomb potential. The Coulomb amplitude is thus

$$f_c(\pi^{\pm}\text{–p}) = -\frac{(\pm\alpha m)}{2k^2 \sin^2(\theta/2)}, \tag{4.66}$$

where the $(\pm)$ sign goes with the $(\pm)$ superscript on the $\pi$ symbol. Owing to the small value of $\alpha$, the Coulomb amplitude contributes appreciably only at small angles (generally $\leq 30°$) in the region of the low energy $\pi^+$–p resonance.† The Coulomb amplitude in the above expression also contains a form factor $F$, which has been set equal to its value at $\theta = 0$. This form factor (see Chapter 11, which describes form factors) has only a second order effect at low $|t|$ values.

It has been shown that we can take the Coulomb amplitude into account, approximately, by adding the Coulomb amplitude to the ordinary (non-spin flip) nuclear amplitude with which it is coherent, since Coulomb scattering does not involve spin flip except to a negligible degree. Hence we obtain

$$f_{\text{total}}(\theta) = f_c(\theta) + f(\theta),$$

$$\frac{d\sigma}{d\Omega} = \lambda\!\!\!^{-2}[|f_c(\theta)+f(\theta)|^2 + |g(\theta, \varphi)|^2]. \tag{4.67}$$

A more detailed treatment of how to more quantitatively include Coulomb effects is given in Chapter 5, § 5.8.

### 4.9. Nuclear interaction scattering

Let us consider the large angle approximation $f_c(\theta) \ll f(\theta)$, and neglect $f_c(\theta)$.

† At 10–30 GeV/$c$ the Coulomb amplitude contributes appreciably only within a small fraction of a degree.

In eqn (4.62), we can replace $\sin^2\theta$ by $1-\cos^2\theta$. Hence we obtain

$$\frac{\mathrm{d}\sigma}{\mathrm{d}\Omega} = \lambda\!\!\!^{-2}\{|a_{0,\frac{1}{2}}|^2+|a_{1,\frac{3}{2}}-a_{1,\frac{1}{2}}|^2+2\,\mathrm{Re}(a^*_{0,\frac{1}{2}}(2a_{1,\frac{3}{2}}+a_{1,\frac{1}{2}})\cos\theta)+ \\ +(|2a_{1,\frac{3}{2}}+a_{1,\frac{1}{2}}|^2-|a_{1,\frac{3}{2}}-a_{1,\frac{1}{2}}|^2)\cos^2\theta\}. \quad (4.68a)$$

Hence

$$\frac{\mathrm{d}\sigma}{\mathrm{d}\Omega} = \lambda\!\!\!^{-2}(A+B\cos\theta+C\cos^2\theta), \quad (4.68b)$$

where

$$A = |a_{0,\frac{1}{2}}|^2+|a_{1,\frac{3}{2}}-a_{1,\frac{1}{2}}|^2, \quad (4.69)$$

$$B = 2\,\mathrm{Re}\{a^*_{0,\frac{1}{2}}(2a_{1,\frac{3}{2}}+a_{1,\frac{1}{2}})\}, \quad (4.70)$$

$$C = 3\{|a^*_{1,\frac{3}{2}}|^2+2\,\mathrm{Re}(a^*_{1,\frac{3}{2}}a_{1,\frac{1}{2}})\}, \quad (4.71)$$

or we can write for the three $\pi^{\pm}$–p scattering processes,

$$\frac{\mathrm{d}\sigma}{\mathrm{d}\Omega}(\pi^+\to\pi^+) = \lambda\!\!\!^{-2}(A^++B^+\cos\theta+C^+\cos^2\theta), \quad (4.72)$$

$$\frac{\mathrm{d}\sigma}{\mathrm{d}\Omega}(\pi^-\to\pi^-) = \lambda\!\!\!^{-2}(A^-+B^-\cos\theta+C^-\cos^2\theta), \quad (4.73)$$

$$\frac{\mathrm{d}\sigma}{\mathrm{d}\Omega}(\pi^-\to\pi^0) = \lambda\!\!\!^{-2}(A^0+B^0\cos\theta+C^0\cos^2\theta). \quad (4.74)$$

Using eqn (4.63) and eqn (4.72–74), we obtain

$$A^+ = |a_{0,\frac{1}{2}}(\tfrac{3}{2})|^2+|a_{1,\frac{3}{2}}(\tfrac{3}{2})-a_{1,\frac{1}{2}}(\tfrac{3}{2})|^2, \quad (4.75a)$$

$$B^+ = 2\,\mathrm{Re}\{a^*_{0,\frac{1}{2}}(\tfrac{3}{2})(2a_{1,\frac{3}{2}}(\tfrac{3}{2})+a_{1,\frac{1}{2}}(\tfrac{3}{2}))\}, \quad (4.75b)$$

$$C^+ = 3\{|a_{1,\frac{3}{2}}(\tfrac{3}{2})|^2+2\,\mathrm{Re}(a^*_{1,\frac{3}{2}}(\tfrac{3}{2})a_{1,\frac{1}{2}}(\tfrac{3}{2}))\}. \quad (4.75c)$$

Similar but more complicated expressions involving $a_{l,j}(\frac{3}{2})$ and $a_{l,j}(\frac{1}{2})$, can be derived for $A^-$, $B^-$, $C^-$ and $A^0$, $B^0$, $C^0$ by inserting the expression for $a_{l,j}$ given in terms of $a_{l,j}(\frac{3}{2})$ and $a_{l,j}(\frac{1}{2})$ by eqn (4.64) and eqn (4.65) into eqn (4.69–71), respectively.

Several conclusions can be drawn from these angular distribution formulae.

(a) If we had only $s$-state scattering, we would have an isotropic angular distribution as expected.

(b) A $\cos^2\theta$ term can be obtained from scattering in a single state only if $J = \frac{3}{2}$.

(c) The $\cos\theta$ term is a result of $s$–$p$ interference and would be absent if there were only $p$-state scattering.

Figures 4.5(a), (b), (c) show [17], [18] the angular distribution coefficients $A$, $B$, and $C$. (See [18].)

As we have already concluded, the total cross-section data strongly supports the assumption of a resonance in the $T = J = \frac{3}{2}$ state, accompanied by much smaller scattering cross-sections corresponding to the other states.

Let us now express the scattering amplitudes in terms of the corresponding phase shifts since the data analyses are most conveniently done in this fashion. We shall also, henceforth, adopt the standard notation for a phase shift as follows, $\delta_{2T,2J}$ is the phase shift for the state with isotopic spin $T$ and angular momentum $J$. For $s$-waves, the second (right) index is omitted and it is, of course, understood that $J = \frac{1}{2}$. In this standard phase shift notation

$$a_{l,J}(T) = \frac{1}{2\mathrm{i}}\,(\mathrm{e}^{2\mathrm{i}\delta_{2T,2J}}-1).$$

Expressing eqns (4.75) in terms of phase shifts, we obtain

$$A^{+} = \sin^2\delta_3+\sin^2(\delta_{33}-\delta_{31}), \tag{4.76a}$$
$$B^{+} = 2\sin\delta_3\{2\sin\delta_{33}\cos(\delta_{33}-\delta_3)+\sin\delta_{31}\cos(\delta_{31}-\delta_3)\}, \tag{4.76b}$$
$$C^{+} = 3\{\sin^2\delta_{33}+2\sin\delta_{31}\sin\delta_{33}\cos(\delta_{33}-\delta_{31})\}. \tag{4.76c}$$

Integrating eqn (4.74a) to obtain the total $\pi^{+}$+p cross-section, the integral of the $B$ term drops out, and we obtain

$$\sigma_{\text{total}} = 4\pi\lambda\!\!\!^{-2}(\sin^2\delta_3+\sin^2\delta_{31}+2\sin^2\delta_{33}). \tag{4.77}$$

Let us now assume that only $\delta_{33} \neq 0$. Then we find

$$A^{+} = |a_{1,\frac{3}{2}}(\tfrac{3}{2})|^2 = \sin^2\delta_{33}, \tag{4.78a}$$
$$B^{+} = 0, \tag{4.78b}$$
$$C^{+} = 3\sin^2\delta_{33}. \tag{4.78c}$$
$$A^{-} = \tfrac{1}{9}\,|a_{1,\frac{3}{2}}(\tfrac{3}{2})|^2 = \tfrac{1}{9}\sin^2\delta_{33}, \tag{4.79a}$$
$$B^{-} = 0, \tag{4.79b}$$
$$C^{-} = \tfrac{1}{3}\,|A_{1,\frac{3}{2}}(\tfrac{3}{2})|^2 = \tfrac{1}{3}\sin^2\delta_{33}, \tag{4.79c}$$
$$A^{0} = \tfrac{2}{9}\,|A_{1,\frac{3}{2}}(\tfrac{3}{2})|^2 = \tfrac{2}{9}\sin^2\delta_{33}, \tag{4.80a}$$
$$B^{0} = 0, \tag{4.80b}$$
$$C^{0} = \tfrac{2}{3}\sin^2\delta_{33}. \tag{4.80c}$$

If we now look at Fig. 4.5 which shows $A$, $B$, and $C$ as functions of energy, it is clear that near the resonant energy (195 MeV in the laboratory system),

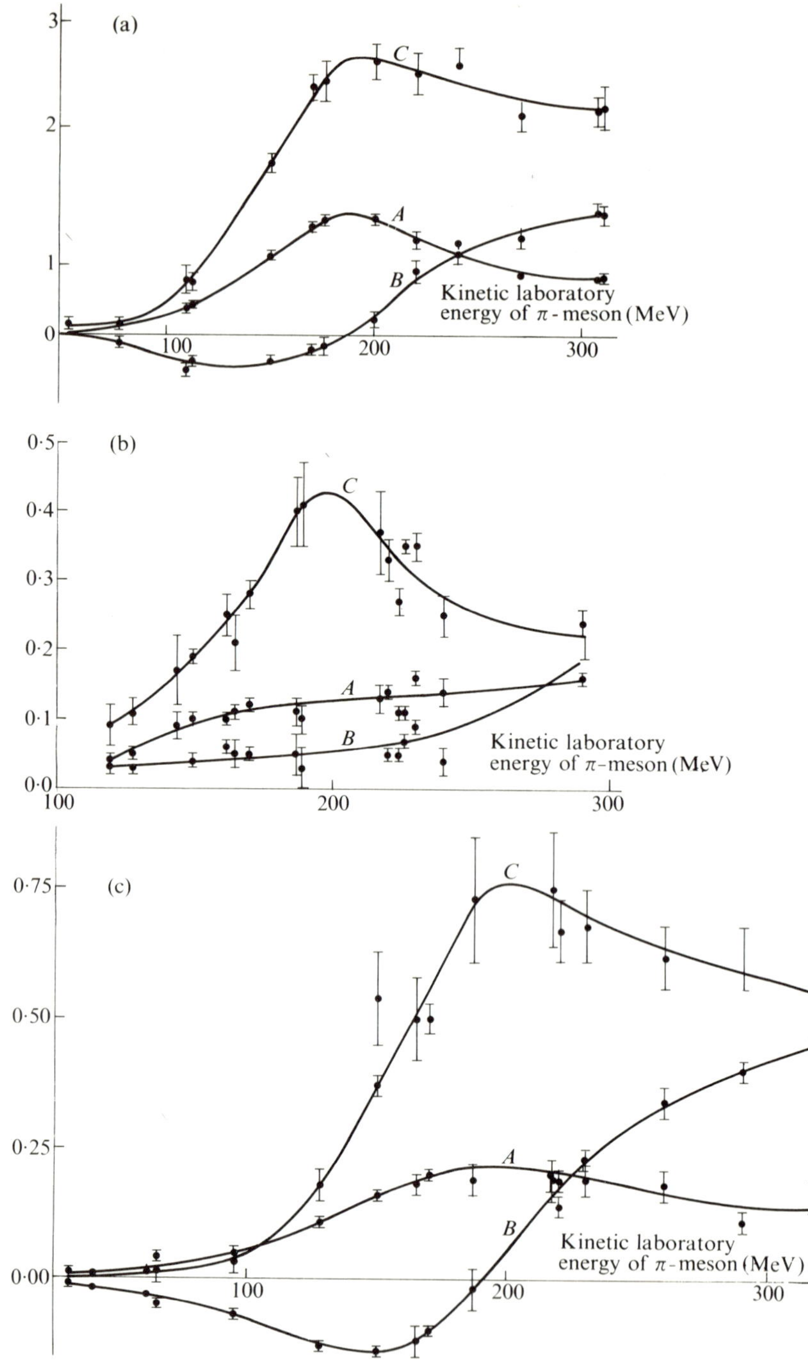


FIG. 4.5. Empirical angular distribution coefficients for
(a) $\pi^+ + p \rightarrow \pi^+ + p$,
(b) $\pi^- + p \rightarrow \pi^- + p$,
(c) $\pi^- + p \rightarrow \pi^\circ + n$.
(After Miyake *et al.* (1962). *Phys. Rev.* **126,** 2188, and Kallen (1964), *Elementary Particle Physics*, Addison-Wesley, Reading, Mass.)

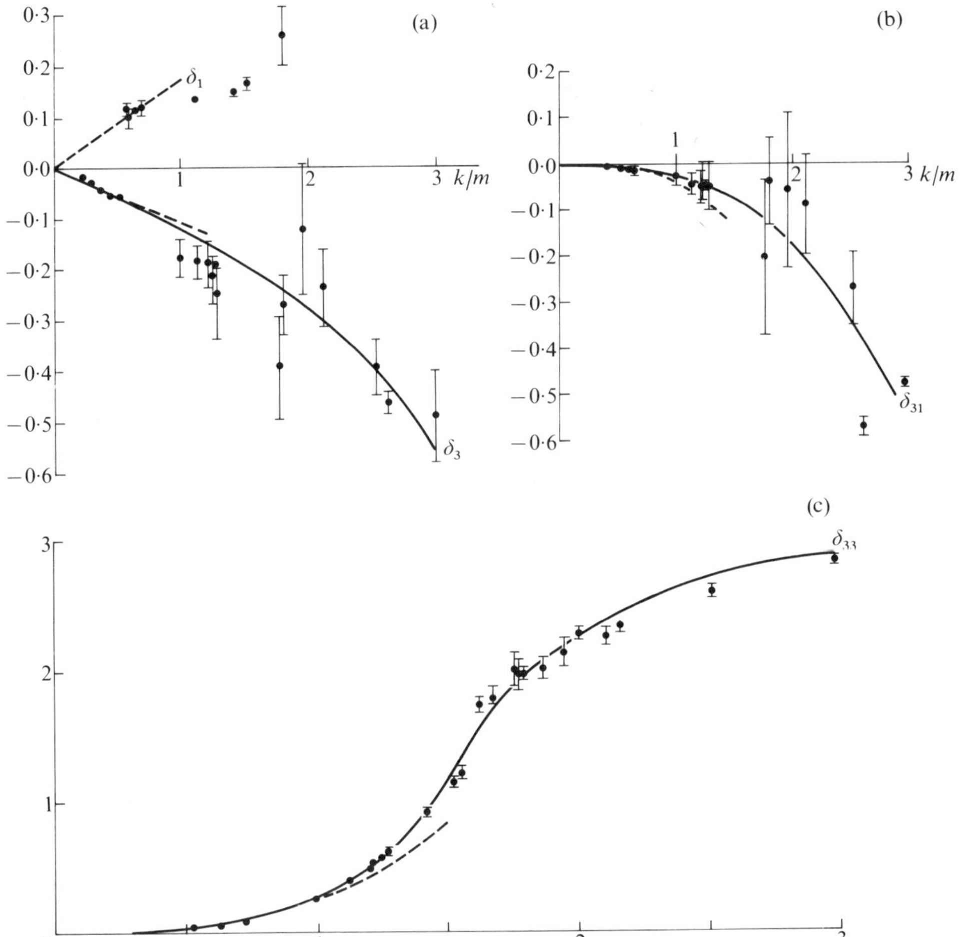


FIG. 4.6. $\pi$–nucleon phase shifts as functions of $k/m$, determined empirically. The dashed lines are the solutions by Miyakawa *et al.*, which have the effective range characteristics appropriate to low energies, but which are incorrect at the higher energies. (After Miyake *et al.* (1962). *Phys. Rev.* **126,** 2188, and Kallen 1964. *Elementary Particle Physics*. Addison-Wesley, Reading, Mass.)

$C^+ \approx 3$, $A^+ \approx 1$, and $B^+ \approx 0$, as expected $C^- \approx 0{\cdot}4$, with an error which includes the expected value of about $\frac{1}{3}$, while $A^-$ has the expected value $\approx 0{\cdot}1$. $B$ in this case is positive, but less than half the value of $A$. $C^0 \approx \frac{3}{4}$, which is clearly consistent with the expected value of about $\frac{2}{9}$. $B^0$ is near zero. As we move away from the resonance, the situation becomes more complicated by $s$–$p$ interference. However, the major reason why the original analysis by Fermi *et al.* did not find the $\delta_{33}$ resonant solution is that there are phase shift ambiguities which allows more than one solution. We shall discuss the most general of these phase shift ambiguities in the next section. Figure 4.6 shows the behaviour of the phase shifts as a function of energy.

## 4.10. Phase-shift ambiguities

### 4.10.1. *Sign ambiguities*

Since the relationships between the differential cross-sections and the phase shifts are complicated and non-linear, it is not surprising that there are several sets of phase shift solutions which fit the differential cross-section data well. From the form of the scattering amplitude it is clear that reversal of the sign of all the phase shifts will leave the differential scattering cross-section unchanged. Therefore there is no way of determining the absolute sign of the nuclear phase shifts from differential cross-section measurements of nuclear scattering alone. However, since the amplitude and absolute phase of the Coulomb scattering amplitude is well-known, it is possible to determine the absolute phases of the nuclear scattering amplitude, by observing the interference between the Coulomb and nuclear scattering at small enough angles so that both amplitudes are comparable.

### 4.10.2. *Yang ambiguity*

After Fermi had calculated his phase shift solutions, Yang pointed out [19] that there existed an ambiguity whereby, in reversing the sign of the quantity $\delta_{31}-\delta_{33}$, one also obtained a good fit to the experimental differential cross-section measurements. Let us demonstrate this in the case of $\pi^+ + \text{p} \to \pi^+ + \text{p}$-scattering (i.e. $T = \frac{3}{2}$ scattering only).

The Yang ambiguity for low energy $\pi$–p scattering can be stated as follows. Given a set of phase shifts of the Fermi type, let

$$\delta_{33}-\delta_{31} \to \delta'_{31}-\delta'_{33}, \tag{4.81}$$

and a new set is derived which also fits the experimental data. Then in eqn (4.62), we see that the coefficient of the spin flip term for $\pi^+ + \text{p}$ scattering gives

$$\begin{aligned} |a_{1,\frac{3}{2}}(\tfrac{3}{2})-a_{1,\frac{1}{2}}(\tfrac{3}{2})|^2 &= \left|\frac{1}{2\text{i}}(\text{e}^{2\text{i}\delta_{33}}-\text{e}^{2\text{i}\delta_{31}})\right|^2, \\ &= \tfrac{1}{4}\,|\text{e}^{\text{i}(\delta_{33}+\delta_{31})}[\text{e}^{\text{i}(\delta_{33}-\delta_{31})}-\text{e}^{\text{i}(\delta_{31}-\delta_{33})}]|^2 \\ &= \tfrac{1}{4}\,|2\text{i}\sin(\delta_{33}-\delta_{31})|^2 = |\sin(\delta_{33}-\delta_{31})|^2 \\ &= |\sin(\delta'_{31}-\delta_{33})|^2. \end{aligned} \tag{4.82}$$

Hence this term satisfies the Yang ambiguity condition. Referring to the first term of eqn (4.62) since $\delta_3 = \delta'_3$, $a_{0,\frac{1}{2}}$ does not change, and the coefficient of the first term of eqn (4.62) becomes

$$\left|\frac{1}{2i}(e^{2i\delta_3}-1)+\frac{1}{2i}(2e^{2i\delta_{33}}-2+e^{2i\delta_{31}}-1)\cos\theta\right|^2 =$$

$$= \tfrac{1}{4}\,|e^{2i\delta_3}-1+(2e^{2i\delta_{33}}+e^{2i\delta_{31}}-3)\cos\theta|^2.$$

The only part of the above term which depends on the Yang substitution is the quantity

$$Q = 2e^{2i\delta_{33}}+e^{2i\delta_{31}} = e^{2i\delta_{33}}(2+e^{-2i(\delta_{33}-\delta_{31})}). \tag{4.83}$$

After the Yang substitution, we obtain

$$Q' = e^{2i\delta'_{33}}(2+e^{+2i(\delta_{33}-\delta_{31})}).$$

We would like to find a $\delta'_{33}$ such that $Q = Q'$, to establish the equivalence of the Yang set of phase shifts. This requires that

$$e^{2i(\delta'_{33}-\delta_{33})} = \frac{2+e^{-2i(\delta_{33}-\delta_{31})}}{2+e^{2i(\delta_{33}-\delta_{31})}}, \tag{4.84}$$

$$e^{2i(\delta'_{33}-\delta_{33})} = \frac{2+e^{-2i\epsilon}}{2+e^{2i\epsilon}}, \tag{4.85}$$

where $\epsilon = \delta_{33}-\delta_{31}$. Regardless of the values of $\delta_{31}$ and $\delta_{33}$, the right-hand side of the above equation has the property that the modulus of the numerator and denominator are identical and, hence, the modulus of the right-hand side is unity, since

$$\left|\frac{2+e^{-2i\epsilon}}{2+e^{+2i\epsilon}}\right| = \frac{\sqrt{\{(2+\cos 2\epsilon)^2+\sin^2 2\epsilon\}}}{\sqrt{\{(2+\cos 2\epsilon)^2+\sin^2 2\epsilon\}}} = 1. \tag{4.86}$$

The modulus of the left-hand side of eqn (4.85) is also unity. Hence, given a particular $\delta_{33}$ and $\delta_{31}$ we can always choose a real $\delta'_{33}$ which will make the phase of the left-hand side of eqn (4.85) equal to the phase of the right-hand side. Hence there is always a Yang solution. In fact, this phase $\delta'_{33}$ is given by the equation

$$(\delta'_{33}-\delta_{33}) = \sin^{-1}\frac{-\sin 2\epsilon}{\{(2+\cos 2\epsilon)^2+\sin^2 2\epsilon\}^{\frac{1}{2}}}. \tag{4.87}$$

For small phase shifts this becomes

$$\delta'_{33}-\delta_{33} = -\frac{2\epsilon}{3} = -\tfrac{2}{3}\delta_{33}+\tfrac{2}{3}\delta_{31} \tag{4.88}$$

or

$$\delta'_{33} = \tfrac{1}{3}\delta_{33}+\tfrac{2}{3}\delta_{31}, \tag{4.89a}$$

but

$$\delta'_{31}-\delta'_{33} = \delta_{33}-\delta_{31}. \tag{4.89b}$$

Hence

$$\delta'_{31} = \tfrac{4}{3}\delta_{33}-\tfrac{1}{3}\delta_{31}. \tag{4.90}$$

Adding eqn (4.89b) to three times eqn (4.89a) we obtain the relationship

$$2\delta'_{33}+\delta'_{31} = 2\delta_{33}+\delta_{31}. \tag{4.91}$$

Now let us consider the question of whether this ambiguity exists in the general case, when we consider $\pi^-+\mathrm{p}\rightarrow\pi^-+\mathrm{p}$ and $\pi^-+\mathrm{p}\rightarrow\pi^0+\mathrm{n}$ scattering, where the $T=\frac{1}{2}$ scattering amplitude enters, also, and introduces interference terms between the two amplitudes ($T=\frac{1}{2}$ and $T=\frac{3}{2}$). The phase shift analysis has demonstrated that in the low energy region the $T=\frac{1}{2}$ $p$-wave phase shifts ($\delta_{11}$ and $\delta_{13}$) have a small effect on the scattering compared to the $s$-wave phase shift $\delta_1$. Hence we can, to a reasonable approximation, assume $\delta_{11}=\delta_{31}=0$. In this case, also, it can easily be shown from eqns (4.62–65) that an exact Yang solution can always be found for a Fermi solution. Even if both $\delta_{11}$ and $\delta_{31}$ are finite but small (which is the actual case), a Yang solution can be found, as was demonstrated to be the case in the computer search for phase shift solutions, which will be described in the latter part of this section.

From eqns (4.89a) and (4.90) we see that, as we obtain an increasing $\delta_{33}$ with increasing energy in the Fermi type solution, accompanied by a small $\delta_{31}$, this is translated into a large $(\delta_{31})_{\mathrm{Yang}}$ $(\approx\frac{4}{3}(\delta_{33})_{\mathrm{Fermi}})$ in the Yang solution and a small $(\delta_{33})_{\mathrm{Yang}}$ $(\approx\frac{1}{3}(\delta_{33})_{\mathrm{Fermi}})$. The Yang solutions were found to behave very peculiarly with energy, showing three resonances ($\delta_{31}=90°$ and $270°$, and $\delta_{33}=90°$) by the time $\delta_{33}$ had reached $120°$. This kind of behaviour would be very difficult to understand from a theoretical point of view. Fermi pointed out that the two sets of phase shifts give different predictions for the polarization of the recoil nucleon and, therefore, in principle, polarization measurements could be used to discriminate between the two sets.

The requirement that the spin flip forward dispersion relations are satisfied by the phase shift solution was later shown [20] to eliminate the Yang solutions.

### 4.10.3. *The Minami ambiguity*

Minami [21] has shown that there is a general exact ambiguity which can be applied to any arbitrary set of phase shifts.

If the pion–nucleon state is characterized by a $T$ (isotopic spin), a total angular momentum $J$, and a parity $P$, the Minami ambiguity states that the transformation $P \to -P$, applied to all $T$ and $J$, leaves the differential scattering cross-section invariant. Hayakawa, Kawaguchi, and Minami [22] (1954) pointed out that this ambiguity had a more general nature. Dyson and Nambu (1956) provided a very general proof of the ambiguity.

In terms of phase shifts, for a particular $T$, $J$, and $l$, this ambiguity states that

$$\begin{aligned} \delta'(T, J, l = J-\tfrac{1}{2})_{\mathrm{M}} &= \delta(T, J, l = J+\tfrac{1}{2}), \\ \delta'(T, J, l = J+\tfrac{1}{2})_{\mathrm{M}} &= \delta(T, J, l = J-\tfrac{1}{2}), \end{aligned} \tag{4.92}$$

for all $T$, $J$, $l$ states, where $\delta'_{\mathrm{M}}$ is the Minami set of phase shifts corresponding to the original set denoted by $\delta$. Hence the prescription for applying the Minami ambiguity to a set of phase shifts is the interchange of the phase shifts for $l = J-\frac{1}{2}$ and $l = J+\frac{1}{2}$. The Minami ambiguity is of much more general interest than the Yang ambiguity, since it generally exists for any arbitrary set of phase shifts. Hence, for the $T = \frac{3}{2}$ state which is the dominant state in low energy $\pi$–p scattering, the Minami ambiguity leads to the following relations in terms of the Fermi set of phase shifts.†

$$\delta'_{\mathrm{M}}(T = \tfrac{3}{2}, S_{\frac{1}{2}}) = \delta(T = \tfrac{3}{2}, P_{\frac{1}{2}}) = \delta_{31}, \tag{4.93a}$$

$$\delta_{\mathrm{M}}(T = \tfrac{3}{2}, P_{\frac{1}{2}}) = \delta(T = \tfrac{3}{2}, S_{\frac{1}{2}}) = \delta_{3}, \tag{4.93b}$$

$$\delta_{\mathrm{M}}(T = \tfrac{3}{2}, D_{\frac{3}{2}}) = \delta(T = \tfrac{3}{2}, P_{\frac{3}{2}}) = \delta_{33}. \tag{4.93c}$$

The only difference between the predictions of the Minami phase shifts and those of the original Fermi set appear in the polarization of the recoil nucleon.‡ This Minami set of phase shifts derived from the Fermi set would exhibit a resonance in a $D_{\frac{3}{2}}$-state instead of the $P_{\frac{3}{2}}$-state. Furthermore, the $S_{\frac{1}{2}}$-state phase shift in the Fermi set would become a $P_{\frac{1}{2}}$-state phase shift. The scattering from a potential of well-defined range at low energies is expected to give phase shifts $\delta$ which vary like

$$\delta \sim p^{2l+1} \sim \eta^{2l+1}. \tag{4.94}$$

† Where the capital letters, $S$, $P$, $D$, ..., stand for $l = 0, 1, 2, \ldots$, in the usual spectroscopic notation, and the subscript represents the angular momentum ($J$).

‡ Even a measurement of $\dfrac{\mathrm{d}\sigma}{\mathrm{d}\Omega}(\theta)$ and the polarization $P(\theta)$ does not lead to a unique selection of a set of phase shifts, since the operation of 'complex conjugation' changes the sign of the polarization, but leaves $\mathrm{d}\sigma\theta/\mathrm{d}\Omega$ the same. Thus some dynamical criteria, the best of which are the (spin flip) dispersion relations, must be used.

The Fermi set shows the following behaviour,

$$\delta_{33} \sim p^{2l+1} \sim \eta^{2l+1}, \tag{4.95a}$$

$$\delta_3 \sim p \sim \eta. \tag{4.95b}$$

However, the Minami set does not exhibit the above behaviour. Although this is a qualitative argument against the Minami set, it had been shown [23] (1958) that the spin flip forward dispersion relations are not satisfied by the Minami set, which are generated from the Fermi set, nor are these spin flip forward dispersion calculations satisfied by the new set of phase shifts obtained by applying the Minami ambiguity to the Yang set. These will be discussed in Chapter 5. By 1958 it had been established that the only set of low energy pion–nucleon phase shifts which satisfied the forward dispersion relations was the Fermi set, which exhibited a resonance in $\delta_{33}$ at about 200 MeV. Hence, it became practically certain that the Fermi set was the only correct set and is the unique solution for the pion–nucleon scattering at low energies up to about 300 MeV.

## 4.11. The behaviour of the $\pi^+$–p total cross-section in the resonance region

High precision (1·6 per cent r.m.s. absolute errors) total cross-section measurements in the $T = J = \frac{3}{2}$ resonance region demonstrated [24] how

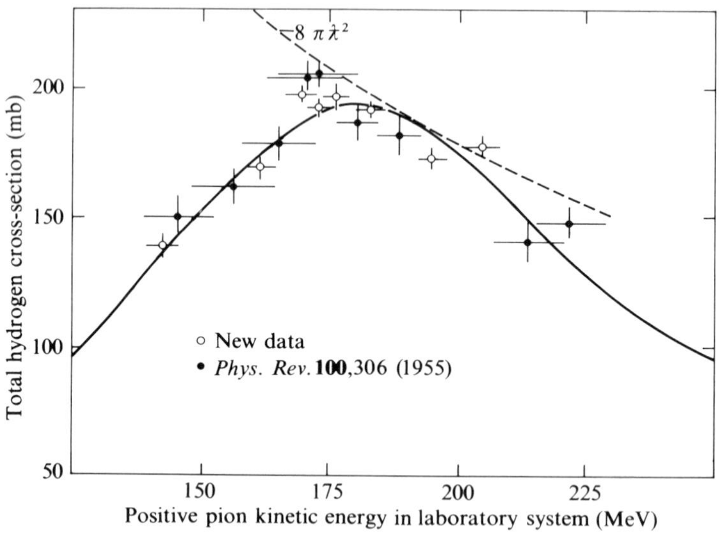


FIG. 4.7. The total cross-section values of hydrogen for positive pions of 143–205 MeV determined in the present work, plotted as a function of pion kinetic energy in the laboratory system. Our earlier measurements in, and near, this energy interval are plotted also for comparison. The solid line represents the contribution of $\alpha_{33}$ alone, taken from a Chew–Low plot.

well the behaviour of the $\pi^+$–p total cross-section can be explained by the $\alpha_{33}$ phase shift (see Fig. 4.7). The solid curve shown represents the contribution of $\alpha_{33}$ alone, determined by a Chew–Low plot through the available data. As we will explain in § 4.12, a Chew–Low plot is essentially a more sophisticated representation of a $T = J = \frac{3}{2}$ state resonance. Figure 4.8 shows this data in comparison with the data of good precision obtained by other groups.

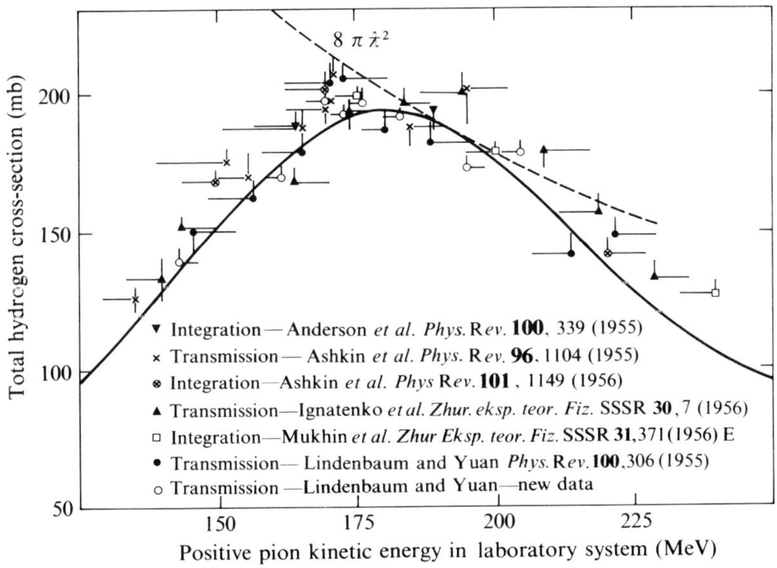


FIG. 4.8. A comparison of the recent $\pi^+ + p$ total cross-section data in the energy region of 125 to 250 MeV, determined by various groups. The solid line represents the contribution of $\alpha_{33}$ alone, taken from a Chew–Low plot.

Although the peak in the $\pi^+$–p total cross-section appears to occur at about 180 MeV in the laboratory system, the downward slope of the $8\pi\lambda\!\!\!^{-2}$ curve causes the resonant energy to occur beyond the peak, at that point where the cross-section data curve is either tangent to, or due to the small contributions of other waves reaches its maximum relative to, the $8\pi\lambda\!\!\!^{-2}$ curve. The solid line (contribution of $\alpha_{33}$ alone) represents the total cross-section data well, to within a few per cent.

From a Chew–Low plot it was concluded (1958) [24] that the resonant energy occurs at $E \sim 190^{+20}_{-10}$ MeV in the laboratory system. De Hoffman *et al.* (1954) had concluded that $E_R$ was about 195 MeV [14]. Roper *et al.* [25] in their extensive phase shift analysis (1965) concluded $E_R$ was about 193 MeV. This later work will be considered in § 4.14.

### 4.12. Chew–Low plot

Low [26] had determined a set of equations which included the pion cloud effects in the nucleon wave function. These equations were applicable to both the static model and the relativistic Yukawa theory, and they exhibited the low energy properties of both in a clear and useful way. Chew and Low [27] investigated this equation in the one-meson approximation for the symmetric pseudoscalar meson theory (with PV coupling) and obtained results for $p$-wave scattering.

A cut-off factor was employed which allowed the maximum energy to be $\omega_{\max}$ (about 1 GeV). They obtained the following relationship

$$\frac{\eta^3}{\omega_t^*} \cot \alpha_{33} = \tfrac{3}{4}(f^2)^{-1} F(\omega_t^*), \qquad (4.96)$$

where $\eta$ is the pion momentum in units of $m_\pi c$. $\omega_t^*$ is the total c.m.s. energy of the pion plus the nucleon kinetic energy,† in units of $m_\pi c^2$, where $m_\pi$ is the pion rest mass and $c$ is the velocity of light. $\alpha_{33}$ is the phase shift in the $T = J = \frac{3}{2}$ state, $f^2$ is the renormalized unrationalized coupling constant, and $F(\omega_t^*) = 1 - \omega_t^* r_\alpha(\omega_t^*)$, with $r_\alpha(\omega_t^*)$ almost constant for $\omega_t^* \ll \omega_{\max}$. This then allows $f^2$ to be determined by extrapolation, as we shall see later. For $\omega_t^* \lesssim \omega_{\text{resonance}}$, we have $\omega_t^* \ll \omega_{\max}$ and the effective range approximation can be made. To a good approximation $r_\alpha$ is a constant. Since $\cot \alpha_{33} = 0$ at the resonance energy, from eqn (4.96) it follows that $F(\omega_r^*) = 0$ (where $\omega_r$ is written for $\omega_{\text{resonance}}$). Thus,

$$F(\omega_t^*) = 1 - \frac{\omega_t^*}{\omega_r^*} \qquad (4.97)$$

and, therefore,

$$\frac{\eta^3}{\omega_t^*} \cot \alpha_{33} = \tfrac{3}{4}(f^2)^{-1}\left(1 - \frac{\omega_t^*}{\omega_r^*}\right). \qquad (4.98)$$

The cut-off model used predicts a low energy resonance in the $T = J = \frac{3}{2}$ state for the observed values of the coupling constant provided $\omega_{\max}$ is large enough, and $\omega_{\max} \sim 1$ GeV leads to a resonance at about the correct energy. The resonance energy ($\omega_r^*$) depends on $\omega_{\max}$ as well as $f^2$ and, therefore, $\omega_r^*$ can replace $\omega_{\max}$ as a parameter in the theory of eqns (4.97) and (4.98). It is apparent that if $\dfrac{\eta^3}{\omega_r^*} \cot \alpha_{33}$ is plotted versus $\omega_t^*$ the theory predicts a straight line.

† The inclusion of the recoil nucleon in the total c.m.s. energy is an attempt to take into account, approximately, the nucleon recoil effects. Notice $\alpha_{33} = \delta_{33}$.

It has been shown [10] that eqn (4.98) results in a variation of the $\pi^+$–p total cross-section generally similar to the one obtained by Brueckner [5], when he fitted a one-level Breit–Wigner resonance formula to the observed energy variation of the $\pi^+$–p total cross-section for a resonance in the $T = J = \frac{3}{2}$ state. We can demonstrate this as follows. In eqn (4.15), we have seen (neglecting the smaller terms) that

$$\frac{\sigma_{\text{total}}}{8\pi\bar{\lambda}^2} = \sin^2\delta_{33} = \frac{\rho^2}{\rho^2+\epsilon^2}, \tag{4.99}$$

where

$$\rho = \left\{\frac{\left(\dfrac{a}{\bar{\lambda}}\right)^3}{1+\left(\dfrac{a}{\bar{\lambda}}\right)^2} b\right\}, \tag{4.100}$$

where $a$ is the channel radius and is of order $\hbar/m_\pi c$, $b$ is a constant related to the width of the resonance, and $\epsilon = (\omega_r - \omega_\pi)$. Let us use natural units, where $a \approx 1$ (i.e. $\hbar - m_\pi = c = 1$), hence

$$\rho = \frac{\eta^3 b}{\omega_\pi^2}. \tag{4.101}$$

Since, from eqn (4.99),

$$\sin\delta_{33} = \frac{\rho}{(\rho^2+\epsilon^2)^{\frac{1}{2}}}, \tag{4.102}$$

we obtain

$$\cos\delta_{33} = \frac{\epsilon}{(\rho^2+\epsilon^2)^{\frac{1}{2}}}, \tag{4.103}$$

then

$$\rho\cot\delta_{33} = \epsilon \tag{4.104}$$

or

$$\frac{\eta^3}{\omega_\pi}\cot\delta_{33} = \frac{\omega_\pi}{b}\epsilon = \frac{\omega_\pi}{b}\omega_r\left(1-\frac{\omega_\pi}{\omega_r}\right), \tag{4.105}$$

which is quite similar in form, but not identical, to the Chew–Low equation. The differences are that, in the Brueckner form, we have $\omega_\pi$ instead of $\omega_t^*$ (nucleon recoil effects are neglected and the result does not have a constant multiplying $\left(1-\dfrac{\omega_\pi}{\omega_r}\right)$ on the right-hand side, but contains $\omega_\pi$ as a factor which, however, is nearly constant at the lower energies, increasing as the energy increases.

A Chew–Low plot (Lindenbaum and Yuan 1958) using the best values of $\delta_{33}$ (available in 1958), obtained from the phase shift analysis of the pion–nucleon scattering, is shown in Fig. 4.9.

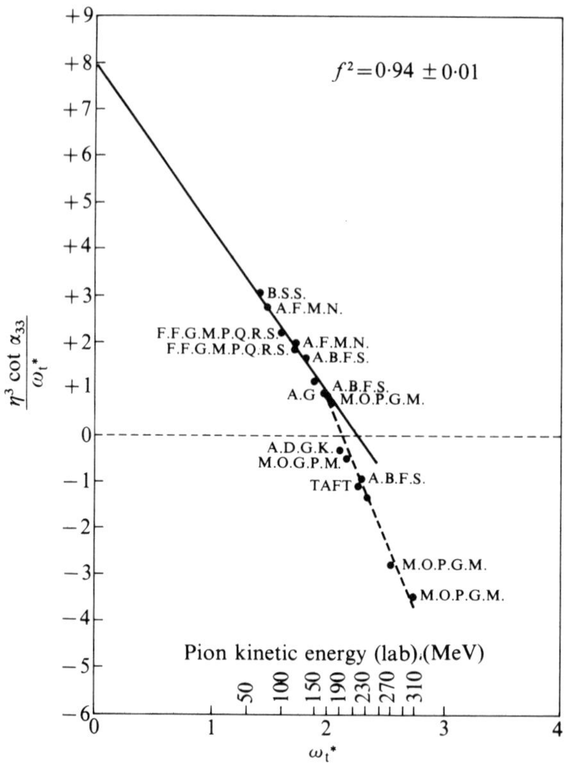


FIG. 4.9. A Chew–Low plot. Points are identified by the initial of author's last names. (Lindenbaum and Yuan. (1958). *Phys. Rev.* **111,** 1380.)

As pointed out, eqn (4.98) predicts a linear plot. We see in the figure that it was found that there was a change in slope in the Chew–Low plot beyond the resonant energy. The observed change in slope in these effective range plots is not too surprising since they are only expected to be straight lines for energies low compared to the effective cut-off energy $\omega_{\max}(\sim 6(m_\pi c^2))$. We might also notice that the resonance energy is slightly above the threshold for producing an additional pion and, hence, two meson states may begin to contribute appreciably.

The change of slope of the Chew–Low plot near the resonance makes the determination of the resonance energy somewhat uncertain. The resonance energy for the Chew–Low plot is determined by the point where the ordinate $(\eta^3 \cot \delta_{33}/\omega_t^*)$ is zero, which is the intercept on the $\omega_t^*$ axis. The best guess for the resonant energy in Fig. 4.9 is the intersection of the broken line with the $\omega_t^*$ axis. At the time the estimate was $\simeq 190^{+20}_{-10}$ MeV pion laboratory system kinetic energy, where the estimate of errors is somewhat uncertain.

The renormalized unrationalized coupling constant $f^2$ is obtained by extrapolating the line to determine the $y$-axis intercept at $\omega_t^* = 0$. From Fig. 4.9 we obtain $f^2 = 0{\cdot}94 \pm 0{\cdot}01$.

It was pointed out by Serber and Lee that the exact solution of the Chew–Low equation for charged and neutral scalar meson theories in the one meson approximation is

$$\frac{\eta^3}{\omega_t^*}\cot\alpha_{33} - \frac{1}{\omega_t^*} = \tfrac{3}{4}(f^2)^{-1}F(\omega_t^*). \tag{4.106}$$

However, as Chew and Low pointed out, the change in $f^2$ due to the inclusion of the $-(1/\omega_t^*)$ term in the ordinate of the plot may be of the same order of magnitude as the second order correction and the errors in the experimental data. The resonant energy for this latter plot (which also showed a break) turned out to be the same as that determined for the Chew–Low plot, but the value derived for the coupling constant was $f^2 = 0{\cdot}107 \pm 0{\cdot}01$, which was somewhat higher.

The tendency for a break (change in slope) in the Chew–Low curve is still present, even if we use the latest data. Furthermore, the values of the coupling constant are larger than the true values. A great deal of this increase in coupling constant can be attributed to the following cause. In the $P_{33}$-state there is, in addition to nucleon exchange which is taken into account in the Chew–Low theory, another exchange interaction which consists of exchange of a pair of pions which are in a $T = J = 0$ state which can be denoted by $\pi\pi_{T=0,J=0}$. This exchange also gives an attraction and it has been shown [28] that it contributes about 25–30 per cent to the coupling constant term in the Chew–Low formula. The contribution of $N^*$ exchange is negligible and only a very small contribution is made by $\rho$ exchange. Therefore, we can correct the Chew–Low formula prediction for the coupling constant for these additional exchanges by approximately replacing $f^2$ by $\frac{5}{4}f_e^2$ in the Chew–Low equation. The corrected effective value of $f_e^2$ then becomes

$$f_e^2 \approx (0{\cdot}94 \pm 0{\cdot}01) \times \tfrac{4}{5} \approx 0{\cdot}075 \pm 0{\cdot}008,$$

which agrees well with the best value of $f^2 = 0{\cdot}081 \pm 0{\cdot}002$ determined from dispersion relations (see Chapter 5).

## 4.13. Measurements of the $\pi^+$–p total cross-sections from 500 MeV to 2 GeV

More detailed subsequent measurements of the $\pi^\pm$–p total cross-sections beyond 500 MeV revealed that there was considerable structure in these

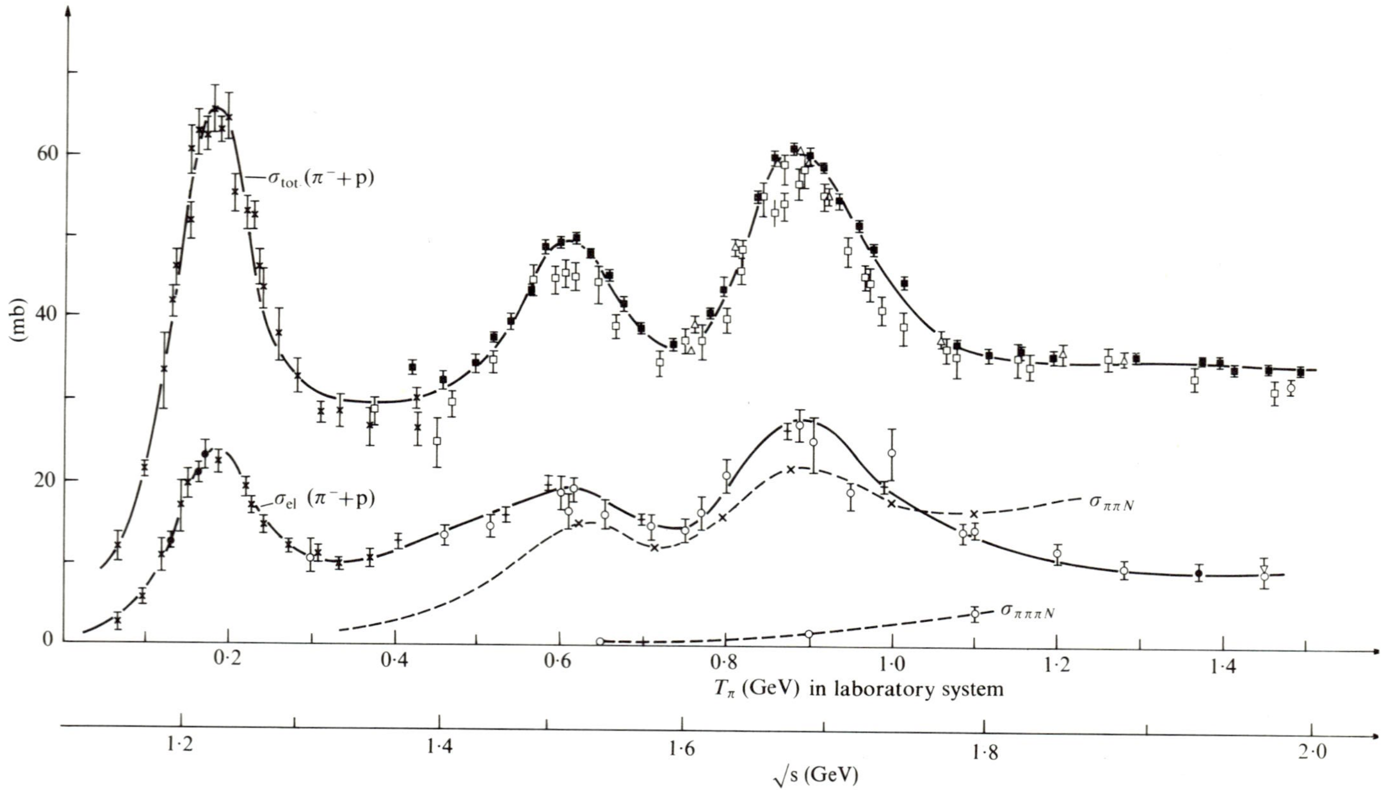


FIG. 4.10.(a) $\sigma_{tot}(\pi^- + p)$, $\sigma_{el}(\pi^- + p)$ (in mb) plotted against the laboratory kinetic energy. The data is from a compilation by Focaeci and Giacomelli, CERN 66–18 (1966). $\sigma_{\pi\pi N}$ is also shown. (Rosenfield and Soding. (1968). *Pion-nucleon scattering*. Wiley Interscience, New York.)

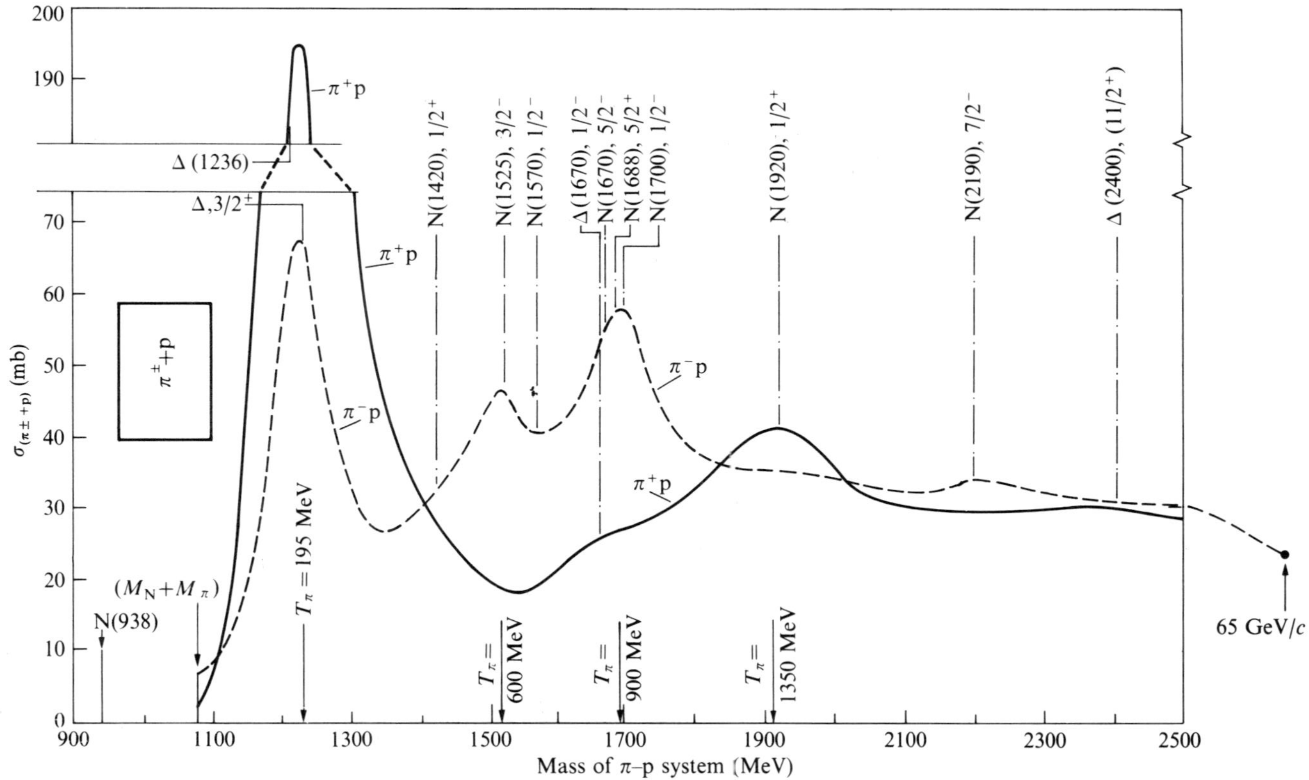


FIG. 4.10.(b) $\sigma_{\text{total}}(\pi^{\pm}-\text{p})$ (in mb) plotted against the mass of the $\pi$–p system (MeV) (i.e. $\approx$ the mass of a resonance, where a resonance peak occurs). Various resonance are indicated. See Particle Data Group tables (August 1970) ((1970). *Phys. Lett.* **B 33,** 1.) for details.

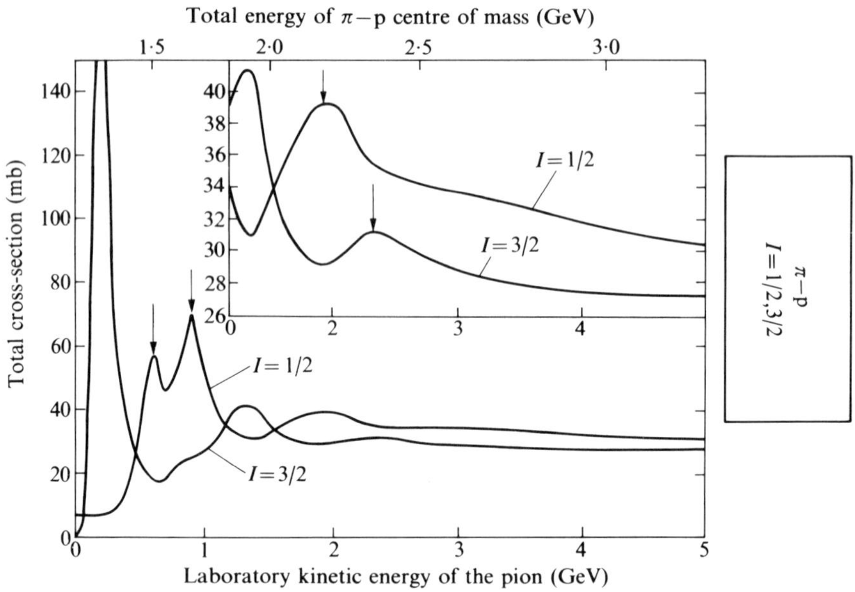


FIG. 4.10.(c) $\sigma_{\text{total}}(I = \frac{3}{2})$ and $\sigma_{\text{total}}(I = \frac{1}{2})$ (where $I = T$ = isotopic spin) plotted in mb as a function of incident pion laboratory energy, and the total energy of the $\pi$–p centre of mass (GeV). The upper right insert shows a more detailed plot for incident laboratory kinetic energy ($\geq$ GeV) (see Particle Data Group tables. (1970). *Phys. Lett.* **B 33,** 1).

cross-sections [29]. Figure 4.10 shows well-defined peaks in the $\pi^-$–p total cross-sections at 600 and 900 MeV. In the $\pi^+$–p cross-section there is a shoulder at about 800–1000 MeV, and a peak at around 1300 MeV. The $\pi^+$–p curve is the pure $T = \frac{3}{2}$ cross-section, (see Fig. 4.10(a) for plot against laboratory energy), but the $\pi^-$–p curve is a mixture of $T = \frac{1}{2}$ and $T = \frac{3}{2}$. Therefore the deduced $T = \frac{1}{2}$ curve is also shown.

In contrast to the low energy $T = J = \frac{3}{2}$ entirely elastic resonance, which is the only important interaction at low energies, these resonance phenomena are quite complex, involving more than one resonance, considerable inelasticity, and background amplitudes. We will defer their analysis until the section on generalized phase shift analysis.

### 4.14. Phase-shift analysis for generalized complex phase shifts

Phase shift analysis, in terms of partial wave amplitudes, can be conveniently considered by plotting once, or twice, the partial wave amplitudes on an Argand diagram.

From eqn (4.57), we have

$$2a_{l,j} = \frac{1}{\mathrm{i}}\,(\mathrm{e}^{2\mathrm{i}\delta_{l,j}}-1) = [\mathrm{e}^{\mathrm{i}\pi/2}-\mathrm{e}^{\mathrm{i}(2\delta_{l,j}+\pi/2)^{2}}]. \tag{4.107}$$

When the scattering is entirely elastic $\delta_{l,j}$ is real, and Fig. 4.11 shows a plot of $2a_{l,j}$ indicating the addition of the two equal modulus vector terms in eqn (4.107) which clearly requires that the locus of $a_{l,j}$ be a circle with its centre at $+\mathrm{i}$, and having a radius of unity. The magnitude of $a_{l,j}$ and its phase are both zero when $\delta_{l,j} = 0$, and increase continuously to a maximum. As $\delta_{l,j}$ increases beyond 90° (the resonant point) the magnitude of $a_{l,j}$ decreases toward zero, which is the value associated with $\delta_{l,j} = 180°$.

The causality requirement forces the resonant amplitude to contain a factor $\dfrac{1}{\omega-\omega_{\mathrm{r}}+\mathrm{i}\Gamma/2}$, which means that the point corresponding to $a_{l,j}$ must move around the circle in a counterclockwise direction as the energy increases and passes through resonance ($\delta_{l,j} = 90°$). When the interaction is not entirely elastic the situation becomes much more complex. In this case the phase shift $\delta_{l,j}$ has to be considered complex. Let us define $\delta_{l,j} = \alpha_{l,j}+\mathrm{i}\beta_{l,j}$. Then

$$\mathrm{e}^{2\mathrm{i}\delta_{l,j}} = \mathrm{e}^{2\mathrm{i}\alpha_{l,j}}\mathrm{e}^{-2\beta_{l,j}} = |\eta_{l,j}|\,\mathrm{e}^{2\mathrm{i}\alpha_{l,j}}, \tag{4.108}$$

where $|\eta_{l,j}| = \mathrm{e}^{-2\beta_{l,j}} = \mathrm{e}^{-2\,\mathrm{Im}\,\delta_{l,j}}$. Due to the unitarity requirement, it follows that $\beta \geq 0$, with the equality holding only for elastic scattering. Obviously,

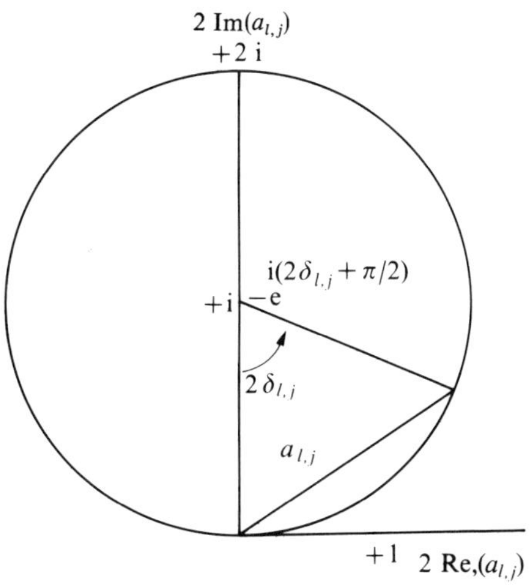


FIG. 4.11

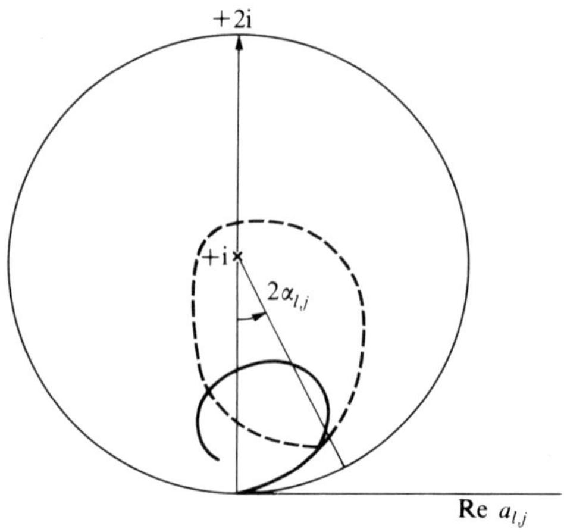


FIG. 4.12

$0 \leq \eta \leq 1$ with the lower limit ($|\eta| = 0$) corresponding to maximum inelasticity and the upper limit corresponding to elastic scattering.

Hence in this general case, we obtain

$$2a_{l,j} = (e^{i\pi/2} - |\eta_{l,j}|\, e^{i2\delta_{l,j}+\pi/2}). \tag{4.109}$$

Figure 4.12 illustrates this general case.

For $|\eta| = 1$ we again have the elastic condition. To consider the case where the inelasticity is appreciable, it is convenient to define $\eta_{l,j}$ as the complex quantity

$$\eta_{l,j} = e^{2i\delta_{l,j}}. \tag{4.110}$$

Then we have

$$2a_{l,j} = (e^{i\pi/2} - \eta_{l,j} e^{i\pi/2}) = e^{i\pi/2}(1-\eta_{l,j}). \tag{4.111}$$

From previous definitions of the cross-section in terms of $a_{l,j}$, it can be shown that for a particular angular momentum state

$$(\sigma_{\text{total}})_{j\atop\text{elastic}} = \pi\lambda\!\!\!^{-2}(j+\tfrac{1}{2})\,|1-\eta_j|^2,$$

$$\sigma_{\text{inel}_j} = \pi\lambda\!\!\!^{-2}(j+\tfrac{1}{2})(1-|\eta|^2). \tag{4.112}$$

Hence,

$$\sigma_{\text{total}_j} = \sigma_{\text{el}_j} + \sigma_{\text{inel}_j} = \pi\lambda\!\!\!^{-2}(j+\tfrac{1}{2})(1-|\eta|^2+|1-\eta|^2) =$$
$$= \pi\lambda\!\!\!^{-2}(2j+1)(1-\text{Re}\,\eta). \tag{4.113}$$

Let us define

$$E_{l,j} = \left(\frac{\sigma_{\text{el}}}{\sigma_{\text{tot}}}\right)_{l,j}, \tag{4.114}$$

where $E$ is related to the elasticity of the process, $\frac{1}{E}$ is a measure of the inelasticity, and $|E| \leq 1$. Now in Fig. 4.12 when there is inelasticity, $|\eta| < 1$ and, in general, $\eta$ is a function of $\alpha$, which is a function of energy. Therefore as shown in Fig. 4.12, we no longer describe a circle with the vector representing $a_{l,j}$ but, instead, a more generalized type of curve. If $E > \frac{1}{2}$ (the dotted line in Fig. 4.12) the general features of elastic scattering are still maintained. The curve described by $a_{l,j}$ crosses the imaginary axis above the centre of the circle, and the real part of the phase shift ($\alpha$) goes through 90°. For this case, $\eta$ generally drops to a minimum value as we go through resonance, and then rises again. For cases with low elasticity $E < \frac{1}{2}$. The curve describing $a_{l,j}$ passes below the centre of the circle for the elastic case. Hence we have a complicated behaviour for the real part of the phase shift. At low energies it is positive and increases to a maximum, which is less than $\pi/4$, well below the resonance. In fact, as the resonant energy is passed the real part of the phase shift goes through zero and changes its sign, becoming negative. Some illustrations of these types of plots for resonances involving $s$-, $p$-, $d$-, and $f$-waves are taken from a phase shift analysis of the then available data in the region of 300–1000 MeV (Bareyre *et al.*) [30] and shown in Fig. 4.13(a)–(c). The corresponding phase shifts (their $\delta$ corresponds to our $\alpha$) are shown in Figs. 4.13(d)–(e) and compared to the experimental data. The analysis was made including $f$-waves up to the two highest energies 950 and 990 MeV, where $g$-waves were also included. However, the $g$-wave phase shifts were found to be less than 2°, and could have been neglected throughout this energy range.

As was previously pointed out, there is generally more than one phase shift solution at a particular energy. It was required that the selected solution be joined smoothly to the solutions found at lower energy and vary smoothly throughout the energy range investigated. This smoothing requirement eliminated all but one solution. Such a smoothing assumption, whether made by inspection, or specifically made by requiring a slowly varying parameterization [25], [31], or by using as a starting point for a solution the theoretical predictions obtained with the partial wave dispersion relations [31], [32], [34], [35] has been found to be required to select a unique set of phase shifts. The smoothing requirement is intimately related to the requirements of microscopic causality or, more precisely, the requirement of analysticity of the scattering amplitudes as embodied in the dispersion relations and, hence, is a reasonable approach to the problem. In Fig. 4.13(a), for the $T = \frac{3}{2}$ state (Bareyre *et al.*) the locus of the $s_{31}$ partial wave is drawn in the complex plane. $s_{31}$ is a repulsive wave up to 550 MeV and, as

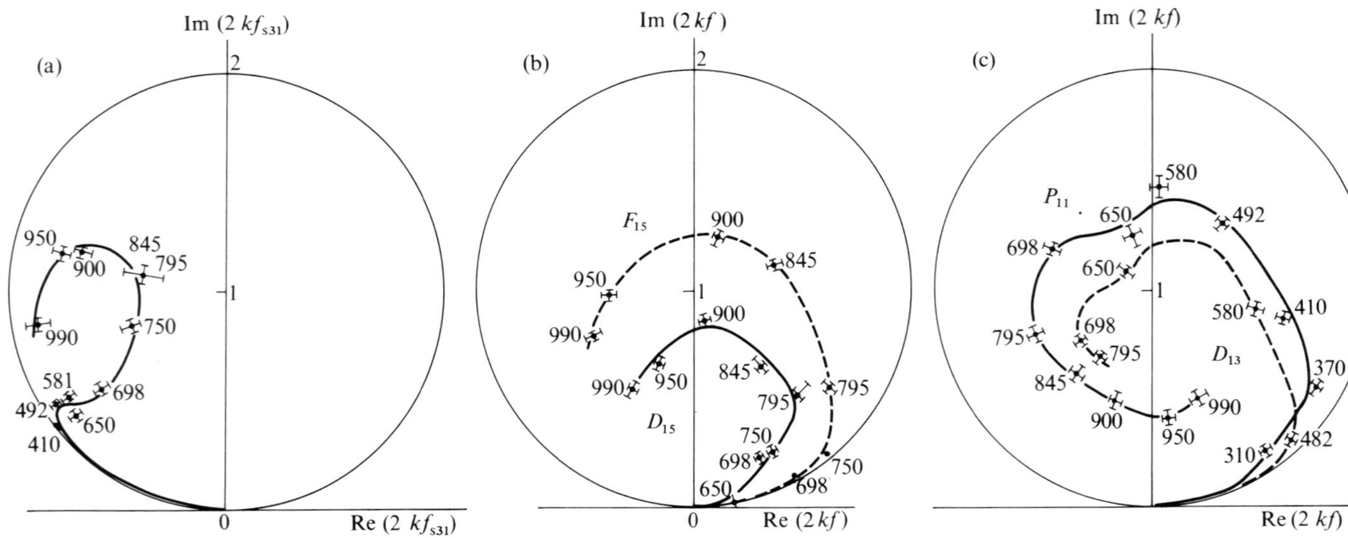


FIG. 4.13. Locus in the complex plane of (a) an $s$-wave (b) $d$- and $f$-waves (c) $d$- and $p$-waves.

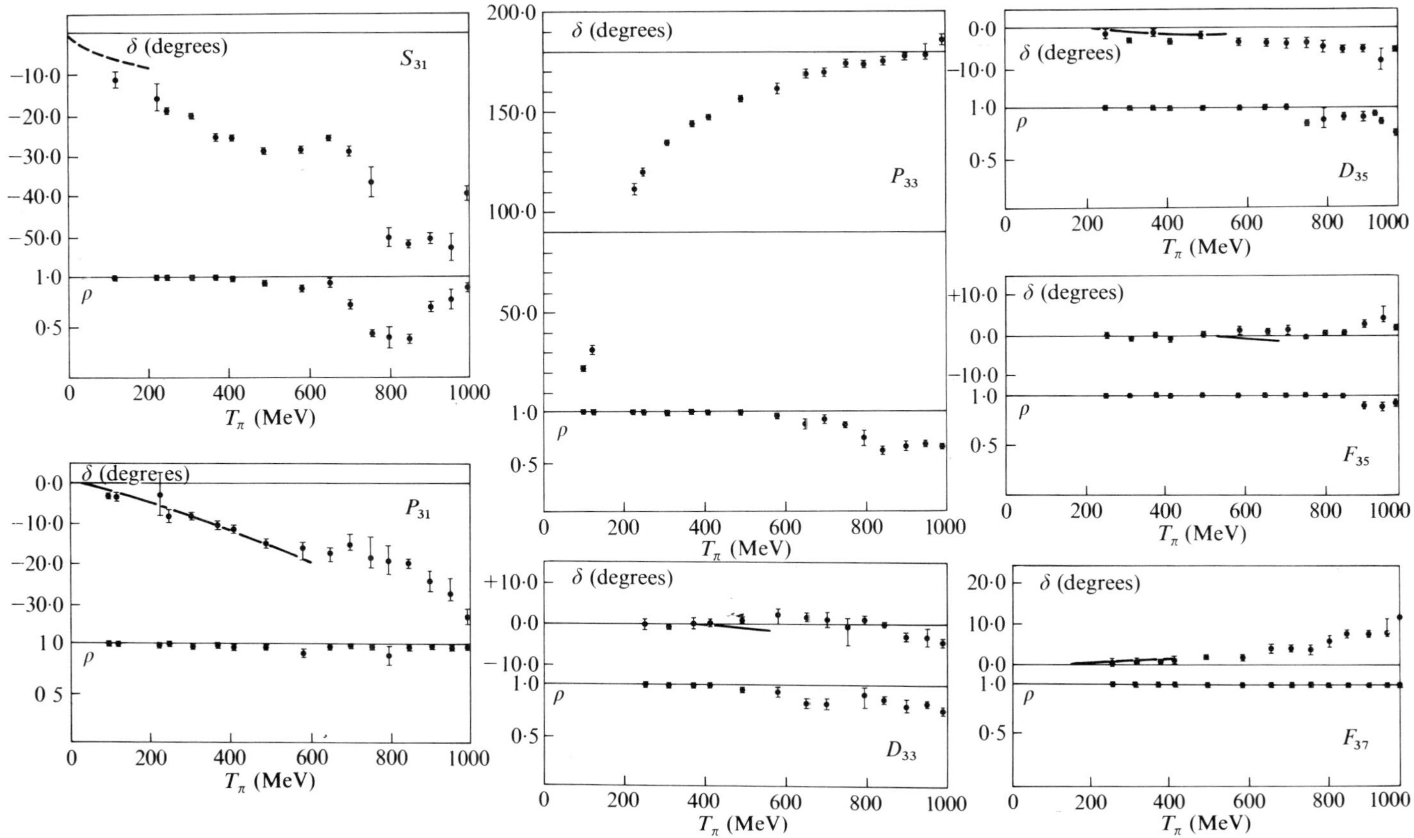


FIG. 4.13.(d) Solution for isotopic spin $I = \frac{3}{2}$. The solid curves represent the predictions of [34], dashed curves refer to [18] of Chapter 5. (For energies less than 310 MeV the values indicated are those of similar published analyses.)

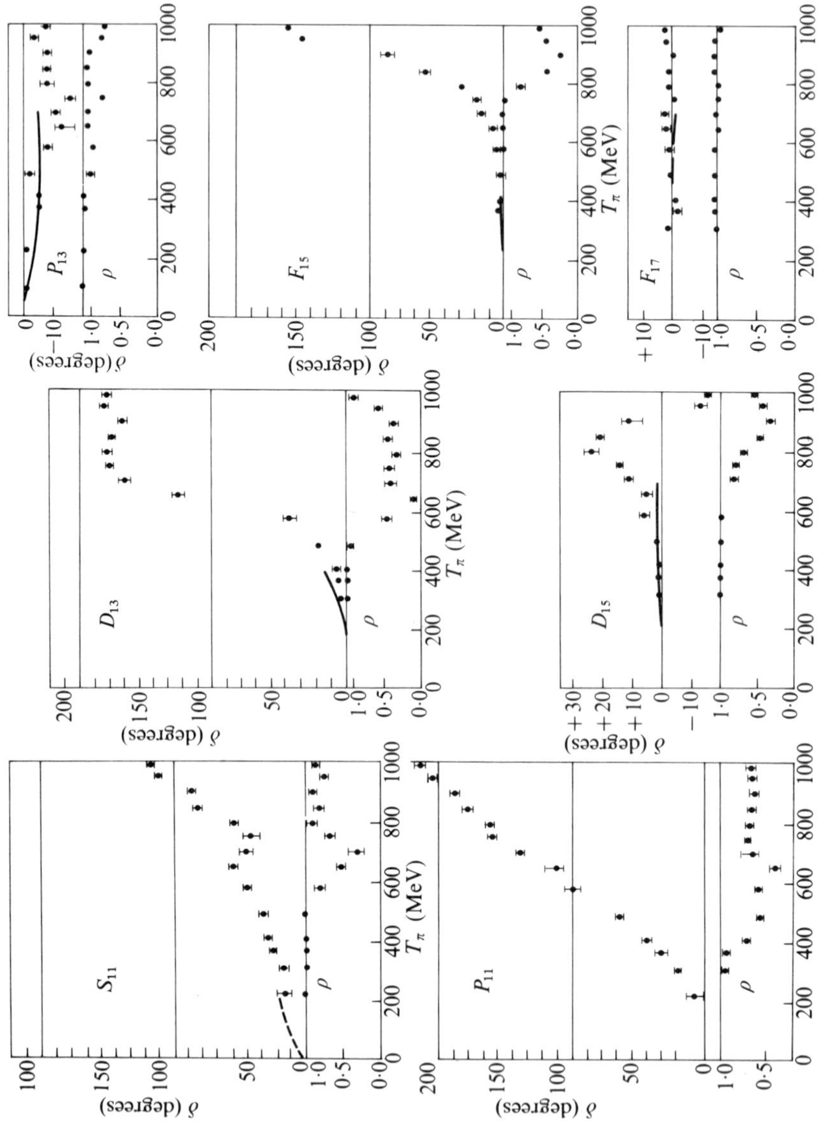


FIG. 4.13.(e) Solution for isotopic spin $I = \frac{1}{2}$ (Fig. 4.13 is from Bayere *et al.* (1965). *Phys. Lett.* **18**, 342).

can be seen, as the energy increases beyond 550 MeV the locus moves in a counter-clockwise direction along a circular curve. This behaviour may be connected with a resonance in the inelastic channel $\pi N^*$. The inelasticity and behaviour of possible background amplitudes may cause this curve never to cross the imaginary axis even though a resonance occurs [34]. For the $T = \frac{1}{2}$ state, it had been shown that the $f_{15}$ wave is responsible for the 900 MeV resonance [34].

The analysis under discussion has shown that the situation in the neighbourhood of the resonance is complicated. The $F_{15}$ wave crosses 90° at around 900 MeV, the width of the resonance is about 110 MeV, and the absorption is obviously strong. The behaviour of the $D_{15}$ wave could, as can be seen from Fig. 4.13, be interpreted as a resonance at an energy of about 915 MeV, which decays mostly into strongly inelastic channels. The phase shift goes through 0° at about 915 MeV. We should recall from our previous discussion that if $E < \frac{1}{2}$, which is the case here, a resonance is characterized by a phase shift, which reaches a maximum and then decreases toward zero (which it passes through at the resonant energy).

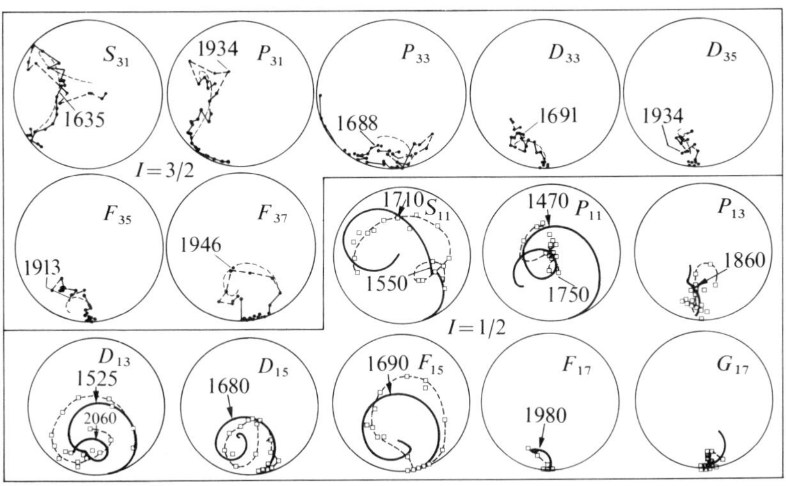


FIG. 4.14. Results of the phase shift analyses of the CERN and Berkeley groups. The CERN I results are the smooth curves (dashed in the $I = \frac{3}{2}$ diagrams). This analysis used dispersion relations to join and smooth the solutions found at different energies. The arrows in the $I = \frac{1}{2}$ diagrams indicate approximate resonance positions. They have been drawn by us. The CERN II solution is shown (as a dot–dash line) only for the $I = \frac{3}{2}$ amplitudes, since the $I = \frac{1}{2}$ are not available. The arrows have been drawn by the authors. The Berkeley solution is shown only for the $I = \frac{1}{2}$ state (as empty squares joined by dashes). (From Particle Data Group. (1970). *Rev. mod. Phys.* p. 163.)

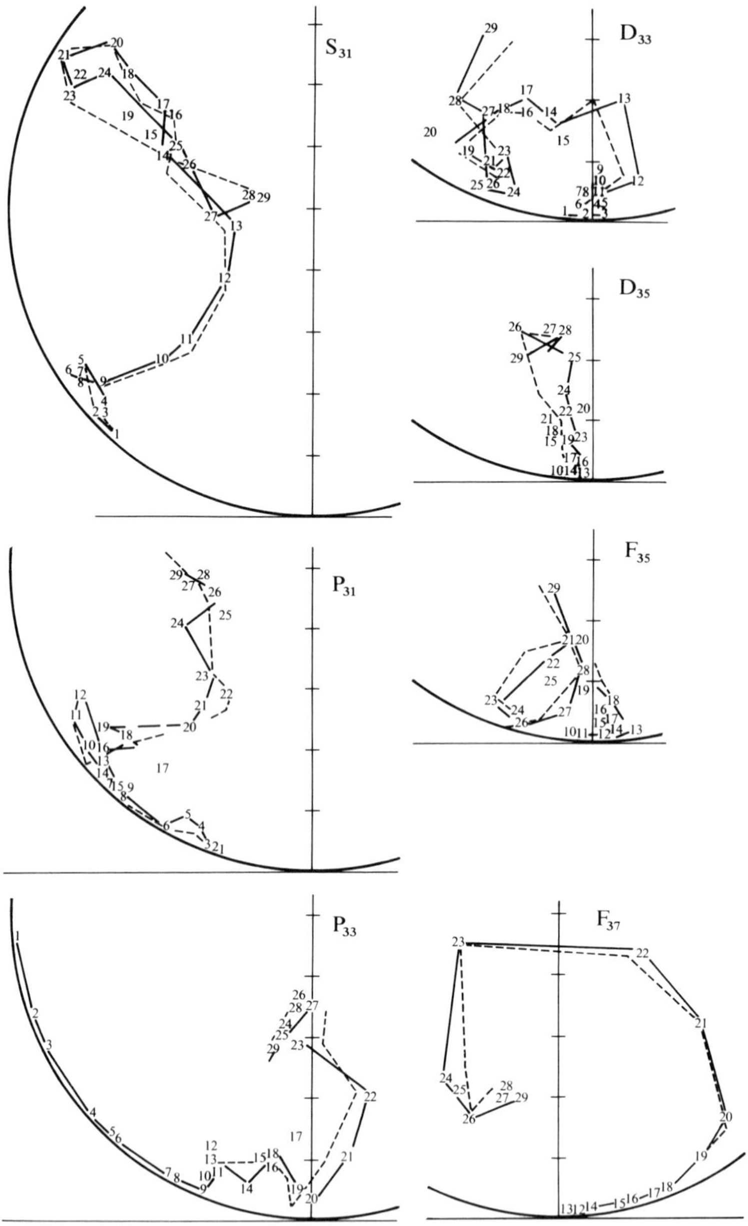

$S_{31}$
$D_{33}$
$D_{35}$
$P_{31}$
$F_{35}$
$P_{33}$
$F_{37}$

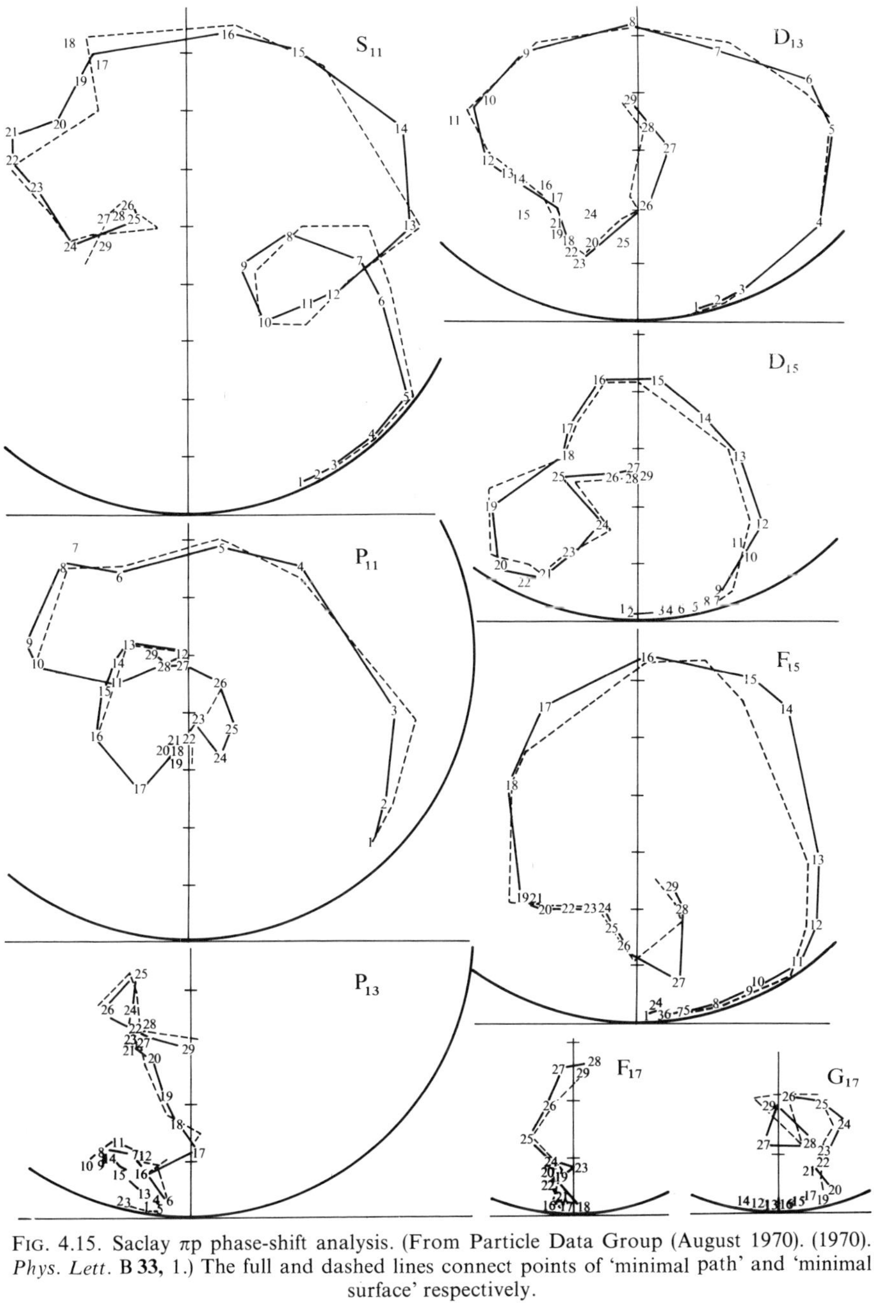


FIG. 4.15. Saclay $\pi p$ phase-shift analysis. (From Particle Data Group (August 1970). (1970). *Phys. Lett.* B **33,** 1.) The full and dashed lines connect points of 'minimal path' and 'minimal surface' respectively.

The $D_{13}$ phase shift shows the usual resonance behaviour and crosses 90° near 630 MeV. It has a width of about 90 MeV and is quite inelastic. As we can see from Fig. 4.13(c), the solution also exhibits a clear-cut resonance behaviour in the $P_{11}$ wave at 600 MeV, with a width of 350 MeV. This $P_{11}$-wave resonance appears to be confirmed by high energy experiments which detect a nucleon-isobar type behaviour at a mass of $\approx$ 1400 MeV. This will be discussed more fully in Chapter 6. Figure 4.13(d) gives the deduced phase shifts for an isotopic spin of $\frac{3}{2}$, while Fig. 4.13(e) gives the phase shifts deduced for isotopic spin of $\frac{1}{2}$.

Since the Argand diagram method illustrates the problem well, we have deferred discussing *a priori* [25] extensive phase shift analysis extending over several energy ranges from 0 to 700 MeV (see [25] which also reviews earlier work in the field and gives the first detailed analysis which justifies a $P_{11}$ resonance at about 585 MeV incident pion laboratory energy).

To obtain a unique phase shift solution over a range of energies, the phase shifts and absorption parameters were parameterized as functions of energy. The Breit–Wigner resonance forms for the amplitudes were put in for states which have the characteristics of a resonance, and a background amplitude was incorporated. The parameterizations used gave the right threshold behaviour for the phase shifts $\delta_{l,j} \sim \eta^{2l+1}$ and approximately correct behaviour for the absorption coefficients $(k-k_0)^{2l+1}$, provided the first important inelastic state after threshold has the same orbital angular momentum as the initial state.

Satisfaction of the dispersion relation which connects the real and imaginary parts of the partial wave amplitude is not guaranteed. However, other analyses [31], [32], [34], [35] have this condition required in the analysis. The available data on total cross-sections, elastic and charge exchange differential cross-sections, and the recoil nucleon polarizations for elastic and charge exchange scattering have been used. The data was analysed in the 0–350 MeV and 0–700 MeV ranges. At a later time, 0–50 MeV and 0–100 MeV ranges were considered. The analysis was carried through *f*-waves. A consistent behaviour for the solutions in all these regions was found at least for the large phase shifts, which was generally continued by later analyses.

Unexpectedly, the 0–350 MeV analysis yielded a negative $P_{11}$ phase shift below 150 MeV, which became large and positive at 350 MeV (32°). Upon extension of the analysis to 700 MeV it was impossible to get good fits to the data unless the $P_{11}$ phase shift was allowed to become large. The best fit occurs when the $P_{11}$ resonates at approximately 585 MeV along with the $D_{13}$ resonance at about 638 MeV. Appreciable reduction of the $P_{11}$ phase shift from the large values (although going through resonance was not necessary to fit the data) caused the forward $\pi$–p cross-section to fall far below the observed values throughout the 300–600 MeV energy region. The resonant behaviour of the $P_{11}$ phase shift has been subsequently observed in other analyses which we have just discussed. The fact that a number of independent analyses, using different criteria, agree in so many features of this complex problem lends weight to the probable validity of the major conclusions.

In regard to the 0–350 MeV region, we should note that *d*- and *f*-wave phases stay below about 3° and, hence, the earlier *s*, *p* analysis, at least up to and through the resonance energy, are justified with reference to the behaviour of large phase shifts. Predictions for the *p*, *d*, and *f* phase shifts, up to moderate energies (400–900 MeV), have been made, using a peripheral method [32], [34] which evaluates the dispersion relation for

$$\mathscr{F}_{l\pm}(s) = f_{l\pm}(s)/q^{2l},$$

where $f_{l\pm}(s)$ is the partial wave amplitude and $q$ is the pion momentum. The poorly-known shorter range parts of the interaction are suppressed by the factor $\frac{1}{q^{2l}}$. A unitary sum is used to estimate the short range parts of the pion–nucleon interaction. This gives reasonable predictions of the non-resonant *p*-, *d*-, and *f*-wave $\pi$–N amplitudes up to about 600 MeV, which agree reasonably with other analyses. A more detailed discussion of this method will be presented in Chapter 5.

We have just described some of the early work in the field of phase shift analyses. Subsequent to this early work, an era of feverish activity in this field ensued, and various degrees of refinement in the techniques were used with the objective of uncovering nucleon resonances. A recent review of this work was given by Donnachie [36] in a rapporteur's Report at the Vienna conference (1968). Table 1 (p. 142 [36]) summarized the twenty conjectured pion–nucleon resonance assignments below 2·2 GeV mass, with the status of the corresponding structure observed in the five most recent phase shift analyses which, as one would expect, do not all agree. Table 2 (p. 143 [36]) shows the proposed structure of the 16 to 18 pion–nucleon resonances, at the time of the Vienna conference, with the masses, widths, and elasticities taken from the CERN and Glasgow groups solutions. This subject is updated, and complete recent references are given by the Particle Data Group (*Rev. mod. Phys.* (1970) **42**, 87). Some recent Argand diagrams illustrating the results of recent phase shift analyses are also given in the above reference and reproduced here in Figs. 4.14, 4.15, and 4.16.

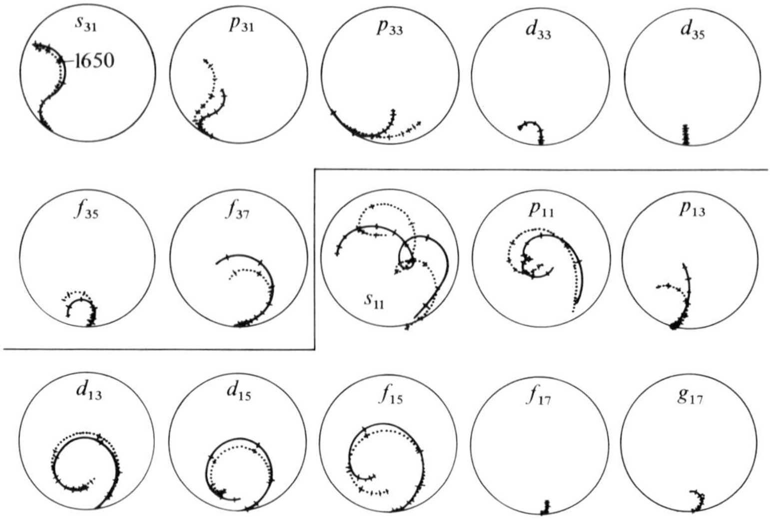


FIG. 4.16. Partial wave analysis results of Davies. The curves plotted here are the the results of the energy-dependent analysis. Davies A (solid curves) is obtained by starting from the best set of phase shifts solution of the Glasgow group. Davies B (dashed curves) is obtained by starting from the CERN 1 phase shifts. Davies B is not shown when it is very close to Davies A. Scale marks of 50 MeV are shown. The first large mark is at $M$ = 1400 MeV, the last large mark at $M$ = 1900 MeV (Davies 1968). (From Particle Data Group (August 1970). (1970). *Phys. Lett.* B **33,** 1.)

In general, use has been made of two methods to find resonance parameters (a) the energy at the point where the total cross-section is a maximum and (b) the energy where the speed of variation in the Argand plot is a maximum. CERN generally refers to only one method—where the absorption is maximum. Davies employs a Breit–Wigner parameterization (Fig. 4.16).

## REFERENCES

1. PAULI, W. (1946). *Meson theory of nuclear forces.* Interscience, New York.
2. WENZEL, G. (1947). *Rev. Mod. Phys.* **19,** 1.
3. MARSHAK, R. E. (1952). *Meson physics.* McGraw-Hill, New York.
4. ANDERSON, H. L., FERMI, E., LONG, E. A., MARTIN, R., and NAGLE, D. E. (1952). *Phys. Rev.* **85,** 934; ANDERSON, H. L., FERMI, E., NAGLE, D. E., and YODH, G. B. (1952). *Phys. Rev.* **86,** 793.
5. BRUECKNER, K. A. (1952). *Phys. Rev.* **86,** 106.
6*a*. ANDERSON, H. L., FERMI, E., MARTIN, R., and NAGLE, D. E. (1953). *Phys. Rev.* **91,** 155.

6*b*. FERMI, E., GLICKSMAN, M., MARTIN, R., and NAGLE, D. E. (1953). *Phys. Rev.* **92,** 161.
7. FERMI, E., METROPOLIS, N., and ALEI, E. F. (1954). *Phys. Rev.* **95,** 1581.
8. CHEW, G. F. (1953). *Phys. Rev.* **89,** 591; (1954). *Phys. Rev.* **95,** 285.
9. LINDENBAUM, S. J. and YUAN, L. C. L. (1953). *Phys. Rev.* **92,** 1578; (1954). **93,** 917. Also *Proceedings of IVth annual Rochester conference.* (1954). Interscience, New York. p. 100.
10. LINDENBAUM, S. J. and YUAN, L. C. L. (1955). *Phys. Rev.* **100,** 306.
11. ASHKIN, J., BLESER, J. P., FEINER, F., GORMAN, J. G., and STERN, M. O. (1954). *Phys. Rev.* **96,** 1104.
12. COOL, R. L., MADANSKY, L., and PICCIONI, O. (1954). *Phys. Rev.* **93,** 249.
13. LINDENBAUM, S. J. and YUAN, L. C. L. (1956). *Phys. Rev.* **103,** 404; (1954). **93,** 1431.
14. DE HOFFMAN, F., METROPOLIS, N., ALEI, E. F., and BETHE, H. A. (1954). *Phys. Rev.* **95,** 1586.
15. CONDON and SHORTLEY. (1935). *Theory of atomic spectra.* Cambridge University Press, London.
16. BLATT, J. M. and WEISSKOPF, V. F. (1952). *Theoretical nuclear physics.* John Wiley and Sons, New York.
17. MIYAKE, K., KINSEY, K. F., and KNAPP, D. E. (1962). *Phys. Rev.* **126,** 2188.
18. KÄLLEN, G. (1964). *Elementary particle physics.* Addison-Wesley, New York.
19. YANG, C. N. (unpublished).
20. DAVIDON, W. C. and GOLDBERGER, M. L. (1956). *Phys. Rev.* **104,** 1119.
21. MINAMI, S. (1954). *Prog. theor. Phys., Osaka.* **11,** 213.
22. HAYAKAWA, S., KAWAGUCHI, M., and MINAMI, S. (1954). *Prog. theor. Phys., Osaka.* **11,** 332.
23. LINDENBAUM, S. J. and STERNHEIMER, R. M. (1958). *Phys. Rev.* **110,** 1174.
24. LINDENBAUM, S. J. and YUAN, L. C. L. (1958). *Phys. Rev.* **111,** 1380.
25. ROPER, L. D., WRIGHT, R. M., and FELD, B. T. (1965). *Phys Rev.* B**138,** 190.
26. LOW, F. E. (1955). *Phys. Rev.* **97,** 1392.
27. CHEW, G. F. and LOW, F. E. (1956). *Phys. Rev.* **101,** 1570.
28. DONNACHIE, A. and HAMILTON, J. (1964). *Phys. Rev.* B**133,** 1053; HAMILTON, J. (1967). *High energy physics*, Vol. 1 (Edited by Burhop). Academic Press, New York.
29. Data by FOCACCI, N. and GIACOMELLI, G. (1966). CERN 66–18; ROSENFELD and SODING (1968). *Pion-nucleon scattering.* Wiley Interscience, New York; Particle Data Group Tables (August 1970).
30. BAREYRE, P., BRICKMAN, C., STIRLING, A. V., and VILLET, G. (1965). *Phys. Lett.* **18,** 342.
31. BRANDSEN, B. H., O'DONNELL, P. J., and MOORHOUSE, R. G. (1964). *Phys. Lett.* **11,** 339.
32. DONNACHIE, A., LEA, A. T., and LOVELACE, C. (1965). *Phys. Lett.* **19,** 146. In a slightly later treatment these authors showed that a strongly repulsive background which they understood in terms of a dispersion relation prediction was responsible for the shift of $S_{31}$ to the left in the complex plane, and when this effect was corrected out the $S_{31}$ curve crossed the imaginary axis at resonance.

33. Duke, P. J., Jones, D. P., Kemp, M. A. R., Murphy, P. G., Prentice, J. D., Thresher, J. J., Atkinson, H. H., Cox, C. R., and Heard, K. S. (1965). *Phys. Rev. Lett.* **15,** 468.
34. Donnachie, A., Hamilton, J., and Lea, A. T. (1964). *Phys. Rev.* **B135,** 515; Donnachie, A. and Hamilton, J. (1965). *Phys. Rev.* **B138,** 678.
35. Auvil, P., Lovelace, C., Donnachie, A., and Lea, A. T. (1964). *Phys. Lett.* **12,** 76.
36. Donnachie, A. (1968). Nuclear resonances. *14th International conference on high energy physics, Vienna.* Published by CERN Scientific Information Service, Geneva.

# 5

# DISPERSION RELATIONS

## 5.1. General characteristics of two-body scattering

In Chapters 1 and 2 we briefly discussed invariance principles including Lorentz invariance. Lorentz invariance with respect to translations leads to conservation of energy and momentum. Lorentz invariance with respect to rotations leads to conservation of angular momentum. Invariance with respect to time reversal leads to the principle of time-reversal invariance. These requirements, and the appropriate Lorentz transformation characteristics of these quantities, are most conveniently expressed in terms of certain invariant functions of the energy–momentum vectors.

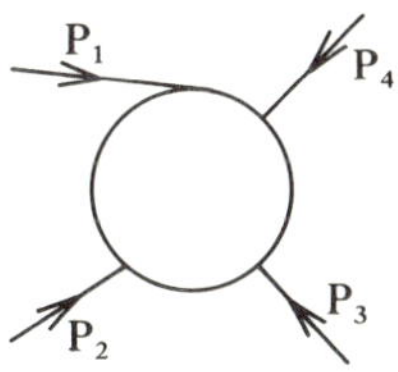


FIG. 5.1. It is generally most convenient to evaluate $s$, $t$, and $u$ in either the c.m.s. or the laboratory system, but since they are Loretz invariants, they will have the same values in any Lorentz frame moving with fixed velocity vector relative to these two frames.

Consider the scattering process $\alpha+\beta \rightarrow \alpha+\beta$, as depicted in Fig. 5.1, where $P_1$, $P_2$, $P_3$, $P_4$ are energy–momentum four-vectors all pointing inward toward the interaction volume in the symmetrical arrangement.† A line can be reversed in direction by replacing particle by anti-particles, and this transformation depends on time, charge conjugation, and parity CPT invariance. Let us define

$$s = (P_1+P_2)^2 = (P_3+P_4)^2, \tag{5.1}$$

$$t = (P_1+P_4)^2 = (P_2+P_3)^2, \tag{5.2}$$

$$u = (P_1+P_3)^2 = (P_2+P_4)^2. \tag{5.3}$$

Only two of $s$, $t$, and $u$ are independent. The third then being determined by the relationship $s+t+u = m_1{}^2+m_2{}^2+m_3{}^2+m_4{}^2$.

Since

$$\mathbf{p}_1 = -\mathbf{p}_2$$

† When the four-vectors $P_3$ and $P_4$ are reversed in direction (in the symmetrical arrangement shown above) all four components change sign, and $P_3$ and $P_4$ represent anti-particles of the corresponding outgoing particles.

then,

$$s = (E_{(1)}^{\text{cms}} + E_{(2)}^{\text{cms}})^2,$$

that is, $s$ is the square of the total centre of mass energy. In the laboratory system, for the case where the target particle is at rest, $\mathbf{p}_1 = \mathbf{p}_l$ and $\mathbf{p}_2 = 0$. Hence,

$$s = (E_1^l + m_2^l)^2 - (p_1^l)^2, \tag{5.4}$$

$$s = (E_1^l)^2 + (m_2^l)^2 + 2E_1^l m_2^l - (p_1^l)^2$$

$$= (m_1^l)^2 + 2E_1^l m_2^l + (m_2^l)^2,$$

or

$$s = (m_1^l)^2 + (m_2^l)^2 + 2m_2^l E_1^l.$$

For the case of elastic scattering

$$m_1 = m_4$$

and

$$m_2 = m_3.$$

We can evaluate $t$ conveniently in c.m.s., and obtain

$$t = -\{|p|^2 \sin^2\theta + |p|^2 (1 - \cos\theta)^2\} = -2\,|p|^2 (1 - \cos\theta). \tag{5.5}$$

If $m_1 = m_3$ and $m_2 = m_4$ (elastic exchange scattering),

$$u = -2\,|p|^2 (1 + \cos\theta).$$

For the special case $m_1 = m_2 = m_3 = m_4$ (elastic scattering with all masses equal) we have, simultaneously,

$$t = -2\,|p|^2 (1 - \cos\theta),$$

$$u = -2\,|p|^2 (1 + \cos\theta).$$

If both $\alpha$ and $\beta$ are spinless, the only vectors we have at our disposal are $\mathbf{p}_1$ and $\mathbf{p}_2$. We must require that the scattering amplitude depend on scalar products of these, which are invariant with respect to rotations, in order to conserve angular momentum. The only possibility is $\mathbf{p}_1 \cdot \mathbf{p}_2$. Hence the scattering amplitude must be of the form

$$f(\mathbf{p}_1 \cdot \mathbf{p}_2) = |p_1|\,|p_2| \cos\theta = f(\cos\theta) = f(t), \tag{5.6}$$

$f(\mathbf{p}_1 \cdot \mathbf{p}_2) = f(t)$ is a scalar under the parity transformation and, therefore, this form of the amplitude also conserves parity. This form of the amplitude is also time-reversal invariant. Therefore, for scattering of the spinless particles, allowing for variation with energy,

$$\frac{d\sigma}{d\Omega}(E) = |f(\cos\theta, E)|^2, \tag{5.7}$$

where the scattering amplitude $f(\cos\theta, E)$ is a complex function. If we wish to express the scattering differential cross-section in an invariant form,† we can use the notation

$$\frac{d\sigma}{dt} = |A(s, t)|^2. \tag{5.8}$$

If one of the particles has spin $\sigma = \frac{1}{2}$, then we can form a second term meeting the invariance conditions, such as $g(\cos\theta, E)(\boldsymbol{\sigma}.\hat{\mathbf{n}})$,

$$\hat{\mathbf{n}} = \frac{\mathbf{p}_1 \times \mathbf{p}_2}{|\mathbf{p}_1 \times \mathbf{p}_2|} \tag{5.9}$$

is the unit vector normal to the scattering plane. This term is invariant with respect to the parity, and time, reversal operations. Terms of the form $\boldsymbol{\sigma}.\mathbf{p}_1$ or $\boldsymbol{\sigma}.\mathbf{p}_2$ change sign under the parity transformation and hence are excluded. Thus the most general scattering amplitude for $\pi$–$p$ scattering, in a particular charge state, is

$$f(\cos\theta, E) + g(\cos\theta, E)\boldsymbol{\sigma}.\hat{\mathbf{n}}. \tag{5.10}$$

Hence,

$$\frac{d\sigma}{d\Omega}(\cos\theta, E) = |f(\cos\theta, E)|^2 + |g(\cos\theta, E)|^2 \sin^2\theta. \tag{5.11}$$

The $\sin\theta$ factor comes from evaluation of $\boldsymbol{\sigma}.\hat{\mathbf{n}}$. This form has been derived, previously, from a more specific point of view in Chapter 4. As we saw, the second term corresponds to flipping the spin of the proton, and is incoherent with the first term (or non-flip scattering). Therefore, the sum of the squares are taken separately. We then have for the invariant form†

$$\frac{d\sigma}{dt} = |A(s, t)|^2 + |B(s, t)\boldsymbol{\sigma}.\hat{\mathbf{n}}|^2.$$

The appropriate relationship connecting the terms is given by the relation

$$\frac{\pi}{k^2}\frac{d\sigma}{d\Omega} = \frac{d\sigma}{dt}.$$

Hence, invariance principles limit the $\pi$–p scattering amplitude to contain only two independent complex functions of $s$ and $t$. As $\theta$ or $t \to 0$, the second or spin flip term $\to 0$ and hence for forward, or near forward, $\pi$–p scattering,

† Since $t$ is a negative quantity in the physical region, $\dfrac{d\sigma}{dt} = -\dfrac{\pi}{k^2}\dfrac{d\sigma}{d\Omega}$. However, since cross-sections are usually defined as positive quantities, we shall, for convenience, drop the negative sign, and consider $d\sigma/dt$ a positive quantity in the physical region. However, $t$ will always be considered a negative quantity in the physical region.

the most general scattering amplitude contains only one complex function $f(s, t)$ or, equivalently, $f(E, \cos\theta)$.

### 5.2. The causality condition

By requiring invariance with respect to the proper Lorentz transformations corresponding to displacement (time and space) and rotations and the improper Lorentz transformations (parity and time-reversal), we have limited the forward $\pi$–p scattering amplitude to one complex function.

So far we have not made use of the causality condition inherent in the special theory of relativity. This is the requirement that signals do not propagate faster than the velocity of light in a vacuum. Therefore, space-like points cannot interfere with each other, and the operators corresponding to space-like points must commute. Einstein had conceived of special relativity as a macroscopic theory, and expressed the point of view that he would not expect it necessarily to hold at submolecular distances, just as the concept of temperature cannot be applied to submolecular distances [1]. However, Kronig (1946) [2] suggested placing on the $S$-matrix the requirement that signals do not travel faster than light in a vacuum. It is this limitation which is generally referred to as 'microscopic causality', in contrast to Einstein's concept which is referred to as 'macroscopic causality'.

### 5.3. Optical dispersion relations

Twenty years earlier Kronig and Kramers had applied [3] the restriction that light shall not propagate with speeds exceeding $c$, when treating the propagation of electromagnetic phenomena in optical media. On the basis of this requirement, they obtained the relationship between the real and imaginary parts of the refractive index of the media. Previously, this relation was obtained by studying specific models in classical electrodynamics. This treatment of the subject, and the general subject of dispersion relations, is developed in considerable detail in a book by Goldberger and Watson [4]. Let us denote the complex refractive index for light, as a function of frequency $\omega$, by

$$n(\omega) = \mathrm{Re}\, n(\omega) + \mathrm{i}\,\mathrm{Im}\, n(\omega). \tag{5.12}$$

Let the absorption coefficient be defined by the statement

$$\text{Intensity} \propto \mathrm{e}^{-\beta x},$$

then

$$\beta(\omega) = \frac{2\omega}{c}\,\mathrm{Im}\, n(\omega). \tag{5.13}$$

The result (5.14), based on classical electrodynamics, was

$$\operatorname{Re} n(\omega) = 1 + \frac{c}{\pi} \mathrm{P} \int_0^\infty \frac{\beta(\omega')}{\omega'^2 - \omega^2} \, \mathrm{d}\omega', \tag{5.14}$$

where P means that the principal value is to be taken at the singularity $\omega = \omega'$. Equivalently, by substituting eqn (5.13) into eqn (5.14), we obtain

$$\operatorname{Re}(n(\omega) - 1) = \frac{2}{\pi} \mathrm{P} \int_0^\infty \frac{\omega' \operatorname{Im} n(\omega')}{\omega'^2 - \omega^2} \, \mathrm{d}\omega'. \tag{5.15}$$

Kramers defined the refractive index for negative $\omega$ as follows

$$n(\omega) = n^*(-\omega). \tag{5.16}$$

Hence

$$\operatorname{Re} n(\omega) = \operatorname{Re} n(-\omega). \tag{5.17a}$$

Re $n(\omega)$ is an even function of $\omega$.

$$\operatorname{Im} n(\omega) = -\operatorname{Im} n(-\omega). \tag{5.17b}$$

i.e., Im $n(\omega)$ is an odd function of $\omega$ and, hence, $\beta(\omega)$ is even. By making use of eqn (5.17b), eqn (5.15) can then be written

$$\operatorname{Re}(n(\omega) - 1) = \frac{1}{\pi} \mathrm{P} \int_{-\infty}^{\infty} \frac{\operatorname{Im}(n(\omega') - 1)}{\omega' - \omega} \, \mathrm{d}\omega'. \tag{5.18}$$

Equation (5.18) is the real part of the following equation

$$\begin{aligned} n(\omega) - 1 &= \frac{1}{\mathrm{i}\pi} \mathrm{P} \int_{-\infty}^{\infty} \frac{\mathrm{d}\omega'(n(\omega') - 1)}{\omega' - \omega} \\ &= \frac{1}{2\pi \mathrm{i}} \int_{\substack{-\infty \\ \eta \to 0(+)}}^{\infty} \frac{(n(\omega') - 1) \, \mathrm{d}\omega'}{\omega' - \omega - \mathrm{i}\eta}. \end{aligned} \tag{5.19}$$

Only if $n(\omega') - 1$ approaches zero faster than $\log \omega'$, does the above integral converge, and the above equations depend on this condition which is, in fact, satisfied.

The characteristics of $n(\omega)$ as stated in eqn (5.19) allow us to define $n(\omega)$ as a function of the complex variable $z$, which is analytic in the entire upper-half

$z$-plane. Thus, we can write

$$n(z)-1 = \frac{1}{2\pi \mathrm{i}} \int_{-\infty}^{\infty} \frac{(n(\omega')-1)\,\mathrm{d}\omega'}{\omega'-z}. \tag{5.20}$$

The limit $z \to \omega + \mathrm{i}\eta$, when $\eta \to 0(+)$ gives us the physical $n$. On the other hand, if we had assumed that $n$ was a function of a complex variable $z$, which is analytic in the entire upper-half $z$-plane, and which approaches unity at infinity, we could, using Cauchy's theorem, obtain eqn (5.20).

Kramers, using arguments similar to the above, demonstrated that, when one considers the propagation of light described by a refractive index which satisfies the above conditions, signals could not be transmitted at speeds greater than $c$. Equations such as (5.18) and (5.19), which relate the real (or dispersive part) of the $n(\omega)$ to functions of the imaginary (or absorptive) part, are, therefore, referred to as dispersion relations.

The refractive index is related to the coherent forward scattering amplitude by the Lorentz relation

$$n(\omega)-1 = \frac{2\pi c^2}{\omega^2} N f(\omega), \tag{5.21}$$

where $f(\omega)$ is the coherent forward scattering amplitude for light of frequency $\omega$, and $N$ is the number of scattering centres per cubic centimetre. Inserting eqn (5.21) into eqn (5.19), we obtain

$$\frac{2\pi c^2}{\omega^2} N \operatorname{Re} f(\omega) = \frac{1}{\pi} \mathrm{P} \int_{-\infty}^{\infty} \frac{2\pi c^2}{\omega'^2} \frac{N \operatorname{Im} f(\omega')\,\mathrm{d}\omega'}{\omega'-\omega}, \tag{5.22}$$

$$\operatorname{Re} f(\omega) = \omega^2 \frac{1}{\pi} \mathrm{P} \int_{-\infty}^{\infty} \frac{\operatorname{Im} f(\omega')\,\mathrm{d}\omega'}{\omega'^2(\omega'-\omega)}$$

$$= \frac{2\omega^2}{\pi} \mathrm{P} \int_{0}^{\infty} \frac{\mathrm{d}\omega' \operatorname{Im} f(\omega')}{\omega'(\omega'^2-\omega^2)},$$

where we have used the fact that $\operatorname{Im} f(\omega)$ is an odd function of $\omega$. If $f(0)$ is finite instead of zero, as was implicitly assumed in the above treatment, $f(\omega)$ could contain a pole at $\omega = \mu$, and we would have to modify the treatment leading to eqn (5.22) and, thereby, obtain

$$\operatorname{Re} f(\omega)-\operatorname{Re} f(0) = \frac{2\omega^2}{\pi}\, \mathrm{P}\int_0^\infty \mathrm{d}\omega' \, \frac{\operatorname{Im} f(\omega')}{\omega'(\omega'^2-\omega^2)}. \tag{5.23}$$

We can express $\operatorname{Im} f(\omega')$ in terms of the total cross-section by making use of eqn (5.13).

The intensity of the wave is reduced in a distance of travel ($x$) by the factor $\mathrm{e}^{-\beta x} = \mathrm{e}^{-(2\omega/c)\operatorname{Im} n(\omega)}$. If the total cross-section is denoted by $\sigma$, then the reduction factor is $\mathrm{e}^{-N\sigma x}$. Hence

$$N\sigma = \frac{2\omega}{c} \operatorname{Im} n(\omega) \tag{5.24}$$

but, from eqn (5.21),

$$\operatorname{Im} n(\omega) = \frac{2\pi c^2}{\omega^2} N \operatorname{Im} f(\omega) \tag{5.25}$$

or

$$\operatorname{Im} f(\omega) = \frac{\omega\sigma}{4\pi c} \tag{5.26}$$

which is the optical theorem relationship. Substituting eqn (5.25) into eqn (5.23), we obtain

$$\operatorname{Re} f(\omega)-\operatorname{Re} f(0) = \frac{\omega^2}{2\pi c^2}\, \mathrm{P}\int_0^\infty \frac{\sigma(\omega')\,\mathrm{d}\omega'}{\omega'^2-\omega^2}, \tag{5.27}$$

which is the Kronig–Kramers dispersion relation.

Hence, the imposition of the condition of causality has led in optics to relations between the real part of the scattering amplitude and integrals over energy of functions involving the total cross-section. As we shall see, similar types of relations can be obtained for pion–nucleon scattering. Their significance is two-fold. First, such relations allow one to check the consequences of the concept of microscopic causality at very small distances and, therefore, to check whether its predictions are valid. Such dispersion relations, if valid and especially when extended to finite $t$, help introduce a new powerful calculational tool which can make use of empirical data. In the classical electrodynamical case which we have just considered, the assumption of causality had the consequence that the scattering amplitude became an analytic function of the frequency. Alternatively, the assumption that the scattering amplitude was an analytic function of frequency would have led to the same physical consequences.

It is interesting to note that what, in physical applications, are referred to as dispersion relations are, in mathematics referred to as Hilbert transforms. This is explained in Eden's book [20].

### 5.4. Pion–nucleon dispersion relations

If the complex scattering amplitude, or matrix elements of the $\pi$–N scattering matrix, can be expressed as analytic functions of the energy in the complex energy plane $\omega$, then the Cauchy integral formula can be used to give so-called dispersion relations, connecting the real and imaginary parts of the scattering amplitudes. The Kramers–Kronig dispersion formula for light was the earliest example of this type of relationship.

The Cauchy formula allows us to represent an analytic function $f(\omega)$, where $\omega$ is a complex variable, as follows

$$f(\omega) = \frac{1}{2\pi \mathrm{i}} \int_{\Gamma_\omega} \frac{f(\omega')}{\omega' - \omega} \, \mathrm{d}\omega', \tag{5.28}$$

where $\Gamma_\omega$ is a closed contour containing $\omega$, bounding a region within which $f(\omega)$ is analytic.

If $\omega$ does not lie within $\Gamma$,

$$\int_{\Gamma} \frac{f(\omega') \, \mathrm{d}\omega'}{\omega' - \omega} = 0 \qquad \text{when } \Gamma \text{ does not contain } \omega. \tag{5.29}$$

A special case arises when $\omega$ itself is on the contour. Then, the interpretation of the Cauchy formula is made in terms of the principal value integral. For example, if $\omega$ is on the real axis and the contour $\Gamma$ is in the upper-half plane with its linear boundary being along the real axis, then we replace $\dfrac{1}{\omega' - \omega}$ by $\dfrac{1}{\omega' - \omega - \mathrm{i}\eta}$ in the integral and take the limit $\eta \to 0(+)$. We then obtain

$$f(\omega) = \frac{1}{\pi \mathrm{i}} \, \mathrm{P} \int_{\Gamma} \frac{f(\omega')}{\omega' - \omega} \, \mathrm{d}\omega', \tag{5.30}$$

where P stands for principal value.

The real and imaginary parts of eqn (5.30) must separately be equal, and this gives us the dispersion relations

$$\mathrm{Re}\, f(\omega) = \frac{1}{\pi}\, \mathrm{P}\int_{-\infty}^{\infty} \frac{\mathrm{Im}\, f(\omega')}{\omega' - \omega}\, \mathrm{d}\omega' \tag{5.31}$$

and

$$\mathrm{Im}\, f(\omega) = -\frac{1}{\pi}\, \mathrm{P}\int_{-\infty}^{\infty} \frac{\mathrm{Re}\, f(\omega')\, \mathrm{d}\omega'}{\omega' - \omega}\, . \tag{5.32}$$

Equation (5.31) is the basis of the usual forward dispersion relations and is of much greater practical value than eqn (5.32), since the imaginary part of the forward amplitude can be much better evaluated experimentally by measuring the total cross-section as a function of energy and utilizing the optical theorem. The real part of the forward scattering amplitude is more difficult to measure and the accuracy attained is much worse.

We will discuss in the next section the conditions necessary to demonstrate the required analytic properties of the scattering amplitude. Bogoliubov and Shirkov [5] have a detailed discussion of this.

Let us for the moment assume that the $\pi$–N forward elastic scattering amplitude in a particular charge state† is analytic in the complex energy plane. Let us further assume that everywhere in the upper-half plane, as $|\omega| \to \infty$, $f(\omega)$ falls off fast enough so that, if we choose as the contour of integration an infinite semicircle bounded by the real axis, the integral over the semicircle is negligible and, of course, the integral along the real axis converges.

If‡

$$|f_{\pm}(\omega)| \leq \frac{\text{Constant}}{|\omega|}, \tag{5.33}$$

as $|\omega| \to \infty$ this is, obviously, true. Then we obtain

$$\mathrm{Re}\, f_{\pm}(\omega) = \frac{1}{\pi}\, \mathrm{P}\int_{-\infty}^{\infty} \frac{\mathrm{Im}\, f_{\pm}(\omega')}{\omega' - \omega}\, \mathrm{d}\omega'. \tag{5.34}$$

† $\pi^- + \mathrm{p} \to \pi^- + \mathrm{p}$ and $\pi^+ + \mathrm{p} \to \pi^+ + \mathrm{p}$ are examples of a particular charge state.

‡ The $\pm$ subscript refers to $\pi^{\pm} + \mathrm{p}$ scattering.

The above integral should be considered in three separate regions, as follows

$$\operatorname{Re} f_{\pm}(\omega) = \frac{1}{\pi} \mathrm{P} \int_{-\infty}^{-\mu} \frac{\operatorname{Im} f_{\pm}(\omega')}{\omega' - \omega} \, \mathrm{d}\omega'$$

$$+ \frac{1}{\pi} \mathrm{P} \int_{-\mu}^{\mu} \frac{\operatorname{Im} f_{\pm}(\omega')}{\omega' - \omega} \, \mathrm{d}\omega'$$

$$+ \frac{1}{\pi} \mathrm{P} \int_{+\mu}^{\infty} \frac{\operatorname{Im} f_{\pm}(\omega')}{\omega' - \omega} \, \mathrm{d}\omega', \tag{5.35}$$

where $\mu$ represents the pion rest mass energy. The third term in eqn (5.35) is evaluatable using the optical theorem,

$$\operatorname{Im} f_{\pm} = \frac{k\sigma}{4\pi}.$$

The first term involves scattering of negative-energy mesons, but can be related to positive energy scattering by making use of the crossing relations, which relate unphysical $\pi^- + \mathrm{p}$ scattering to physical $\pi^+ + \mathrm{p}$ scattering and vice versa, as follows

$$\langle p_{\mathrm{f}}, q_{\mathrm{f}} | \, T_+ \, | p_{\mathrm{i}}, q_{\mathrm{i}} \rangle = \langle p_{\mathrm{f}}, -q_{\mathrm{i}} | \, T_- \, | p_{\mathrm{i}}, -q_{\mathrm{f}} \rangle, \tag{5.36a}$$

$$\langle p_{\mathrm{f}}, q_{\mathrm{f}} | \, T_- \, | p_{\mathrm{i}}, q_{\mathrm{i}} \rangle = \langle p_{\mathrm{f}}, -q_{\mathrm{i}} | \, T_+ \, | p_{\mathrm{i}}, -q_{\mathrm{f}} \rangle. \tag{5.36b}$$

The middle term requires special attention, since it ranges over the unphysical region ($|\omega| < \mu$), and is not directly relatable to physical scattering processes. Fortunately, in the case of $\pi$–p scattering, it has been shown that the unphysical region integral vanishes, except at one point (the neutron intermediate state) and that, using the symmetrical pseudoscalar meson theory, the entire contribution of the unphysical region can be expressed in terms of one parameter $f^2$, the pseudovector coupling constant, which is an experimentally evaluable parameter. Therefore, as we shall see later, the pion–nucleon forward dispersion relations can be expressed entirely in terms of experimentally measurable quantities. At this point we should note that the assumption we made in eqn (5.33) is not borne out by experiments. According to high-energy experiments and cosmic ray investigations, the total cross-sections appear to approach constants at high energy (as $\omega \to \infty$) and hence, approximately, $f(\omega) \propto \omega$.

We shall see in § 5.13 that we can expect one general theoretical restriction, given by the Froissart bound, which states that

$$f(\omega) < \omega \log^2\omega \tag{5.37}$$

There is a procedure for making a dispersion relation more convergent. Let us assume that $f(\omega)$ is bounded at infinity by a finite order polynomial,

$$f(\omega) \leq \text{constant } \omega^n. \tag{5.38}$$

Then consider the function

$$\frac{f(\omega)}{(\omega-\omega_0)^{n+1}}, \tag{5.39}$$

where $\omega_0$ is finite. Assuming $\text{Im } \omega_0 \leq 0$, this function will be analytic in the entire upper-half plane and will have a negligible contribution from the infinite semicircle, hence we can write a dispersion relation for it,

$$\text{Re } f(\omega) = \frac{(\omega-\omega_0)^{n+1}}{\pi} \text{P} \int_{-\infty}^{\infty} \frac{\text{Im } f(\omega')\, d\omega'}{(\omega'-\omega)(\omega'-\omega_0)^{n+1}} + \\ + \text{Re } f(\omega_0) + \ldots + \frac{\text{Re } f^{(n)}(\omega_0)}{n(\omega-\omega_0)^n}, \tag{5.40}$$

where $f^{(n)}$ represents the $n$th derivative. The precise knowledge of the functions of the scattering amplitude, required for the above equation, is not easily obtained. The usual method of improving the convergence of the integrals in a dispersion relation is the method of making additional subtractions at various values of $\omega$. Equation (5.34) represents an unsubtracted dispersion relation.

The following simplified treatment will illustrate the general features of the method that one must use in actual cases to make sure the integral over the infinite semicircle approaches zero, and that convergence of the principal value integrals is strong enough. Let us write eqn (5.34) and use the optical theorem to obtain

$$\text{Re } f_{\pm}(\omega) = \frac{1}{\pi} \text{P} \int_{-\infty}^{\infty} \frac{\text{Im } f(\omega')\, d\omega'}{\omega'-\omega} = \frac{1}{4\pi^2} \text{P} \int_{-\infty}^{\infty} \frac{k\sigma(\omega')\, d\omega'}{(\omega'-\omega)}. \tag{5.41}$$

Let us evaluate eqn (5.41) for a particular energy $(\omega_{0_1})$ where, for example, an experimental measurement of $\text{Re } f(\omega_{0_1})$ has been, or can, in principle, be made, or is known for other reasons. Then, if we subtract eqn (5.34), evaluated at $\omega_{0_1}$, from the original eqn (5.34), we obtain

$$\operatorname{Re} f(\omega) - \operatorname{Re} f(\omega_{0_1}) = (\omega - \omega_{0_1}) \frac{1}{4\pi^2} \mathrm{P} \int_{-\infty}^{\infty} \frac{k'\sigma(\omega')\,\mathrm{d}\omega'}{(\omega'-\omega)(\omega'-\omega_{0_1})}. \tag{5.42}$$

Hence we have placed another power of $\omega'$ in the denominator of the principal value integral (by using our knowledge of $\operatorname{Re} f(\omega_{0_1})$) and, hence, improved its convergence. The same improved convergence was obtained for the infinite semicircle integral and, hence, it may be negligible after the subtraction whereas before, it may have diverged. The right-hand side of eqn (5.42) will converge if $\sigma(\omega')$ decreases faster than $1/\log \omega'$ as $\omega' \to \infty$. We can, of course, repeat the subtraction procedure as many times as necessary to obtain the desired convergence.

Making another subtraction at $\omega_{0_2}$, we obtain a doubly subtracted dispersion relation of the form

$$\operatorname{Re} f(\omega) - \frac{\operatorname{Re} f(\omega_{0_1})(\omega - \omega_{0_2})}{\omega_{0_1} - \omega_{0_2}} - \frac{\operatorname{Re} f(\omega_{0_2})(\omega_{0_1} - \omega)}{\omega_{0_1} - \omega_{0_2}}$$
$$= (\omega - \omega_{0_1})(\omega - \omega_{0_2}) \frac{1}{4\pi^2} \mathrm{P} \int_{-\infty}^{\infty} \frac{k'\sigma(\omega')\,\mathrm{d}\omega'}{(\omega'-\omega)(\omega'-\omega_{0_1})(\omega'-\omega_{0_2})}. \tag{5.43}$$

It will converge if the total cross-section is bounded to increase more slowly at high energies than $\omega/\log \omega$. Therefore, the observed experimental behaviour of the total cross-sections if continued to higher energies, would mean that eqn (5.40) would be a suitable convergent dispersion relation. It is, therefore, clear that the usefulness of the forward dispersion relation can depend strongly on our knowledge of asymptotic bounds on the total cross-sections, which we shall consider in § 5.13.

## 5.5. Proof of the forward dispersion relations

The establishment of the required analytic properties of the scattering amplitude, so that use may be made of Cauchy's theorem, and thereby a proof of dispersion relations obtained, has presented a difficult and subtle problem.

The pion–nucleon forward dispersion relations were originally derived by Goldberger and co-workers [6] using heuristic arguments which yield the forward dispersion relations as identities, which follow from representing the scattering amplitude as a sum over a complete set of intermediate states.

The dispersion relations for neutral meson–nucleon scattering were first derived by Karplus and Ruderman [7], but the required analyticity was assumed. It has been shown by Bogoliubov [8], for an arbitrary angle, and Symanzik [9], for forward scattering, that the dispersion relations can be derived from the basic axioms of modern quantum field theory. The principle of microscopic causality is necessary for their derivation. To see the relationship between microscopic causality and analyticity properties, consider the following simplified situation.

Let us consider an $f(\omega)$ which is the Fourier transform of a function of time ($G(t)$), hence

$$f(\omega) = \int_{-\infty}^{\infty} G(t)\mathrm{e}^{\mathrm{i}\omega t}\,\mathrm{d}t. \tag{5.44}$$

The requirement of 'causality' can be introduced by requiring that $G(t) = 0$ for $t < 0$. Assuming the integral converges for real $\omega$, its convergence will be even stronger in the upper-half plane, since for $t > 0$

$$\mathrm{e}^{\mathrm{i}(\omega+\mathrm{i}|\eta|)t} = \mathrm{c}^{\mathrm{i}\omega t}\mathrm{c}^{-|\eta|t},$$

for $t < 0$, $G(t)$ vanishes and, therefore, the increasing exponential factor causes no trouble.

The Titchmarsh theorem states that for an $f(\omega)$ which is the Fourier transform of a function $G(t)$, which vanishes for negative values of $t$ and satisfies the condition of quadratic integrality,

$$\int_{-\infty}^{\infty} \mathrm{d}\omega\, |f(\omega)|^2 < \infty. \tag{5.45}$$

The function of $f(\omega)$ is the boundary value of a function $f(\omega+\mathrm{i}\eta)$, $\eta \to 0(+)$, which is analytic in the upper-half plane and satisfies the relationship

$$\int_{-\infty}^{\infty} |f(\omega+\mathrm{i}\eta)|^2\,\mathrm{d}\omega \qquad < \infty \text{ for all } \eta > 0. \tag{5.46}$$

The above are necessary and sufficient conditions for the transform relationship

$$f(\omega) = \frac{1}{\mathrm{i}\pi}\,\mathrm{P}\int_{-\infty}^{\infty} \frac{f(\omega')}{\omega'-\omega}\,\mathrm{d}\omega'. \tag{5.47}$$

The first of these conditions is satisfied by our $f(\omega)$, as evidenced by eqn

(5.44). Let us make it a requirement that the second condition (eqn (5.45)) is also satisfied by our $f(\omega)$. Then eqn (5.47) holds for our $f(\omega)$. Equating the real parts of both sides of eqn (5.47), we obtain

$$\operatorname{Re} f(\omega) = \frac{1}{\pi} \mathrm{P} \int_{-\infty}^{\infty} \frac{\mathrm{d}\omega' \operatorname{Im} f(\omega')}{\omega' - \omega}. \tag{5.48}$$

Equating the imaginary parts of both sides of eqn (5.48), we obtain

$$\operatorname{Im} f(\omega) = \frac{-1}{\pi} \mathrm{P} \int_{-\infty}^{\infty} \frac{\mathrm{d}\omega' \operatorname{Re} f(\omega')}{\omega' - \omega}. \tag{5.49}$$

The requirement for quadratically integrable functions can be eliminated by performing the required number of subtractions on the original function $f(\omega)$, using the procedures previously described. For example, performing a subtraction in eqn (5.48) at $\omega_{0_1}$ gives us the equation

$$\operatorname{Re} f(\omega) - \operatorname{Re} f(\omega_{0_1}) = (\omega - \omega_{0_1}) \frac{1}{\pi} \mathrm{P} \int \frac{\operatorname{Im} f(\omega')\, \mathrm{d}\omega'}{(\omega' - \omega)(\omega' - \omega_{0_1})}. \tag{5.50}$$

We can obtain the Kramers–Kronig dispersion relations for light, from the above, by letting $f(\omega)$ be the forward scattering amplitude for light, and using the optical theorem, which gives

$$\operatorname{Im} f(\omega) = \frac{\omega\sigma(\omega)}{4\pi},$$

Then taking $\operatorname{Im} f(\omega)$ as an odd function of $\omega$ and letting $\omega_{0_1} = 0$, we obtain

$$\operatorname{Re} f(\omega) - \operatorname{Re} f(0) = \frac{\omega^2}{2\pi^2} \mathrm{P} \int_{0}^{\infty} \frac{\sigma(\omega')\, \mathrm{d}\omega'}{(\omega')^2 - \omega^2}. \tag{5.51}$$

To summarize, we have illustrated how the introduction of a causality requirement allows us to obtain necessary analyticity properties and, thus, obtain dispersion relations for a scattering amplitude.

### 5.5.1. *Fundamental length*

One final point worth considering is failure of causality at some very small distance ($l_0$) corresponding to a so-called fundamental length. We can approximately describe this situation by assuming $\sigma(t)$ vanishes only for

$t < -l_0$. Then we can substitute $t - l_0$ for $t$ in eqn (5.44) and obtain

$$f(\omega) = e^{-il_0\omega}F_0(\omega), \tag{5.52}$$

where $F_0(\omega)$ is now a function without an essential singularity at infinity, and dispersion relations would hold for this function. But the $e^{-il_0\omega}$ factor has an essential singularity at infinity and, therefore, $f(\omega)$ must be multiplied by the factor $e^{+il_1\omega}$, where $l_1 > l_0$ is positive, and obtain a function for which the dispersion relations hold. Therefore, we can have modified dispersion relations of the form

$$e^{il_1\omega}f(\omega) = \frac{1}{\pi i} \mathrm{P} \int_{-\infty}^{\infty} \frac{e^{il_1\omega'}f(\omega')}{\omega' - \omega} \, d\omega'. \tag{5.53}$$

There are mathematical difficulties when we attempt to give up locality in any space–time region with a sharp boundary, which causes locality to fail in the other regions of space–time. However, in physical models a gradual transition between regions is reasonable, and could be expected to circumvent these mathematical difficulties. It is also well known that no one has yet been able to formulate a field theory containing a fundamental length in a consistent manner. Thus, any treatments of a fundamental length are necessarily model dependent. Although these are some difficulties which prevent a rigorous model-independent treatment of the effect of the fundamental length, reasonable treatments of its consequences (which we shall discuss in § 5.8) show that its effect is expected to allow the consequences of the dispersion relations to remain valid provided $\omega \ll \frac{1}{l_0}$ (i.e., for this case the added exponential is near unity). At $\omega \gtrsim \frac{1}{l_0}$, the dispersion relations break down completely (i.e., the added exponential factor departs from unity and approaches zero). This is one of the important consequences of the dispersion relations, namely, they allow an experimental test of whether such a fundamental length exists, within which the principle of microscopic causality is violated. Obviously, the higher the energy ($\omega_{\text{max}}$) of the test, the smaller the limit on the fundamental length investigated. It will be shown later in this chapter that it has been experimentally demonstrated that, if a fundamental length exists in some simple models which will be considered, $l_0 < 10^{-15}$ cm ($= 10^{-2}$ fm).

### 5.5.2. *Proof of the dispersion relations*

The proof of the dispersion relations reduces to a proof of the necessary analytic properties of the scattering amplitude in the complex $s$-plane for a fixed value of momentum transfer ($t$). It has been shown that these required analytic properties can be derived from the basic principles of a local relativistic quantum field theory satisfying the following requirements:

(a) *Lorentz invariance of the theory.*

(b) *The microscopic causality condition* which requires that, if two space-time points $(x_1, t_1)$ and $(x_2, t_2)$ are separated by a finite space-like distance, they cannot influence one another. If

$$l^2 = (t_2 - t_1)^2 - (x_2 - x_1)^2 < 0,$$

the separation is space-like. Even if $l^2 > 0$ (i.e. the separation is time-like) an event at $(x_1, t_1)$ can only influence a future event at $(x_2, t_2)$, but not a past one. That is to say, if $t_2 < t_1$, causality prevents $(x_1, t_1)$ from influencing $(x_2, t_2)$. The causality condition is imposed by requiring that operators which correspond to physically measurable quantities shall commute for any two points separated by a finite space-like distance. Since all such operators are constructed from field operators, this requirement is met if all boson field operators commute, and all fermion field operators anti-commute. If $\phi(s, t)$ describes the pion field, then we require

$$[\phi(x_1, t_1), \phi(x_2, t_2)] = 0$$

for space-like separation ($l^2 < 0$). This causality condition is, therefore, referred to as the requirement of *local commutativity*, and is the basis of all local field theories.

(c) *Mass spectrum conditions.* The existence of a spectrum of positive energy-stable particles is assumed, which satisfies the following relationship,

$$\text{total energy} = |\sqrt{(m_\alpha^2 + P_\alpha^2)}|.$$

(d) *The polynomial boundedness condition.* As $s \to \infty$, the scattering amplitude $T(s, t)$ can be bounded by a polynomial of finite power,

$$\text{i.e. } T(s, t) < \text{constant } |s|^n \quad \text{as} \quad |s| \to \infty,$$

where $n$ is a positive integer.

(e) *Unitarity.* The $S$-matrix is unitary (i.e. $SS^+ = 1$). This assumption allows us to use the optical theorem in the forward dispersion relations.

Based on the above conditions, which essentially represent the L.S.Z. formalism, Bogoliubov *et al.* [8] derived the necessary analytic properties for the scattering amplitude for an arbitrary angle, and Symanzik [9] derived the necessary analytic properties for the forward scattering amplitude. Recently, [10] it has been shown that the analyticity properties of the two-particle scattering amplitudes, necessary for proof of dispersion relations, were also rigorously derivable from the general set of axioms of Wightman for a local relativistic quantum field theory. The assumption that one-particle states are created from the vacuum by Wightman fields and that certain mass spectral conditions are satisfied, allowed a rigorous proof of the desired analyticity properties of the two-particle scattering amplitudes required to allow a proof of the dispersion relations.

**5.6. Evaluation of the pion–nucleon forward dispersion relations**

In eqns (5.36a) and (5.36b), we saw the relationship of scattering of negative frequency pions by protons to the scattering of positive frequency pions. If we represent the scattering amplitude in terms of its real (or dispersive) and imaginary (or absorptive) parts for $\omega > 0$ by

$$f_{\pm}(\omega) = D_{\pm}(\omega) + \mathrm{i}A_{\pm}(\omega), \tag{5.54}$$

then it can be shown (from CPT invariance) that the connection between scattering of positive and negative frequency pions by protons is represented by

$$f^{*}_{\mp}(-\omega) = f_{\pm}(\omega), \tag{5.55a}$$

which gives

$$D_{\pm}(\omega) = D_{\mp}(-\omega), \tag{5.55b}$$

$$A_{\pm}(\omega) = -A_{\mp}(-\omega). \tag{5.55c}$$

The most convenient way to proceed is to write the forward dispersion relations for the symmetric and anti-symmetric combinations of scattering amplitudes.

$$D^{+}(\omega) = \tfrac{1}{2}[D_{-}(\omega) + D_{+}(\omega)]$$

$$= \frac{1}{\pi}\,\mathrm{P}\int_{-\infty}^{\infty} \frac{A^{+}(\omega')}{\omega' - \omega}\,\mathrm{d}\omega'. \tag{5.56}$$

$$D^{-}(\omega) = \tfrac{1}{2}[D_{-}(\omega) - D_{+}(\omega)] = \frac{\mathrm{P}}{\pi}\int_{-\infty}^{\infty} \frac{A^{-}(\omega')\,\mathrm{d}\omega'}{\omega' - \omega}, \tag{5.57}$$

where

$$A^{+} = \tfrac{1}{2}(A_{-}(\omega) + A_{+}(\omega)), \tag{5.58}$$

and

$$A^{-} = \tfrac{1}{2}(A_{-}(\omega) - A_{+}(\omega)). \tag{5.59}$$

From eqn (5.55),

$$D^+(\omega) = D^+(-\omega), \tag{5.60}$$

$$D^-(\omega) = -D^-(-\omega), \tag{5.61}$$

$$A^+(\omega) = -A^+(-\omega), \tag{5.62}$$

$$A^-(\omega) = A^-(-\omega). \tag{5.63}$$

In the physical region, the imaginary parts of the forward scattering amplitudes are related to the total cross-sections by the optical theorem relationship

$$A = \frac{k\sigma}{4\pi}, \qquad \text{where } \sigma \text{ is the total cross-section.} \tag{5.64}$$

Hence,

$$A^+ = \frac{k}{8\pi}(\sigma(\pi^- + \mathrm{p}) + \sigma(\pi^+ + \mathrm{p})), \tag{5.65}$$

$$A^- = \frac{k}{8\pi}(\sigma(\pi^- + \mathrm{p}) - \sigma(\pi^+ + \mathrm{p})). \tag{5.66}$$

Hence these symmetric and anti-symmetric combinations have simple evenness and oddness characteristics when $\omega \to -\omega$, and this is the reason for their usefulness. Using eqns (5.55), (5.60), (5.61), (5.62), and (5.63) in eqns (5.56) and (5.57), we obtain

$$D^+(\omega) = \frac{\mathrm{P}}{\pi}\int_0^\infty A^+(\omega')\left(\frac{1}{\omega'-\omega} + \frac{1}{\omega'+\omega}\right)\mathrm{d}\omega'$$

$$= \frac{2}{\pi}\mathrm{P}\int_0^\infty \frac{\omega' A^+(\omega')\,\mathrm{d}\omega'}{(\omega')^2 - \omega^2}, \tag{5.67}$$

$$D^-(\omega) = \frac{\mathrm{P}}{\pi}\int_0^\infty A^-(\omega')\left(\frac{1}{\omega'-\omega} - \frac{1}{\omega'-\omega}\right)\mathrm{d}\omega'$$

$$= \frac{2\omega}{\pi}\mathrm{P}\int_0^\infty \frac{A^-(\omega')\,\mathrm{d}\omega'}{(\omega')^2 - \omega^2}. \tag{5.68}$$

For constant, or near constant, total cross-sections at high energies, $D^+(\omega)$ is not a valid dispersion relation since $A^+(\omega) \sim \omega$. Therefore, at high energies the principal value integral would diverge like $\omega'$. Therefore we must make a subtraction. This is conveniently done at $\omega = \mu$. Evaluating eqn (5.67) for $\omega = \mu$ (the pion rest energy), we obtain

$$D^{+}(\mu) = \frac{2}{\pi}\,\mathrm{P}\int_{0}^{\infty} \frac{\omega' A^{+}(\omega')\,\mathrm{d}\omega'}{(\omega')^2-\mu^2}\,. \tag{5.69}$$

Subtracting eqn (5.69) from eqn (5.67), we obtain

$$\begin{aligned} D^{+}(\omega)-D^{+}(\mu) &= \frac{2}{\pi}\,\mathrm{P}\int_{0}^{\infty} \omega' A^{+}(\omega')\left(\frac{1}{(\omega')^2-\omega^2}-\frac{1}{(\omega')^2-\mu^2}\right)\mathrm{d}\omega' \\ &= (\omega^2-\mu^2)\,\frac{2}{\pi}\,\mathrm{P}\int_{0}^{\infty} \frac{\omega' A^{+}(\omega')\,\mathrm{d}\omega'}{(\omega'^2-\omega^2)(\omega'^2-\mu^2)} \\ &= \frac{2k^2}{\pi}\,\mathrm{P}\int_{0}^{\infty} \frac{\omega' A^{+}(\omega')\,\mathrm{d}\omega'}{(\omega'^2-\omega^2)(\omega'^2-\mu^2)}\,. \end{aligned} \tag{5.70}$$

All quantities appearing in the above equations refer to the laboratory system, in which the proton is at rest. Since $A^{-}(\omega')$ now appears, experimentally, to rise more slowly with increasing energy than $\dfrac{\omega}{\log \omega'}$ (there are theoretical reasons for expecting this behaviour—the Pomeranchuk theorem), eqn (5.69) is expected to be a valid convergent dispersion relation without further subtraction. This subject will be discussed in detail later. In the original work, and in many subsequent works, an additional subtraction was introduced into $D^{-}(\omega)$ to desensitize the results to the unknown high-energy behaviour. However, it makes the higher energy prediction more sensitive to the error in $D^{-}(\mu)$. Thus we can write

$$D^{-}(\mu) = \frac{2\mu}{\pi}\,\mathrm{P}\int_{0}^{\infty} \frac{A^{-}(\omega')\,\mathrm{d}\omega'}{(\omega'^2-\mu^2)}\,, \tag{5.71a}$$

or, equivalently,

$$\frac{\omega}{\mu}\,D^{-}(\mu) = \frac{2\omega}{\pi}\,\mathrm{P}\int_{0}^{\infty} \frac{A^{-}(\omega')\,\mathrm{d}\omega'}{\omega'^2-\mu^2}\,, \tag{5.71b}$$

Subtracting eqn (5.71b) from eqn (5.68), we obtain

$$D^{-}(\omega)-\frac{\omega D^{-}(\mu)}{\mu} = \frac{2\omega(\omega^2-\mu^2)}{\pi} \mathrm{P}\int_0^{\infty} \frac{A^{-}(\omega')\,\mathrm{d}\omega'}{(\omega'^2-\omega^2)(\omega'^2-\mu^2)}. \tag{5.72}$$

Equations (5.70) and (5.72) show considerably more convergence than eqn (5.42) after one subtraction, but this is due to the symmetry relations between positive and negative $\omega$ scattering, which increases the effective convergence (or effective number of subtractions) in the equivalent integral over $\omega$ from 0 to $\infty$. In the sense of eqns (5.42) and (5.43), from the point of view of convergence, eqn (5.70) has the equivalent of two subtractions, and eqn (5.72) has the equivalent of three subtractions.

Let us now consider the non-physical region. In eqns (5.68), (5.70), and (5.72), the part of the principal value integral where $\omega$ lies in the range $0 \leq \omega \leq \mu$ corresponds to a non-physical region, since the conservation of energy requirement leads to an imaginary momentum for the pion. This can be seen as follows. If

$$\omega^2 = \mu^2+k^2 < \mu^2,$$

then $k^2 < 0$, and

$$k = \mathrm{i}\,|k|.$$

In the non-physical region, conservation of energy, momentum, and charge make the entire contribution from this region vanish, except for the intermediate state consisting of a neutron. The exact treatment of the non-physical region, which involves obtaining scattering amplitudes, which are analytically continuable in the upper-half plane and which yield convergent principal value integrals with negligible contribution from the infinite semi-circles, is quite involved. The historical treatment outlined here is merely for illustrative purposes, and is heuristic in nature.

The non-zero terms in the amplitudes $A_{\pm}(\omega)$ for $\omega > 0$, which can satisfy energy, momentum, and charge conservation are single nucleon states of the form

$$A_{+}(\omega) = \sum |\langle \mathrm{n}|\, j_{-}(0)\,|\mathrm{p}\rangle|^2\, \delta(M-E_n+\omega), \tag{5.73}$$

where $j_{\pm}(x)$ are the currents associated with the charged meson fields, $|\mathrm{p}\rangle$ is a proton state, $|\mathrm{n}\rangle$ is a neutron state, $M$ is proton total energy, and $E_n$ is the neutron total energy. The argument of the $\delta$-function vanishes only for $E_n = M+\mu^2/2M$, for which $\omega = \dfrac{\mu^2}{2M}$. The corresponding value of

$$k^2 = \left(\frac{\mu^2}{2M}\right)^2 - \mu^2,$$

which is obviously negative. The symmetrical pseudoscalar meson theory can then be used to relate this matrix element to an observable coupling constant. The result is independent of details of the pion–nucleon interaction and depends primarily on parity and angular momentum considerations. We obtain

$$\langle \mathrm{n}|\, j_{-}(0)\, |\mathrm{p}\rangle = \mathrm{i}g\sqrt{2}\,\frac{\boldsymbol{\sigma}.\mathbf{k}}{2M}, \tag{5.74}$$

where $g$ is the renormalized coupling constant of the symmetrical pseudoscalar theory (Gaussian units), with $\hbar = c = 1$. The equivalent pseudovector coupling constant $\left(f = \dfrac{\mu}{2M}g\right)$ is generally employed. We have found that $f^2$ is only accurate to terms of order $\left(\dfrac{\mu}{M}\right)^2$. The result, for $0 < \omega < \mu$, is

$$\begin{aligned} A_{+}(\omega) &= \frac{2\pi g^2 k^2}{(2M)^2}\,\delta\!\left(\omega - \frac{\mu^2}{2M}\right) \\ &= -\frac{2\pi}{\mu^2} f^2 k^2\, \delta(\omega - \mu^2/2M). \end{aligned} \tag{5.75}$$

In a more formal treatment [9] it was shown that the scattering amplitude may be analytically continued from arguments on the real axis ($\omega' \geq \mu$) onto the entire cut $\omega'$-plane, with the physical cuts from $-\infty$ to $-\mu$ and from $+\mu$ to $+\infty$. It was shown that the scattering amplitude then possesses a pair of poles at $\omega' = \pm\mu^2/2M$. These poles are just the ones we have been considering but when restricting ourselves to $\omega' > 0$, we considered only the contribution of the pole at $\omega' = \mu^2/2M$. Therefore, the whole non-physical region reduces to a point which is the pole at $\omega'$. Evaluating the non-physical region as indicated (i.e. substituting an expression like eqn (5.74)† in the principal-value integrals of eqns (5.68), (5.70), and (5.72) and using the optical theorem relationships, we obtain the following results for:

(a) *once subtracted* $D^{+}$,

$$D^{+}(\omega) = D^{+}(\mu) + \frac{f^2 k^2}{M\left\{1 - \left(\dfrac{\mu}{2M}\right)^2\right\}\left\{\omega^2 - \left(\dfrac{\mu^2}{2M}\right)^2\right\}} + \frac{k^2}{4\pi^2}\,\mathrm{P}\int_{+\mu}^{\infty} \frac{\mathrm{d}\omega'\,\omega'((\sigma_{-}(\omega') + \sigma_{+}(\omega'))}{k'(\omega'^2 - \omega^2)}; \tag{5.76}$$

† Correcting to order $\left(\dfrac{\mu}{2M}\right)^2$.

(b) *unsubtracted forward dispersion relations for $D^-$*,

$$D^-(\omega) = \frac{2f^2\omega}{\{\omega^2-(\mu/2M)^2\}} + \frac{\omega}{4\pi^2}\,\mathrm{P}\int_{+\mu}^{\infty} \mathrm{d}\omega' k' \,\frac{(\sigma_-(\omega')-\sigma_+(\omega')}{(\omega'^2-\omega^2)}; \tag{5.77}$$

(c) *once subtracted $D^-$*

$$D^-(\omega) = \frac{\omega}{\mu}D^-(\mu) \frac{-2f^2k^2\omega}{\{\omega^2-(\mu^2/2M)^2\}\mu^2\left\{1-\left(\dfrac{\mu}{2M}\right)^2\right\}} +$$

$$+\frac{k^2\omega}{4\pi^2}\,\mathrm{P}\int_{\mu}^{\infty}\frac{\mathrm{d}\omega'}{k'}\frac{\{\sigma_-(\omega')-\sigma_+(\omega')\}}{\omega'^2-\omega^2}. \tag{5.78}$$

The dispersion relations for $D_\pm$ can be obtained by taking sums and differences, respectively, of the $D^+$ and $D^-$ relations. If we use the once subtracted relations both for $D^+$ and $D^-$, we obtain for the individual charge state relations,

$$D_\pm(\omega) = \tfrac{1}{2}(1+\omega/\mu)D_\pm(\mu) + \frac{1}{2}\left(1-\frac{\omega}{\mu}\right)D_\mp(\mu)$$

$$\pm\frac{2f^2k^2}{\mu^2(\omega\mp\mu^2/2M)(1-\mu^2/4M^2)}+$$

$$+\frac{k^2}{4\pi^2}\,\mathrm{P}\int_{\mu}^{\infty}\frac{\mathrm{d}\omega'}{k'}\frac{\sigma_\pm(\omega')}{\omega'-\omega}+\frac{\sigma_\mp(\omega')}{\omega'+\omega} \tag{5.79}$$

Only observable quantities are involved in eqns (5.76)–(5.79), and we have defined $\hbar = c = 1$ (natural units), $\omega$ as the total pion energy in the laboratory system, $\mu$ as the pion rest energy, and $k = |\sqrt{(\omega^2-\mu^2)}|$. $M$ is the nucleon mass and $f^2$ is the renormalized unrationalized pseudovector interaction coupling constant. In these units, where $\hbar = c = l$, we can still fix one of the following—a mass, a length, or a time. The units of $D$ and $A$ are those of length.

## 5.7. Early application of the pion–nucleon forward dispersion relations

### 5.7.1. *Discrimination against some phase-shift sets*

One of the first applications [11], [12] of the forward $\pi$–N dispersion relations was the discrimination between various sets of phase-shift solutions obtained in analyses of $\pi$–N scattering experiments. A method employed [11] was to calculate the predicted real part of the forward scattering amplitude for $T = \frac{3}{2}$ and $T = \frac{1}{2}$ isotopic spin states using in the dispersion relations the experimental data for $\pi^{\pm}$–p total cross-sections up to 1·9 GeV, and assuming that they were constant thereafter. Thus they obtained $D_{\pm}(k)$. Since

$$D_3(k) = D_+(k) \tag{5.80a}$$

and

$$D_1(k) = \tfrac{3}{2}D_-(k) - \tfrac{1}{2}D_+(k), \tag{5.80b}$$

(where the 3 subscript refers to $T = \frac{3}{2}$, and the 1 subscript refers to $T = \frac{1}{2}$ states), it is straightforward to calculate these quantities. This was more convenient since the phase-shift analyses of the experiments assumed charge independence, and were given in terms of the $T = \frac{3}{2}$ and $T = \frac{1}{2}$ states. Up to terms including $p$-waves,

$$D_3(k) = (k/2k_{\text{cms}}^2)(\sin 2\alpha_3 + 2\,\text{Sin}\, 2\alpha_{33} + \text{Sin}\, 2\alpha_{31} + \ldots), \tag{5.81a}$$

$$D_1(k) = (k/2k_{\text{cms}}^2)(\sin 2\alpha_1 + 2\,\text{Sin}\, 2\alpha_{13} + \text{Sin}\, 2\alpha_{11} + \ldots), \tag{5.81b}$$

where $k_{\text{cms}}$ is the pion wave number in c.m.s. and $k$ is the pion wave number in the laboratory system. The connection between c.m.s. and laboratory quantities is given by

$$k_{\text{cms}} = k\left(1 + \frac{2\omega}{M} + \frac{\mu^2}{M^2}\right)^{-1/2}, \tag{5.82a}$$

$$\frac{D(k)}{k} = \frac{D_{\text{cms}}(k_{\text{cms}})}{k_{\text{cms}}}. \tag{5.82b}$$

Figures 5.2(a) and 5.2(b) show the results of the calculation using the two then-available solutions for the $s$-wave scattering lengths $a_3$ and $a_1$.†

The most significant feature of the behaviour of $D_3(k)$ is the change of sign above 180 MeV, almost forcing a resonant solution in the $T = \frac{3}{2}$ state. As can be seen, there was reasonable agreement with the results deduced from the experimental data. These old calculations should not be taken too seriously as they give too low a resonance energy, but they show the general features of the problem. The behaviour of $D_1(k)$ was found to be in agreement with the values determined from the experiment, but was less convincing due to the small magnitude of $D_1(k)$ and the large errors caused by the great uncertainties in experimentally determining the smaller $T = \frac{1}{2}$ phase shifts.

By using techniques similar to the above, and requiring that any set of phase shift solutions should satisfy the forward dispersion relations, we are

† The solid line is shown for the usually assumed solutions for $a_3$ and $a_1$.

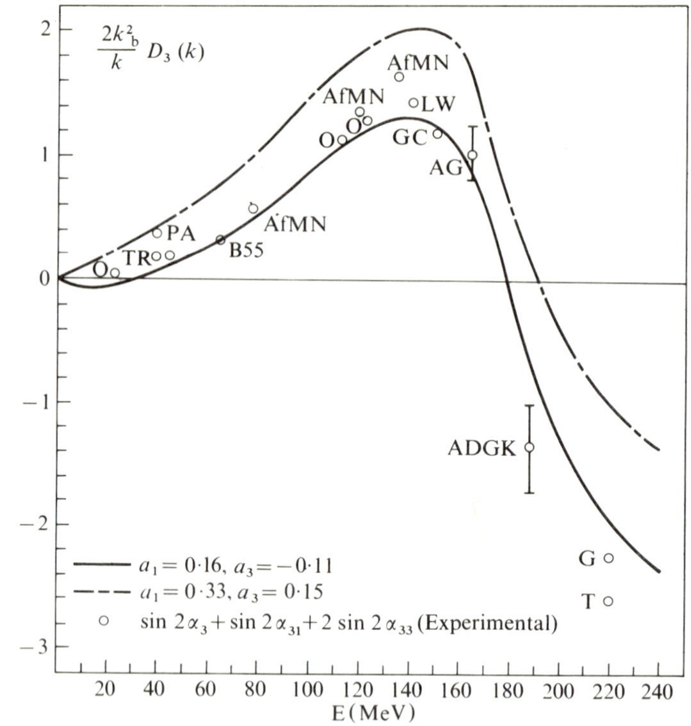


FIG. 5.2.(a) Comparison of $(2k_b^2/k)D_3(k)$ calculated from causality conditions with $\sin 2\alpha_{31} + 2 \sin 2\alpha_{33}$ obtained from phase-shift analysis. The abscissa represents the laboratory kinetic energy in MeV. Sources for data are in references (11–27) of [11] of this chapter. $k_b$ is the pion wavenumber in the c.m.s. (Anderson *et al.* [11].)

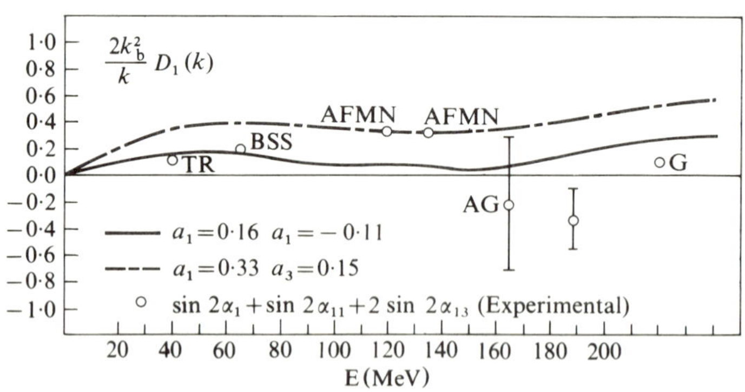


FIG. 5.2.(b) Comparison of $(2k_b^2/k)D_1(k)$ calculated from causality conditions with $\sin 2\alpha_{11} + \sin 2\alpha_{13}$ obtained from phase-shift analysis. The abscissa represents the laboratory kinetic energy in MeV. Sources for data are in references (11–27) of [11] of this chapter. $k_b$ is the pion wavenumber in c.m.s. (Anderson *et al.* [11].)

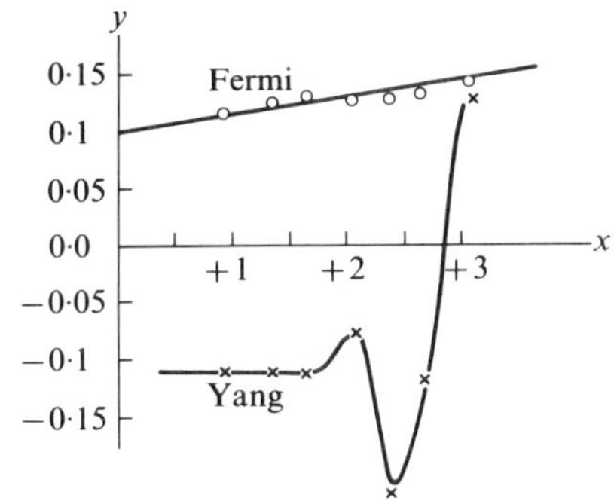


FIG. 5.2.(c) Plot of $y = \frac{1}{2}x\{\mathrm{Re}(\alpha_3) - \gamma I_3(\gamma) + \frac{1}{3}I_3(-\gamma)\}$ against $x = \gamma/(1+\gamma_B/\gamma)$ to test dispersion relations and obtain a value of the coupling constant by extrapolation. (From Davidon and Goldberger. (1956). *Phys. Rev.* **104,** 1119.)

able to reject four of six possible solutions which, within experimental error, fit the differential scattering data well. However, there still remain two sets of phase shifts, a Fermi- and a Yang-type, referred to previously.

Another important early application of the $\pi$–N forward dispersion relations was the accurate determination of the value of the renormalized unrationalized pseudoscalar coupling constant ($f^2$). This subject will be treated in a subsequent section.

### 5.7.2. *Differences between the Fermi and Yang sets of phase shifts*

As previously discussed, the Fermi set are large in the $T = J = \frac{3}{2}$ state, while the Yang set are large in the $T = \frac{3}{2}$, $J = \frac{1}{2}$ state. Both of these sets of phase shifts give essentially identical results for the magnitude and sign of the forward scattering amplitude and, therefore, cannot be distinguished by the forward dispersion relation predictions. However, the sign of the spin flip amplitude is different for the Fermi and Yang sets, and the polarization of the recoil nucleon is also different for the two sets.

### 5.7.3. *The spin-flip dispersion relations and resolution of the Fermi-Yang ambiguity*

The spin flip amplitudes vanish in the forward direction, since they contain a sin $\theta$ term. However, by taking the derivative of the spin flip amplitudes with respect to sin $\theta$, evaluated at $\theta = 0$, we can obtain finite quantities for which forward dispersion relations can, and have been, obtained [13]. These relations are referred to as spin flip dispersion relations. The derivative of the spin flip amplitude in the forward direction can be written as $\frac{1}{\mu}\bar{\eta}a^2$, where $\bar{\eta}$ is the c.m.s. momentum in units of $\mu c$, and $a$ is dimensionless, and is generally finite, as $\bar{\eta} \to 0$. Just as the dispersion relations for the ordinary amplitude determine the sign of this amplitude, the spin flip dispersion relations determine the sign of the spin flip amplitude, and were used to discriminate between the Fermi and Yang solutions, which differ from one another in the opposite sign of the spin flip amplitude. This still left the Minami ambiguity, and it was subsequently shown [15] that the spin flip dispersion relations can also be used to eliminate this ambiguity as well, therefore leaving the Fermi set as the only solution.

The method used applied the fact that the spin flip amplitude in the forward direction is given by $\frac{1}{\mu}(\bar{\eta}^2 a_{3,1})$, where $\eta$ is the centre of mass momentum in units of $\mu c$, $a$ is dimensionless and, in general, approaches a finite non-zero limit as $\bar{\eta} \to 0$, and

$$a_{3,1} = \sum_{l=1}^{\infty} \frac{l(l+1)}{4\mathrm{i}\bar{\eta}^3} (\mathrm{e}^{2\mathrm{i}\delta_{l+}} - \mathrm{e}^{2\mathrm{i}\delta_{l-}})_{3,1}, \tag{5.83}$$

where $\delta_{l_+}$ and $\delta_{l_-}$ are the phase shifts for orbital angular momentum $l$ and total angular momentum $J = l+\frac{1}{2}$ and $l-\frac{1}{2}$, respectively. The subscripts 3 and 1 refer to total isotopic spin $T = \frac{3}{2}$ and $\frac{1}{2}$, respectively.

It was shown [14] that, to a good enough approximation,† the spin flip dispersion relations lead to a relationship of the form

$$y = f^2 + cx, \tag{5.84}$$

where $x$, $y$, and $c$ are defined by

$$x = \gamma(1+\gamma_{\mathrm{B}}/\gamma)^{-1}, \tag{5.85a}$$

$$y = \tfrac{1}{2}x(\mathrm{Re}(a_{\frac{3}{2}} - \gamma[I_3(\gamma) + \tfrac{1}{3}I_3(-\gamma)]), \tag{5.85b}$$

$$c \equiv \frac{1}{3\pi} \int_1^{\infty} \mathrm{d}\gamma' \, \frac{\mathrm{Im}(a_3)}{\gamma'}, \tag{5.85c}$$

where $f^2$ is the coupling constant squared and $c$ can also be considered a constant to be determined by the data. $\gamma$ is the laboratory total energy of the pion in units of $\mu c^2$, the constant $\gamma_{\mathrm{B}} = \frac{\mu}{2M}$, the constant $M$ is the nucleon

† The small contributions from the $T = \frac{1}{2}$ state are negligible.

mass, and $I_3(\gamma)$ is the following principal value integral

$$I_3(\gamma) = \frac{1}{\pi} \mathrm{P} \int_1^\infty \mathrm{d}\gamma' \, \frac{\mathrm{Im}(a_3)}{\gamma'(\gamma'-\gamma)}. \tag{5.86}$$

The values of $a_3$ were obtained from the energy dependence of the phase shifts [11]. Figure 5.2(c) shows plots of $y = f^2 + cx$ (eqn (5.84)). The upper curve is for the Fermi set and the lower curve is for the corresponding Yang set obtained from the Fermi set.

The Fermi set shows the expected straight line behaviour, and extrapolates to the expected value of $f^2$ ($\approx 0{\cdot}1$) and, hence, satisfies the spin flip dispersion relations. The Yang set, on the other hand, extrapolates to a negative (unphysical) value of $f^2$, and deviates grossly from a straight line behaviour at the higher energies and, hence, does not satisfy the spin flip dispersion relations.

### 5.7.4. *Resolution of the Minami ambiguity*

In the works described in the preceding section, the authors did not consider the Minami ambiguity, which is an exact ambiguity true under all conditions, whereas the Yang ambiguity is true only under certain special conditions. However, it was soon pointed out and shown [15] that the Minami ambiguity can be resolved by requiring satisfaction of the spin flip dispersion relations. The corresponding Minami phase shifts ($\delta'_{\mathrm{M}}$) were calculated, according to eqns (4.92) and (4.93), for the Fermi set previously selected. It was also pointed out that there existed a new set of phase shifts ($\alpha'$) which can be obtained by applying the Minami transformation to the Yang set. This new set would also satisfy the experimental total cross-section and differential scattering data and the ordinary forward dispersion relations. However, since the Minami transformation changes the polarization predictions, it was concluded that this ambiguity should be distinguishable by the spin flip dispersion relations.† The solid line in Fig. 5.3(a) shows the resulting plot of eqn (5.84) for the Minami phase shifts (determined from the Fermi set).

It is clear that $y$ becomes infinite as $x \to 0$ and this curve cannot be extrapolated to give a finite value of $f^2$. Moreover, eqn (5.84) predicts a linear behaviour of $y$ with constant slope $c$, which is not satisfied in Fig.

† Polarization measurements, as well as $\frac{\mathrm{d}\sigma}{\mathrm{d}\Omega}(\theta)$, would not lead to selection of a unique set of phase shifts since 'complex conjugation' changes the sign of $P(\theta)$ while leaving $\frac{\mathrm{d}\sigma}{\mathrm{d}\Omega}(\theta)$ the same.

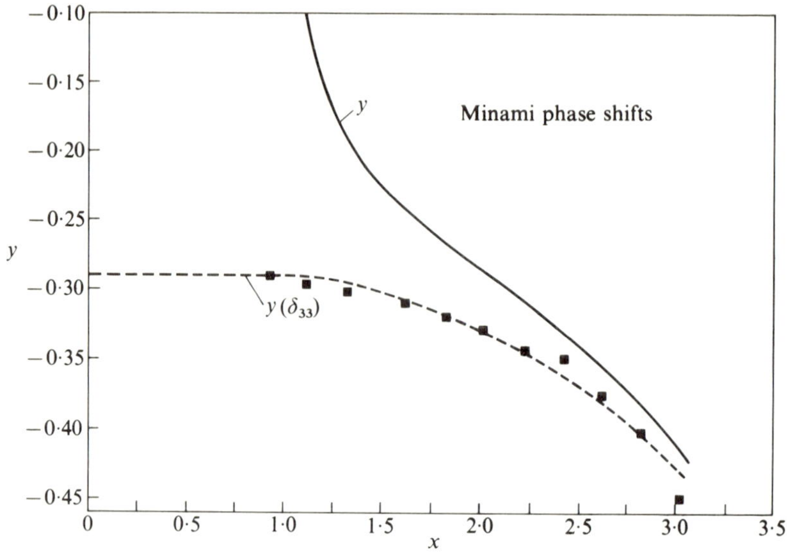


FIG. 5.3.(a) Plot of $y$ against $x$ for the Minami phase shifts. The solid curve represents $y$. The dashed curve marked $y(\alpha_{33})$ gives the contribution to $y$ due to $\alpha_{33}$ alone. (From Lindenbaum and Sternheimer [15].)

5.3(a). We can, therefore, conclude that the Minami phase shifts are incompatible with the requirements of the dispersion relations. The divergence results essentially from the fact that the Minami set gives a $p$-wave phase shift $\delta_{\mathrm{M}}(P_{\frac{1}{2}}, T = \frac{3}{2})$ which behaves as $\bar{\eta}$ as $\bar{\eta} \to 0$, instead of $\bar{\eta}^3$, as expected for a $p$-wave phase shift, scattered from a short range potential.† It appears that the dispersion relations essentially predict that a $p$-wave phase shift must increase at least as rapidly as $\bar{\eta}^3$ at low energies. The dashed line in Fig. 5.3(a) shows the contribution of $\delta_{33}$ alone, denoted as $Y(\delta_{33})$. Modifying the behaviours of $\delta_m(P_{\frac{1}{2}}, T = \frac{3}{2})$ for very low energies to obtain a finite intercept at $x = 0$ does not lead to a sensible solution with the expected coupling constant and linear behaviour. Hence, the Minami set has to be rejected.

In Fig. 5.3(b), the same plot (eqn (5.84)) is shown for the $\alpha'$ phase shift set (the Minami set derived from the Yang set). Again it is clear that this Minami set has to be rejected. Hence it was demonstrated for the first time that no other phase shift solution but the Fermi set (which went through a resonance

† It should be remarked that the Minami phase shifts had been considered a physically less plausible set than the Fermi set, due to the fact that the momentum dependence of the various phase shifts at low energies were not the usual ones expected for a short range potential interaction and, also, because a $d$-wave resonance was hard to understand physically, whereas a $p$-wave resonance followed from general strong coupling meson theory for an odd parity pion.

in $\delta_{33}$ at $E_\pi \sim 195$ MeV) could satisfy the experimental total cross-section and scattering data, the $\pi$–N forward dispersion relations, and the spin flip forward dispersion relations. This made it almost certain that the Fermi set of phase shifts is the only correct set, and is also the best solution for the pion–nucleon scattering at low energies up to about 300 MeV.

5.7.5. *Determination of the pion–nucleon coupling constant*

Since the $\pi$–p forward dispersion relations involve $f^2$, it is obvious that they can be used to determine this quantity. The difficulty, however, is finding a method for doing this which is not very sensitive to the unknown quantities or errors in the known quantities.

The early determination [16] of $f^2$ used the following procedure. Starting with eqn (5.77), the unsubtracted $D^-$

$$D^-(\omega) = \frac{2f^2\omega}{\left\{\omega^2 - \left(\dfrac{\mu^2}{2M}\right)^2\right\}} + \frac{\omega}{4\pi^2}\,\mathrm{P}\int_{+\mu}^{\infty} \frac{\mathrm{d}\omega' k'[\sigma_-(\omega') - \sigma_+(\omega')]}{(\omega'^2 - \omega^2)}, \tag{5.87}$$

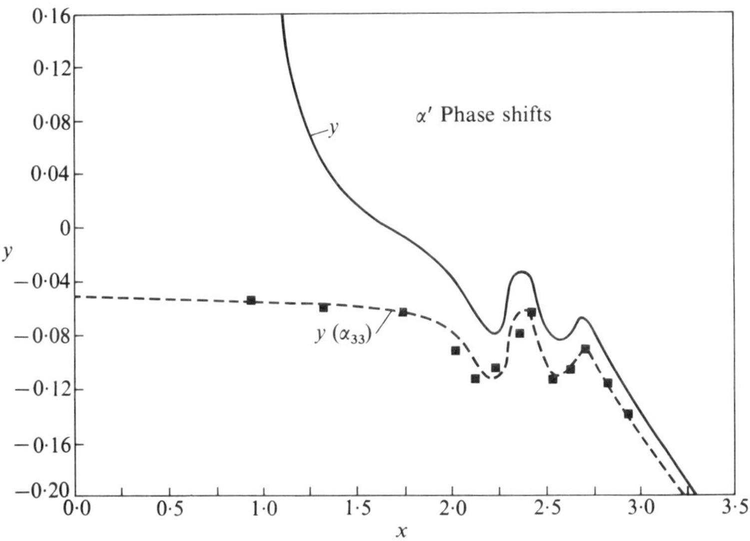


FIG. 5.3.(b) Plot of $y$ against $x$ for the $\alpha'$ phase shifts. The solid curve represents $y$. The dashed curve, marked $y(\alpha_{33})$, gives the contribution to $y$ due to the phase shift $\alpha_{33}$, alone. (From Lindenbaum and Sternheimer [15].)

and then, using the identity

$$\frac{1}{\omega'^2-\left(\dfrac{\mu^2}{2M}\right)^2} = \frac{1}{\omega'^2}+\frac{\left(\dfrac{\mu^2}{2M}\right)^2}{\omega'^2\left\{\omega'^2-\left(\dfrac{\mu^2}{2M}\right)^2\right\}}, \tag{5.88}$$

the above equation for $D^-$ can be rewritten as

$$\left\{\frac{D^-(\omega)}{\omega}-\frac{\omega^2}{4\pi^2}\,\mathrm{P}\int_{\mu}^{\infty}\frac{\mathrm{d}\omega' k'}{\omega'^2}\frac{[\sigma_-(\omega')-\sigma_+(\omega')]}{(\omega'^2-\omega^2)}\right\}\left(\omega^2-\left(\frac{\mu^2}{2M}\right)^2\right)$$

$$= 2f^2+\frac{\omega^2-\left(\dfrac{\mu^2}{2M}\right)^2}{4\pi^2}\;\mathrm{P}\int_{\mu}^{\infty}\frac{\mathrm{d}\omega' k'}{\omega'^2}[\sigma_-(\omega')-\sigma_+(\omega')]. \tag{5.89}$$

Provided we assume that the integral on the right-hand side converges (the Pomeranchuk theorem), the left-hand side of the above equation is then a linear function of $\omega^2$. Unfortunately, Re $D^-(\omega)$ has not been measured directly at low energies, due to experimental difficulties, however, its value was estimated on the basis of phase-shift analyses using the relation between the scattering amplitude and phase shifts, as described in Chapter 4, § 6, and, hence the first term in the bracket on the left-hand side was obtained. The second term in the bracket on the left-hand side was calculated from the measured values of the total cross-section and reasonable assumptions about the high energy behaviour. Fortunately, the integral in the bracket is not very sensitive to the assumed energy behaviour at very high energies. After evaluation of the left-hand side, the value of $2f^2$ is determined by extrapolating the straight line relationship to the point

$$\omega^2 = \left(\frac{\mu^2}{2M}\right)^2.$$

The result obtained was $f^2 \approx 0{\cdot}08$. Although there were some indications of discrepancies in the early work, it was subsequently [17] shown that, within the experimental errors on the data, the prediction of the $\pi$–p forward dispersion relations are consistent with the data, using a value of $f^2 \approx 0{\cdot}08$. A more recent, and more accurate, value of $f^2$ obtained using the dispersion relation is [18] $f^2 = 0{\cdot}081 \pm 0{\cdot}002$. A reasonable consistency between the low energy experimental data and the $\pi$–p forward dispersion relations has been attained and, in fact, more accurate values of low energy parameters have been obtained using dispersion relations.

## 5.8. Experimental test of the validity of the pion–nucleon forward dispersion relations

### 5.8.1. *Background*

The previous major application of the pion–nucleon forward dispersion relations was made in the low energy region of 0 to several hundred MeV incident pion kinetic energy. As they had been successfully applied to the phase-shift analyses and determinations of the coupling constant etc., it was generally assumed that the $\pi$–p forward dispersion relations were valid at least up to a few hundred MeV of incident pion energy. There were several points of doubt which could be raised as to the validity of these earlier conclusions.

(a) The uncertainty of the effects of unforeseen asymptotic behaviour. Neither the experimental evdience for, or the theoretical understanding of, asymptotic behaviour was in good shape at this time.

(b) The forward scattering amplitude was not measured directly (except in isolated cases) in these experimental checks, and was deduced from the best-fitting set as a function of the energy of a number of non-unique phase shift analyses.

(c) The low energy region was overwhelmed by the $T = J = \frac{3}{2}$ resonance, and the results were sensitive to the low energy parameters which were deduced from these same data.

(d) If the causality condition failed within a fundamental length corresponding to a few hundred MeV incident energy, the large effect of the low energy resonances and low energy parameters could tend to mask its effects in the region where the most extensive experimental checks were made.

Therefore, it became clear [19] that a critical test of the pion–nucleon forward dispersion relations required a direct precision measurement of the $\pi^{\pm}$–p forward elastic scattering amplitudes as well as total cross-sections to permit evaluation of the dispersion relation energy integral. The insensitivity of the asymptotic behaviour had also to be demonstrated. The following advantages are obtained by utilizing the highest available energies.

(a) Additional subtractions can be introduced at the highest energy to virtually eliminate sensitivity to even highly implausible and unphysical behaviour of the high energy total cross-sections.

(b) At high energies, there is insensitivity to errors in the low energy parameters and total cross-section data.

(c) As discussed in § 5.8 (where the effects of failures of the microscopic causality condition, i.e., introduction of a fundamental length, is treated) the demonstration of the validity of the pion–nucleon forward dispersion relation to an incident maximum energy ($\omega_{max}$) strongly implies that, if there is a fundamental length $l$, then

$$l < \frac{1}{\omega_{max}}.$$

Therefore the forward dispersion relations are expected to be valid for all energies $\omega \ll \omega_{max}$.

### 5.8.2. *Experimental measurement of the* π–N *forward elastic scattering amplitude*

As we have seen previously for π–p forward elastic scattering, only one complex function is required to represent the forward (and near forward) scattering amplitude as the spin flip amplitude vanishes.

As shown previously, [19] for the one complex amplitude case, the results of the Coulomb scattering added to the nucleon scattering results in the following expression

$$\frac{d\sigma}{dt}(\pi^{\pm}-p) = \left| \pm \frac{\mathscr{F}}{|t|} e^{2i\delta} + \text{Re } A_N(s,t) + i \text{ Im } A_N(s,t) \right|^2, \tag{5.90}$$

where $\delta$ is the relative phase introduced between the nuclear and Coulomb phase shifts by the long range Coulomb interaction. $\mathscr{F}$ is essentially the point-charge constant times the equivalent form factor,† and $A_N$ is an invariant nuclear amplitude which, for convenience, we have normalized such that

$$|A_N(s,t)|^2 = \frac{d\sigma}{dt_{nuclear}}(s,t). \tag{5.91}$$

The usually-employed invariant amplitude $F(s,t)$, defined in terms of the $T$-matrix elements, has a different normalization to that of our invariant amplitude.

The scattering, or reaction, amplitude $T_{fi}$, for a transition from an initial to final state, is given in terms of the $S$-matrix element as

$$S_{fi} = \delta_{fi} + i(2\pi)^4\, \delta(\textstyle\sum p_{r_f} - \sum p_{r_i}) T_{fi}, \tag{5.92}$$

† There are also some small radiative correction terms contained in $\mathscr{F}$.

where $(p_0, \mathbf{p})$ is the energy–momentum four-vector

$$\delta(\sum p_{\mathrm{f}} - \sum p_{\mathrm{i}}) = \prod_{i=1}^{4} \delta(\sum p_{r_i} - \sum p_{r_f}) \tag{5.93}$$

The amplitude $T_{\beta\alpha}$ is made invariant by choosing a covariant normalization, such as

$$\langle p_{\mathrm{f}} \mid p_{\mathrm{i}} \rangle = 2p_0(2\pi)^3\, \delta(\mathbf{p}' - \mathbf{p}). \tag{5.94}$$

For two-particle elastic scattering, we can write the matrix elements of the $T$-matrix as $F(s, t)$, and following the methods of Eden [20] we can obtain

$$\frac{\mathrm{d}\sigma}{\mathrm{d}\Omega} = \frac{1}{64\pi^2 s}|F(s, t)|^2 \tag{5.95a}$$

or, if spins are present,

$$\frac{\mathrm{d}\sigma}{\mathrm{d}\Omega} = \frac{1}{64\pi^2 s} \sum_{\text{spins}} |F(s, t)|^2 = \sum_{\text{spins}} |f(s, t)|^2. \tag{5.95b}$$

Thus

$$\frac{1}{8\pi s^{\frac{1}{2}}} F(s, t) = f(s, t), \tag{5.95c}$$

since also,

$$\frac{\mathrm{d}\sigma}{\mathrm{d}\Omega} = |f(s, t)|^2 = \frac{p_{\mathrm{cms}}^2}{\pi} |A_{\mathrm{N}}(s, t)|^2 \tag{5.96a}$$

$$f(s, t) = \frac{p_{\mathrm{cms}}}{\pi^{\frac{1}{2}}} A_{\mathrm{N}}(s, t). \tag{5.96b}$$

From the relation

$$\frac{\mathrm{d}\sigma}{\mathrm{d}\Omega} = \frac{1}{64\pi^2 s}|F(s, t)|^2 = \frac{p_{\mathrm{cms}}^2}{\pi}|A_{\mathrm{N}}(s, t)|^2, \tag{5.97}$$

we obtain

$$A_{\mathrm{N}}(s, t) = \frac{F(s, t)}{8 s^{\frac{1}{2}} \pi^{\frac{1}{2}} p_{\mathrm{cms}}} = \frac{\pi^{\frac{1}{2}} f(s, t)}{p_{\mathrm{cms}}}. \tag{5.98}$$

Both are invariant amplitudes, but differ in normalization. In our natural units, $p_{\mathrm{cms}} = k$. Hence, in c.m.s.,

$$\frac{\mathrm{d}\sigma}{\mathrm{d}\Omega} = |f|^2 = \frac{p_{\mathrm{cms}}^2}{\pi}\frac{\mathrm{d}\sigma}{\mathrm{d}t} = \frac{p_{\mathrm{cms}}^2}{\pi}|A_{\mathrm{N}}|^2. \tag{5.99}$$

$\mathscr{F}$ is the point-charge constant times the equivalent form factor, which to a good approximation can be expressed as the product of the two individual particle form factors. If the pion form factor is assumed equivalent to the proton form factor† at low $|t|$, where Coulomb effects are important,

$$\mathscr{F} \approx 2e^2\sqrt{\pi}\left\{G_E^2(t) - \frac{t}{4M^2}G_M^2(t)\left(\frac{1}{1-t/4M^2}\right)\right\}. \tag{5.100}$$

As we shall see in Chapter 13, for $-t < 0{\cdot}2\,(\text{GeV}/c)^2$, several experiments demonstrate that $A_\text{N}$ can be represented by [19]–[21]

$$A_\text{N} = \exp(a/2+(b/2)t+(c/2)t^2), \tag{5.101}$$

where the quadratic term is small compared to the linear term

$$b \sim 8\text{–}10\,(\text{GeV}/c)^2$$

and

$$c \sim 2\text{–}3\,(\text{GeV}/c)^{-4}.$$

Therefore, it follows from the above formulae and the observed values of $\frac{\mathrm{d}\sigma}{\mathrm{d}t}(t \approx 0)$ that the known real Coulomb amplitude, which is infinite at $t \approx 0$ and drops as $1/|t|$, will become equal to the slowly varying nuclear amplitude at $t \approx 0{\cdot}003$, and the Coulomb amplitude will then rapidly become unimportant at higher $|t|$.

In the region where the Coulomb amplitude is comparable with the nuclear amplitude, we may expect to measure by coherent interference both the value and sign of the real part of the nuclear scattering amplitude. Since $t \approx 0{\cdot}003$ corresponds to an impact parameter of about 3·5 fm, which is well outside the range of nuclear force, we can expect, by observing these interference effects at small $|t|$, to measure the real part of the nuclear scattering amplitude at large enough distances so that a reliable extrapolation to its final value at $t = 0$ can be made.

The imaginary part of the forward scattering amplitude can be determined from high-precision total cross-section measurements using the optical theorem, which depends only on unitarity and gives the relationship in these

† This assumption is supported by the general similarity of $\pi$–p and p–p elastic scattering curve slopes at low $|t|$. In any event, the form factor effect is quite small and would appreciably affect the results only if the pion form factor corresponded to a much bigger pion radius than the proton radius. This latter possibility is ruled out by the behaviour of the data and the fact that the nuclear force range is limited by the pion mass which, according to the experiments, is the lightest nuclearly interacting particle. Hence the uncertainty in the form factor introduces a negligible error.

The general treatment of electromagnetic form factors is considered in Chapter 11. The expression we use here is essentially the first term with different units of eqn (11.20). The second term is negligible at small $|t|$. There are also some small radiative correction terms which apply to $\mathscr{F}$.

units,

$$\text{Im } A_N(t = 0) = \frac{1}{4\sqrt{\pi}}\sigma. \tag{5.102}$$

Of course, observation of the differential elastic scattering near $t = 0$ also determines the imaginary part of the ordinary scattering amplitude near $t = 0$, and we can use this additional information to check the consistency with the $t = 0$ value determined for total cross-section measurements. It is convenient to define the dimensionless ratio

$$\alpha = \frac{\text{Re } A_N}{\text{Im } A_N}. \tag{5.103}$$

As we shall see later, the experimental data is very consistent with the assumption that, at small $|t|$, $\alpha$ is a constant. We shall assume and justify this point later. Using the above, we can express the differential cross-sections as

$$\frac{d\sigma}{dt}(\pi^{\pm}\text{-p}) = \frac{\mathscr{F}}{|t|^2} \mp \frac{2\mathscr{F}}{|t|}\,\text{Im}(A_{N\pm})(\alpha_{\pm}\cos 2\delta \pm \sin 2\delta) +$$
$$+(1+\alpha_{\pm}^2)(\text{Im } A_{N\pm})^2 + \text{multiple scattering corrections.} \tag{5.104}$$

The original calculation for $\delta$ was made by Bethe [22] and gave

$$\delta = \frac{e^2}{hc\beta}\log\frac{1{\cdot}06}{ka\theta}, \tag{5.105}$$

where $a$ is the effective radius, $k$ the momentum, and $\theta$ is the scattering angle (either laboratory or c.m.s. quantities can be used, since $ka\theta$ is an invariant). Although it was originally referred to as non-relativistic, it turned out that the result is, within the uncertainties of the calculations, also relativistically correct.

Over the years, a number of attempts [23]–[27] have been made to improve the Bethe formula (i.e. make it relativistic and derive it from a more fundamental relativistic quantum electrodynamic field theory [19], [27] point of view). Where sizeable differences have appeared, they have turned out to be errors† in the calculations. Three calculations agree within 15 per cent with the above result. In a recent comprehensive analysis of the problem, from the point of view of relativistic quantum field theory, West and Yennie [27] demonstrated that the Bethe formula was correct within the limits of possible

† In [24] magnetic effects were not originally taken into account. In reference [23*a*] when making the infra-red approximation, a form factor effect was neglected [23*b*].

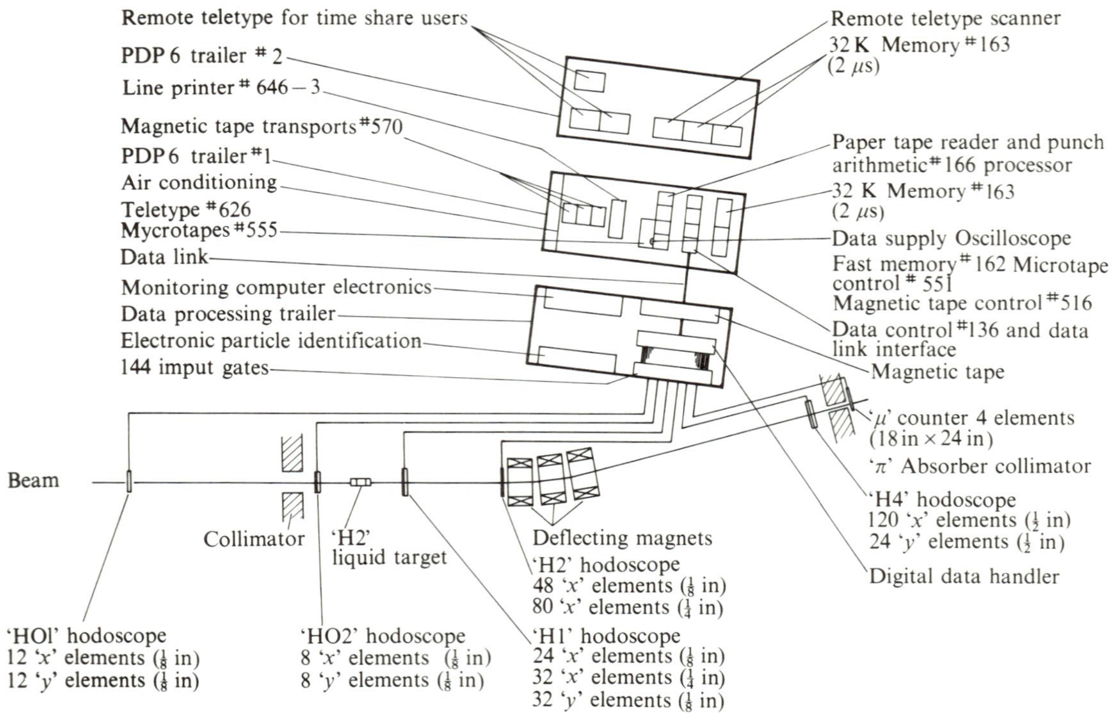


FIG. 5.4. Experimental arrangement of [29]. A simplified sketch of the counter hodoscope on-line computer system for studying small angle scattering, utilizing a high resolution, high-data-rate capacity magnetic spectrometer. Hodoscopes $HO_1$ and $HO_2$ measure the location and direction of the particle before scattering in the hydrogen target. Hodoscopes $H_1$ and $H_2$ determine its location and direction after scattering, and reject events not coming from the target. Hodoscope $H_4$, after the magnets, then determines its momentum. The digital data handling equipment in the $40' \times 10'$ trailer near the hodoscopes. The on-line PDP-6 computer system is contained in the other $40' \times 10'$ trailers. The apparatus is about $400'$ long. (From Lindenbaum [34] and Foley *et al.* [29].)

higher order corrections, which would introduce corrections of less than twice the spread in values ($\Delta\delta \lesssim 0{\cdot}005$).

The first demonstration of the existence of sizeable real parts of the $\pi^{\pm}$–p forward scattering amplitudes at high energies occurred in 1964, in a series of preliminary measurements [28]. Owing to the size of the systematic errors associated with the analysis to obtain the real part of the forward scattering amplitude, these results did not allow a conclusive check of the forward dispersion relations. The accurate measurement of the real part of the forward scattering amplitude at high energies is an extremely difficult technical problem. Since the whole Coulomb interference effect occurs in a few milliradians ($\ll 1°$ near the forward direction), where the steep $|t|$ dependence of the Coulomb and multiple scattering and the incident beam are present, high angular and momentum resolutions, high and accurately known detector efficiency, and the ability to measure and analyse an enormous number of events (mostly in the Coulomb region) are all required to obtain an accurate measurement of the real part of the forward scattering amplitude. An over-all absolute accuracy of 1 per cent, or better, was required and obtained in the new differential scattering measurements [29], which were run at the Brookhaven AGS in 1965–66.

A highly accurate knowledge of the imaginary part of the forward scattering amplitude is also required in the analysis of the small angle scattering experiments and to allow accurate calculations of the dispersion relation predictions. An absolute precision of 0·3 per cent was obtained in the associated total cross-section measurements [29*b*]. Only the on-line computer techniques originally developed several years earlier (1962) by the author's group made experiments of this precision possible. The experimental arrangement is shown in Fig. 5.4.

About 400–500 counter hodoscopes were employed in a magnetic spectrometer system which allowed $10^{12}$ possible counter combinations. Three million trigger selected events per hour were handled and recorded by the digital data handler system, and about 10 per cent of these were analysed automatically by the on-line computer. Typical differential cross-sections are shown in Fig. 5.5.

Equation (5.101) was substituted for $A_{\mathrm{N}}$ into eqn (5.104), and $\alpha$ and $b$ were allowed to be free parameters. Since the quadratic term contributed very little in the $t$ range investigated and the value of $\alpha$ is very insensitive to the value of $c$, $c$ was fixed at its average value determined from previous experiments (at higher $|t|$). The values of $\mathrm{Im}\, A_{\mathrm{N}}(t = 0)$ were determined from precision total cross-section measurements (Chapter 8). The solid curves in Figs. 5.5(a) and 5.5(b) are the best fits to the data. The dashed curves are the

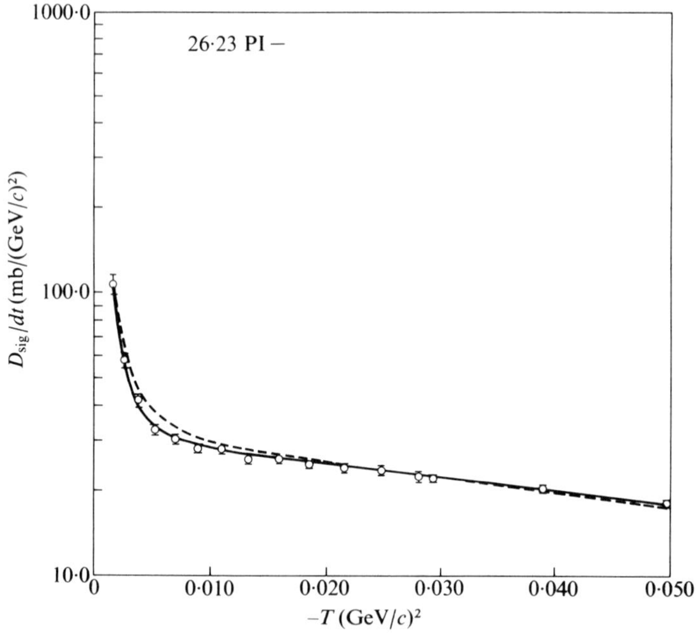


FIG. 5.5.(a) A typical $\pi^-+p$ differential elastic scattering cross-section. The solid line is the best fit allowing for a real amplitude ($\alpha \neq 0$). The dotted line is the best fit assuming the real amplitude is zero (i.e. $\alpha = 0$). The parametric forms used, and all the details of the fit, are explained in [29] and discussed in this chapter. (From Foley *et al.* [29].)

best fits with $\alpha = 0$, a constraint which is obviously ruled out by the $\chi^2$ values. The interference is destructive for $\pi^-+p$ and constructive for $\pi^++p$, demonstrating that the real part of the amplitude is always negative. The value of $\alpha$ is very insensitive to the minimum value of $|t|$ used in the fits. This shows that the parameterization used (including the assumption of no $|t|$ dependence of $\alpha$) is reasonable. A minimum cut-off at $|t| = 0{\cdot}001$ (GeV/$c$)$^2$ was selected for the final determinations. That the value of $\alpha$ determined then corresponds to the value of $\alpha$ at $t = 0$ is well justified, since the measurements have been made to $|t|$ values which correspond to impact parameters of about 6·5 fm, which is well outside the range of nuclear force.†

† That is to say, in order to produce even a 5 per cent change (which is small compared to the error) in the value of $\alpha$, in going from $|t| = 0{\cdot}001$ to $|t| = 0$, would require that the $b$ parameter for the real amplitude be increased by a factor of 10 compared to the $b$ parameter for the imaginary amplitude, which is not to be seriously expected since this would correspond to scattering from a range of interaction of 3 fm and the largest nuclear force range corresponding to the lightest known nuclear particle, the pion, is about 1 fm.

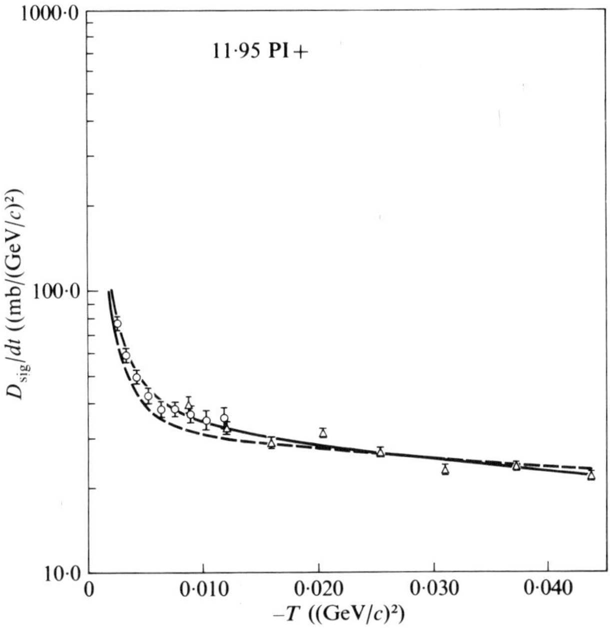


FIG. 5.5.(b) A typical $\pi^+ + \text{p}$ differential elastic scattering cross-section. The solid line is the best fit allowing for a real amplitude ($\alpha \neq 0$). The dotted line is the best fit assuming the real amplitude is zero (i.e. $\alpha = 0$). (From Foley *et al.* [29].)

### 5.8.3. *Comparison of experimental measurements and the* $\pi$–N *forward dispersion relations*

The values of $\alpha_{\pm}$ computed according to eqn (5.104) are compared in Fig. 5.6 to the dispersion relation predictions (using eqns (5.76)–(5.77). The solid points represent measurements with the apparatus rearranged to cover a larger angular range with consequently worse (about 50 per cent) angular resolution. The good agreement demonstrates the insensitivity of the results to the resolution. The errors shown are those obtained from the least-squares fit. In addition, there is an over-all systematic error of $\pm 0{\cdot}02$ due to uncertainties in the total cross-section measurements, inelastic background subtraction, efficiency, and multiple scattering corrections.

Figure 13.30 shows the results of fitting the high precision total cross-section measurements to two of three variations of a five parameter power law, in which for all cases it is assumed $\sigma_{\pi^-} - \sigma_{\pi^+} \to 0$ as $p_{\text{lab}} \to \infty$. Allowing an additional parameter to provide for a finite cross-section difference at infinity gave results consistent with zero difference within the errors.

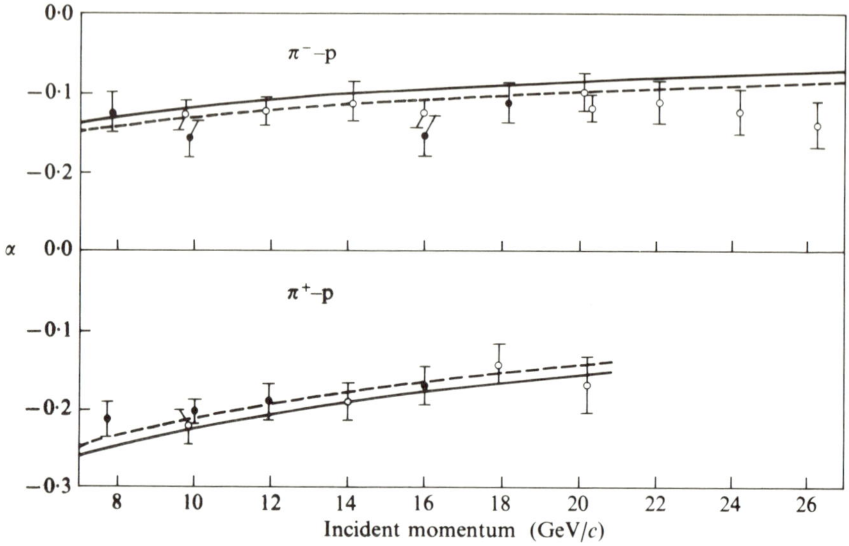


FIG. 5.6. Experimental values of $\alpha = \dfrac{\mathrm{Re}\, f}{\mathrm{Im}\, f}$. The solid lines are the results of the dispersion relation fits. The dashed line is the result of shifting the dispersion relation results by the additional systematic scale error to be applied to $\alpha$. As is discussed in the text, this error is such as to move $\alpha_+$ and $\alpha_-$ in opposite directions. (From Foley *et al.* [29].)

Fit I† (of Figs. 13.3 and 13.30) was used in the dispersion relation calculations shown above 8 GeV in Figs. 5.6–5.7. Fit III‡ would give very similar results. Fit II,§ which is similar to a three pole $P$, $P'$, $\rho$ Regge fit, gave much too large a theoretical value of $D^-$ compared to the experimentally determined values due to its slow convergence of the difference and, therefore, was rejected. However, we should note that with two standard deviations on the exponent of the difference in Fit II, it could be brought close to agreement with the experimental values of $D^-$, within the systematic error on $D^-$. As can be seen in Figs. 5.6–5.7, within the systematic scale error there is a

† Individual power law fits to $\sigma(\pi^\pm\text{–p})$ of the form $a+\dfrac{b_\pm}{p_\pm^n}$.

‡ Similar individual power law fits to $\sigma(T = \frac{3}{2})$ and $\sigma(T = \frac{1}{2})$ of the form $\sigma_{\frac{3}{2},\frac{1}{2}} = a+\dfrac{b_{\frac{3}{2},\frac{1}{2}}}{p^n_{\frac{3}{2},\frac{1}{2}}}$.

§ $\sigma_{\pi^-}+\sigma_{\pi^+} = a+\dfrac{b}{p^n}$,

$$\sigma_{\pi^-}-\sigma_{\pi^+} = \frac{c}{p^n}.$$

generally good agreement between the experimental results and Fit I used in the dispersion relations. Within the total cross-section errors, this extrapolation can be adjusted to give an even better agreement between the predictions and the data.

Since all important systematic errors, including the uncertainty in $\delta$, tend to make about equal and opposite contributions to $\alpha_-$ and $\alpha_+$, they cancel to a large extent in the experimentally determined $D^+$. Furthermore, the calculated $D^+$ is very insensitive to the differences between Fit I, Fit II, and Fit III while, on the other hand, $D^-$ is extremely sensitive to the systematic errors and fit differences. Hence a more critical test of the validity of the forward dispersion relations can be made by comparing the experimental results with the $D^+$ dispersion relation, which is shown in Fig. 5.7(a). We see that the agreement is good. If the total cross-section obeys the Froissart bound (which follows from axiomatic field theory and unitarity—see Chapter 13), and the data is certainly consistent with this, the singly subtracted $D^+$ dispersion integral converges reasonably well. However, an additional subtraction was made in $D^+$ at 20 GeV/$c$, in order to greatly reduce the sensitivity of the integral to the high energy cross-section behaviour beyond the measured incident momentum range. The result after this additional subtraction is also shown in Fig. 5.7(a), which fits the data excellently.

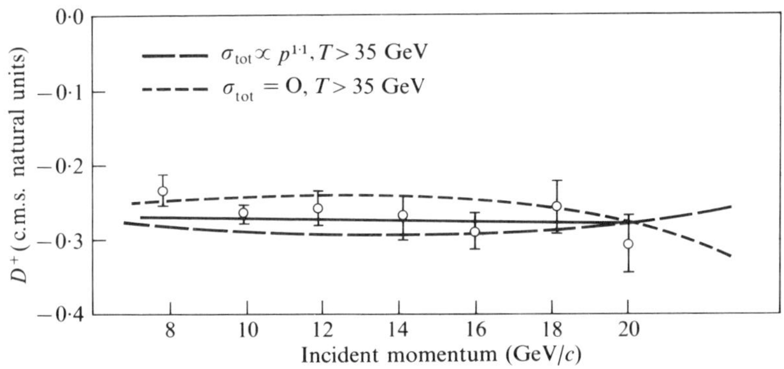


FIG. 5.7.(a) Values of $D^+$ in natural units in c.m.s. The solid curve is the predicted result of the singly-subtracted dispersion relation integrals and, also, is the predicted result for a doubly subtracted dispersion relation, when the second subtraction is made at 20 GeV/$c$ using the value of $D^+$ predicted by the singly subtracted dispersion relation. The total cross-sections at higher energies are allowed to follow Fits I or III (based on the experimental data). The dotted and dashed curves are the results obtained from the doubly subtracted dispersion relation integral, using the assumed total cross-section behaviours shown in the figure. (From Foley *et al.* [29], also [34].)

This doubly subtracted $D^+$ dispersion relation is virtually independent of the high-energy behaviour of the total cross-sections. This is demonstrated by Fig. 5.7(a), which shows the effect of drastic (and what most of us would consider unphysical) assumed changes in the behaviour of the high-energy total cross-sections.

The two assumptions made are that, for $p > 35$ GeV/$c$:

(a) The total $\pi^{+}$–p cross-sections both vanish.

(b) Both total cross-sections increase faster than linearly (i.e. as $P^{1 \cdot 1}$). Since these totally unreasonable changes introduce only small changes in the predicted values of $D^+$, it is clear that the agreement of the data and the prediction of the $D^+$ dispersion relation are virtually independent of any acceptable changes in the asymptotic cross-section behaviour. Hence, the validity of the doubly subtracted $D^+$ dispersion relation has been established to beyond 20 GeV/$c$. Thus, it follows that it has at least been proved that $F(s, t \simeq 0)$ must be analytic at least for $s$ smaller than the value of $s$ corresponding to 20 GeV/$c$ incident pions. It is clear that probably considerably more about the analyticity properties of scattering amplitudes has been demonstrated than the minimum, easily justifiable conclusions stated above.

As previously demonstrated, the singly subtracted $D^+$ dispersion relation is valid provided the asymptotic behaviour of the high-energy total cross-sections does not change considerably from the behaviour of the good parametric fits obtained for the data. As would be expected, $D^+$ is mostly sensitive to the average level of the total cross-section at higher energies, and it has been concluded that the constant at infinity, deduced in the parametric fit, gives good agreement, and a change of about 20 per cent in its value would lead to disagreement.

Figure 5.7(b) shows a comparison of the data and the calculated results for $D^-$, using Fit I (simple individual power law fits). As we have previously explained, the systematic scale errors in $D^-$ add. Their effect on the comparison is shown by the dashed curve, which represents a displacement of the solid curve (Fit I) downward by the systematic scale error, which is equivalent to a change of scale of the experimental points for comparison purposes. It is clear that there is reasonable agreement between the $D^-$ prediction for Fit I and the data, within the systematic error. However, as discussed previously, $D^-$ is very sensitive to the exact power of convergence of the differences cross-section

$$\sigma^- = \frac{\sigma^- - \sigma^+}{2} = \frac{\text{constant}}{p^n}.$$

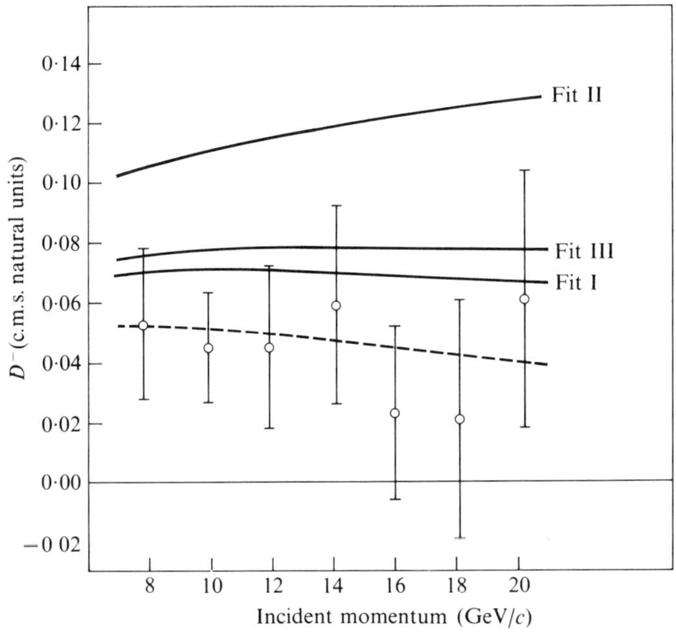


FIG. 5.7.(b) Values of $D^-$ in the centre of mass natural units. The solid curves are the results of the unsubtracted dispersion relation integrals discussed in the text. The dotted curve represents a shift of Fit I by the value of the systematic scale error. (Foley *et al.* [29], also [34].)

For example, for Fit II, the derived $D^-$ does not agree with the data. This subject will be discussed further in Chapter 13. For the present, though, we can conclude that a satisfactory fit to the total cross-sections can be found, which gives predicted values of $D^-$ using the unsubtracted dispersion relation, which agrees with the experimental data, within error. One subtraction introduced in $D^-$ will completely densensitize it to any reasonable changes in asymptotic behaviour, and the prediction will be in good agreement with the data, hence, at least the singly subtracted $D^-$ is a valid dispersion relation, virtually independent of asymptotic behaviour.

### 5.8.4. *Limits on a fundamental length*

The dispersion relations we have employed so far have been derived assuming microscopic causality holds down to infinitesimal distances, and that points separated by a space-like distance cannot interfere with each other (i.e. boson (fermion) field operators at such points commute (anti-commute)). Some time ago, it was shown [30] that if we relax this requirement and assume an acausal region for space-like points separated by an absolute distance $<|l|$, where $|l|$ is a so-called 'fundamental length', then the Cauchy dispersion integrals hold for a newly defined function

$$F(\omega) = f(\omega)e^{i|l|\omega}, \tag{5.106}$$

where, as before, $f(\omega) = D(\omega)+iA(\omega)$. Oehme evaluated the result for neutral pions, and it is straightforward, using the crossing symmetry relation between positive and negative frequency cross-sections, to derive the result for charged pions. Doing so, subtracting at $\omega = \mu$, and assuming $l \ll \frac{1}{\mu}$ we obtain for $D^+$.

$$D^+(\omega)\cos|l|\,\omega - A^+(\omega)\sin|l|\,\omega = D^+(\mu)+\frac{f^2k^2}{M\left\{1-\left(\frac{\mu}{2M}\right)^2\right\}\left\{\omega^2-\left(\frac{\mu^2}{2M}\right)^2\right\}}+$$

$$+\frac{2k^2}{\pi}\mathrm{P}\int_0^\infty \omega'\,d\omega'\,\frac{[A^+(\omega')\cos|l|\,\omega'+D^+(\omega')\sin\omega'\,|l|}{(\omega'^2-\omega^2)(\omega'^2-\mu^2)}. \tag{5.107}$$

When $\omega l \to 0$, these relations reduce to the ordinary forward dispersion relations. However, when $\omega l$ becomes an appreciable fraction of 1, they change drastically.

It appears that even a fundamental length ($|l|$) of about $10^{-16}$ cm results in sizeable changes in the predicted $D^+$ values from eqn (5.107) which disagree with the experimental data values, which we should remember are almost free of experimental scale error. It is obvious that unforeseen changes in asymptotic behaviour could also change the predicted $D^+$. However, an additional subtraction in $D^+$ would completely densensitize the predicted values of $D^+$ to changes introduced by unforeseen asymptotic behaviour, or probably even for a fundamental length such that $\frac{1}{l} \gg 20$ GeV. However, if there were a fundamental length $|l|$, such that $\omega_l = \frac{1}{|l|}$ occurred within, or below, our experimental observation range, due to the structure of eqn (5.107) (particularly the left-hand side) we would have drastic changes in both the magnitude, and energy behaviour, of $D^+$ (i.e. the dispersion relations would change drastically and give completely different predictions) and therefore we can, on the basis of the above, conclude that if a fundamental length exists, $|l| < 10^{-15}$ cm.

There is one uncertainty in the above procedure, since two points can be space-like even when separated by a large spatial difference. The form of the above changes in the forward dispersion relations strongly implies that the most important contributions for the acausal region come from the space-like region which is physically close, since the changes depend on the product $\omega |l|$, indicating that high energy (i.e. accessibility to short distances) is necessary to observe the effects. If we attempt to remove this uncertainty we need a particular model. A recent model [31] for avoiding this uncertainty expressed the changes in the predicted values for the causal $D_{\pm}$ due to an acausal region characterized by $|l|$ in terms of the values of $D_{\pm}(\Omega)$, where $\Omega = \frac{1}{|l|}$ is the laboratory energy corresponding to the fundamental length. Using this procedure again, we can exclude a fundamental length occurring at energies which correspond to those within, or below, the range of measurement, as it would drastically change the dispersion relations. If a fundamental length at higher energies is assumed, and its effects in the present energy range are calculated, the procedure is somewhat uncertain, since we must estimate by extrapolation $D_{\pm}(\Omega)$ as well as $\sigma_{\pm}(\Omega)$, and we cannot be sure that the characteristics of these (especially the former) will not change drastically at the energy which corresponds to the fundamental length. If we ignore these uncertainties and assume we can detect $\sim 0{\cdot}04$ in $D^+$, we can conclude that $|l| \lesssim 3 \times 10^{-16}$ cm. Similar considerations can be applied to $D^-$, but due to its extreme sensitivity to asymptotic behaviour and the sensitivity of the $D^-$ data to systematic scale errors, it is safer to restrict these analyses to $D^+$.

In summary, it has been shown directly that $|l| < 10^{-15}$ cm for the models considered, independent of reasonable uncertainty in asymptotic behaviour. If we assume the characteristics exhibited by the parametric fits of the total cross-section data (at least for the sum) continue at higher energy, we find $|l| < 3 \times 10^{-16}$ cm to $10^{-16}$ cm, depending upon the model employed. This then implies that $F(s, t \approx 0)$ is analytic for all $s$ smaller than the value of $s$ corresponding to 60–200 GeV/$c$ incident pions.

No one has been able to obtain the analyticity properties of the scattering amplitudes from basic axioms, without including local commutativity. Attempts to violate local commutativity anywhere lead to its violation everywhere, or else unacceptable physical consequences occur. Therefore, the models we have just discussed cannot be made rigorous. However, Oehme [30*b*] recently analysed this problem and concluded that simple physical models of fundamental length effects can be expected to be justified on 'dimensional grounds'. It will be shown in Chapter 13 that the results of the experiments just described, and the charge-exchange measurements at high energies, are consistent with the prediction of charge independence, and the analysis of the asymptotic behaviour of total cross-sections will be discussed.

### 5.9. Forward dispersion relations for p–p and p̄–p scattering

The general form of the p–p scattering amplitude satisfying conservation of angular momentum, parity, and time-reversal invariance contains five complex functions of $(s, t)$. Only two of these vanish in the forward direction, leaving three remaining terms, the ordinary singlet, the ordinary triplet, and one spin flip term. If we assume the scattering amplitude terms are analytic, and we attempt to write forward dispersion relations, there are individual relationships for each of these amplitudes [32]. Furthermore, the negative frequency parts of the integral must be related by crossing symmetry to the p̄–p forward scattering amplitudes of which there are also three.

In order to evaluate these relations in a way which will allow comparison with experiment, the assumption of spin independence is usually made—this then results in an equivalent single complex amplitude. However, there is still, unfortunately, an extensive non-physical region which cannot be evaluated uniquely. If we assume that the Pomeranchuk theorem is approached faster than $1/\ln \omega$, the resultant $D^-$ converges without subtraction. If the total cross-sections do not increase faster than $\omega^{1-\epsilon}$, where $\epsilon$ is a finite (positive) fraction, then the resultant $D^+$, with one subtraction, is convergent.

The assumption of spin independence results in the usual relation between the optical theorem and the total cross-section. This assumption is necessary, also, to allow a unique and accurate determination of the real part of the forward scattering amplitude by analysis of Coulomb–nuclear interference.

Owing to the much higher intensities and ease of handling protons, the p–p small angle scattering has been more thoroughly studied than the $\pi$–p scattering.

Figure 5.8 shows the comparison between the experimental results [33]–[34], deduced for the real part of the p–p forward scattering amplitudes, and the predictions of dispersion relation calculations.

The solid line represents a calculation [35] using a singly subtracted forward dispersion relation for $D^+$ and an unsubtracted forward dispersion relation for $D^-$. The unphysical region contribution was estimated by assuming that the known poles at $\rho$ and $\omega$ predominate, with the usual assumptions about their coupling strengths. The $\eta$ and $\pi$ pole amplitudes were found to be negligible.

The dotted lines [36] represent the estimated range of uncertainty introduced by lack of knowledge of the exact contribution of the non-physical region. In this calculation a second subtraction was taken using experimental data at 24 GeV. This calculation is very insensitive to the high-energy asymptotic behaviour of the total p–p and p̄–p cross-sections.

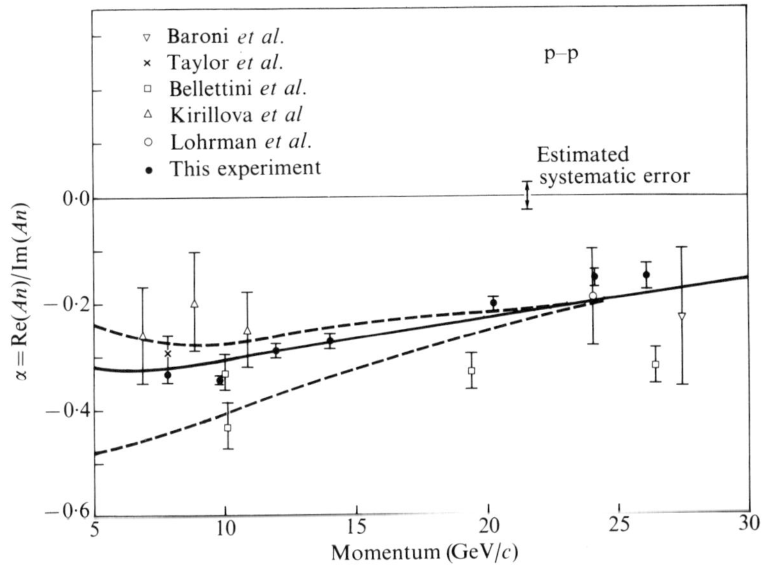


FIG. 5.8. $\alpha$, the ratio of the real to imaginary part of the forward p–p scattering amplitude (assuming spin independence) compared to other data, and forward dispersion relations calculations. (Foley *et al.* [33] and Lindenbaum [34].)

The data (except for the high-momentum points of Belletini *et al.*)† fit the predictions rather well. The significance of this is not obvious, in view of the many assumptions made in the p–p calculations. Nevertheless, it at least implies that p–p forward scattering amplitudes are, to a large degree, spin independent and that, in spite of the extensive non-physical region, the high-energy forward dispersion relation predictions are reasonable. Forward dispersion relations can be obtained for p̄–p using analogous methods to those employed for p–p. The predicted $\alpha_{\bar{p}-p}$ is near zero in the energy range considered, and an experimental determination of $\alpha_{\bar{p}-p}$ at $\approx 12$ GeV/$c$ agrees quite well with the prediction [34].

## 5.10. Fixed momentum transfer dispersion relations and their analyticity domains

Although eqns (5.28)–(5.32) have so far been applied only to the derivation of the pion–nucleon forward dispersion relations, they are quite generally valid for any other fixed values of $t$, provided only that $f(\omega)$ is analytic in the

† The shape of the curve which was originally obtained is consistent with the other work considered, but there is a scale shift. On re-checking the point at 10 GeV, it came in lower in agreement in previous work. Thus, a scale shift may explain this discrepancy.

entire upper-half complex energy plane for these values of $t$. The following are two well-known domains.

(a) *The Mandelstam analyticity domain*

The maximum domain of analyticity is that assumed by the Mandelstam representation [37] which assumes the full cut-plane analyticity in both $s$, $t$. This domain can only be demonstrated in perturbation theory.

(b) *The Lehman analyticity domain*

Lehman [38] starting from axiomatic quantum field theory was able (1958) to demonstrate† that the invariant scattering amplitude $F(s, t(\cos\theta))$ is analytic for all $\cos\theta$ inside an ellipse (the Lehman ellipse) in the complex ($\cos\theta$) plane. Where $s$ is in the physical region and considered fixed, the focii of the ellipse are at $\cos\theta = \pm 1$. The length of the semi-major axis depends on the masses. For example, for pion–pion scattering,

$$b(s) = \left(1+\frac{64\mu^2}{k^2 s}\right)^{\frac{1}{2}}.$$

We should note that the imaginary part of $F(s, t)$ is analytic in a somewhat larger ellipse. As $s \to \infty$, the ellipse shrinks and approaches the real line $|\cos\theta| \leq 1$ and the end of the major axis moves toward $t = 0$ according to the following,

$$t_{\text{end}}(s) \sim s^{-1}.$$

5.10.1. *Enlargement of the analyticity domain*

A number of authors subsequently enlarged the analyticity domain. The largest domain was obtained by Martin [39]. He combined the unitarity condition with the L.S.Z. formalism of axiomatic field theory and obtained a much larger analyticity domain. The domain has a size which can be characterized [40] by

$$|t| \leq 2k^2\{b(s)-1\},$$

where $b(s)$ is the semi-major axis of the Lehman ellipse. We can see from the foregoing that, as $s \to \infty$, $|t| \leq R$, where $R$ is a constant.

For the $\pi$–$\pi$ scattering, it has been shown by Martin that $F(s, t)$ satisfies a dispersion relation for $s$, when $t$ is in the range $-28\mu^2 < t < 4\mu^2$. The lower

† With suitable conditions on the masses of the particles.

limit is caused by the inelastic threshold at $16\mu^2$ in $s$ and $u$. For $\pi$–K scattering similar results have been attained. For $\pi$–N scattering the nucleon pole causes some difficulties and the results are less satisfactory.

As a function of two variables, the analyticity domain of $F(s, t)$ is much extended for real $s$ and $t$, and approaches the boundary of the Mandelstam spectral functions $\rho$ except for small indentations (see Chapter 4 of Eden's book [20]). However, this domain is still much smaller than the Mandelstam domain.

## 5.11. The Mandelstam representation

Mandelstam [37] (1958) conjectured that the pion–nucleon, and other two-body elastic scattering amplitudes, $F(s, t)$ could be represented by relativistic dispersion relations which were a function of two complex variables $s$ and $t$, the real parts of which are the Lorentz invariants $s$ and $t$. In order to justify these dispersion relations he needed an extended domain of analyticity, which is referred to as the Mandelstam domain.

Two-particle scattering can be represented by the diagram

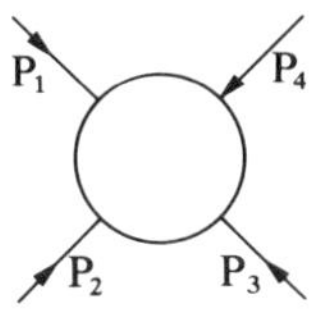


The diagram represents three channels.

$$
\begin{array}{lll}
\text{I.} & s \text{ channel} & 1+2 = 3+4, \\
\text{II.} & t \text{ channel} & 1+\bar{4} = \bar{2}+3, \\
\text{III.} & u \text{ channel} & 1+\bar{3} = \bar{2}+4.
\end{array} \tag{5.108}
$$

For the sake of illustration it is most convenient to consider the case of two-particle scattering, where both particles have equal masses. The Mandelstam diagram for the equal mass case is shown in Fig. 5.9. Kibble [41] gives a general prescription for constructing such a diagram. The three physical regions I, II, III are indicated by the shaded area and have the following characteristics.

$$
\begin{array}{lllll}
\text{I.} & s \text{ channel} & s > 4m^2 & t \le 0 & u \le 0, \\
\text{II.} & t \text{ channel} & t > 4m^2 & s \le 0 & u \le 0, \\
\text{III.} & u \text{ channel} & u > 4m^2 & s \le 0 & t \le 0.
\end{array} \tag{5.109}
$$

The cross-shaded areas are those regions where the Mandelstam spectral functions are non-zero.

Using the axioms of quantum field theory [42], it has been shown that $F(s, t)$, the scattering amplitude, can be analytically continued from region I to region II to region III. Hence, we can describe the three distinct processes in channel I, channel II, and channel III, by the boundary values of the same

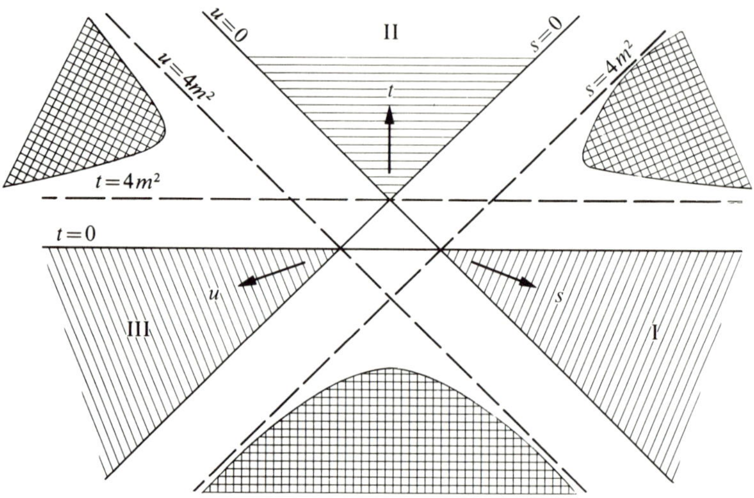


FIG. 5.9. The Mandelstan diagram for the equal mass case (not drawn to scale for illustrative purposes only).

analytic function. The relationships between I, II, and III, which we obtain by employing analytic continuation, are just those referred to as crossing symmetry.

Mandelstam's proposed double dispersion relation representation was based on the evaluation of the fourth-order Feynman diagram for scattering. He evaluated Im $F(s', t)$ and showed, explicitly, that it is analytic in the complex $t$-plane cut from $4m^2$ to $\infty$. Thus, this diagram gives single and double dispersion relations similar to those obtained in potential scattering. Mandelstam then based his representation on the assumption that the resulting double dispersion relations can be generalized to represent the full amplitude as follows.

$$F(s,t) = \frac{1}{\pi^2}\iint\limits_{4m^2}^{\infty} \mathrm{d}s'\,\mathrm{d}t'\,\frac{\rho_{12}(s',t')}{(s'-s)(t'-t)}$$

$$+\frac{1}{\pi^2}\iint\limits_{4m^2}^{\infty} \mathrm{d}t'\,\mathrm{d}u'\,\frac{\rho_{23}(t',u')}{(t'-t)(u'-u)}$$

$$+\frac{1}{\pi^2}\iint\limits_{4m^2}^{\infty} \mathrm{d}u'\,\mathrm{d}s'\,\frac{\rho_{31}(u',s')}{(u'-u)(s'-s)}\,. \tag{5.110}$$

The spectral function ($\rho$) is zero outside the cross-shaded areas in Fig. 5.9.

When considering the scattering of identical bosons, we can demonstrate from crossing symmetry that $\rho_{12}$, $\rho_{23}$, $\rho_{31}$ are the same function ($=\rho$). The number of subtractions required is uncertain in this representation. For fixed real $t$ ($<4m^2$), the dispersion relation is convergent with two subtractions. However, for $t > 4m^2$, there is no way of knowing how many subtractions are required, and one cannot be sure that an infinite number are not required [43]. It is generally assumed, however, that a finite number will suffice.

With one subtraction at $s_0$, $t_0$, and $u_0 = 4m^2-s_0-t_0$, the Mandelstam representation has the form

$$F(s,t) = F(s_0,t_0)+\frac{s-s_0}{\pi}\int\limits_{4m^2}^{\infty}\frac{\mathrm{d}s'\rho_1(s')}{(s'-s_0)(s'-s)}+$$

$$+\frac{(t-t_0)}{\pi}\int\limits_{4m^2}^{\infty}\frac{\mathrm{d}t'\rho_2(t')}{(t'-t_0)(t'-t)}+\frac{u_0-u}{\pi}\int\limits_{4m^2}^{\infty}\frac{\mathrm{d}u'\rho_3(u)}{(u'-u_0)(u'-u)}+$$

$$+\frac{(s-s_0)(t-t_0)}{\pi^2}\iint\limits_{4m^2}^{\infty}\frac{\mathrm{d}s'\,\mathrm{d}t'\rho(s',t')}{(s'-s_0)(s'-s)(t'-t_0)(t'-t)}+$$

$$+\text{cyclic permutations.} \tag{5.111}$$

For a more detailed and thorough treatment of the Mandelstam representation see Eden's book [20], Chap. 4.

### 5.12. Partial wave dispersion relations

Dispersion relations have been developed for the partial wave amplitudes

$$f_{l\pm}^{(T)} = \frac{\exp(2i\,\delta_{l\pm}^{(T)}-1)}{2ik}. \tag{5.112}$$

Since each partial wave amplitude corresponds to a particular phase shift, these partial wave dispersion relations are particularly useful in conjunction with phase-shift analyses. The partial waves have a unique value of angular momentum ($J$) and a unique value of parity ($P$). Thus requiring the conservation of these two quantities is straightforward when using partial wave amplitudes. The unitarity requirement, also, can be expressed easily, and the peripheral approximation is simple to make. A peripheral method can be employed to treat higher waves ($l \geq 1$) approximately. It is convenient in this method to replace the partial wave amplitude by the reduced amplitude defined as [44]

$$\mathscr{F}_{l\pm}^{(T)}(s) = \frac{f_{l\pm}^{(T)}(s)}{k^{2l}}. \tag{5.113}$$

In other words, the partial wave amplitude has been reduced by a factor $k^{2l}$. We should note that, for small $k$,

$$\delta_l^{(T)} \propto k^{2l+1}.$$

This effect can be understood in terms of the centrifugal potential. Hence, for small $k$,

$$\mathscr{F}_{l_\pm}(s) \propto \frac{2i\,\delta_{l_\pm}^{(T)}}{2ik(k^{2l})} = \text{constant}. \tag{5.114}$$

Thus, as $k \to 0$, there is no difficulty with the definition of $\mathscr{F}_{l_\pm}(s)$. The advantage of this method is that contributions to the dispersion integral from the neighbourhood of $s = -M^2$, relative to those for $s = +M^2$, are suppressed by a factor of $\left(\frac{\mu}{M}\right)^{2l}$. The result is that contributions from the cut $-\infty \leq s \leq 0$ are small, for $l \geq 1$, and can be neglected. For $l \geq 1$, the pole at $s = 0$ is removed. We shall treat the subject briefly here, and the reader is referred to Hamilton [44] for the detailed treatment. The major point is that, at not too high values of $k$, the short-range parts of the interaction are increasingly strongly suppressed as $l$ increases, and we can get good results from considering the few known long-range interactions. Following these considerations and applying Cauchy's theorem we obtain the partial wave dispersion relation

$$\mathscr{F}_{l_\pm}^{(T)}(s) = \mathscr{F}_{l_\pm}^{(T)\prime}(s)+\frac{1}{\pi}\,\mathrm{P}\int_{(M+\mu)^2}^{\infty}\frac{\mathrm{Im}\,\mathscr{F}_{l_\pm}^{(T)}(s')\,\mathrm{d}s'}{s'-s}, \tag{5.115}$$

where $\mathscr{F}_{l_\pm}^{(T)\prime}(s)$ is the total contribution of all the unphysical cuts and we have, as usual, approached the physical cut in $s$ from above.

### 5.12.1. *Unitarity*

The partial wave total elastic cross-sections can be shown to be

$$\sigma_{\mathrm{el}_{l\pm}}^{(T)}(s) = \frac{\pi}{k^2}(J+\tfrac{1}{2})\,|1-\exp(2\mathrm{i}\,\delta_{l_\pm}^{(T)})|^2. \tag{5.116}$$

The total inelastic cross-section is

$$\sigma_{\mathrm{inel}_{l\pm}}^{(T)}(s) = \frac{\pi}{k^2}(J+\tfrac{1}{2})\{1-|\exp(2\mathrm{i}\,\delta_{l_\pm}^{(T)})|^2\}. \tag{5.117}$$

Unitarity requires $\sigma_{\mathrm{inel}} \geq 0$.

The total cross-section is (adding eqns (5.116) and (5.117))

$$\sigma_{l_\pm}^{(T)} = \frac{\pi}{k^2}(2J+1)\{1-\mathrm{Re}(\exp(2\mathrm{i}\,\delta_{l_\pm}^{(T)}))\}. \tag{5.118}$$

From the above relations we can write the optical theorem for the particular partial wave considered,

$$(J+\tfrac{1}{2})\mathrm{Im}\,f_{l_\pm}^{(T)}(s) = \frac{q}{4\pi}\,\sigma(T,l,J).$$

Let us define the parameter $\rho = \sigma_{l_\pm}^{(T)}/\sigma_{\mathrm{el}_{l\pm}}^{(T)}$, $\rho = 1/E$, where $E$ was defined in § 4.14. Obviously, $\rho \geq 1$, for purely elastic scattering $\rho = 1$, and for black body sphere scattering $\rho = 2$. When inelastic scattering becomes appreciable, there is no reliable theory to estimate it, and experimental values must be employed when available.

We can now express the unitarity condition in simple terms as a function of $\rho$ and $\mathscr{F}_{l_\pm}^{(T)}$ as follows,

$$(\mathrm{Re}\,\mathscr{F}_{l_\pm}^{(T)})^2+(\mathrm{Im}\,\mathscr{F}_{l_\pm}^{(T)})^2 = \frac{1}{\rho_{l_\pm}^{(T)}k^{2l+1}}\mathrm{Im}\,F_{l_\pm}^{(T)}.$$

Using these methods most of the pion–nucleon partial waves have been predicted up to those energies where the effects of inelastic interactions become large.

The analysis of $s$-wave and $p$-wave pion–nucleon scattering at energies $\lesssim 1$ GeV shows that there are four processes in pion–nucleon interactions which have medium or long range† ($R \gtrsim 3\times 10^{-14}$ cm).

These four processes are

(a) N exchange,

(b) N* exchange,

(c) $(\pi, \pi)$ exchange,

(d) $\rho$ exchange.

Figure 5.10 illustrates these process.

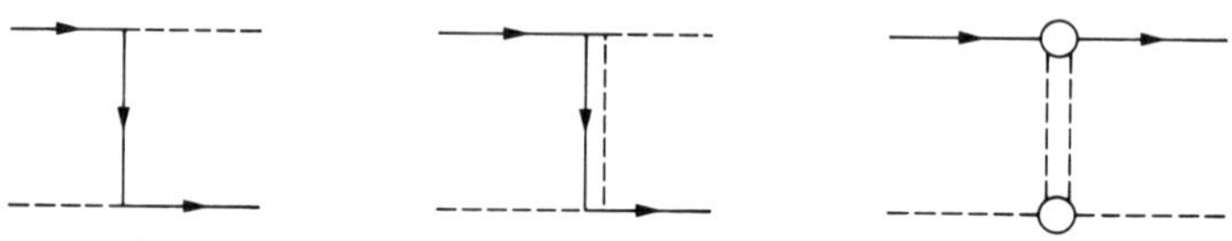

(a) (b) (c) or (d)

FIG. 5.10.

An analysis of the complete system of cuts in the pion–nucleon partial wave amplitudes $f_{l\pm}^{(T)}(s)$, or $F_{l\pm}^{(T)}(s)$, which are responsible for these processes, is given in reference [44].

N-exchange is due to the cut $(M-\mu^2/M)^2 \leq s \leq M^2+2\mu^2$. The nucleon isobar ($T = \frac{3}{2}, J = \frac{3}{2}$) contribution to the cut $0 \leq s \leq (M-\mu)^2$ gives the N*-exchange contribution. The $(\pi, \pi)_0$-exchange and $\rho$-exchange come from the cut which is the arc of the circle $|s| = M^2-\mu^2$. N-exchange is of long range and provides the major contribution to formation of the N* resonance and is, thus, the major cause of the resonance behaviour of $(\frac{3}{2}, \frac{3}{2})$ amplitudes. It has the largest range force. $(\pi, \pi)_0$-exchange is the only other important one in forming the N* resonance contributing about 25 per cent of the total.

N*-exchange is of medium range and is relatively unimportant, except for the $(\frac{1}{2}, \frac{1}{2})$ state, (the nucleon ground state) where it contributes an (comparable with N-exchange) attractive interaction. However, $(\pi, \pi)$ $T = 0$, $J = 0$, which is of the medium range, and $\rho$ exchange, which is of shorter range, are also necessary to obtain enough attraction to obtain the nucleon ground state.

† Exchanges of higher mass nucleon isobars N*, higher mass pionic isobars, nucleon–antinucleon pairs, multi-particle exchanges, etc., all give rise to possible shorter range interactions.

5.12.2. *Low-energy s-wave scattering*

For analysing $s$-wave scattering, the ordinary partial wave charge combination amplitudes $f_{0+}^{\pm}(s)$ are used, since the suppression factors in the reduced amplitude are only useful for $l \geq 1$. Thus $s$-wave scattering suffers more from uncertainties in short range interactions

There is considerable evidence, both from partial wave dispersion relations analyses of low-energy pion–nucleon $s$-wave scattering and from particle production experiments at higher energies, that there is a strong attraction (but probably not a resonance) in a low-energy $\pi$–$\pi$ ($T = 0, J = 0$) state, which we denote by $(\pi, \pi)_{0,0}$. The analyses of $s$-wave scattering also requires some $\rho$-exchange for their explanation.

5.12.3. *Medium-energy p-, d-, and f-wave scattering*

Using the foregoing methods, a reasonably good prediction for the $p$-, $d$-, and $f$-waves has been obtained up to moderate energies ($\sim$600 MeV). Qualitative predictions for $p$- and $d$-waves have been obtained up to about 1 GeV, and for $f$-waves up to about 1·5 GeV. It can be seen from simple semi-classical arguments that the range of validity of the peripheral method in partial wave dispersion relations limits the energy to which it can be applied.

The impact parameter $b$ can be defined by

$$b^2k^2 = l(l+1),$$

where $k$ is the wave number in the c.m.s. For $d$-, or higher, waves $b$ can, with a reasonable approximation, be interpreted in the classical sense. As stated previously, the short-range interactions, which are relatively unknown and neglected in the peripheral method, are of typical range $2 \times 10^{-14}$ cm. Hence, to prevent the wave function from penetrating down to those distances, we must aim for impact parameters several times larger (let us say 4–5 times larger, or about $(0{\cdot}8–1{\cdot}0) \times 10^{-13}$ cm). This is satisfied for $d$-waves only for laboratory energies less than about 800 MeV, but extends to about 1500 MeV for $f$-waves. For $p$-waves there is less suppression of the short-range interactions and more uncertainty due to these effects. However, Donnachie *et al.* subtracted the partial wave dispersion relation at $s = (M+\mu)^2$, by using the known $p$-wave scattering lengths for subtraction constants and, thus, further suppressed short-range interactions. With this additional procedure the $p$-wave has been estimated to about 900 MeV. Only the results for $P_{11}$ are sensitive to the short-range interactions.

Taking into account the effects of unitarity allows approximate correction for the effects of the unknown short-range interactions provided these are small. This is done in terms of a unitary sum rule. For a given $l$, the long-range parts of the interaction become less important more rapidly with increasing energy than the medium-, or short-, range parts.

Using the peripheral method of analysis with only the previously mentioned four long- and medium-range interactions treated, Donnachie and Hamilton [45] (1964) found that there were four strongly attractive amplitudes $P_{33}$, $D_{13}$, $F_{15}$, and $F_{37}$, and that these all exceeded the defined unitary limit. Hence the partial wave resonated at an energy somewhat below that at which, in their calculation, they reached the unitary limits. The estimated resonance energies were compatible with unique identification with the experimentally discovered resonances at 195, 610, 900, and 1410 MeV, respectively. Hence, this method predicts the most characteristic features of the pion–nucleon interactions at those energies. It was also shown, in an analysis of inelastic effects, that these only modify the results slightly. Thus, even though inelastic decay channels are important for $D_{13}$, $F_{15}$, and $F_{37}$, the major causes of the resonance at those energies appear to be the four basic interactions which occur in the elastic channels.

Using these partial wave dispersion relations, approximate predictions have been made for non-resonant small phase shifts. These form useful starting points for detailed phase-shift analyses of the entire available body of experimental data. Requiring that the partial wave dispersion relations are satisfied is a very useful smoothing requirement, which allows us to select a unique solution from several available phase shifts which fit the experimental data.

To summarize, the use of partial wave dispersion relations is a most useful technique, both as a practical aid in phase-shift analysis and in clarifying which exchange interactions are responsible for pion–nucleon interactions in the $\lesssim 1$ GeV incident energy range.

## 5.13. Bounds on the scattering amplitude

### 5.13.1. *Froissart bound*

In order to determine the maximum number of subtractions required to make a dispersion relation convergent, we must obtain bounds on the asymptotic behaviour of the scattering amplitudes at high energies. The first, and most famous, of these bounds was established by Froissart [46] in 1961. He showed that, by assuming the analyticity of the Mandelstam representation and the requirements of unitarity, we could obtain a bound on the forward scattering amplitude and the total cross-section.

The Froissart bound on $F(s, t)$ is

$$|F(s, 0)| < C_0 s \log^2(s/s_0). \tag{5.119}$$

The corresponding bound on the total cross-section is (via the optical theorem)

$$\sigma < C \log^2(s/s_0). \tag{5.120}$$

Crudely speaking, such bounds occur because the analyticity and unitary requirements allow us to deduce that the nuclear interaction exhibits a finite range (i.e. falls off exponentially), while the probability for an interaction within this range is bounded say, by a term of the form $s^N$.

With the Froissart bound only two subtractions are required in forward dispersion relations for convergence. This can be seen, as follows, for the two-body interactions where particle (1) is incident on particle (2) at rest in laboratory system. We have from eqn (5.4) (as $s \to \infty$)

$$s \to 2m_2^l \omega_1^l, \tag{5.121}$$

where $\omega_1^l$ is the total energy of particle (1) in the laboratory system. We then have for the integral term in eqn (5.42), which is a singly subtracted dispersion relation,

$$\frac{\mathrm{P}}{4\pi^2}\int_{-\infty}^{\infty} \frac{k'\sigma(\omega')\,\mathrm{d}\omega'}{(\omega'-\omega)(\omega'-\omega_{0,1})}, \tag{5.122}$$

Inserting the Froissart bound, we obtain

$$\frac{\mathrm{P}}{4\pi^2}\int_{-\infty}^{\infty} \frac{k'C\log^2\left(\dfrac{\omega'}{\omega_0}\right)\mathrm{d}\omega'}{(\omega'-\omega)(\omega'-\omega_{0,1})}, \tag{5.123}$$

which diverges. However, for the doubly subtracted forward dispersion relations, eqn (5.43), we obtain for the integral term, after inserting the Froissart bound,

$$\frac{\mathrm{P}}{4\pi^2}\int_{-\infty}^{\infty} \frac{k'C\log^2\left(\dfrac{\omega'}{\omega_0}\right)\mathrm{d}\omega'}{(\omega'-\omega)(\omega'-\omega_{0,1})(\omega'-\omega_{0,2})},$$

which converges well $\left(\text{approximately like } \dfrac{1}{\omega}\right)$. At this time it is worth clarifying a common point of confusion. Equation (5.76) for the singly subtracted $D^+$ converges well when the cross-sections follow the Froissart bound and, in fact, has the degree of convergence corresponding to two subtractions. This is a special effect due to the properties of the crossing symmetry relations for $\pi$–N, which relate the negative energies for $\pi^- + \mathrm{p}$

scattering to the positive energies for $\pi^+ + \mathrm{p}$ scattering. These relations have the effect of increasing the effective convergence of the resultant combined integral from 0 to $\infty$. Hence, when we talk generally about the number of subtractions required to obtain convergence of forward dispersion relations with a particular bound, such as the Froissart bound, we talk of two subtractions as necessary (i.e. counting the number of additional individual powers of $\omega'$ introduced into the denominator) and, in general, each subtraction introduces one additional power of $\omega'$. However, for the $\pi$–p case, due to crossing symmetry we have introduced in $D^+$ two additional powers of $\omega'$ in the denominator by performing only one subtraction. Similarly, the unsubtracted $D^-$ has the equivalent of one subtraction due to the crossing symmetry effects.

### 5.13.2. *The Greenberg–Low bound*

Until recently, the maximum analyticity domain established from axiomatic field theory was the Lehman ellipse. Using this considerably smaller analyticity domain than the Mandelstam representation, Greenberg and Low [47] derived the Greenberg–Low bound which had one more power of the energy

$$|F(s, 0)| < C_0 s^2 \log^2 s.$$

Hence,

$$\sigma_{\text{total}} < Cs \log^2 s.$$

Thus the Greenberg–Low bound would require three subtractions. This would mean, for example, that two subtractions in $D^+$ would be required for convergence, instead of the single subtraction required with the Froissart bound.

The Greenberg–Low bound was subsequently improved by Martin and Eden [48], so that two subtractions were sufficient to assure validity of the forward dispersion relations. They obtained

$$|F(s, 0)| < \frac{C_0 s^2}{\log^2 s},$$

$$\sigma < \frac{Cs}{\log^2 s}.$$

### 5.13.3. *Establishment of the Froissart bound*

With the enlarged domain of analyticity which Martin [39] obtained, he was able, by also requiring that unitarity be satisfied, to demonstrate the validity of the Froissart bound. The derivation is based on the fact that the domain of convergence of the partial wave series for the imaginary part of the amplitude is, for fixed $s$, the extended Lehman ellipse domain established by Martin. The partial wave series is

$$\operatorname{Im} F(s, \cos\theta) = \frac{8\pi s^{\frac{1}{2}}}{k}\sum_{0}^{\infty}(2l+1)\operatorname{Im} f_l(s)P_l(\cos\theta). \tag{5.124}$$

Within this domain of convergence, unitarity limits the value of each partial wave, since unitarity requires that

$$0 \leq |f_l(s)|^2 \leq \operatorname{Im} f_l(s) \leq 1. \tag{5.125}$$

From axiomatic field theory [10] we have polynomial boundedness, as $s \to \infty$,

$$F(s, t) < s^N, \tag{5.126}$$

when $|t| <$ constant.† Thus, since all terms are positive, each term must be less than $s^N$, as $s \to \infty$,

$$\frac{s^{\frac{1}{2}}}{k}(2l+1)\operatorname{Im} f_l(s)P_l(\cos\theta) < s^N, \tag{5.127}$$

where the ellipse with focii at $\pm 1$ extends to a positive point which, for the case of equal masses, is

$$\cos\theta = 1+\frac{2t_0}{s-4m^2} \tag{5.128}$$

and $t_0 = 4m^2$ is the nearest threshold. Therefore, putting

$$t_0' < t_0, \tag{5.129}$$

we obtain

$$(2l+1)\operatorname{Im} f_l(s)P_l\left(1+\frac{t_0'}{2k^2}\right) < s^N. \tag{5.130}$$

For $l$ large and $x > 1$,

$$P_l(x) > \frac{C}{\sqrt{2l+1}}(1+(2x-2)^{\frac{1}{2}})^l. \tag{5.131}$$

Thus

$$\operatorname{Im} f_l(s) < \frac{C_1 s^N(1+(t_0'/k^2)^{\frac{1}{2}})^{-l}}{\sqrt{(2l+1)}}. \tag{5.132}$$

Hence, as $s \to \infty$,

$$\operatorname{Im} f_l(s) < C_2 \exp(-2l(t_0/s)^{\frac{1}{2}})+N\log s. \tag{5.133}$$

† This constant is $4\mu^2$ for the case of $\pi$–$\pi$ scattering.

Thus, from eqn (5.125) and eqn (5.133) it is clear that both the real and imaginary parts of $f_l(s)$ suffer an exponential decrease with increasing $l$, and for $l > Cs^{\frac{1}{2}} \log s$ they become negligible for large $s$. For $0 < |t| < t_0$ we can neglect the series terms for which $l > Cs^{\frac{1}{2}} \log s$, and this is also true for $\cos \theta$ inside the analyticity domain discussed.

Using eqn (5.125) we can obtain an analogous result for the invariant amplitude, namely,

$$F(s, t) \rightarrow \frac{8\pi s^{\frac{1}{2}}}{k} \sum_{l=0}^{L} (2l+1) f_l(s) P_l(\cos \theta), \tag{5.134}$$

as $s \rightarrow \infty$, for $-1 \leq \cos \theta \leq +1$ (the physical region).

From eqn (5.125) and eqn (5.134),

$$\operatorname{Im} F(s, 0) < \frac{8\pi s^{\frac{1}{2}}}{k} \sum_{0}^{Cs^{\frac{1}{2}} \log s} (2l+1) \tag{5.135}$$

or $\sim 2(Cs^{\frac{1}{2}} \log s)^2$ as $s \rightarrow \infty$ but, by the optical theorem,

$$\sigma_{\text{total}} = \frac{1}{2ks^{\frac{1}{2}}} \operatorname{Im} F(s, 0). \tag{5.136}$$

Hence

$$\sigma_{\text{total}} < \frac{1}{2ks^{\frac{1}{2}}} \times 2C^2 s \log^2 s, \tag{5.137}$$

$$\sigma_{\text{total}} < \text{constant} \times \log^2 s, \tag{5.138}$$

which is just the Froissart bound.

Roughly speaking, the bound arises from the existence of the domain of analyticity giving the interaction a definite range in terms of the number of partial waves. It includes those up to

$$l < Cs^{\frac{1}{2}} \log s. \tag{5.139}$$

Now, since the impact parameter range included is such that, for large $s$,

$$s^{\frac{1}{2}} b \propto l, \tag{5.140a}$$

$$b \propto \frac{l}{s^{\frac{1}{2}}} = C \log s. \tag{5.140b}$$

If the bound is considered to be given by the condition that all partial waves are considered absorbed in the circle $\pi b^2$, we then obtain

$$\sigma < \text{Constant} \times \log^2 s, \tag{5.141}$$

which is the Froissart bound. As we shall see in Chapter 13, the highest energy measured total cross-sections appear to be varying slowly at high energies and approaching constants. Thus they seem to satisfy the Froissart bound closely. It is interesting that the Froissart bound gives a result close to the intuitive expectation of a constant cross-section at infinity, but we should be careful not to carry the analogy too far. However, the bound is now seen to be rigorously derivable from the basic principles of axiomatic field theory and unitarity.

So far we have been considering bounds on collisions involving bosons of zero spin. However, these bounds have been generalized to cases involving particles with spin. The result is that the Froissart bound also applies for particles with spin (Yamamoto [49] and Kornille [50]).

### 5.13.4. *Finite angle scattering bounds*

Let us first consider the upper bound. Kinoshita, Loeffel, and Martin [51] assumed the Mandelstam representation, and obtained the bound

$$|F(s, t)| < Cs(\log s)^{\frac{3}{2}}, \quad \text{for fixed } t < 0. \tag{5.142}$$

Thus only two subtractions are required for convergent dispersion relations. For $t > 0$, we are in the unphysical region. Jin and Martin [52] demonstrated from the Froissart bound that

$$|F(s, t)| < s^{2-\epsilon}, \quad \text{as } s \to \infty, \tag{5.143}$$

with $\epsilon > 0$ for $t < 4\mu^2$. Thus only two subtractions are required for convergent dispersion relations in this region.

In relation to forward amplitude lower bounds, Jin and Martin [52] and Sugawara [53] have shown that

$$|F(s, 0)| > \frac{1}{s^{2+\epsilon}}, \quad \text{with } \epsilon > 0, \quad \text{as } s \to \infty, \tag{5.144}$$

if $F$ does not oscillate. If $F$ is oscillating then the least upper bound of $F$ will satisfy the above. We cannot use the optical theorem to obtain a limit on the total cross-section as $s \to \infty$, since $F(s, 0)$ may be purely real in this limit. Therefore, obtaining the bound from the elastic cross-section must be considered from the point of view of axiomatic field theory [52].

Using the narrowest allowed forward peak, it can be shown that

$$\sigma_{\text{total}} > \sigma_{\text{elastic}} > \frac{1}{s^6 \log^2 s}. \tag{5.145}$$

Sugawara [53] showed that, if $F(s, 0)$ at threshold is sufficiently small and

$|F(s, t = 0)| > \text{constant} > 0$, this bound could be much improved, and obtained

$$\sigma_{\text{total}} > \frac{C}{s^2 \log^2 s}.$$

A summary of bounds, the assumptions of their derivation, and references is contained in a table from a paper by Martin (Berkeley Conference (1966)) and in reference [20], Chapter 6.

## REFERENCES

1. EINSTEIN, A. (1923). *Geometry and experiment*. (Out of print). See also reference [31].
2. KRONIG. (1946). *Physica*. **12,** 543.
3. KRONIG. (1926). *J. opt. Soc. Am.* **12,** 547; KRAMERS, H. A. (1927). *Atti del congress internationale de Fisica convention.*
4. GOLDBERGER, M. and WATSON, K. (1964). *Collision theory*, Wiley, New York.
5. BOGOLIUBOV, N. N. and SHIRKOV, D. V. (1959). *Introduction to the theory of quantised fields*. Interscience, New York.
6. GOLDBERGER, M. L. (1955). *Phys. Rev.* **99,** 979; GOLDBERGER, M. L., MIYAZAWA, H., and OEHME, R. (1955). *Phys. Rev.* **99,** 986.
7. KARPLUS, R. and RUDERMAN, M. (1955). *Phys. Rev.* **98,** 771.
8. BOGOLIUBOV, N. N. and VLADIMIROV, V. S. (1958). *Izv. Akad Nauk SSSR*. **22,** 15.
9. SYMANZIK, K. (1957). *Phys. Rev.* **105,** 743.
10. HEPP, K. (1964). *Helv. phys. Acta*. **37,** 639.
11. ANDERSON, H. L., DAVIDON, W., and KRUSE, U. (1956). *Proceedings VIth Annual Conference High Energy Physics*. Interscience Publishing Inc., New York; and (1955). *Phys. Rev.* **100,** 339.
12. KARPLUS, R. and RUDERMAN, M. (1955). *Phys. Rev.* **98,** 771.
13. OEHME, R. (1955). *Phys. Rev.* **100,** 1503; (1956). **102,** 1174. SALAM, A. (1956). *Nuovo Cim.* **3,** 424.
14. DAVIDON, W. C. and GOLDBERGER, M. L. (1956). *Phys. Rev.* **104,** 1119.
15. LINDENBAUM, S. J. and STERNHEIMER, R. M. (1958). *Phys. Rev.* **110,** 1174.
16. HABER-SCHAIM, U. (1956). *Phys. Rev.* **104,** 1113.
17. SCHNITZER, H. J. and SALZMAN, G. (1959). *Phys. Rev.* **113,** 1153.
18. HAMILTON, J. and WOOLCOCK, W. S. (1963). *Rev. mod Phys.* **35,** 737.

19*a*. LINDENBAUM, S. J. (1964). *Rapporteur's Report, Proceedings International Conference on High Energy Physics, Dubna.*

19*b*. LINDENBAUM, S. J. (1965). *Plenary session review lecture, Oxford International Conference on Elementary Particles.*

20. EDEN, R. J. (1967). *High energy collisions of elementary particles*. Cambridge University Press, London.

21*a*. FOLEY, K. J., LINDENBAUM, S. J., LOVE, W. A., OZAKI, S., RUSSELL, J. J., and YUAN, L. C. L. (1963). *Phys. Rev. Lett.* **10,** 376, 543; (1963). **11,** 425, 503.

21*b*. FOLEY, K. J., GILMORE, R. S., JONES, R. S., LINDENBAUM, S. J., LOVE, W. A.,

OZAKI, S., WILLEN, E. H., YAMADA, R., and YUAN, L. C. L. (1965). *Phys. Rev. Lett.* **15,** 45.
22. BETHE, H. A. (1958). *Ann. Phys.* **3,** 190.
23. SOLOVIEV, L. D. (1965). *Zh. eksp. teor. Fiz.* **49,** 292.
24. RIX, J. and THALER, R. M. (1966). *Phys. Rev.* **154,** 1357.
25. ISLAM, M. M. (1967). *Phys. Rev.* **162,** 1426.
26. LOCHER, M. P. (1967). *Nucl. Phys.* **B2,** 525.
27. WEST, G. B. and YENNIE, D. R. (1968). *Phys. Rev.* **172,** 1413.
28. FOLEY, K. J., GILMORE, R. S., JONES, R. S., LINDENBAUM, S. J., LOVE, W. A., OZAKI, S., WILLEN, E. H., YAMADA, R., and YUAN, L. C. L. (1965). *Phys. Rev. Lett.* **14,** 862; (1964). *Proceedings of the International Conference on High Energy Physics, Dubna.*
29*a*. FOLEY, K. J., JONES, R. S., LINDENBAUM, S. J., LOVE, W. A., OZAKI, S., PLATNER, E. D., QUARLES, C. A., and WILLEN, E. H. (1967). *Phys. Rev. Lett.* **19,** 193; (1969). *Phys. Rev.* **181,** 1775; *b* (1967. *Phys. Rev. Lett.* **19,** 330.
30. OEHME, R. *a* (1955). *Phys. Rev.* **100,** 1503; *b Festchrift for Gregor Wenzel.* (1970). Chicago University Press, Chicago. In this latter paper an updated discussion of the entire subject is presented.
31. BLOKHINTSEV, D. I. (1966). *Soviet Phys. Usp.* **9,** 405.
32. GOLDBERGER, M. L., NAMBU, Y., and OEHME, R. (1957). *Ann. Phys.* **2,** 226.
33. FOLEY, K. J., JONES, R. S., LINDENBAUM, S. J., LOVE, W. A., OZAKI, S., PLATNER, E. D., QUARLES, C. A., and WILLEN, E. H. (1967). *Phys. Rev. Lett.* **19,** 857.
34. LINDENBAUM, S. J. (1969). *Pion–nucleon scattering.* Wiley Interscience, New York.
35. SÖDING, P. (1964). *Phys. Lett.* **8,** 285.
36. LEVINTOV, I. I. and ADELSON VELSKY, G. M. (1964). *Phys. Lett.* **13,** 185.
37. MANDELSTAM, S. (1958). *Phys. Rev.* **112,** 1344; (1959). **115,** 1741, 1752.
38. LEHMANN, H. (1958). *Nuovo Cim.* **10,** 579.
39. MARTIN, A. (1966). *Nuovo Cim.* **42,** 930; (1966). **44,** 1219; *CERN Report* TH702.
40. SOMMER, G. (1964). *CERN pre-print 64/1591/5-Th 731.*
41. KIBBLE, T. W. (1961). *Phys. Rev.* **117,** 1159.
42. BROS, J., EPSTEIN, H., and GLASER, V. (1965). *Com. math. Phys.* **1,** 240.
43. MANDELSTAM, S. (1963). *Nuovo Cim.* **30,** 1127, 1148.
44. HAMILTON, J. (1967). *High energy physics*, Vol. I. (edited by Burhop). Academic Press, New York. p. 193–338.
45. DONNACHIE, A. and HAMILTON, J. (1964). *Phys. Rev.* **B133,** 1053.
46. FROISSART, M. (1961). *Phys. Rev.* **123,** 1053.
47. GREENBERG, W. and LOW, F. E. (1961). *Phys. Rev.* **124,** 2047.
48. MARTIN, A. (1964). *CERN Pre-print 64/1699/5, Th630 (NC 1966);* EDEN, R. J. (1966). *Phys. Lett.* **19,** 695.
49. YAMAMOTO, K. (1963). *Nuovo Cim.* **27,** 1277.
50. CORNILLE, H. (1964). *Nuovo Cim.* **31,** 1101.
51. KINOSHITA, T., LOEFFEL, J. J., AND MARTIN, A. (1964). *Phys. Rev.* **B135,** 1464.
52. JIN, Y. S. and MARTIN, A. (1964). *Phys. Rev.* **B135,** 1369 and **B135,** 1375.
53. SUGAWARA, M. (1965). *Phys. Rev. Lett.* **14,** 336.

# 6

# PION PRODUCTION

## 6.1. Nucleon–nucleon collisions near threshold

THE discovery of the pion in cosmic rays, by observing positive pions decay at rest in nuclear emulsion, (Chapter 3) led to its detection and investigation at cyclotrons operating in the several hundred MeV energy range.

Nucleon–nucleon collisions allow much clearer interpretation of the observed results and are, therefore, of much greater value than nucleon–nucleus collisions. For a nucleon–nucleon collision a straightforward application of kinematics shows that the threshold energy for single pion production would correspond to $s = (2M+\mu)^2$, where $M$ is the proton mass and $\mu$ is the pion mass. Thus

$$\begin{aligned} s = (2M+\mu)^2 &= M^2+M^2+2M(M+\text{K.E.}_{\cdot\text{lab}}) \\ = 4M^2+\mu^2+4M\mu &= 2M^2+2M^2+2M\text{K.E.}_{\cdot\text{lab}}. \end{aligned} \tag{6.1}$$

Hence, $\text{K.E.}_{\cdot\text{lab}}$ (the kinetic energy of the incident proton in the laboratory system) is

$$\text{K.E.}_{\cdot\text{lab}} = \frac{\mu^2}{2M}+2\mu. \tag{6.2}$$

The ratio of the second term to the first is about $\dfrac{4M}{\mu} \approx 25$ and, hence, the first term is negligible. Thus the threshold is, approximately, at a laboratory system kinetic energy of $2\mu$. For charged pion production this threshold is 290 MeV and, thus, was reachable by the Chicago ($\sim$ 440 MeV), Columbia ($\sim$ 385 MeV), converted Berkeley ($>$400 MeV), and Carnegie Tech. ($>$400 MeV) F.M. proton cyclotron beams.

Double pion production would have a threshold of

$$\begin{aligned} s = (2M+2\mu)^2 &= M^2+M^2+2M(M+\text{K.E.}_{\cdot\text{lab}}) \\ = 4M^2+4\mu^2+8M\mu &= 4M^2+2M\text{K.E.}_{\cdot\text{lab}} \end{aligned}$$

or,

$$(\text{K.E.})_{\text{lab}} = \frac{2\mu^2}{M}+4\mu \approx 600 \text{ MeV}, \tag{6.3}$$

which was an energy range attainable in the later Russian ($\approx$ 660 MeV) and European (CERN $\approx$ 600 MeV) F.M. cyclotrons and the proton synchrotons—Cosmotron (3 GeV), Bevatron (6 GeV), Birmingham (1·0 GeV), and Saclay (3·0 GeV).

p+p collisions are directly observable by allowing the incident protons to strike liquid hydrogen targets. n+p collisions require a deuteron (or heavier) target. Collisions with the neutron can be treated as quasi-free, with the proton being considered, in first approximation, as a spectator proton. The following is a list of all p–p, and n–p, single pion production reactions consistent with charge conservation,

$$
\begin{aligned}
&\text{(a)}\ \mathrm{p}+\mathrm{p} \rightarrow \mathrm{p}+\mathrm{n}+\pi^{+},\\
&\text{(b)}\ \mathrm{p}+\mathrm{p} \rightarrow \mathrm{p}+\mathrm{p}+\pi^{0},\\
&\text{(c)}\ \mathrm{p}+\mathrm{p} \rightarrow \mathrm{D}+\pi^{+},\\
&\text{(d)}\ \mathrm{p}+\mathrm{n} \rightarrow 2\mathrm{n}+\pi^{+},\\
&\text{(e)}\ \mathrm{p}+\mathrm{n} \rightarrow \mathrm{p}+\mathrm{p}+\pi^{-},\\
&\text{(f)}\ \mathrm{p}+\mathrm{n} \rightarrow \mathrm{p}+\mathrm{n}+\pi^{0},\\
&\text{(g)}\ \mathrm{p}+\mathrm{n} \rightarrow \mathrm{D}+\pi^{0}.
\end{aligned}
\tag{6.4}
$$

They have all been observed [1*a*]. The corresponding n–n reactions are related to (a), (b), and (c) by the principle of charge symmetry. Reactions (c) and (g), involving deuteron formation, dominate pion production near threshold ($\sim$ 300–400 MeV). This is due to the large final state interaction of the two slow nucleons (*S*-state) in the deuteron state, when the meson is produced near threshold. The relatively large cross-sections for these reactions and their energy behaviour near threshold were explained on the basis of deuteron formation [2], [3]. The dominance of the deuteron final state, for the two nucleons, was shown by a large peak in the $\pi^+$ spectrum (in (c)) corresponding to the mass of the deuterons for the recoiling system. The deuteron has also been detected directly.

Deuteron formation would be expected, and is observed, to become much less important as the energy increases [4]. Above threshold the production of pions increases rapidly with increasing c.m.s. momentum, due to increased phase-space for production, and the apparent dominance of *p*-wave production. In the period when these investigations were performed, there were a number of phenomenological analyses made [2], [3] which explained the data primarily in terms of the final 2-nucleon state enhancement effects, and dominant *p*-wave and, to some extent, *s*-wave production.

The true nature of the interactions are obscured at these low energies just above threshold by a combination of threshold effects, which give large increases of cross-sections with increasing available pion c.m.s. momentum and the final state enhancement effects. Nevertheless, it was clear that *p*-wave pion production was favoured, except at very low energies, where the relatively small *s*-wave production became important. The results primarily

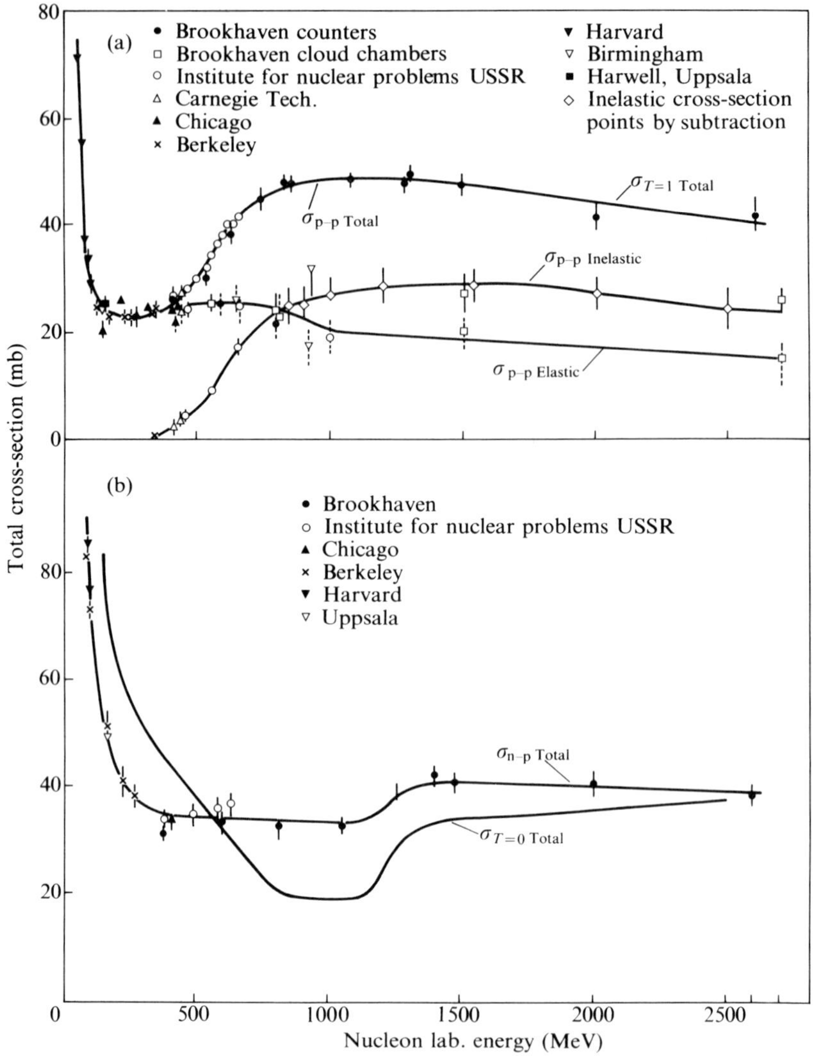


FIG. 6.1.(a) The p–p total, the elastic, and the inelastic cross-section as a function of energy. All curves are empirical. The open diamond points have been obtained by a subtraction of the elastic cross-section curve from the total cross-section curve, and estimated errors are attached. In the region beyond 300–400 MeV, the errors on elastic points have been made by broken lines so that they may be distinguished from inelastic or total cross-sections. Since p–p contains only the $T = 1$ state, the cross-sections are equal to the corresponding ones for the $T = 1$ state (Lindenbaum. (1957). *A. Rev. nucl. Sci.* **7**, 317).

FIG. 6.1.(b) The total n–p cross-section is shown as a function of energy. The $T = 0$ total cross-section deduced from the total p–p, and total n–p, is shown also. All curves are empirical. (Lindenbaum. (1957). *A. Rev. nucl. Sci.* **7**, 317).

followed what we would expect for near-threshold production of a pseudoscalar pion, with gradient coupling to the nucleon. The resultant cross-section for pion production, for example, in p+p collisions ($\sigma_{\mathrm{p-p\,inelastic}}$) increased rapidly with increasing energy above threshold, until about 800 MeV (as shown in Fig. 6.1), where the inelastic cross-section of about 25 mb appeared to become saturated. This phenomena will be discussed much more completely in § 6.4.

We should note, at this time, that pion production by protons incident on heavy nuclei can occur below the nucleon–nucleon threshold given above. This can be understood, since a lower threshold corresponds to interactions with a group of nucleons, the threshold decreasing as the total mass of the conglomerate increases. It is easy to see that, as the number of nucleons in the nucleus becomes large, the threshold approaches approximately $\mu$, instead of about $2\mu$.

Analyses of pion production by an incident nucleon, in nuclei, have been understood on the basis of the Fermi gas model, in which the production process is pictured as a single nucleon–nucleon encounter, or a series of such encounters, inside the nucleus. However, the individual target nucleons have a momentum distribution corresponding to a Fermi gas at low temperatures. This is, essentially, the effect of the Pauli exclusion principle for nucleons constrained to the limited nuclear volume resulting from the short-range strong interaction. The exact momentum distribution is different for different models such as, for example, the shell model. However, in all models where individual nucleon–nucleon collisions are considered, the nucleons in the target nucleus must have a ground-state momentum distribution. This internal momentum of the nucleus drastically increases the effective c.m.s. energy available in the nucleon–nucleon system for those cases where the internal momentum is opposite to the incident momentum (and, similarly, decreases the available c.m.s. energy when the two momenta are in the same direction). Thus the effective threshold is considerably lowered in heavy nuclei. We should be careful, in applying these internal momentum considerations, to realize that these procedures are only approximate, especially in regard to high-momentum tails of the distribution, and that, under no circumstances, can the absolute threshold be improved.

The kinetic energy threshold for a nucleon incident on a target nucleus (A) is given by

$$\approx \mu-(M_{\mathrm{A}}+M_{\mathrm{nucleon}}-M'_{(\mathrm{A}+1)})+\frac{p^2}{2M'_{(\mathrm{A}+1)}}, \tag{6.5}$$

where $M'_{(\mathrm{A}+1)}$ is the lowest attainable mass state of the resulting A+1 nucleons (and $c = 1$), and $p$ is the incident momentum in the laboratory system.

In treating pion production in nuclei, the possible effects of several collisions inside the nucleus of the incident nucleon, target recoil nucleon, and the produced pions must also be taken into account. These effects are small in light nuclei such as beryllium and carbon, where the average mean free path of pions and nucleons inside the nucleus are of the order of the nuclear radius or larger and, hence, these nuclei are reasonably transparent. On the other hand, in heavy nuclei, these secondary interaction effects become quite important. The analysis of investigations of pion production at low energies can be understood in terms of internal momentum distributions and, in fact, is good evidence for determining internal momentum distributions.

In early investigations [5], it was shown that the predicted energy distribution of positive and negative mesons, emitted at 90°, is quite sensitive to the internal momentum distribution of the target nucleon.

Three momentum distributions were considered. (1) The Fermi degenerate gas model which has a fixed upper momentum limit, (2) the Gaussian, and (3) the Chew–Goldberger distribution, which has a higher momentum tail than the Gaussian distribution. It was found that the Gaussian distribution gave the best fit to the data.

### 6.1.1. *High-energy pions produced by cosmic rays*

Production of pions by cosmic ray nucleons with energies greater than, or equal to, several hundred MeV had been studied for some time. A number of simple theories, which should perhaps, more appropriately, be referred to as models, had been proposed for their explanation.

Heisenberg [8] (1949) considered high-energy pion production as similar to the generation of turbulence in a liquid.

One of the early models (by Lewis, Oppenheimer, and Wouthuyzen) [6] used symmetric scalar, or pseudoscalar, theory, with pseudovector coupling, to describe the meson fields. Pion production was considered as a radiation due to the change of the pion fields in the collision. The changes in the field leading to the radiation were attributed to changes in isotopic spin direction and, in the pseudoscalar case, the change in nucleon (angular momentum) spin. For a review of these, and improved and related theories, see reference [7].

## 6.2. The statistical theories

The most frequently applied of the early theories, and the one of most subsequent interest, was the statistical theory proposed in 1950 by Fermi [9].

The basic idea was that, for the strongly interacting particles, the interaction is considered strong enough that, when two high energy particles collide, all the kinetic energy of the two incident particles (nucleons) is localized in a state of statistical equilibrium in a characteristic volume ($\Omega$), which is approximately the volume extent of the meson clouds surrounding the nucleons.

An observer in the centre of mass system would conclude that the meson cloud of each of the incident particles would have the shape of a Lorentz contracted sphere. That is to say, the cross-section of the 'nucleon' normal to the direction of motion, would be, approximately, a circle of radius equal to the pion Compton wavelength $\frac{\hbar}{\mu c}\left(=\frac{1}{\mu}\right)$. However, along the direction of motion, the dimension would appear Lorentz-contracted by the factor $\gamma = \frac{E_{\text{cms}}}{2M}$, where $E_{\text{cms}}$ is the total energy of the two nucleons in the centre of mass system, and $M$ is proton mass. Thus, when the particles collide and penetrate each other, the total energy passes into a volume

$$\Omega \approx \frac{\Omega_0}{\gamma} = \frac{2M\Omega_0}{E_{\text{cms}}} = \frac{8\pi M}{3E_{\text{cms}}\mu^3}, \tag{6.6}$$

where $\Omega_0 = \frac{4}{3}\frac{\pi}{\mu^3}$ is the volume of a sphere of radius approximately equal to the Compton wavelength of a pion. The generalization to the case where the incident particle is different from the target particle is trivial. The strong interactions are then supposed to rapidly set up a statistical equilibrium between all possible strong interaction final states, which can be produced in many possible ways. The electromagnetic and weak interaction strengths are assumed to be too weak to affect the strong interaction statistical equilibrium. This equilibrium state is such that the final state particles can be thought of as a free particle gas in some large normalization volume ($V$). Thus the cross-section for each final state ($n$) will be proportional to the probability that the $n$ particles will be simultaneously contained in the volume. This factor can be expressed by $\left(\frac{\Omega}{V}\right)^{n-1}$. The exponent $n-1$ is due to the fact that the position of $\Omega$ inside $V$ is arbitrary, or equivalently, it can be seen to follow from dimensional reasons, since the original number of degrees of freedom of the final state ($3n$) is reduced to $3(n-1)$ by the requirements of three-momentum conservation. It is also clear that the cross-section for each final state will be proportional to the density (with respect to the total energy $E$)

of final states in phase-space. The relation between the cross-section and these factors can be more formally [10a] developed using the $S$-matrix theory which gives†

$$\sigma_{\Phi} = \frac{(2\pi)^2\omega'_1\omega'_2}{\sqrt{[(p'_1p'_2)^2 - m_1'^2m_2'^2]}} \sum_{f\epsilon\Phi} \delta^4(P_f - p'_1 - p'_2)\,|M_{fi}|^2, \tag{6.7}$$

where $\sigma_{\Phi}$ is the cross-section of the reaction leading to any state out of a set $\Phi$ of final states.

$P_f$ is the total four-momentum in the final state; $\sum$ represents a summation over discrete variables and integration over continuous variables. There is a uniquely defined initial state describing the collision of two elementary particles with masses $m'_k$ and four-momenta‡

$$p'_k = (\omega'_k, \mathbf{p}'_k).$$

The matrix elements $M_{fi}$ are related to the $S$-matrix elements by

$$S_{fi} = I_{fi}\text{ (unity operator)} + \mathrm{i}\,\delta^4(P_f - p'_1 - p'_2)M_{fi}. \tag{6.8}$$

The $M_{fi}$ with the appropriate invariant normalization are the usual $T$-matrix elements. The above formulae apply rigorously for any theoretical calculation, or model, which can be used to obtain the unknown $M_{fi}$. At the minimum we assume that the $M_{fi}$ have all the usually accepted invariance properties that correspond to the known strong-interaction conservation laws, such as energy, linear momentum, total angular momentum, total isotopic spin ($T$), $T_Z$ (electric charge conservation), total baryon number ($B$), and total strangeness ($S$) conservation.

In effect, the statistical theory described above allows us to make rather simple assumptions about the matrix elements.

To illustrate the general characteristics of the method, we neglect the effects of spin, charges, etc. and take into account only the conservation of total four-momentum. $\sigma_{\Phi}$ will be Lorentz invariant if

$$M_{fi} = \left(\frac{m'_1m'_2m_1m_2\ldots m_n}{\omega'_1\omega'_2\omega_1\omega_2\ldots\omega_n}\right)^{\frac{1}{2}} M'_{fi}, \tag{6.9}$$

where $m_1\ldots m_n$ and $\omega_1\ldots\omega_n$ are masses and energies, respectively, in the final state, and $M'_{fi}$ is a function only of Lorentz invariant quantities.

† Kretzchmar uses a different normalization, and different notation, in his $S$-matrix treatment than had been used previously. In order that the reader can conveniently refer to his article for further details we have followed his treatment here.

‡ The scalar product of the two four-vectors $p_1p_2$ is defined as $p_1p_2 = p_{10}p_{20} - \mathbf{p}_1\cdot\mathbf{p}_2$.

The Fermi assumption was that

$$|M'_{\mathrm{fi}}|^2 \propto (2\pi)^{-3(n-1)}\Omega^{n-1}. \tag{6.10}$$

The $M_{\mathrm{fi}}$ contains the phase-space integral and the particular weight factor for the transition. The statistical assumption is that the weight factors (i.e. $T$-matrix elements for each transition) are the same. The above assumptions allow a simple summation of eqn (6.7) and we obtain

$$\sigma(E, m_1, \ldots, m_n) \propto \left(\frac{\Omega}{V}\right)^{n-1}\left(\frac{V}{8\pi^3}\right)^{n-1} \rho(E, 0, m_1, \ldots, m_n)$$

$$\propto \Omega^{n-1}\rho(E, 0, m_1, \ldots, m_n), \tag{6.11}$$

where $V$ is the large arbitrary normalization volume referred to previously and which, as one would expect, cancels out. $\rho(E, 0, m_1, \ldots, m_n)$ stands for the phase-space integral. In the early work only the following form of phase-space integral was considered

$$\rho_{\mathrm{F}}(E, \mathbf{P}, m_1, \ldots, m_n) = \int \mathrm{d}\mathbf{p}_1 \int \mathrm{d}\mathbf{p}_n\, \delta\left(E - \sum_{i=1}^{n} \sqrt{(m_i^2 + \mathbf{p}_i^2)}\right) \times \delta^3\left(\boldsymbol{P} - \sum_{i=1}^{n} \mathbf{p}_i\right). \tag{6.12}$$

This corresponds to the situation $M_i \approx M'_i$, which is the non-relativistic limit.

More generally, [10*b*] Neuman and Sirvastava and Sudarshan demonstrated

$$\rho_{\mathrm{S}}(E, \boldsymbol{P}, m_1, \ldots, m_n) = \int \mathrm{d}\mathbf{p}_1 \int \mathrm{d}\mathbf{p}_n \frac{m_1 m_2 \ldots m_n}{\omega_1 \omega_2 \ldots \omega_n} \times$$

$$\times\, \delta\left(E - \sum_{i=1}^{n} \sqrt{(m_i^2 + \mathbf{p}_i^2)}\right) \delta^3\left(\boldsymbol{P} - \sum_{i=1}^{n} \mathbf{p}_i\right).$$

More detailed treatments are given by Kretzschmar [10*a*]. Modifications of the statistical theory are described in reference [10*c*].

The production of pions at cosmic ray energies seemed to be consistent, in general, with the Fermi statistical theory. Thus, it was generally anticipated that at the new multi-GeV laboratory accelerators (Cosmotron (3 GeV) and Bevatron (6 GeV)) the predictions of the Fermi statistical theory would be valid, and Fermi and others (Christian and Yang) made predictions for the characteristics of pion production to be expected in this energy range.

### 6.3. The experimentally observed production of pions of energy $\gtrsim$ 1 GeV

The nature of pion production in the multi-GeV region was revealed by a series of counter experiments [11] at the 3·0 GeV Cosmotron, which determined the energy spectra of charged pions from (p+Be) and (p+p).† The observations were at 32° in the laboratory system, and employed a counter telescope with internal magnetic deflection, to determine momentum. Time-of-flight was used to separate pions from other particles.

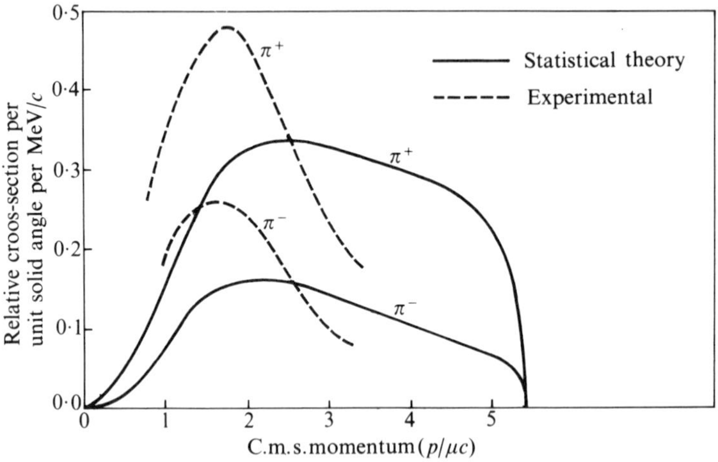


FIG. 6.2. Comparison of the experimental results for the production of pions in Be and H (positive pion spectrum only) with the statistical theory predictions. ([11]. (1956). *Phys. Rev.* **103**, 404.)

Figure 6.2 shows how the first observations at 2·3 GeV disagreed strikingly with the Fermi statistical theory predictions by exhibiting an unexpected low-energy peak. Measurements at 1·0 GeV showed an even greater discrepancy with the statistical theory predictions, since the low-energy peak was even more prominent experimentally. In trying to understand the discrepancy we can examine the single parameter in the statistical theory, namely, the volume. It is obvious from eqn (6.10) that varying the volume parameter can affect the multiplicity distribution, larger $\Omega$ emphasizing higher multiplicities, and smaller $\Omega$ emphasizing smaller multiplicities. Other than this effect on the multiplicity distribution, within a particular multiplicity, the momentum spectra and correlations (like $Q$-value distribution) are independent of $\Omega$. In fact, we can easily see that a momentum

† A poly-ethylene target and carbon target were alternated to determine the difference and, thus, the pion production in hydrogen.

spectrum, in the reaction channel yielding $m_1...m_n$, can be obtained by essentially leaving out one integration over momentum in $\rho(E, 0, m_1, ..., m_n)$. If two integrations over momentum are left out a $Q$-value distribution is obtained.

The multiplicity distribution deduced experimentally from these, and cloud chamber, experiments at 2·3 GeV [12*a*] and the fact that at 1 GeV, owing to the low available c.m.s. energy, we have almost exclusively single pion production, made it impossible to reconcile the statistical theory predictions with these experiments. The observed pion spectra were transformed to the nucleon–nucleon† centre of mass system and compared to the $\pi^+ + p$ total interaction cross-section, as a function of c.m.s. energy. It was noted by the author that the general similarity of the positive and negative pion energy spectra in the centre of mass system, produced by 1·0 and 2·3 GeV protons, and the $\pi^+ + p$ total interaction cross-section, is the most striking feature of these results, and suggests a very intimate relationship between the production and scattering processes. It was further reasoned that excitation of one, or both, nuclei to the $T = J = \frac{3}{2}$ state isobar was a likely explanation of the observed results, and the first crude version of the isobar model was proposed.

Diffusion chamber experiments with a 1·0–2·2 GeV incident neutron spectrum [12*b*] were able to observe the multiplicity distribution directly at these energies. The predictions of the statistical theory with a volume $\Omega(\approx h/\mu c)$ gave far too many single pion production events, and too few double, and triple, production events. An increase of $\Omega$, by a factor of 13, was required to obtain agreement with the experimental multiplicity distribution.

### 6.4. The isobar model

In 1957, Lindenbaum and Sternheimer [13] proposed the isobar model for pion production in nucleon–nucleon and pion–nucleon collisions. The basic assumption was that observed dominance of low energy pion–nucleon interactions by the $T = J = \frac{3}{2}$ isobaric state can also determine the characteristics of pion production. In particular, a series of resonant isobaric levels were presumed characteristic of the internal structure of the nucleon.

Hence, we could conceive of the pion production in a nucleon–nucleon collision as follows: As a result of a collision, there is a transfer of kinetic energy between nucleons, which raises one, or both, to one of their set of

† Assuming that the beryllium spectra represent, predominantly, the effects of individual nucleon–nucleon collisions. The observed similarity (within the errors) of the $\pi^+$ energy spectra for beryllium and hydrogen supports this assumption.

isobaric levels. The lifetime of the isobar is long enough† to allow them to separate before decay by pion emission and, therefore, final state interactions between the decay products of one isobar and the other nucleon, or isobar, in the collision are small. It was further initially assumed that the dominant isobar affecting pion production, in the 1·0 to 3·0 GeV incident energy range, is the $T = J = \frac{3}{2}$ state, which was the only identified isobar at the time. However, the model was shortly thereafter extended [14] to include the newly-discovered $T = \frac{1}{2}$ state isobars.

The isobars were characterized by a variable mass $m_{\mathrm{I}}$, which was just the total energy of the isobar decay products in the isobar centre of mass system. Production of a virtual isobar state, and its subsequent decay, was assumed to explain the pion–nucleon scattering and, thus, the production and scattering processes could be related.

It was shown that it was reasonable to expect that we could express

$$\mathrm{d}\sigma_{\mathrm{single}}/\mathrm{d}m_1 = \alpha_{\mathrm{s}} F(\bar{E}, m)\sigma(m) a_{\mathrm{s}}(\bar{\theta})\,\mathrm{d}\bar{\Omega}, \tag{6.13}$$

and

$$\frac{\mathrm{d}^2\sigma_{\mathrm{double}}}{\mathrm{d}m_1\,\mathrm{d}m_2} = \alpha_{\mathrm{d}} F(\bar{E}, m_1, m_2)\sigma(m_1)\sigma(m_2) a_{\mathrm{d}}(\bar{\theta})\,\mathrm{d}\bar{\Omega}, \tag{6.14}$$

where $F(\bar{E}, m_1)$ and $F(\bar{E}, m_1, m_2)$ are kinematical phase-space factors, which are determined by the detailed mechanism of isobar formation. The simplest assumption used is that they are the ordinary two-body phase-space factors, for forming one or two isobars of mass $m_1$ or $m_1$ and $m_2$, respectively. $a_{\mathrm{s}}(\theta)$ and $a_{\mathrm{d}}(\theta)$ were the angular distribution factors, which were also considered arbitrary, depending on the detailed mechanism of isobar formation. The two cases considered in the calculation were isotropic isobar production and decay, and forward and backward direction production and isotropic decay. A combination of these two cases was considered adequate to represent the available data. $\alpha_{\mathrm{s}}$ and $\alpha_{\mathrm{d}}$ were constants, the ratio of which determined the ratio of single to double pion production (in the case where only the $T = J = \frac{3}{2}$ isobar is considered). They could be estimated also from the observed cross-sections.

Some of the most important and interesting distributions, such as the

† The long enough lifetime assumption was recognized to be questionable, since from the observed width, the order of magnitude of the lifetime of the $T = J = \frac{3}{2}$ state, gave $\sim 10^{-23}$ s, which meant the isobar could still be partly within the interaction volume (radius $\sim 10^{-13}$ cm) at the time of decay. However, due to the considerable relative velocity of the two recoiling bodies in the interaction, the interaction of a decay product of one isobar with that of the other was expected, in general, to be small compared to the resonant interaction which characterizes isobar formation and decay. As the incident energy increases, the lifetime objection disappears due to the Lorentz time dilation factor.

c.m.s. energy distribution, are almost entirely determined by the isobar formation factors ($\sigma(m_1)$ and $\sigma(m_2)$), and are not influenced much by the other factors. Four-momenta and isotopic spin were rigorously conserved. The former determined the appropriate kinematics, and the latter determined the charge-state distribution.

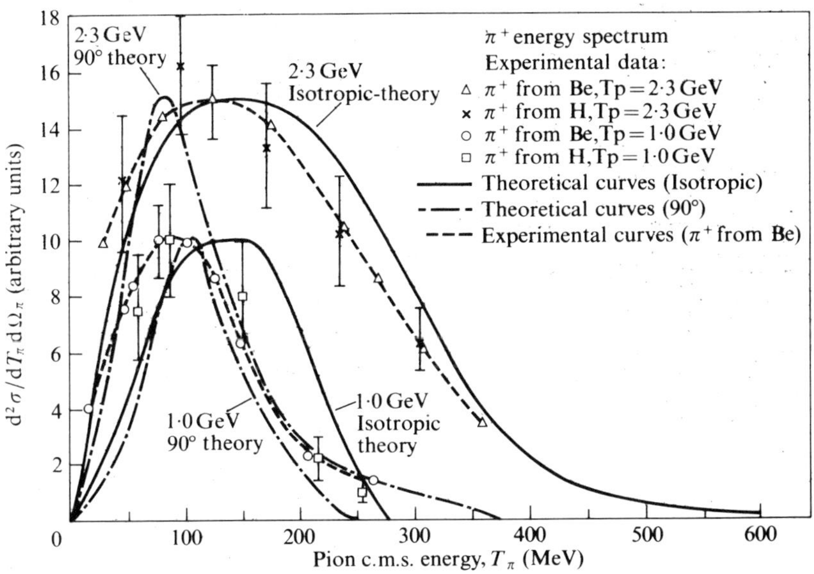


FIG. 6.3. The $\pi^+$ energy spectra from p–p and p–Be collisions at $T_p$ = 1·0 and 2·3 GeV [11] are given, and compared to calculations [13] based on an isobaric nucleon model (discussed in text). Results for isotropic, and only forward-and-backward, emission of isobars in the c.m.s. observed at ~90° (labelled 90° theory) are shown. The experimental data corresponds to pion emission angles in the c.m.s. of ~60° to 75° at 1·0 GeV and ~73° to 105° at 2·3 GeV. An analysis of the cloud chamber experiments [4] indicates isotropic pion emission at 2·3 GeV and a considerable amount of forward–backward peaking at 1·0 GeV which, therefore, implies reasonably good agreement of the calculations with the data. The cloud chamber data (not shown), although of considerably lower statistical accuracy and momentum resolution, is in general agreement with the counter data. (From Lindenbaum. (1957). *A. Rev. nucl. Sci.* **1**, 317.)

The isobar model fits the data on p–p and n–p pion production very well, and the sharp peaks introduced into the pion production by the $\sigma(m)$ factors were strikingly borne out by the appropriate peaks in the data.

Figure 6.3 shows that the pion energy spectra, previously shown to be in disagreement with the statistical theory, were well fitted by the isobar model.

The model was extended [14] to include pion–nucleon and nucleon–antinucleon collisions as well. The calculation of pion production in pion–nucleon interactions was also considered on the basis of exciting only nucleon isobars, since, at that time, the evidence for pion isobars was not clearly established. However, the possibility of later introducing these pion isobar effects was suggested. Considerable success in fitting the details of pion–nucleon interactions, in the range of 1·0 to 1·4 GeV, was attained with these limited calculations.

The general behaviour of the p–p and n–p total cross-section shown in Fig. 6.1 were also understood on the basis of the isobar model. The p–p cross-section, which represents the $T = 1$ cross-section identically, exhibits an inelastic cross-section, which starts to rise just above the single isobar production threshold, and levels off at about 800 MeV, where the probability for single isobar production has levelled off. This increase in the inelastic cross-section is reflected in the total cross-section behaviour.

This is understandable since a $T = \frac{3}{2}$ isobar and a $T = \frac{1}{2}$ recoil nucleon, can combine to form a $T = 1$ initial state. However, in the $T = 0$ state, a single $T = \frac{3}{2}$ isobar and a $T = \frac{1}{2}$ recoil nucleon cannot combine to form a $T = 0$ state, but two $T = \frac{3}{2}$ isobars can form a $T = 0$ state. Thus, single pion production is forbidden in a $T = 0$ state, and it is only at the double pion production threshold that we can expect to see a rise in the inelastic cross-section. As seen in Fig. 6.1, the $T = 0$ total cross-section exhibits its major rise where the double isobar production occurs with good probability. Diffusion chamber experiments, comparing n–p and p–p pion production, confirmed that the p–p pion production has a much higher single to double pion production ratio. It was also determined from counter experiments that the $T = 0$ inelastic cross-section is very low for single pion production, and high for double pion production [15].

The isobar model was extended to include the higher $T = \frac{1}{2}$ isobar states, which were postulated to decay either directly to the ground state via single pion decay, or via a series of cascade steps involving single pion emission between levels at each step. Thus, (starting from the continuum) up to four pions could be emitted from each nucleon. It was proposed that the isobar model could be further generalized to include pionic isobars, strange meson, and strange baryon, isobars. In a sense, this was the forerunner of complicated modern particle spectroscopy.

We have just reviewed the experimental observations which indicated the dominance of isobar production in pion production and the success of the isobar model in explaining the experimental results. There were also some attempts made to modify the statistical theory in various ways to incorporate the important facts connected with the dominance of isobar production. One way was to allow isobars, as a particle, to be created in the statistical process [16], another way was to attempt to consider final state interactions between the created particles [17]. Although these efforts were in the right direction, it became increasingly clear experimentally (as we shall see later in this chapter) that isobar production (including pion isobars) in the separate moving centres was a very important part of the process. Thus, more specific dynamical models, such as the isobar model and the subsequent one-pion exchange and one-meson exchange model were more relevant for explaining the detailed characteristics of the events, exhibiting clear evidence for isobar structure.

There have been a number of proposed variations and refinements of the isobar model [18*a*], [18*b*], and it has also been applied to cosmic ray phenomena [18*c*]. However, the great progress recently made in experimentally discovering resonances and the treatment of interactions via peripheral models have incorporated many of the features of the isobar model in them, within a framework which also allows detailed angular distributions and angular correlation predictions. We shall discuss these models in the subsequent section.

### 6.5. One-pion exchange model

In the isobar model, the detailed mechanism for production of the isobars was not postulated. Thus, the model was quite successful in determining energy spectra, charge ratios, etc., but contained no specific information for detailed prediction of the angular distribution of the isobars.

The introduction of the one-pion exchange model (OPEM) (See Ferrari and Selleri 1962) [19] provided a detailed mechanism for isobar production. The relationship of the one-pion exchange model to the isobar model is nicely explained in a review by Bertocchi and Ferrari [20] in which they state that the one-pion exchange model 'might be considered an isobar model with a specific mechanism of isobar production'. The review develops the interrelationships between the various models, and we shall follow some of their lines of reasoning in this section.

For simplicity, let us consider the following one-pion exchange diagram for single pion production in pion–nucleon collisions ($\pi+\mathrm{N}\rightarrow\pi'+\pi''+\mathrm{N}'$). Such diagrams were first considered in some early work by Chew and Low (1959) [21]. The authors were mainly concerned with extracting the $\pi$–$\pi$ elastic scattering cross-section from the observed inelastic interaction considered above. There are many Feynman diagrams which can contribute to the single production of an additional pion, by the pion incident on the nucleon. However, the particular diagram shown in Fig. 6.4 corresponds to the exchange of a single pion, which is, of course, the lightest possible strongly interacting particle exchange, and this contains a singularity in the variable $t$ nearest to the physical region. Hence, it might be expected to be the most important. Defining the four-momentum transfer squared to the nucleon by the usual Lorentz invariant $(-t)$, this, diagram has the singularity $\frac{1}{(\mu^2-t)}$, where $\mu$ is the mass of the exchanged particle (pion). In the physical region $t$ is always negative and, thus, the singularity occurs in the non-physical region when $t = \mu^2$. Since $\mu$ is smallest for the pion, this is the nearest singularity.

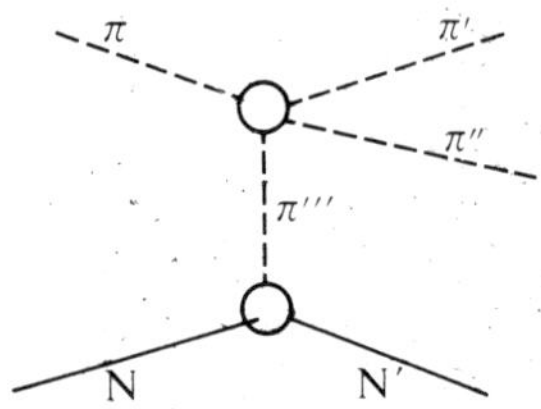

FIG. 6.4.

At the singularity in the non-physical region, we can neglect all other diagrams owing to the infinite denominator of this diagram and thus compute the process purely on the basis of this diagram. If $\mu^2$ is small enough, we can hope to extrapolate from the physical region to the pole. To correspond to the normally used notation for these problems, we shall employ $\Delta^2 = -t$, where $\Delta^2$ is the four-momentum transfer squared. The $T$-matrix element, is computable in the usual way, owing to the diagram. Following the Feynman rules, we can write

$$T = \frac{1}{(\mu^2+\Delta^2)} V(1)V(2), \tag{6.15}$$

where the $V(1)$ and $V(2)$ are the contributions from vertex (1) and vertex (2), respectively. Generally, these vertices can be represented by form factors (which are functions of $t$) which multiply the spinors, calculated from the Feynman graph of lowest order. We can avoid considering the detailed structure of the upper vertex, and represent it by matrix elements for pion–pion scattering

$$T_{\pi-\pi}(p_\pi, p_{\pi'''}, p_{\pi'}, p_{\pi''}), \tag{6.16}$$

where the various $p$ terms are the appropriate four-momenta. Thus,

$$T = G\frac{\{\bar{u}(\mathbf{p}_{N'})\gamma_5 u(\mathbf{p}_N)\}F(t)T_{\pi-\pi}F'(t)}{(\Delta^2+\mu^2)}, \tag{6.17}$$

where $G$ is the renormalized pion–nucleon coupling constant, $F(t)$ and $F'(t)$ are the relevant form factors for the proper vertex part ($\pi\pi$N) and the pion propagators. The definitions are such that as

$$\Delta^2 \to -\mu^2,$$

$$F(\Delta^2) \quad \text{and} \quad F'(\Delta^2) \to 1, \tag{6.18}$$

and the off-shell matrix element $T_{\pi-\pi}$ approaches the on-shell matrix element, describing a physical pion scattering from a physical pion target.

By employing the relationship between matrix element and cross-section, doing summation over nucleon spins, integrating over internal degrees of freedom of $\pi'$ and $\pi''$, and, finally, multiplying the resultant cross-section by $(\Delta^2+\mu^2)^2$, we obtain

$$\lim_{t\to\mu^2}\left\{(\Delta^2+\mu^2)^2\frac{d\sigma}{d\Delta^2}\right\} = \frac{1}{8\pi}\frac{1}{\Phi^2}\left(\frac{G^2}{4\pi}\right)\Delta^2 k\omega\sigma_1(\omega)\,d\omega^2, \tag{6.19}$$

where $\omega^2 = (p_{\pi'}+p_{\pi''})^2$ = mass squared of the final state two-pion system, $\sigma_1(\omega)$ is the $\pi$–$\pi$ scattering cross-section for a total c.m.s. energy $\omega$, $k$ is the c.m.s. momentum, which in the $\pi$–$\pi$ scattering case is

$$k = (\omega^2/4-\mu^2)^{\frac{1}{2}},$$

$\Phi$ is the invariant flux factor in the initial state. In the laboratory system

$$\Phi = \mathbf{p}_{\text{inc}} M \text{ (mass of target)}.$$

Thus, if we have enough statistics, we can select events contained in a small interval of the variable $\omega$. For such an interval, the right-hand side is a function of $t$ only. We can then plot the left-hand side as a function of $t$, going down to as close to $t = 0$ as is practical. Then we can (at least in principle) hope to extrapolate the left-hand side to the unphysical point $\Delta^2 = -\mu^2$. This, in the limit, gives the value of the right-hand side, for which all quantities other than $\sigma_1(\omega)$ are known and, thus, this quantity can be determined.

There have been a number of investigations (Anderson *et al.* (1961) [22], Carmony and Van de Walle (1962) [23]) of this extrapolation method. We illustrate (Fig. 6.5(a)) the work by Carmony and Van de Walle. They extrapolated the function†

$$\mathrm{G}(\Delta^2, s) = (1+\Delta^2)^2\frac{\partial^2\sigma}{\partial s\,\partial(\Delta^2)} \tag{6.20}$$

to $\Delta^2 = -1$.

† Note that $s = \omega^2$ and $\mu = 1$.

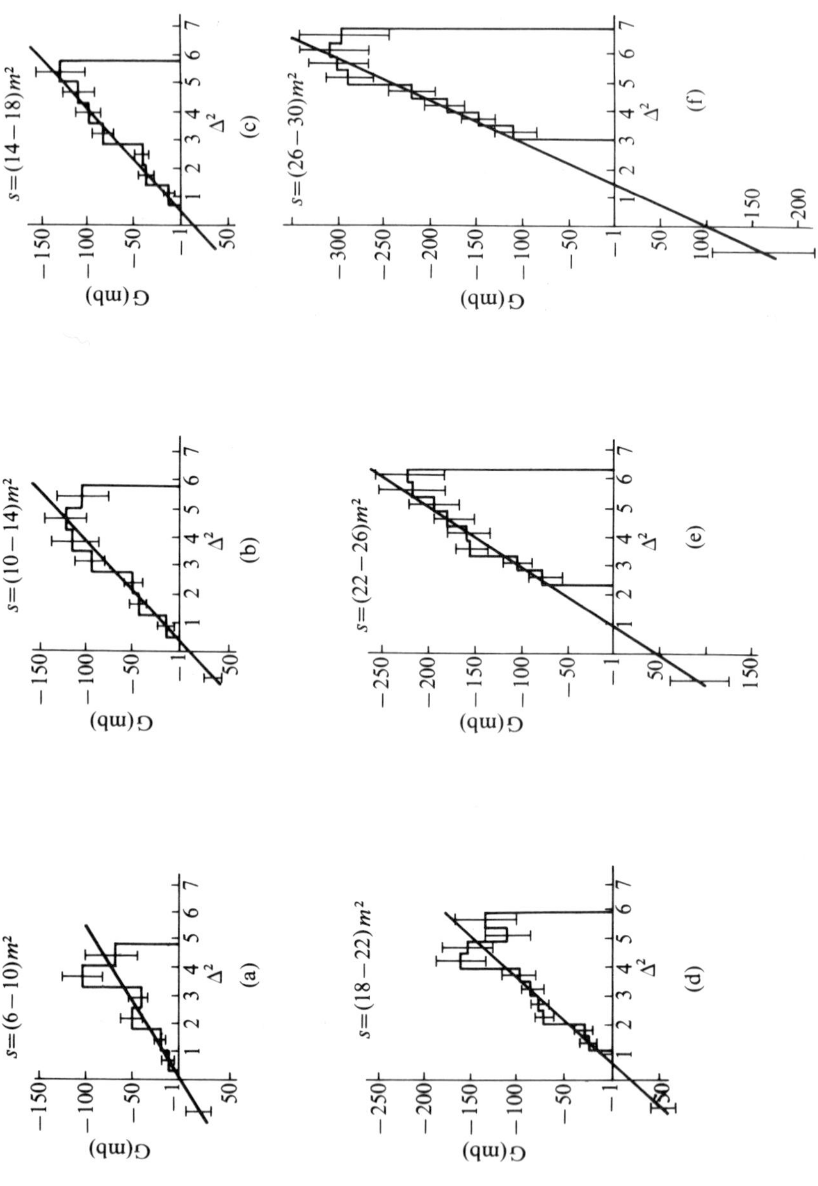


FIG. 6.5.(a) Extrapolation of $G(\Delta^2, s)$ to $\Delta^2 = -1$. (Carmony and Van de Walle. (1962). *Phys. Rev.* **127**, 959.)

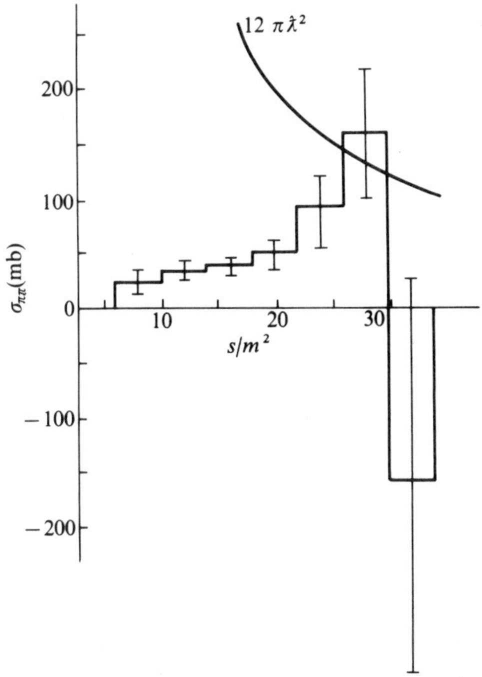


FIG. 6.5.(b) Total cross-section scattering for $\pi$ meson–$\pi$ meson. (Carmony and Van de Walle. (1962). *Phys. Rev.* **127,** 959.)

A number of measurements were determined at values of $s$, ranging from $6m^2$ to $30m^2$, $\Delta^2$ ranging from $+1$ to $+7$. The extrapolation procedure is somewhat uncertain and, thus, the extrapolated values show sizeable errors. From this extrapolated data we can deduce the behaviour of the $\pi$–$\pi$ total cross-section, which is shown in Fig. 6.5(b).

Although there are large errors in the procedure, especially at large values of $s$, it is interesting to note that the peak at $E \approx 725 \pm 25$ MeV, and width $\Gamma \approx 110$ MeV, corresponds to the $\rho$ meson, which we will discuss later. $12\pi\bar{\lambda}^2$ is the expected value of the $\pi$–$\pi$ cross-section at a $J = 1$ resonance, which is, thus, further evidence for the $\rho$-meson.

Original interest in the extrapolation procedure was later replaced with interest in applying this graph directly in the physical region via the so-called pole approximation (Goebel (1958) [24*a*], Drell (1960) [24*b*], Bonsignori and Selleri (1960) [25]. Salzman and Salzman (1960, 1961) [26]) which assumes that, if the pole is close to the physical region, then the major

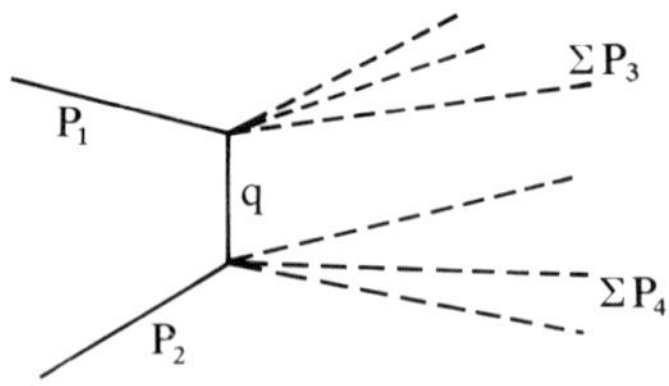


FIG. 6.6.

features of interaction in the physical region for small $\Delta^2$ are likely to be dominated by the $T$-matrix element derived from this diagram.

We can generalize the one-pion exchange graph, by allowing a number of particles to come out from each vertex. This generalized case can then be represented by Fig. 6.6. In Fig. 6.6. $\sum P_3$ and $\sum P_4$ represent the total four-momentum of the particles eminating from each vertex. We follow Bertocchi and Ferrari in this respect.

For this general case, the $T$-matrix will be

$$T = \frac{T_1(q, P_1, \sum P_3).\, T_2(-q, P_2, \sum P_4)}{\Delta^2 + \mu^2} F'(\Delta^2), \tag{6.21}$$

where $T_1$ and $T_2$ represent the interaction at the two respective vertices and $-q$ represents that the exchanged particle travelling from vertex 2 to 1 is equivalent to its anti-particle travelling from vertex 1 to vertex 2. This leads to the following expressions for the cross-section $\frac{d\sigma}{d\Delta^2}$, after integration over the internal degrees of freedom has been performed.

First, for the case containing one three-particle nucleon–nucleon pion vertex (as in Fig. 6.4), the result is

$$\frac{d\sigma}{d\Delta^2} = \frac{1}{8\pi}\frac{1}{\Phi^2}\left(\frac{G^2}{4\pi}\right)\frac{\Delta^2 F^2(\Delta^2) F'^2(\Delta^2) k(\Delta^2, \omega)}{(\Delta^2 + \mu^2)^2}\, \omega\sigma_1(\Delta^2, \omega)\, d\omega^2. \tag{6.22}$$

For the case of a graph containing two multi-particle vertices of mass $\omega_1$ and $\omega_2$, respectively, (as in Fig. 6.6) the result is

$$\frac{d\sigma}{d\Delta^2} = \frac{1}{16\pi^3}\frac{1}{\Phi^2}\frac{F'^2(\Delta^2)}{(\Delta^2 + \mu^2)^2} k_1(\Delta^2, \omega_1)\omega_1\sigma_1(\omega_1) k_2(\Delta^2, \omega_2)\omega_2\sigma_2(\omega_2)\, d\omega_1^2\, d\omega_2^2. \tag{6.23}$$

In the above formulae the fact that the exchanged meson q is off the mass shell, for physical region inelastic processes, has been allowed for in the presence of the form factors ($F$ and $F'$) and also the dependence of the cross-sections $\sigma_1$ and $\sigma_2$, and the flux factors $k_1\omega_1$, $k_2\omega_2$ on $\Delta^2$.

The 'pole approximation' consists in assuming that for small values $\Delta^2 \lesssim \mu^2$, these quantities can be represented by their on-shell value (i.e. $\Delta^2 \approx -\mu^2$). Thus, $\Delta^2$ can be put equal to $-\mu^2$ everywhere, except in the rapidly varying factor $\Delta^2(\Delta^2+\mu^2)^{-2}$. Thus, we obtain a formula like the Chew–Low formula in the physical region.

Bertocchi and Ferrari noticed the similarity of formulae (6.22) and (6.23) to the formulae for the isobar model, and concluded that the one-pion exchange model represented an improvement, rather than a replacement of, the isobar model, and could easily account for the success of the latter in describing pion–nucleon and nucleon–nucleon collisions. These authors further stated that if we integrate over $\Delta^2$, the characteristic dominance of low momentum transfer effects is removed, and the basic shape of the spectra is given by the cross-section function, and is not influenced much by the difference in the phase-space factors in the two cases. 'Therefore, the mass spectra look similar in the two cases. We can say that the one pion exchange model contains the isobar model; the same statement obviously holds for any model which explicitly considers the formation of resonances. In addition, the one pion exchange model predicts a very peculiar $\Delta^2$ dependence which appears to be in good agreement with experiment' (Bonsignori and Selleri 1960). To this we could perhaps add that the generality allowed in the transfer of four-momentum and quantum numbers, between the two excited isobaric centres in the isobar model (and its extensions), allows us to accommodate much more generalized exchanges, including multi-particle (both strange and non-strange) exchanges, but on the other hand, we lose the specific predictions of angular distributions and cross-sections obtained for the one-pion exchange model.

### 6.6. Pion production in nucleon–nucleon and pion–nucleon collisions

Following the early work, continuing experimental investigations established that pion production in the 1–30 GeV incident energy region in nucleon–nucleon and pion–nucleon collisions was, to a large extent, occurring by the formation of two-body intermediate states, each of which could be characterized by a series of isobaric levels or resonant states.

When an inelastic interaction can be characterized in the intermediate state by two well-defined bodies (or particles, isobars, etc.), we often refer to these inelastic processes as quasi-two-body processes. Furthermore, if there is no change in quantum numbers other than those characterized by the quantum numbers of the vacuum, those processes are often referred to as quasi-elastic and, as we shall see later, often behave like elastic scattering in their general characteristics. In addition to an increasing number of nucleon isobars in both the $T = \frac{1}{2}$ and $T = \frac{3}{2}$ states, which were identified, a considerable number of pionic isobars, or new heavier non-strange mesons, were discovered, and their quantum numbers deduced (see Particle Data Group tables).

### 6.6.1. *The 2π, J = 1 resonance (ρ-meson)*

Let us look at the data of Carmony and Van de Walle, in the physical region, and ignore the Chew–Low extrapolation procedure. When we plot the $\pi$–$\pi$ invariant mass quantity $(s/m^2)$ the results shown in Fig. 6.7 clearly indicate a peak at about $27m^2$. This corresponds to total energy of $725 \pm 25$ MeV with a $\Gamma \approx 100$ MeV. The general behaviour of $\pi$–$\pi$ interactions observed in the experiments is quite similar to that deduced by extrapolating to the pole. Thus, the pole approximation seems valid. If this peak is a resonance, we have evidence for its formation directly in the observable laboratory quantities. The work of other authors has previously treated evidence for such an interaction with analyses restricted to the physical region. The peak lies very close to the $12\pi\lambda\!\!\!^{-2}$ curve, which would correspond to the value of the cross-section at a resonance, in a state with $J = 1$. The spin determines the angular distribution and thus, using methods similar to those used in $\pi$–p scattering but allowing for the fact that both particles are spinless, we can demonstrate that [27]†

$$\frac{\mathrm{d}\sigma}{\mathrm{d}\Omega} = 4\pi\lambda\!\!\!^{-2}\left|\sum_j \sqrt{(2j+1)}\,\mathrm{e}^{\mathrm{i}\delta_j}\sin\delta_j Y_j^0(\theta,\phi)\right|^2. \tag{6.24}$$

The experimental angular distribution for the $\pi$–$\pi$ differential scattering was deduced from the data. The result obtained was for the interval $27 \leq s \leq 33$, which was chosen in order to allow reasonable statistics,

$$\left(\frac{\mathrm{d}\sigma}{\mathrm{d}(\cos\theta)}\right) = ((26{\cdot}4 \pm 2{\cdot}4)\cos^2\theta - (1{\cdot}0 \pm 1{\cdot}4)\cos\theta + 6{\cdot}9 \pm 0{\cdot}7) \times 10^{-27}\ \mathrm{cm}^2. \tag{6.25}$$

In eqn (6.25), if we keep just *s*- and *p*-waves, we obtain

$$\frac{\mathrm{d}\sigma}{\mathrm{d}\Omega} = \lambda\!\!\!^{-2}\,|\mathrm{e}^{\mathrm{i}\delta_0}\sin\delta_0 + 3\mathrm{e}^{\mathrm{i}\delta_1}\sin\delta_1\cos\theta|^2 = \lambda\!\!\!^{-2}(\sin^2\delta_0 +$$

$$+ 6\sin\delta_0\sin\delta_1\cos(\delta_0 - \delta_1)\cos\theta + 9\sin^2\delta_1\cos^2\theta). \tag{6.26}$$

† Kallen's book [27] is an excellent general reference for this, and related, material.

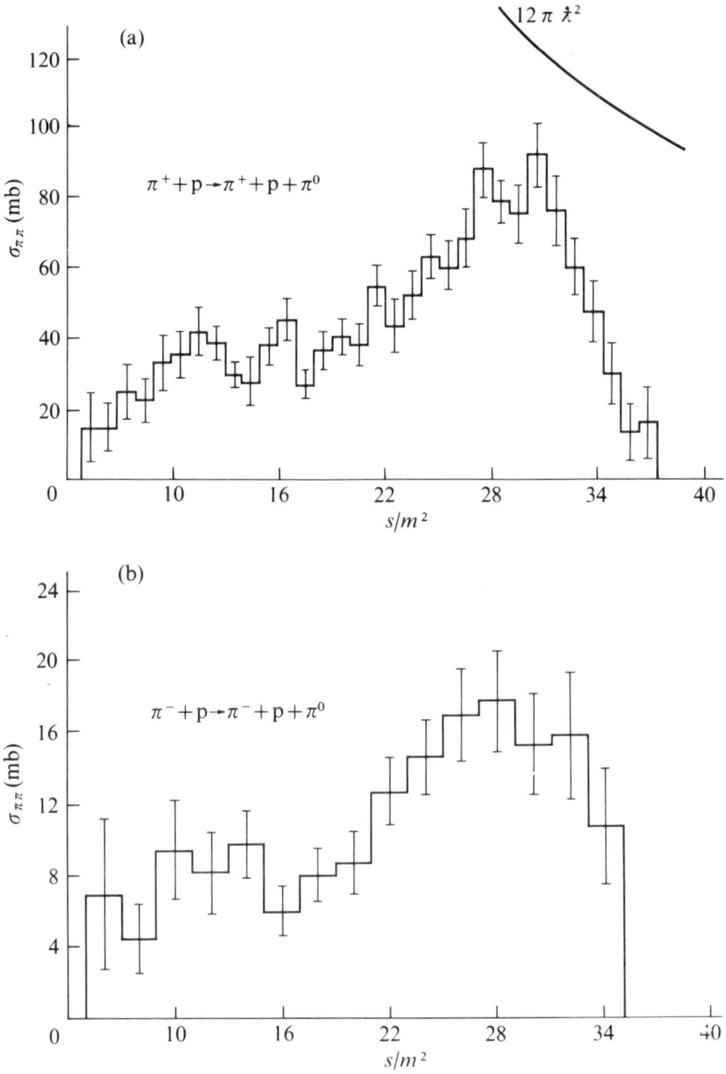


FIG. 6.7. The data used in Fig. 6.5(b) plotted against $(s/m^2)$, instead of the previously-used extrapolation procedure.

When integrated, we obtain

$$\sigma_{\text{tot}} = 4\pi\lambda\!\!\!^{-2}(\sin^2\delta_0+3\sin^2\delta_1). \tag{6.27}$$

The average value of $\lambda\!\!\!^{-2}$ corresponds to an $s_{\pi-\pi} \approx 30m^2$. Hence,

$$\frac{s}{m^2} = 30 = \left(2\sqrt{\left(1+\frac{1}{m^2\lambda\!\!\!^{-2}}\right)}\right)^2 = 4\left(1+\frac{1}{m^2\lambda\!\!\!^{-2}}\right) \tag{6.28}$$

or

$$\frac{(30-4)}{4}m^2 = \frac{1}{\lambda\!\!\!^{-2}} \tag{6.29a}$$

or

$$\lambda\!\!\!^{-2} = \frac{4}{(30-4)m^2} \approx 3\times10^{-27}\ \text{cm}^2. \tag{6.29b}$$

Hence, the predicted $\frac{d\sigma}{d\Omega}$ for *s*- and *p*-waves only is, from eqn (6.26),

$$\frac{d\sigma}{d\Omega} \approx (3\sin^2\delta_0+18\sin\delta_0\sin\delta_1\cos(\delta_0-\delta_1)\cos\theta+27\sin^2\delta_1\cos^2\theta)\times$$
$$\times10^{-27}\ \text{cm}^2. \tag{6.30}$$

Comparing eqns (6.25) and (6.30), we observe that the $\cos^2\theta\sin^2\delta_1$ terms are the same, if $\sin\delta_1 \approx 0{\cdot}9\pm0{\cdot}1$, indicating a resonance with $J = 1$. However, the *s*-wave term is larger than the theoretically allowed limit. This is assumed to be attributable to a strong physical background, in addition to the $\rho$ formation. That this background has a sizeable value can be seen in Fig. 6.7.

The smallness of the $\cos\theta$ term is what we would expect for a small $\delta_0$, which is also consistent with the deduced behaviour of the $\sigma_{\pi-\pi}$. This $2\pi$ meson resonance has become known as the $\rho$-meson.

### 6.6.2. *Isotopic spin of the* $2\pi$, $J = 1$ *resonance* ($\rho$-*meson*)

The isotopic spin of a two-pion $J = 1$ resonance can be deduced purely from the requirement that the two pions obey Bose statistics. Thus, their total wave function must be symmetric under an interchange of both their space coordinates, and isotopic spin coordinates. This is equivalent to the generalized Pauli principle for bosons.

For $J = 1$ the space wave function must be anti-symmetric under space interchange. Thus, the isotopic spin wave function must also be anti-symmetric under exchange of isotopic spin coordinates.

For a two-pion state, the possibilities for $T$ are only 0, 1, or 2. Only $T = 1$ has the required anti-symmetric properties, since even $T$ states are symmetric. Thus, if $J = 1$ for a two-pion resonance state, $T$ must also be equal to

1. In these above experiments the $\rho$-meson shows up in both the $\pi^+, \pi^-$ and $\pi^-, \pi^0$ states. Thus (owing to the latter) $T$ cannot be zero. After many subsequent experiments in which the $\rho$ has been detected under a variety of conditions, it has been found to be a charge triplet $\rho^{0,\pm}$, and thus describeable only by $T = 1$. Furthermore, the charge ratios of various states in $\rho$ production are consistent with $T = 1$ only. In investigations of the reaction

$$\pi^+ + \text{p} \rightarrow \text{n} + \pi^+ + \pi^+,$$

in which the two pions must be in a pure $T = 2$ state there is no evidence for the $\rho$. Accepting that $T = 1$, we can turn the previous argument around and state that the generalized Pauli exclusion principle requires that $J$ be odd. Hence, $J = 1, 3, 5, \ldots$. Although there are errors in the determination of the angular distributions, there is still no question that the only possibility is $J = 1$. The parity is, obviously, negative. This can be seen since the effective intrinsic parity of the combination of two pions is positive, while the parity of a $J = 1$ wave function is negative, thus leading to a negative parity for the $\rho$. The $G$-parity is positive since the $\rho$ decays via strong interactions into two (an even number of) pions.

### 6.6.3. *The $\omega$-meson*

A resonance involving three pions was observed [28] in the proton–anti-proton annihilation reaction

$$\text{p} + \bar{\text{p}} \rightarrow 2\pi^+ + 2\pi^- + \pi^0. \tag{6.31}$$

Figure 6.8 shows the invariant mass distributions for three-pion groups of total charge $Q = 0, 1, 2$.

The smooth curves represent the predictions if production were assumed proportional to available phase-space only. It is clear that there is, within reasonable allowance for statistics, a smooth behaviour everywhere, except for a sharp peak at about 790 MeV, in the $Q = 0$ $(\pi^+\pi^-\pi^0)$ combination. Therefore, this resonance (which is denoted as the $\omega$-meson) has $T = 0$. There have been many confirmations of this $\omega$-particle in various reactions, including $\pi$–p and $\bar{\text{p}}$–p interactions. From the characteristics of the decay angular distribution, and other properties of the reaction, it has been definitely established that the spin of the $\omega$ is one, that the parity is negative, that the $G$-parity is negative, and that the exact mass is $183{\cdot}7 \pm 0{\cdot}4$. The width ($\Gamma$) is $12{\cdot}7 \pm 1{\cdot}2$ MeV.

### 6.6.4. *The $\eta$-meson*

A study [29] of the reaction $\pi^+ + \text{d} \rightarrow \text{p} + \text{p} + \pi^+ + \pi^- + \pi^0$ led to the mass distribution for the pion triplets $(\pi^+\pi^-\pi^0)$ shown in Fig. 6.9. The solid curve

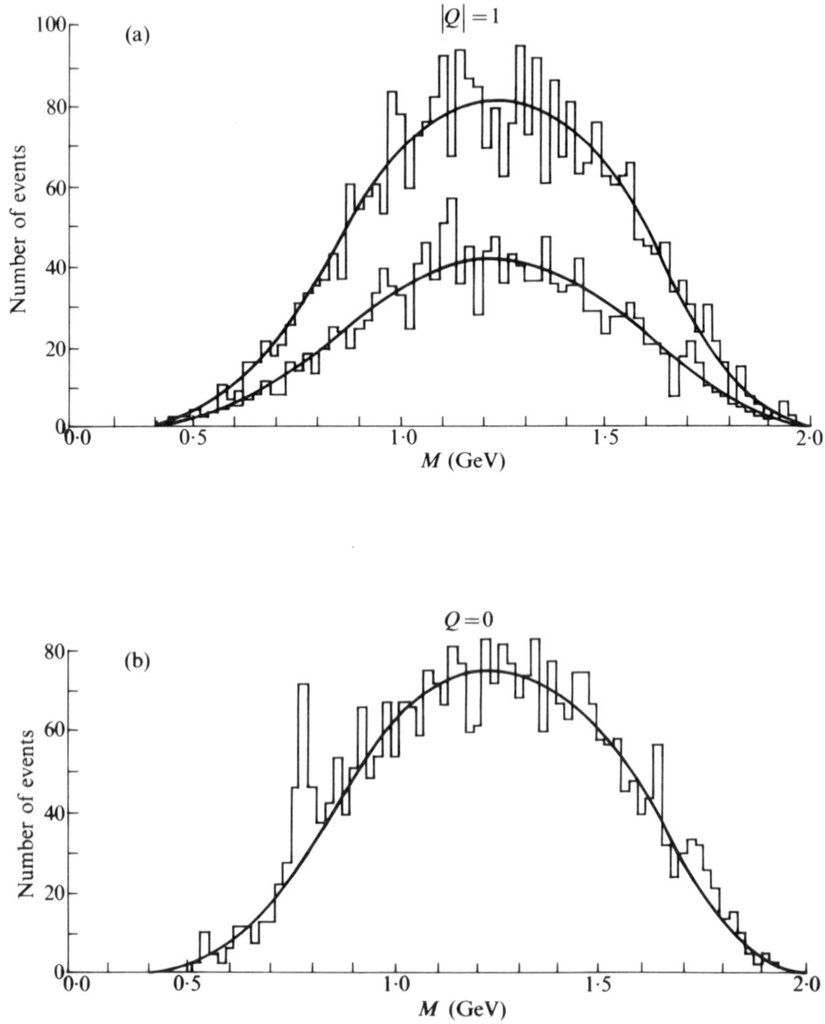


FIG. 6.8. Effective mass distribution of three pions in the reaction $\bar{p}+p \rightarrow n+(\pi^{+}+\pi^{-})+\pi^{0}$ showing a peak only in the $Q = 0$ charge state. (Maglic, Alvarez, Rosenfeld, and Stevenson. (1961). *Phys. Rev. Lett.* **7**, 198.)

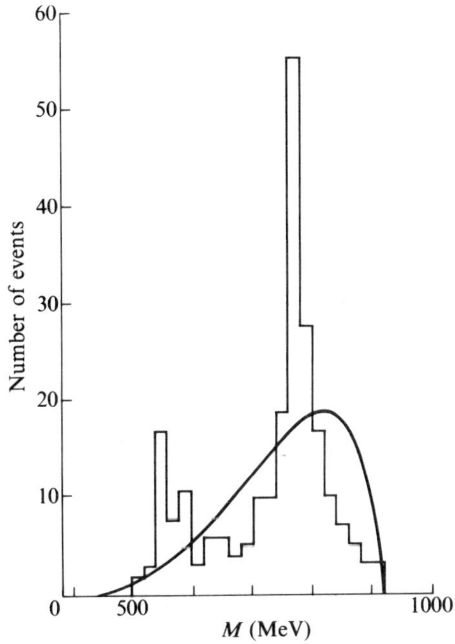


FIG. 6.9. The mass distribution of the neutral $\pi$-meson triplet in the reaction $\pi^+ + d \rightarrow p + p + \pi^+ + \pi^- + \pi^0$. (Pevsner *et al.* (1961). *Phys. Rev. Lett.* **7**, 421.

represents the phase-space distribution. The well-defined $\omega$ peak was found to be accompanied by an $\omega$-like peak with a much lower mass ($\approx$ 550 MeV). This meson was denoted as $\eta$. It was, subsequently, confirmed in many reactions including [30]

$$K^- + p \rightarrow \Lambda + \eta. \quad \downarrow (\pi^+ + \pi^- + \pi^0) \tag{6.32}$$

Since no charged counterparts have been found for the $\eta$, its isotopic spin must be zero, as the only allowed possibilities for the reaction of eqn (6.32) are that the $\eta$ have $T = 0$ or $T = 1$. The $\eta$ mass has been found to be $548{\cdot}8 \pm 0{\cdot}6$ MeV. The width $\Gamma = 2{\cdot}63$ keV$\pm 0{\cdot}64$ keV (i.e. the decay is via the electromagnetic interaction). The parity is positive, the $G$-parity is positive, and the spin is zero.

### 6.6.5. *The Dalitz plot*

It is often of interest to observe the behaviour of final-state interactions by determining the differential cross-sections as a function of an invariant energy (or effective mass) for two or more final-state particles in a reaction. Let us consider the reaction

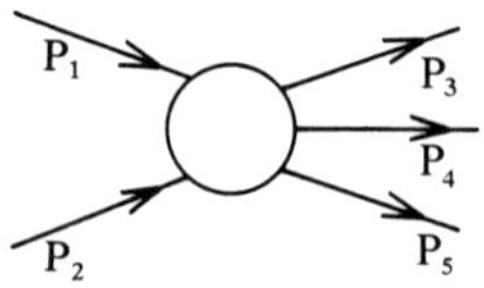


as a function of the invariant energy $s_{34} = (\mathrm{P}_3+\mathrm{P}_4)^2$. Then we can obtain†

$$\frac{\mathrm{d}\sigma}{\mathrm{d}s_{34}} = \frac{1}{2\lambda(s, m_1^2, m_2^2)(2\pi)^5} \int \mathrm{d}\mathbf{p}_3\, \mathrm{d}\mathbf{p}_4\, \mathrm{d}\mathbf{p}_5\, \delta\left(\sum_1^5 \mathbf{p}_\nu\right) \times$$
$$\times\, \delta(s_{34}-(\mathbf{p}_3+\mathbf{p}_4)^2) \prod_3^5 \delta(\mathbf{P}_\nu^2 - m_\nu^2)\theta(P_{\nu 0}) \sum |T|^2, \quad (6.33)$$

$s = s_{12}$ (the invariant energy of the incident particles). For spinless particles, $T$ will be a function of five independent invariant energies (or momentum transfers) chosen from $s_{ii'} = (p_i+p_i')^2$; $i, i' = 1, 2, \ldots 5$. If some particles have spin then $T$ will also be a function of scalar products of spin and momentum $(\gamma_\mu P_i^\mu)$.

In particular, in a three-body final state such as

$$\pi+\mathrm{N} \rightarrow \pi+\pi+\mathrm{N}, \quad (6.34)$$

suppose we suspect a resonance between different pairs of final particles. For example, 3+5, or 4+5 could form $\mathrm{N}^{*+}$, whereas 3+4 could form a $\rho$. In such cases, $T_{35}$, $T_{45}$, and $T_{34}$, respectively, would show a sharp dependence on $s_{35}$, $s_{45}$, or $s_{34}$, respectively, with a peak in the neighbourhood of the resonance. Thus, we would like to look for departures from the phase-space integral prediction (i.e. all $T$ elements are equal to a constant). Hence, as Dalitz proposed (1953) [31] we can employ an $x$–$y$ coordinate system in which the $x$-axis represents $s_{35}$ (or $m_{35}^2$), and the $y$-axis represents $s_{45}$ (or $m_{45}^2$) as in Fig. 6.10. By inserting two delta functions and $|T|$ = constant in eqn (6.33), we can obtain, within a fixed closed curve of the type shown,

$$\frac{\mathrm{d}^2\sigma(s, s_{35}, s_{45})}{\mathrm{d}s_{35}\,\mathrm{d}s_{45}} = \text{Constant } f(s).$$

† From general relativistic kinematics, it can be shown that

$$\frac{\mathrm{d}\sigma}{\mathrm{d}t_{ij}} = \frac{1}{2\lambda}(2\pi)^{3n-4}\sum_n \int \mathrm{d}q_1 \ldots \int \mathrm{d}q_N\, \delta(s_{ij}-(q_i+q_j)^2)\prod_1^\nu \delta(d_\nu^2 - m_\nu^2)\theta(\mathbf{q}_{\nu 0}) \times$$
$$\times\, \delta\left(p_1+p_2-\sum q_\nu\right)\sum |\langle \ldots T \ldots \rangle|^2,$$

where

$$\lambda(x, y, z) = (x^2+y^2+z^2-2xy-2yz-2zx)^{\frac{1}{2}}.$$

(See reference [20], Chapter 5 (Chapter 3)).

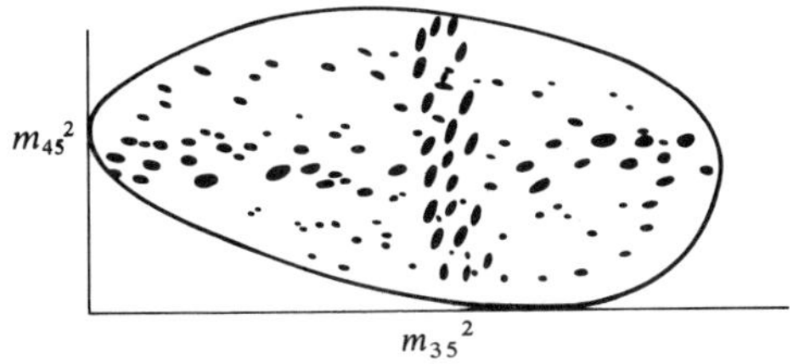


FIG. 6.10. A Dalitz plot for $1+2 \rightarrow 3+4+5$. Each event is represented by a dot and the density of dots is proportioned to $d\sigma/ds_{ii'}$. The clustering of points around $m_{35}$ suggests a resonance between these two particles at this value of invariant mass. Similarly, there appears to be a resonance between 4 and 5 at a mass of $m_{45}$.

Thus, in the Dalitz plot for a particular incident channel $(s)$, the density of points from a random sample of events should be a constant. A clustering of points in the vertical band corresponding to a small range of $s_{35}$ thus corresponds to a resonance of $3+5$, with an invariant mass of $m_{35}^2$ as shown. Similarly, a clustering of points in the band corresponding to a small range of $s_{45}$ corresponds to a resonance of $4+5$ with a mass $m_{45}^2$.

For the case of identical final-state particles, we can use variables that take account of the symmetry. The matrix element has also been allowed to vary [32] with angle, when the resonant intermediate particle is pseudoscalar, or has a spin different from zero. The density distribution of points in diagrams of these types gives us a picture of the behaviour of the matrix elements and can be used to restrict choices of spin and parity of the decaying particles. Although these diagrams are of visual presentation value, any desired investigation of resonance behaviour of matrix elements can be done highly quantitatively, and directly, by analytical means.

### 6.6.6. *Further development of the one-pion exchange model and other peripheral models*

Although the one-pion exchange model was able to qualitatively explain the forwardly collimated angular distribution of pion production in nucleon–nucleon and pion–nucleon collisions in the neighbourhood of 1–3 GeV, it soon becomes obvious that even the sharp forward collimation of one-pion exchange was not enough to explain the experimental results, even at modest $t$ values ($|t| \gtrsim 10\ \mu^2$). In this regard, it was found [33] (De Prato 1961) that the predicted single pion production in nucleon–nucleon collisions is about a factor of 2 or 3 larger than experimentally observed, at energies of a few GeV. The lack of agreement with experiment becomes more marked as the energy increases. In other words, experiments were indicating that even the rapid decrease of the factor $\dfrac{1}{(\Delta^2+\mu^2)^2}$ was not sufficient. Ferrari and Selleri

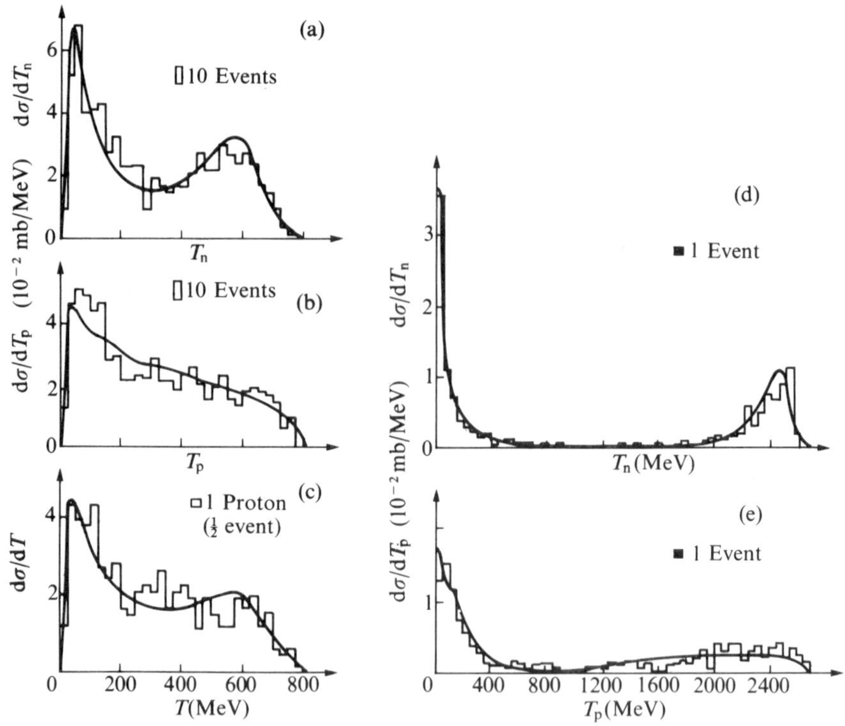


FIG. 6.11. Example of laboratory kinetic energy spectra fitted by the one-pion exchange model with form factors. (Taken from Ferrari and Selleri (1963), which also contains the experimental bibliography.) (a) Neutrons from $pp \rightarrow pn\pi^+$, (b) protons from $pp \rightarrow pn\pi^+$, (c) protons from $pp \rightarrow pp\pi^0$ (at an incident energy of 970 MeV), (d) neutrons, and (e) protons from $pp \rightarrow pn\pi^+$ (at 2·85 GeV), where only events in which the system $p\pi^+$ is in the region of the resonance $N^{*++}$ have been selected. (From Bertocchi and Ferrari, [20].)

(1961, 1962, and 1963) [34] introduced unknown $\Delta^2$-dependent form factors, at the vertices, to take account of the virtuality of the intermediate particle. These form factors were fitted to experiment. Figure 6.11 illustrates some good fits to the experiments which were obtained with this approach.

Figure 6.12 shows the comparison of a calculation with an unmodified one-pion exchange model, and a calculation with a form factor included. The reaction can be represented by the diagram on the facing page.

It is clear that one-pion exchange (unmodified) is in qualitative, but not quantitative, agreement with the data. However, the one-pion exchange with form factors agrees quite well. The one-pion exchange model with form factors can fit the nucleon–nucleon and pion–nucleon production data up to about 4 GeV, but fails at higher energies by predicting too many high $|t|$

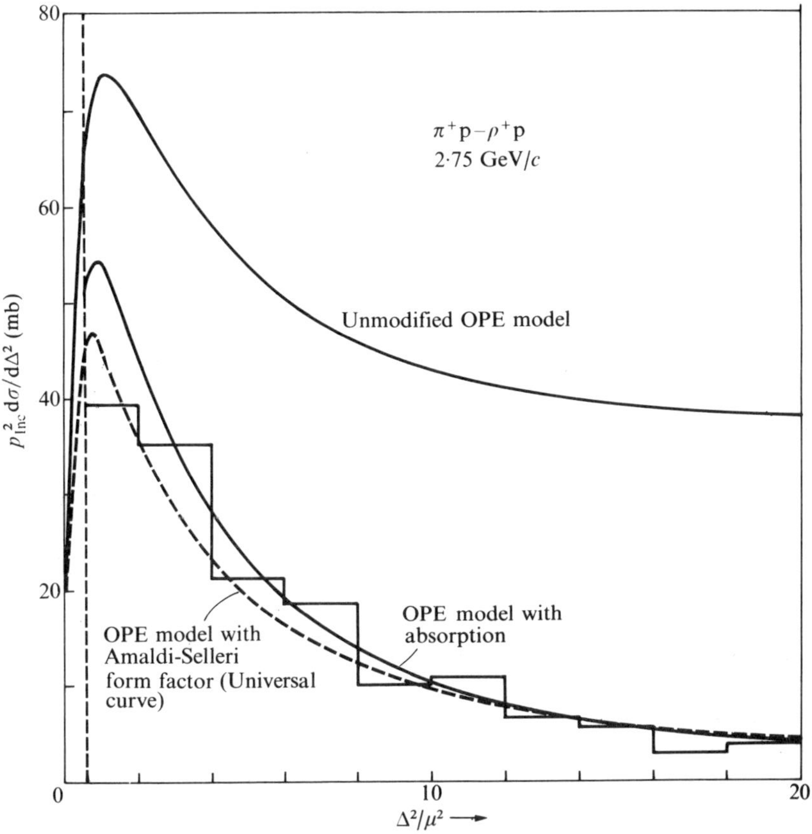


FIG. 6.12. Comparison of theory and experiment for the differential cross-section of the reaction, $\pi^+p+\rho^+p$ at 2·75 GeV/$c$. The histogram represents the data. The upper solid curve is the cross-section predicted by the unmodified OPE. model. The dashed curve is given by the cross-section multiplied by the square of the form factor. The lower solid curve is that predicted by the OPE. model, modified to include absorptive effects as described in §§ 5 and 6. The abscissa is $\Delta^2$ in units of $\mu^2 = m_\pi^2 = 0{\cdot}0195\ (\text{GeV}/c)^2$. (After review by Lindenbaum [38].)

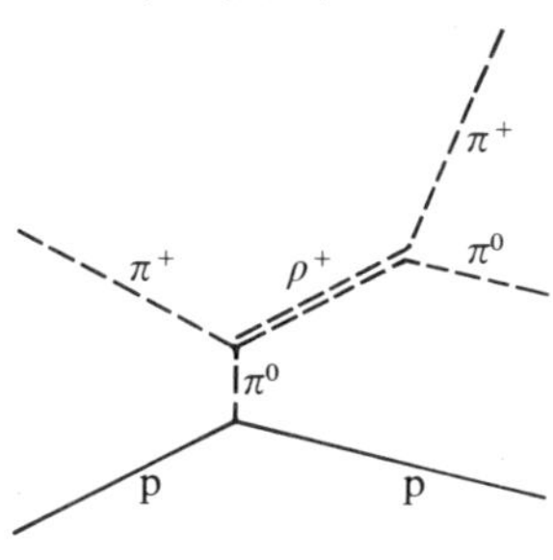

events. If the one-pion exchange model is broadened to the OME (one-meson exchange) model so as to include vector meson exchange, such as $\rho$, the lack of agreement of prediction with experiment is much worse. Much stronger form factor suppression effects are required for the vector meson exchange than for pion exchange and, furthermore, the predicted cross-section behaviours with energy do not agree with experiments in that too large a vector meson exchange cross-section is predicted at the higher energies. This difficulty with vector exchange calculations exists, even in the absorption model, which we shall discuss later.

### 6.6.7. *Treiman and Yang test*

If the only process occurring is one-pion exchange, some definite and unique characteristics are required in the resultant angular correlations in the process.

Consider the following diagram.

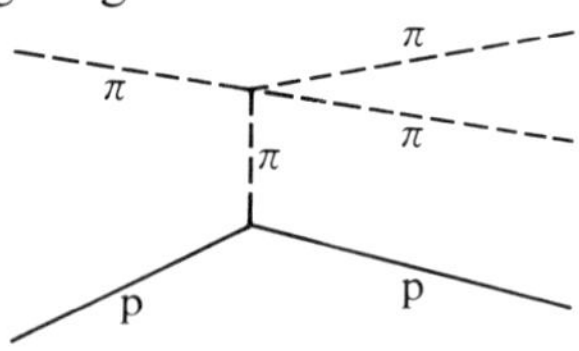


It was pointed out (Treiman and Yang 1962) [36] that, if only a pion is exchanged between the two vertices, the magnitude, but not the direction, of the momentum transfer can be communicated between the two vertices. Thus, we conclude that in the rest system of the incident pion, there should be no correlation between the production plane of the pions and the scattering plane of the nucleons. Actually the experimental results (Hagopion *et al.* 1966) [37] showed such a correlation. Thus, other additional exchanges, such as vector exchange, are required, or diffraction scattering effects (absorption model) must be added. We should perhaps note, at this point, that the increased difficulty with the one-pion exchange model, as the energy increases, is that the low partial waves assumed in this model no longer correspond to peripheral interactions as the energy increases. In fact, such low partial waves (short-range interaction) would be expected to be heavy contributors to higher-order multiple production processes and, thus, absorption corrections are likely to be required.

### 6.6.8. *Peripheral model with absorption*

The experimental results indicated that in a variety of quasi-two-body interactions, at a variety of energies, the general $t$ dependence of the production process was considerably more collimated than the one-meson exchange models. The difference between prediction and experiment increase both with increasing energy, and with increasing spin $J$ of the exchanged object.

The situation is illustrated in Fig. 6.13 which shows quasi-two-body states formed in $\pi^+ + p$ interactions at 8 GeV/$c$.

The three interactions across the top

$$\begin{aligned}\pi^+ + p \rightarrow\ & p + \rho \\ & N^* + \rho \\ & N^* + f \end{aligned} \tag{6.35a}$$

all proceed via one-pion exchange. The next four cases,

$$\begin{aligned}\pi^+ + p \rightarrow\ & N^{*++} + \pi^0 \\ & p + A_1 \\ & p + A_2 \\ & N^* + \omega, \end{aligned} \tag{6.35b}$$

all can proceed via $\rho$-exchange. The last graph in the lower right-hand corner shows elastic scattering.

The $\pi$-exchanges are very similar in $t$ behaviour to elastic scattering and, in general, even the $\rho$-exchange cases are generally similar in $t$ dependence to the elastic scattering behaviour. This represents a serious problem for the one-meson exchange model, since vector exchange predicts a much more slowly decreasing cross-section with increasing $|t|$, than for one-pion exchange. As we recall, in order to get the one-pion exchange prediction to fit the experimental results, arbitrary form factors had to be introduced at the vertices. In the $\rho$-exchange case it was not possible to fit the data except at lower energies ($\lesssim$ 4 GeV) with the form factor approach.

The similarity to the elastic scattering of these particle exchange reactions implies that some quasi-elastic diffraction process similar to elastic diffraction scattering is probably responsible for the angular distribution. The elastic diffraction scattering is essentially due to sizeable absorption in the low partial waves, which fall within the range of interaction. Similarly, we can expect that the competition of many other channels leads to an absorption of the low partial waves.

A number (Sopkovich 1962) [39], (Dar *et al.* 1964) [40], (Durand and Chui 1964–65) [41], and (Jackson *et al.* 1965) [35] of formulations of the absorption approach have been made. This approach generally involves consideration of a re-scattering of the initial- and final-state particles, due to initial- and final-state interactions. The diagram which describes the one-meson exchange model, with absorption, is the following (p. 195).

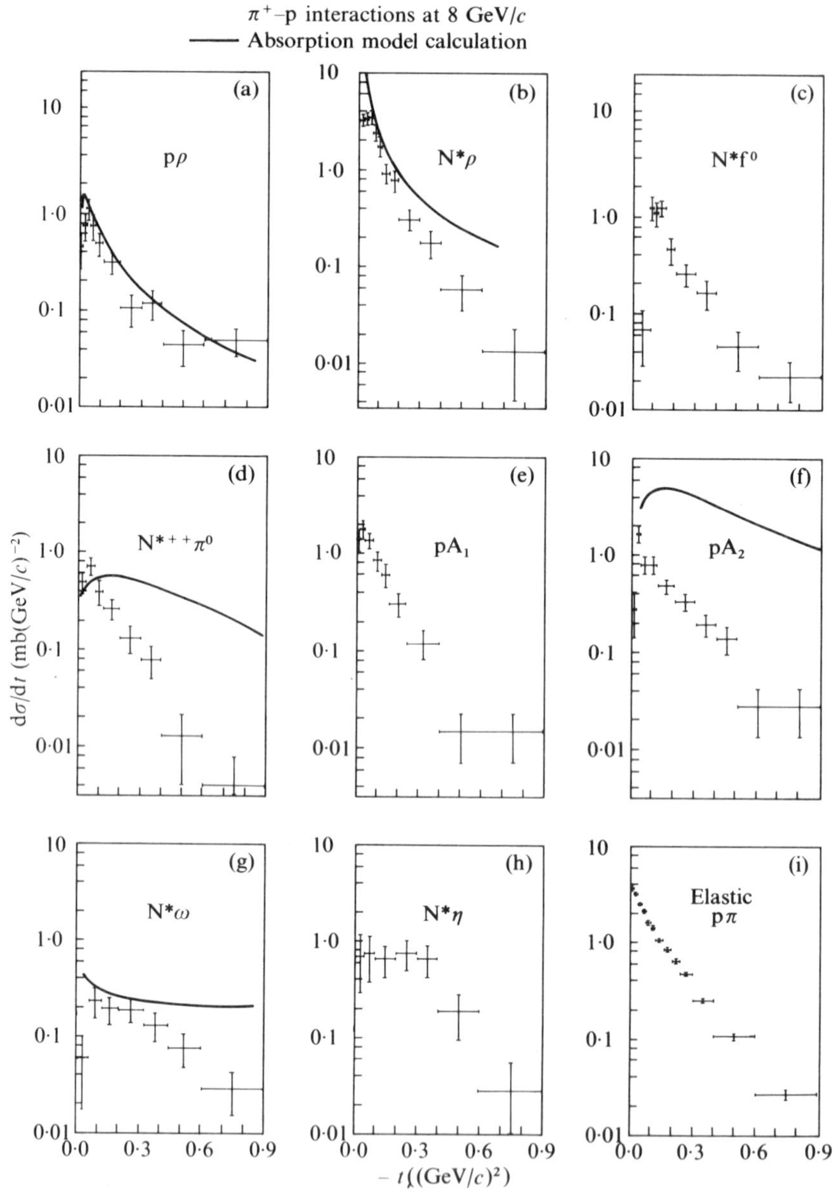


FIG. 6.13. 8 GeV/$c$ $\pi^+$–p inelastic channels and their comparison with predictions of the OME ($\rho$ or $\pi$) model, with absorption effects included. The graph was presented by the Aachen–Berlin–CERN collaboration at the Oxford Conference (1965). The figure is from Lindenbaum (Oxford, 1965).

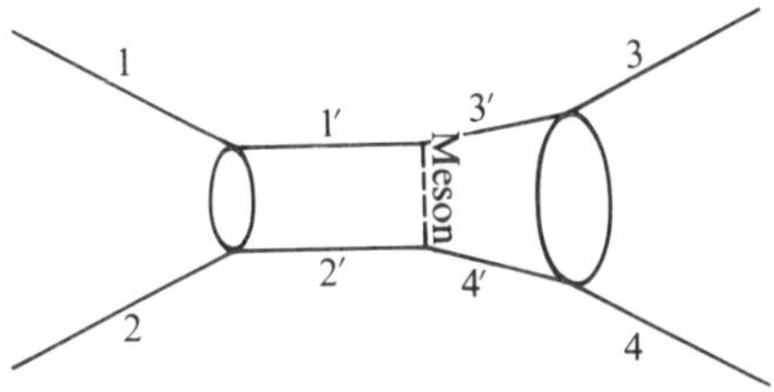


This re-scattering effect must be accompanied by a large absorption in the low partial waves. This then results in a strong forward collimation of the reaction products, similar to that in elastic scattering, which, as we saw, is required to explain the data.

The elastic scattering, in the initial state, is known experimentally for many important cases. Of course, the final-state elastic scattering is, generally, not known experimentally. Often we assume the final-state re-scattering is similar to the initial-state.

Figure 6.13 shows the results of one-meson exchange calculations with absorption. It is clear that, in the pion exchange cases, the agreement is quite reasonable. In fact, it is generally found to work well for pion exchange. However, as can be seen, one-meson exchange with absorption fails quite badly for vector exchange in $\pi^+ + \mathrm{p}$ interactions at 8 GeV/$c$.

Reasonable agreement has been obtained, also, for vector meson exchange at lower energies ($E \lesssim 3$ GeV) but there is a systematic tendency toward failure for the vector meson exchange cases as the energy increases.

In conclusion we found that one-meson exchange models, with absorption, can explain both the $s$ and the $t$ dependence of pion exchange reactions at higher energies ($\gtrsim 4$ GeV), but are grossly contradicted for $\rho$ (vector) exchange at the higher energies.

This latter effect may well be related to the general difficulty in making sure that vector exchanges do not violate unitarity requirements. It is known that a simple pole exchange gives rise to an $s$ dependence of the cross-section, as follows,

$$\frac{\mathrm{d}\sigma}{\mathrm{d}t}(s, t) = f(t)\left(\frac{s}{s_0}\right)^{2(J-1)} \tag{6.36}$$

and, hence, for vector exchange will, at high enough energies, eventually exceed the unitary limit. Appropriate damping of the vector exchange cross-sections is required. One way of doing this is to consider Regge-ized exchanges. This will help since, for the Regge case,

$$J(t) \approx J - m\,|t|. \tag{6.37}$$

Thus, sufficient damping of $J \geq 1$ exchanges is provided to ensure that unitarity requirements are met.

It is interesting to note that the sharp relative growth of effect of a higher spin exchange with increasing $s$ leads us to expect the dominance of higher spin ($J \geq 1$) exchanges at very high energy over spin-zero exchange, such as pion exchange, even though the pole corresponding to the vector exchange is, owing to its greater mass, much further removed from the physical region. Even the severe Regge damping effects do not change this conclusion.

The unique relationship between angular correlations of the decaying resonance state, or scattering at the upper vertex, predicted by the one-meson exchange models is modified by the absorption but, fortunately, the absorption model calculations, in general, agree with the observed experimental angular correlation, whereas the one-meson exchange calculations do not. Although there have been some direct attempts (called $K$-matrix models) [42] to ensure meeting unitarity requirements, the approximation required to work these out often defeats their original purpose. Hence, the satisfactory treatment of vector meson exchange at high energies by one-meson exchange models is still, basically, unsolved.

### 6.6.9. *The status of high-energy production of nucleon and pionic isobars*

A number of (hydrogen and deuterium) bubble chamber investigations of nucleon–nucleon interactions have analysed the various production processes in considerable detail, by separation of single, double, and higher multiplicity final states [43*a*].

In the energy region $\lesssim 3$ GeV, the bubble chamber is very well suited for making these separations. In this energy region it is observed that single pion production in $p+p$ interactions is dominated by the process of nucleon isobar formation, $p+p \rightarrow N+N^*_{\frac{3}{2},\frac{3}{2}}$, which accounts for more than 50 per cent of the single pion production. The most important process in this category is $p+p \rightarrow n+N^{*++}$, which can be understood on the basis of the isobar model, owing to the very favourable Clebsch–Gordan coefficients, which predict that three-quarters of the single isobar production should be in this channel. This effect, and the angular distribution of isobar production, are adequately explained by the one-pion exchange model which, as noted previously, is intimately related to (and contains) the isobar model for these interactions.

Even in double pion production we find no evidence that pionic resonances are formed to any appreciable degree. Double $T = J = \frac{3}{2}$ isobar production is the dominant identified mode, but is not as dominant in double production, as single isobar formation is in single pion production. This can be attributed to the fact that only one $N^{*++}$ can be formed, otherwise we would violate charge conservation. Thus, the Clebsch–Gordan coefficients, for observable double isobar production, are smaller and reduce the cross-section.

Furthermore, unidentified double production from excitation of a single nucleon to the $T = \frac{1}{2}$ isobars at 1400 MeV, 1520 MeV, and 1690 MeV could, as predicted by the isobar model, compete with double $N^{*++}$ excitations. These higher isobars could, as envisaged in the isobar model, decay to a lower isobar level like $N^*_{\frac{3}{2},\frac{3}{2}}$, by emitting a pion, and then the $N^*_{\frac{3}{2},\frac{3}{2}}$ could decay to the ground state by emission of a second pion. These higher $T = \frac{1}{2}$ isobaric states could also decay by single pion emission all the way to the ground state. Below 3 GeV, there are a number of counter investigations [43*b*] which measure, at a fixed angle, the momentum spectra of the pions, and the protons produced in p–p inelastic collisions. These experiments were usually performed by the equivalent of one to a few counter telescopes placed at various angles to an external proton incident on a hydrogen target. The system contains momentum-selecting bending magnets and, thus, the momentum spectra are obtained. The fixed angle in the laboratory system generally corresponds to a small range of angles in the centre of mass system for the high-momentum part of the spectrum. The angular range increases as one proceeds to the low-momentum cut-off.

Although the complete reaction in individual charge states is not seen as in the bubble chamber case, the much higher data rate handling capabilities of counter techniques, and the high intensity of the proton beam allowed high statistics and high angular, and momentum, resolution determination of the pion and nucleon spectra.

The results for single and double pion production are, unfortunately, not directly uniquely separable as in the bubble chamber case. However, in the case of p+p interactions in which a $\pi^-$ meson is produced, single production is ruled out, since charge cannot be conserved and, thus, double or higher multiplicity production is required.

The data from the counter experiments can be explained by either the isobar model (choosing a suitable isobar angular distribution), or by the one-pion exchange model, where the general characteristics of angular distribution are also explainable. The pion production mechanism in n–p collisions shows a richer production of two $N^*_{3,3}$ isobars.† This was attributed to the fact that single isobar production is forbidden in the $T = 0$ state. However, an n–p collision can give rise to an $N^{*++}_{3,3} + N^{*-}_{3,3}$ pair in the $T = 0$ state, which has favourable Clebsch–Gordan coefficients and, hence, is enhanced.

† The notation now employed is $N_{2T,2J}$ to correspond to a paper which we will discuss shortly.

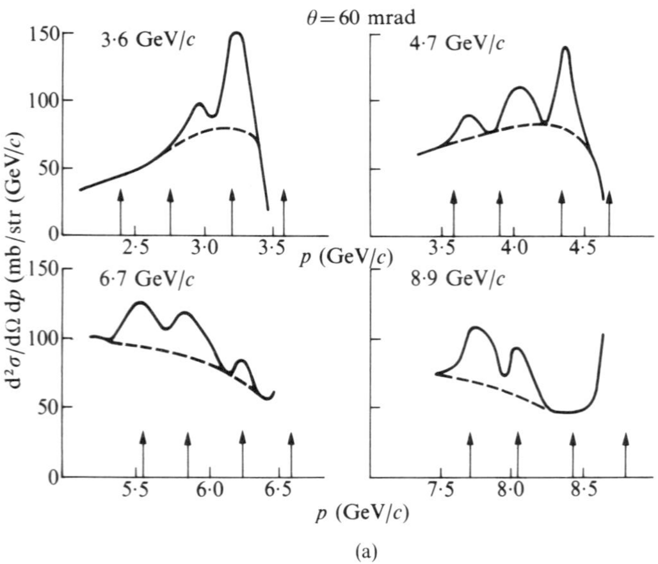


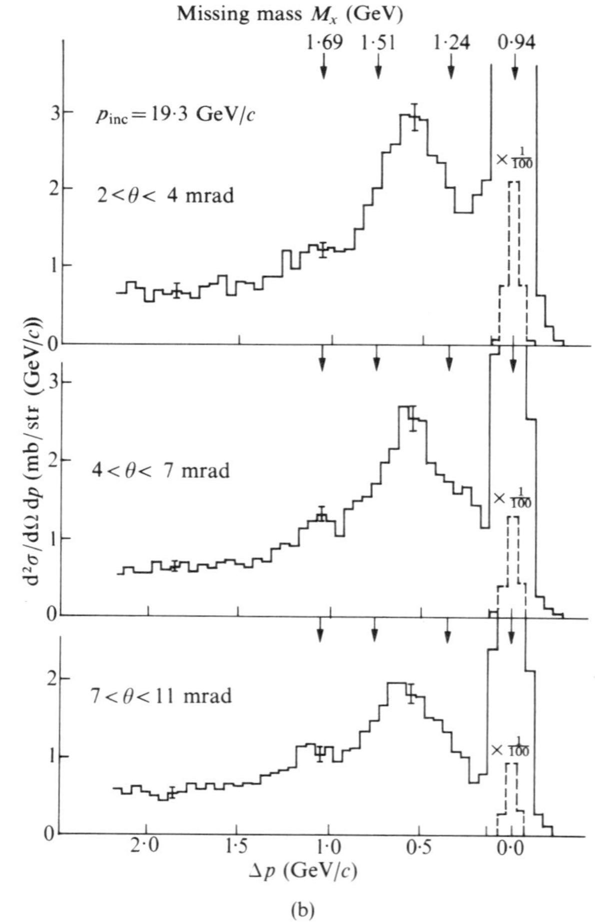


FIG. 6.14. Examples of proton laboratory momentum spectra at a fixed angle in the laboratory system, measured in the CERN experiments of inelastic proton–proton scattering. (a) Spectra at fixed angle and different incident momenta (indicated at top of curves) showing the disappearance of the 33-resonance peak (Cocconi *et al.* 1964) (b) Spectra at fixed incident momentum and various angular ranges, in the very small angle region, showing the enhancement of the 1 400 MeV peak (Belletini *et al.* 1965). The vertical arrows indicate the positions corresponding to the excitation of the isobars $N^*_{15}$, $N^*_{13}$, $N^*_{33}$, and to the position of the elastic peak (from left to right). The elastic peak is not shown in (a), where the dashed lines indicate the supposed behaviour of the non-resonant background. In (b) the momentum scale is measured (toward the left) in terms of the distance from the elastic peak (Berrocchi and Ferrari [20]).

Beyond 3 GeV the separation of individual charge states, and the reconstruction of the kinematics, becomes increasingly difficult with increasing energy. The most significant information on pion production in nucleon–nucleon collisions has, therefore, come from counter and spark chamber experiments, at the highest available energies.

As pointed out in the original isobar model work, in the process $p+p \rightarrow p+N^{*+}$, the observation of the momentum spectra of the inelastically scattered protons will, by conservation of energy and momentum. show peaks corresponding to the resonance cross-section behaviour in the formation of the $N^{*+}$. This was observed in early experiments [43*b*] in the $\lesssim 3$ GeV region, and this technique was used by Cocconi and co-workers [44] at CERN in the incident proton energy range of about 4–26 GeV. Momentum spectra were observed at a fixed laboratory angle (60 mrad) and some of the results are depicted in Fig. 6.14.

For the process which we are considering ($p+p \rightarrow p_{\text{recoil}}+N^{*+}$) it is clear that the isobar production kinematics (except for spin effects) are fully determined once the laboratory direction angles, and the momentum, of the inelastically scattered recoil proton are determined.

An equivalent way of expressing the results of a fixed laboratory angle momentum spectrum is as a doubly-differential spectrum, in terms of the variables $t$ (the momentum transfer between the incident proton and the inelastically scattered proton) and the excited isobar mass $M^*$. In Fig. 6.14 the excited isobars appear as local peaks above the background in the fixed laboratory angular recoil proton-momentum spectrum. The background is expected to vary slowly in the neighbourhood of the isobar peaks and, thus, approximate subtraction can be attempted.

In these experiments at low incident momentum (3·6–4·7 GeV/$c$), which at this fixed angle correspond to low $|t|$, a strong production of $N^{*+}_{3,3}$ is observed. As the incident energy (and consequently the momentum transfer) observed, is increased, the $N^*_{33}$ production drops rapidly, and beyond 8 GeV/$c$ is not significantly observed above background. The higher mass, $T = \frac{1}{2}$, isobars $N^{**}_{13}$ and $N^*_{15}$ are clearly observed throughout the 6–19 GeV/$c$ incident energy range. The momentum transfer varies from $|t| \approx 0{\cdot}1$ to $|t| \approx 1\ (\text{GeV}/c)^2$ over this incident energy range.

Subsequent experiments [44*c*] using digitized sonic spark chamber magnetic spectrometer systems studied very small angle (a few mrads) $p+p$ production of isobars. These results are shown in Fig. 6.14(b), and are in the very small momentum transfer region. In this region, the previously-observed isobars at 1·24 and 1·52 GeV were relatively suppressed and, instead, there appeared a large peak corresponding to an isobar mass of 1400 MeV. The different $t$ dependences of the various isobar peaks in $p+p$ interactions were also observed by Anderson *et al.*, [45] Foley *et al.*, [46] and Ankenbrandt *et al.* [47].

Subsequent observation of production of this isobar in $\pi^{\pm}$–p, as well as p–p interactions, make it likely that this $N^*_{11}$ is associated with the previously-observed resonant behaviour of the $P_{\frac{1}{2}}$ wave in pion–nucleon scattering (Roper resonance), although other possible kinematical accident effects like the so-called Deck effect will be considered also.

### 6.6.10. *Inelastic pion–nucleon interactions*

The most extensive bubble chamber studies of $\pi^{\pm}$–p inelastic interactions have been performed in hydrogen bubble chambers in the region of 1–10 GeV. In this energy range, considerable progress has been made in separating various quasi-two-body interactions. The principal reactions observed are listed in Table 6.1 (Bertocchi and Ferrari [20]), and their cross-section behaviour is given in Figs. 6.15(a) and (b) [48].

TABLE 6.1

| Observed reaction | Quasi-two-body final states observed |
|---|---|
| (1a) $\pi^+ + p \rightarrow \pi^+ + \pi^0 + p$ | $\pi^0 + N^{*++}$, $\rho^+ + p$ |
| (1b) $\pi^- + p \rightarrow \pi^- + \pi^0 + p$ | $\rho^- + p$ |
| (2a) $\pi^+ + p \rightarrow \pi^+ + \pi^+ + n$ | |
| (2b) $\pi^- + p \rightarrow \pi^+ + \pi^- + n$ | $\rho^0 + n$, $f^0 + n$ |
| (3a) $\pi^+ + p \rightarrow \pi^+ + \pi^+ + \pi^- + p$ | $\pi^+ + \pi^- + N^{*++}$, $\rho^0 + \pi^+ + p$, $\rho^0 + N^{*++}$, $A_2^+ + p$ |
| (3b) $\pi^- + p \rightarrow \pi^- + \pi^- \pi^+ + p$ | $\rho^0 + \pi^- + p$, $\pi^- + \pi^- + N^{*++}$, $A_2^- + p$ |
| (4a) $\pi^+ + p \rightarrow \pi^+ \pi^+ \pi^- \pi^0 + p$ | $\pi^+ + \pi^- + \pi^0 + N^{*++}$, $\omega + \pi^+ + p$, $\omega + N^{*++}$ |
| (4b) $\pi^- + p \rightarrow \pi^- \pi^+ \pi^- \pi^0 + p$ | $\omega + \pi^- + p$, $\pi^- + \pi^- + \pi^0 + N^{*++}$ |
| (5a) $\pi^+ + p \rightarrow \pi^+ \pi^+ \pi^+ \pi^- + n$ | |
| (5b) $\pi^- + p \rightarrow \pi^- \pi^+ \pi^- \pi^+ + n$ | |

We should notice that as soon as the energy is high enough to allow appreciable phase-space for the reaction, the cross-section rapidly rises for each inelastic channel in Figs. 6.15(a) and (b) to a substantial value of about 1 to 2 mb, and drops slowly, thereafter, probably at least partly due to increasing competition from other new channels, which are continually opening.

As shown in Fig. 6.15(b), the principal identified quasi-elastic components often contribute a substantial part ($\geq$ one half) of the total value of the channel cross-section. These effects are even more clearly seen in $\pi^+ + p$ interactions at 8 GeV, where quasi-two-body states are most clearly dominant. Some general features which have been noted in analysis of these

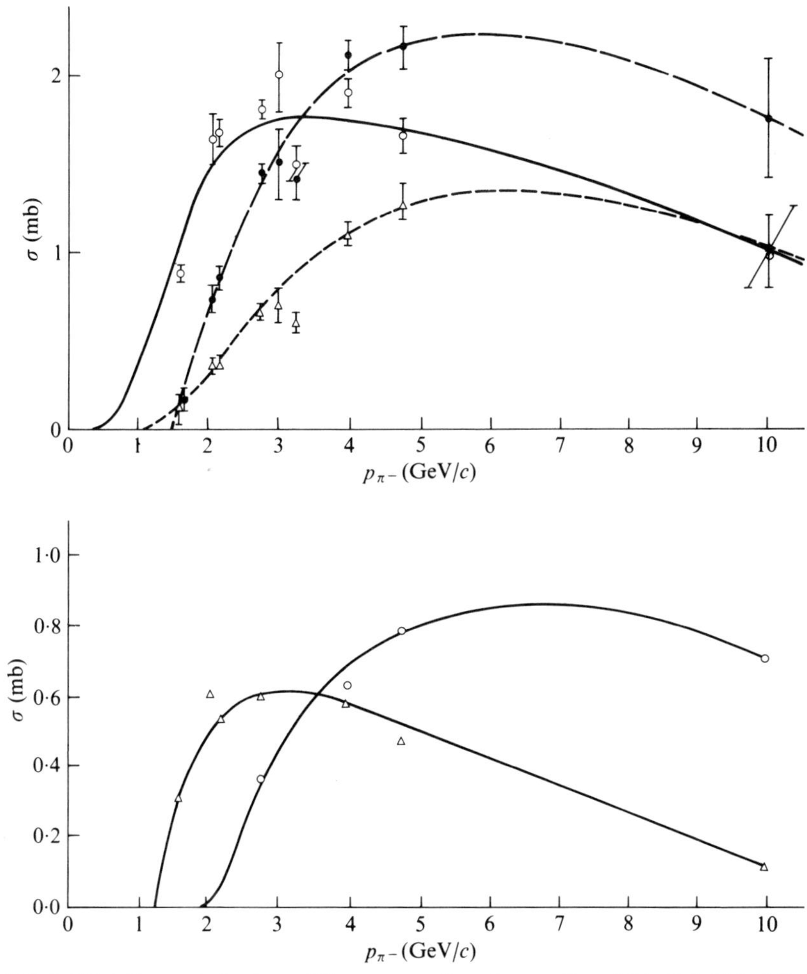


FIG. 6.15.(a) Energy behaviour of the cross-sections for reactions (3b), (4b), (5b) of Table 6.1 ([48]—see this reference for the experimental bibliography).

○ $\pi^- p \rightarrow \pi^- p \pi^+ \pi^-$

● $\pi^- p \rightarrow \pi^- p \pi^+ \pi^- \pi^\circ$

△ $\pi^- p \rightarrow \pi^- n \pi^+ \pi^+ \pi^-$

(References [20] and [48].)

FIG. 6.15.(b) Energy behaviour of the cross-sections for the principle quasielastic channels in the reaction $\pi^- p \rightarrow \pi^- \pi^+ \pi^- p$ from SOBB collaboration ([48] and [20]).

○ $\pi^- p \rightarrow p \pi^- \rho^\circ$

△ $\pi^- p \rightarrow N^{*++} \pi^- \pi^+$

(References [20] and [48].)

investigations are:

(a) In general, nucleon isobar formation is strongly dominant in nucleon–nucleon collisions, which lead to one, or two, nucleon isobars. However, very little, or no, pionic isobar formation is observed.

(b) On the other hand, in pion–nucleon collisions pionic isobars, $\rho$, $\omega$, $A_2$, etc. occur frequently, but in analysing multi-pion final states, it is concluded that double pionic isobar production (of $\rho$, $\omega$ mesons, etc.) is very unlikely. Thus, it seems that, in the collision of the two incident particles, the baryon is primarily excited to baryon isobars, and the pion is primarily excited to pionic isobars.

(c) It seems that $\pi^+ + p$ reactions are favoured over analogous $\pi^- + p$ reactions.

This effect seems to be associated with the fact that exchange of a singly charged particle can lead to formation of the $N^{*++}_{3,3}$ isobar with its associated, very favourable, isospin factors. In the $\pi^- + p$ case, this isobar cannot be formed due to charge conservation. However, we can form the analogous $N^{*-}_{3,3}$, but this would require a double charge exchange which is generally observed experimentally to be a low cross-section process. From the point of view of dominance of single meson exchanges, there is no known doubly charged meson or meson isobar (or resonance). This lack of observation of a doubly charged meson is underlined by the lack of any two-body-inelastic final state in reaction (2a) of Table 6.1 ($\pi^+ + p \rightarrow \pi^+ + \pi^+ + n$).

The missing-mass technique for observing nucleon isobars, previously described, for the reaction

$$p + p \rightarrow N^{*+} + p, \tag{6.38}$$

has been used, also, in studying the reactions

$$\pi^{\pm} + p \rightarrow N^{*+} + \pi^{\pm}, \tag{6.39}$$

and comparing these results to reaction (6.38). In reaction (6.39) the pion is detected, whereas, in studying the reaction (6.38), the proton is detected. Typical results for various incident momenta are shown in Fig. 6.16(a). There is clear evidence for the formation of the $N^*_1(1400)$ in all three reaction channels,† as well as clear evidence for the $N^*_1(1520)$, and $N^*_1(1690)$. There is also some evidence for formation of $N^*_{3,3}(1220)$ at low $|t|$, but owing to its being between the $N^*_1(1400)$ and the elastic scattering peak, it is more difficult to separate. The separation of the isobar peak from background is

† The notation used is $N^*_{2T,2J}$, with the second index omitted if $J$ is not specified, or if resonances with several different $J$ values overlap. In addition, when relevant, the charge of the isobar is indicated to the right of the superscript.

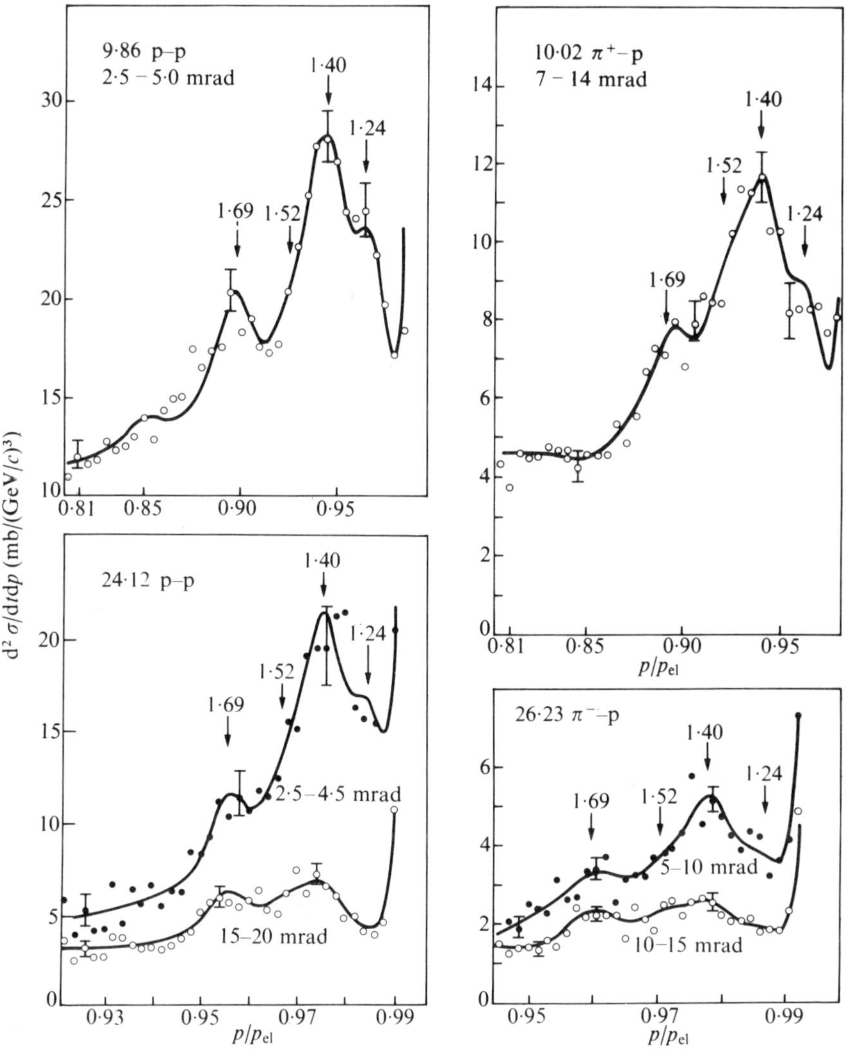


FIG. 6.16.(a) Typical inelastic momentum spectra, $\frac{d^2\sigma}{dp\,dt}$, plotted against the ratio of particle momentum to that of elastically scattered particles at the same angle. The arrows show the expected locations of the isobars. The solid lines are least-squares fits described in the text. (From [46].)

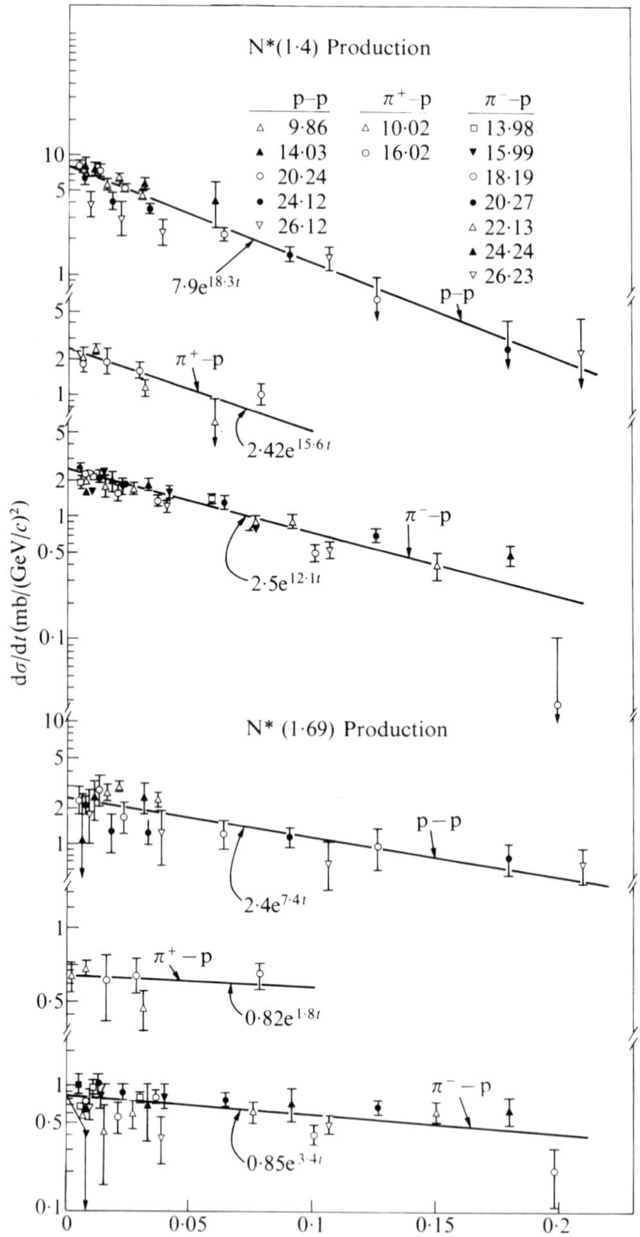


FIG. 6.16.(b) The differential cross-section $d\sigma/dt$ plotted against $t$ for N*(1400) and N*(1690) production in the reactions $\pi^{\pm}+p \rightarrow \pi^{\pm}+N^*$ and $p+p \rightarrow p+N^*$. The lines are least-squares fits to the data (Foley *et al.* (1967). *Phys. Rev. Lett.* **19**, 397).

somewhat arbitrary, and uncertain, in these types of experiments. However, following some simple assumptions, this was done. There is no evidence for an $s$ dependence of the $N_1^*(1400)$, $N_1^*(1520)$, and the $N_1^*(1690)$.

In Fig. 6.16(b), the $t$ dependence of all the above-discussed data and previous data is plotted and fitted by a simple exponential in $t$.

The major conclusions are as follows. Isobar and background production, in all three channels, seem to be the same but scaled in cross-section proportional to the elastic differential scattering cross-section. The $t$ dependence of the $N_1^*(1400)$ is much steeper than the elastic scattering, while the $t$ dependence of $N_1^*(1520)$ and $N_1^*(1690)$ is considerably less steep than elastic scattering. To the extent that we can determine it, the $t$ dependence of the $N_{3,3}^*$ is also much steeper than elastic scattering. The $p+p$ data agree with the data of previous investigations [44], [45], [47] and, in general, with the conclusions reached.

It had been suggested that the $N_1^*(1400)$ peak observed in the p–p case might be a kinematical effect—the Deck-effect [49]—due to the following diagram.

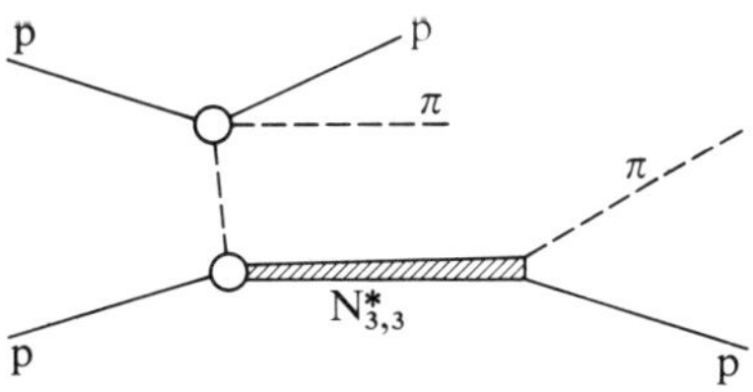


At the lower vertex, the proton disassociates into an $N_{3,3}^*$ and virtual pion, which is scattered elastically from the nucleon at the upper vertex causing it to become a real pion. It was shown that such a process gave a peak around a mass of 1400 MeV for the $N_{3,3}^*+\pi \rightarrow N+2\pi$ combination.

In the incident pion case the diagram would become as follows.

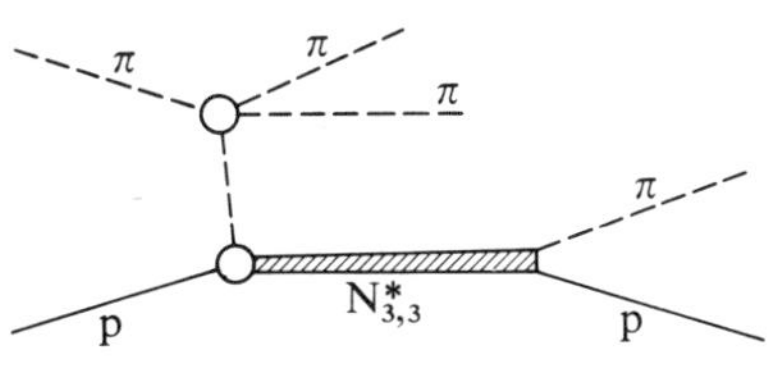


Since only elastic scattering takes place at the upper vertex, quite low momentum transfers are favoured, and the initial mass distribution of the $N_{3,3}^*$ is not changed substantially by the elastic scattering. Hence, the kinematics relating the virtual pion to the $N_{3,3}^{**}$ are not much disturbed. It is, of course, possible that this mechanism is responsible, but the fact that

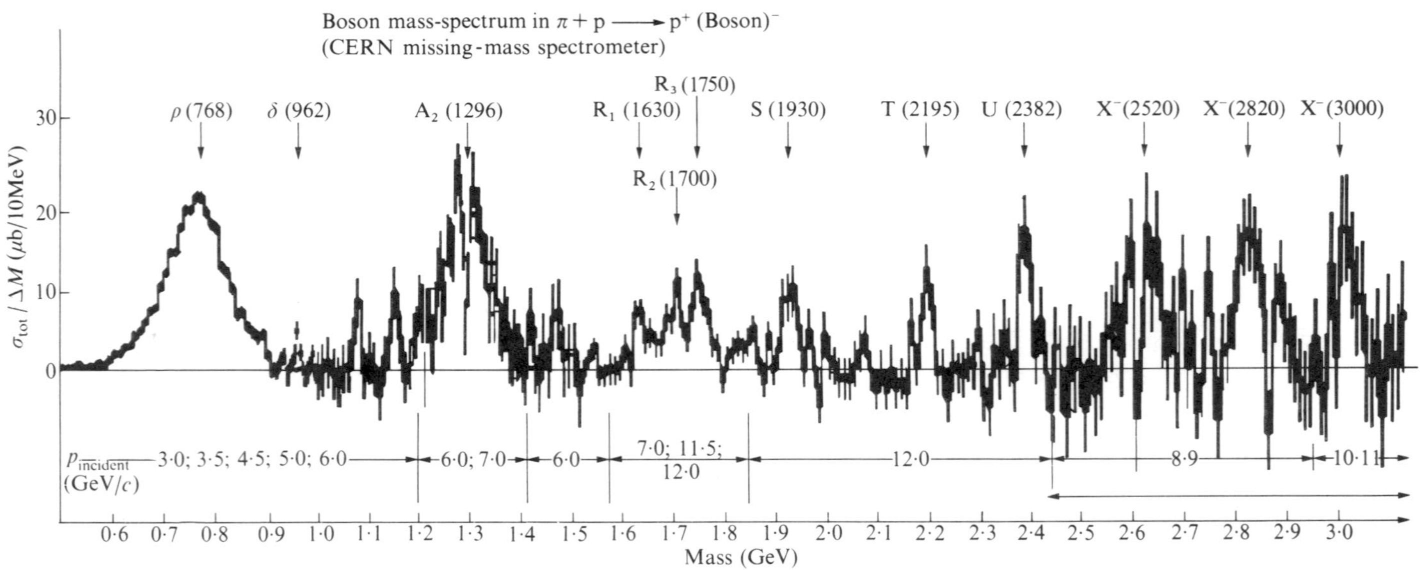


FIG. 6.17. Compiled missing-mass spectrum in the reaction $\pi^{-} + p \rightarrow X^{-} + p$ based on results of the CERN missing-mass, and CERN boson spectrometers. The last four peaks have been scaled by a factor of four, as indicated. The arrows at the top indicate where the authors believe boson resonance states occur. (From Maglic. (1969). Lund Conference.) Recent experiments have disagreed with the split $A_2$, and the status of the higher mass regions R, etc. is now open to question.

all three channels of $\pi^{\pm}$–p and p–p all show the same effect, with the same $s$ and $t$ dependence, strongly enhances the probability of the simple explanation that the $N_1^*(1400)$ is a genuine isobar or, owing to its broadness, perhaps two or more overlapping isobaric levels. We should note that in the p–p case the elastic scattering at the upper vertex is $\pi$–p elastic scattering. In the $\pi^{\pm}$–p case the elastic scattering at the upper vertex is $\pi$–$\pi$ scattering, and different isotopic spin state mixtures are involved in the $\pi^{+}+$p and the $\pi^{-}+$p case. Thus, we would not necessarily expect that, for this mechanism, the results would exhibit the very strong similarity which they do. Finally, we should notice that the approximately constant cross-section behaviour with increasing $s$, of the processes involving $T = \frac{1}{2}$ nucleon isobars, can be roughly understood on the basis of quasi-elastic scattering.

In the case of elastic scattering, we observed a large, more or less constant, cross-section, which we understood as due to elastic diffraction scattering accompanying the absorption of the incident beam. This kind of mechanism has also been considered to be partly operative for high-energy quasi-elastic scattering, which is defined as an inelastic scattering in which the baryon target has no change in the most important quantum numbers, but is increased in mass by a fraction of its rest mass. Thus, all $T = \frac{1}{2}$ nucleon isobars of $Q = +1$ would be preferential candidates for this mechanism. These are just the nucleon isobars which we have discovered exhibit approximately constant production cross-sections with increasing incident energy, which agrees with the foregoing hypothesis. Related phenomenological analyses [50] (diffraction dissociation†) come to a similar conclusion for the formation of $N_1^*$ isobars.

The missing-mass reaction

$$\pi+\mathrm{p} \rightarrow \mathrm{p}+\mathrm{X}, \tag{6.40}$$

where the recoil is detected, has been studied by a CERN group [51]. In the early version of their work, only the recoil proton at wide angles was identified, and its momentum determined by range. In later work observation of some of the decay products in the forward direction were also made, for some of the spectra. Typical results obtained are shown in Fig. 6.17. The usual peaks were observed and in addition splits were observed in the $A_2$ meson. Furthermore, some new meson peaks (referred to as R, S, T, U, etc.) were identified. The characteristics of some of these mesons will be discussed

† In so-called 'diffraction dissociation', as long as the nucleon is transformed, in an inelastic collision, to a $T = \frac{1}{2}$ state of different mass, a certain amount of angular momentum transfer, spin flip, and even parity change can occur, and the process is still favoured in the sense discussed.

in Chapters 9 and 10 and further discussion of particle production will be given in Chapters 7–9.

6.6.11. $\bar{p}$–p *collisions*

In $\bar{p}$–p collisions which lead to meson production, unaccompanied by the annihilation of the nucleon–anti-nucleon pair, the mechanism of exciting one, or both, to the $T = J = \frac{3}{2}$ isobar $N^*_{33}$ or $\overline{N}^*_{33}$ (the anti-isobar), which was predicted for the isobar model, was observed [20], [52], [53] and the energy spectra showed the predicted behaviour.

One-pion exchange calculations, [20], [52], [53] using the same form factors as nucleon–nucleon collisions, yielded predictions of cross-sections which were too large by a factor of two, and angular distributions which were too broad compared to the experimental results for single pion production. However, double pion production was satisfactorily described by the same mechanism as for the p–p interactions, Svensson (1965) [54] reported the absorption model calculations for double pion production in the channel

$$\bar{p}+p \rightarrow N^{*++}+\overline{N^*}^{--}.$$

The total cross-section was over-estimated, but the observed angular correlations were explained.

REFERENCES

1*a*. LINDENBAUM, S. J. (1957). *A. Rev. nucl. Sci.* **7,** 317.

1*b*. GELL-MANN, M. and WATSON, K. M. (1954). *A. Rev. nucl. Sci.* **4,** 219.

2. ROSENFELD, A. H. (1954). *Phys. Rev.* **96,** 139; MORRISON, A. W. (1956) *CERN Symposium Proceedings, Geneva.* Vol. II. pp. 347, 353, 357.

3. BRUECKNER, K. A. (1951). *Phys. Rev.* **82,** 598; WATSON, K. M. and BRUECKNER, K. A. (1951). *Phys. Rev.* **83,** 1.

4. MESHCHERIAKOV, M. G. and NEGANOV, B. S. (1955). *Dokl. akad. Nauk. SSSR.* **100,** 677; MESHCHERIAKOV, M. G., ZRELOV, V. P., NEGANOV, B. S., VZOROV, I. K., and SHABUDIN, A. F. (1956). *CERN Symposium Proceedings, Geneva.* Vol. 2. p.p. 347, 353, 357; MESH KOVSHKI, A. G., PLIGIN, IU. S., SHALOMOV, IA. IA., and SHEBANOV, V. A. (1956). *CERN Symposium Proceedings, Geneva.* Vol. 2. p. 362; SIDOROV, V. M. (1956). *CERN Symposium Proceedings, Geneva.* Vol. 2. p. 366; BAIUKOV, IU. D., KOZODAEV, M. S., and TIAPKIN, A. A. (1956). *CERN Symposium Proceedings, Geneva.* Vol. 2. p. 398; MORRIS, T. W., FOWLER, E. C., and GARRISON, J. D. (1956). *Phys. Rev.* **103,** 1472; FOWLER, W. B., SHUTT, R. P., THORNDIKE, A. M., and WHITTEMORE, W. L. (1956). *Phys. Rev.* **103,** 1479; BLOCK, M. M., HARTH, E. M., COCCONI, V. T., HART, E., FOWLER, W. B., SHUTT, R. P., THORNDIKE, A. M., and WHITTEMORE, W. L. (1956). *Phys. Rev.* **103,** 1484; FOWLER, W. B., SHUTT, R. P., THORNDIKE, A. M., WHITTEMORE, W. L., COCCONI, V. T., HART, E., FOWLER, E. C.,

GARRISON, J. D., and MORRIS, T. W. (1956). *Phys. Rev.* **103,** 1489. BATSON, A. P., CULWICK, B., RIDDIFORD, L., and WALKER, J. (1956). *CERN Symposium Proceedings, Geneva.* Vol. 2. p. 340; HUGHES, I. S., MARCH, P. V., MUIRHEAD, H., and LOCK. W. O. (1956). *CERN Symposium Proceedings, Geneva.* Vol. 2. 344.

5. HENLEY, E. M. and RUDERMAN, M. A. (1953). *Phys. Rev.* **92,** 1036.
6. LEWIS, H. W., OPPENHEIMER, J. R., and WOUTHUYSEN, S. A. (1948). *Phys. Rev.* **73,** 127.
7. LEWIS, H. W. (1952). *Rev. mod phys.* **24,** 241.
8. HEISENBERG, W. (1949). *Nature.* **164,** 65; *Z. Phys.* **126,** 569.
9. FERMI, E. (1950). *Prog. theor. Phys., Osaka.* **5,** 570; (1953). *Phys. Rev.* **92,** 452.

10*a*. KRETZSCHMAR, M. (1961). *A. Rev. nucl. Sci.* **11,** 1.

10*b*. NEUMAN, M. (1959). *Anais Acad. bras. Cienc.* **31,** 361, 487; SRIVASTAVA, P. P. and SUDARSHAN, G. (1958). *Phys. Rev.* **110,** 765.

10*c*. HAGEDORN, R. (1960). *Nuovo Cim.* **15,** 434; FAST, G., HAGEDORN, R., and JONES, L. W. (1963). *Nuovo Cim.* **27,** 856.

11. YUAN, L. C. L. and LINDENBAUM, S. J. (1954). *Phys. Rev.* **93,** 1431; (1956). *Phys. Rev.* **103,** 404.

12*a*. MORRIS, T. W., FOWLER, E. C., and GARRISON, J. D. (1956). *Phys. Rev.* **103,** 1472; FOWLER, W. B., SHUTT, R. P., THORNDIKE, A. M., and WHITTEMORE, W. L. (1956). *Phys. Rev.* **103,** 1479; BLOCK, M. M., HARTH, E. M., COCCONI, V. T., HART, E., FOWLER, W. B., SHUTT, R. P., THORNDIKE, A. M., and WHITTEMORE, W. L. (1956). *Phys. Rev.* **103,** 1484; FOWLER, W. B., SHUTT, R. P., THORNDIKE, A. M., WHITTEMORE, W. L., COCCONI, V. T., HART, E., FOWLER, W. B., GARRISON, J. D., and MORRIS, T. W. (1956). *Phys. Rev.* **103,** 1489.

12*b*. FOWLER, W., SHUTT, R. P., THORNDIKE, A. M., and WHITHEMORE, W. L. (1954). *Phys. Rev.* **95,** 1026.

13. LINDENBAUM, S. J. and STERNHEIMER, R. M. (1957). *Phys. Rev.* **105,** 1874.
14. LINDENBAUM, S. J. and STERNHEIMER, R. M. (1957). *Phys. Rev.* **106,** 1107; (1958). *Phys. Rev.* **109,** 1723; (1961). *Phys. Rev.* **123,** 333.
15. DZHELEPOV, V. P., GOLOVIN, B. M., KAZARINOV, I. M., and SEMENOV, M. M. (1956). *CERN Symposium Proceedings, Geneva.* Vol. 2. 115. See also reference [1].
16. BELENKI, S. Z. and NIKISHOV, A. I. (1955). *Zh. eksp. teor. Fiz.* **28,** 744.
17. KOVACS, J. S. (1956). *Phys. Rev.* **101,** 397.

18*a*. BERGIA, S., BONSIGNORI, F., and STANGHELLINI, A. (1960). *Nuovo Cim.* **16,** 1073; *b* OLSSON, M. G., and YODH, G. B. (1966). *Phys. Rev.* **145,** 1309, 1327.

18*c*. PETERS, B. (1962). *Proceedings of the International Conference on High Energy Physics, Geneva. CERN.* 623; PALA, Y. and PETERS, B. (1964). *Mat.-fys. Medd.* **33,** No. 15.

19. FERRARI, E. and SELLERI, F. (1963). *Nuovo Cim. Sup.* **24,** 453; (Also see (1963)). *Nuovo Cim.* **27,** 1450.
20. BERTOCCHI, L. and FERRARI, E. (1967). *High energy Physics.* Vol. II (Edited by Burhop). Academic Press, New York. p. 72.
21. CHEW, G. F. and LOW, F. E. (1959). *Phys. Rev.* **113,** 1640.

22. ANDERSON, J. A., BANG, V. X., BURKE, P. G., CARMONY, D. D., and SCHMITZ, N. (1961). *Phys. Rev. Lett.* **6,** 365.
23. CARMONY, D. D. and VAN DE WALLE, R. T. (1962). *Phys. Rev.* **127,** 959.
24*a*. GOEBEL, C. (1958). *Phys. Rev. Lett.* **1,** 337.
24*b*. DRELL, S. D. (1960). *Phys. Rev. Lett.* **5,** 278, 342.
25. BONSIGNORI, F. and SELLERI, F. (1960). *Nuovo Cim.* **15,** 465.
26. SALZMAN, F. and SALZMAN, G. (1960). *Phys. Rev. Lett.* **5,** 377; (1960). *Phys. Rev.* **120,** 599; (1961). **121,** 1545.
27. KALLEN, G. (1964). *Elementary particle physics.* Addison-Wellsley, New York.
28. MAGLIC, B. C., ALVAREZ, L. W., ROSENFELD, A. H., and STEVENSON, M. L. (1961). *Phys. Rev. Lett.* **7,** 178.
29. PEVSNER, A., KRAEMER, R., NUSSBAUM, M., RICHARDSON, C., SCHLEIN, P., STRAND, R., TOOHIG, P., BLOCK, M., ENGLER, A., GESSAROLI, R., and MELTZER, C. (1961). *Phys. Rev. Lett.* **7,** 421.
30. BASTIEN, P. L., BERGE, J. P., DAHL, O. I., FERRO-LUZZI, M., MILLER, D. H., MURRAY, J. J., ROSENFELD, A. H., and WATSON, M. B. (1962). *Phys. Rev. Lett.* **8,** 114.
31. DALITZ, R. H. (1953). *Phil. Mag.* **44,** 1068.
32. STEVENSON, M. L., ALVAREZ, L. W., MAGLIC, B. C., and ROSENFELD, A. H. (1962). *Phys. Rev.* **125,** 687.
33. DA PRATO, G. (1961). *Nuovo Cim.* **22,** 123.
34. FERRARI, E. and SELLERI, F. (1961). *Phys. Rev. Lett.* **7,** 387; (1962). *Nuovo Cim. Sup.* **24,** 453; (1963). *Nuovo Cim.* **27,** 1450.
35. JACKSON, J. D. (1965). *Rev. mod. Phys.* **37,** 484; JACKSON, J. D., DONOHUE, J. T., GOTTFREID, K., KEYSER, R., and SVENSSON, B. E. Y. (1965). *Phys. Rev.* B **139,** 428.
36. TREIMAN, S. B. and YANG, C. N. (1962). *Phys. Rev. Lett.* **8,** 140.
37. HAGOPIAN, V. W. and SELOVE, W. (1966). *Phys. Rev.* **145,** 1128.
38. LINDENBAUM, S. J. (1965). Strong interactions at high energy. *Proceedings of the Oxford International Conference on Elementary Particles.*
39. SOPKOVICH, N. J. (1962). *Nuovo Cim.* **26,** 186.
40. DAR, A. (1964). *Phys. Rev. Lett.* **13,** 91; DAR, A., KUGLER, M., DOTHAN, J., and NUSSINOV, S. (1964). *Phys. Rev. Lett.* **12,** 82; DAR, A. and TOBOCMAN, W. (1964). *Phys. Rev. Lett.* **12,** 511.
41. DURAND, L. I. and CHIU, Y. T. (1964). *Phys. Rev. Lett.* **12,** 399; (1965). *Phys. Rev.* B **137,** 1530; (1965). *Phys. Rev.* B **139,** 646.
42. ARNOLD, R. C. (1964). *Phys. Rev.* B **136,** 1388; DIETZ, K. and PILKULN, H. (1964). *Nuovo Cim.* **39,** 928; TREFIL, J. (1966). *Phys. Rev.* **148,** 1452; COURANT, H., SMITH, G. A., FOWLER, E. C., KRAYBILL, H., SANDWEISS, J., and TAFT, H. (1961). **123,** 2160.
43*a*. FICKINGER, W. J., PICKUP, E., ROBINSON, D. K., and SALANT, E. O. (1962). *Phys. Rev.* **125,** 2082; PICKUP, E., ROBINSON, D. K., and SALANT, E. O. (1962). *Phys. Rev.* **125,** 2091; HART, E. L., LOUTTIT, R. I., LÜERS, D., MORRIS, T. W., WILLIS, W. J., and YAMAMOTO, S. J. (1962). *Phys. Rev.* **126,** 747; Also see [20].

43*b*. BUGG, D. V., OXLEY, A. J., ZOLL, J. A., RUSHBROOKE, J. G., BARNES, V. E., KINSON, J. B., DODD, W. P., DORAN, G. A., and RIDDIFORD, L. (1964). *Phys. Rev.* B **133,** 1017; CHADWICK, G. B., COLLINS, G. B., DUKE, P. J., FUJII, T., HIEN, N. C., KEMP, M. A. R., and TURKOT, F. (1962). *Phys. Rev.* **128,** 1823; MELISSINOS, A. C., YAMANOUCHI, T., FAZIO, G. G., LINDENBAUM, S., and YUAN, L. C. L. (1962). *Phys. Rev.* **128,** 2373; REAY, N. W., MELISSINOS, A. C., REED, J. T., YAMANOUCHI, T., and YUAN, L. C. L. (1966). *Phys. Rev.* **142,** 918.

44*a*. COCCONI, G., DIDDENS, A. N., LILLETHUN, E., and WETHERELL, A. M. (1961). *Phys. Rev. Lett.* **6,** 231.

44*b*. COCCONI, G., DIDDENS, A. N., LILLETHUN, E., MANNING, G., TAYLOR, A. E., WALKER, T. G., and WETHERELL, A. M. (1961). *Phys. Rev. Lett.* **7,** 450.

44*c*. BELLETTINI, G., COCCONI, G., DIDDENS, A. N., LILLETHUN, E., SCANLON, J. P., SHAPIRO, A. M., and WETHERELL, A. M. (1965). *Phys. Lett.* **18,** 167.

45. ANDERSON, E. W., BLESER, E. J., COLLINS, G. B., FUJII, T., MENES, J., TURKOT, F., CARRIGAN, R. A., Jr., EDELSTEIN, R. M., HIEN, N. C., MCMAHON, T. J., and NADELHAFT, I. (1966). *Phys. Rev. Lett.* **16,** 855.

46. FOLEY, K. J., JONES, R. S., LOVE, W. A., LINDENBAUM, S. J., OZAKI, S., PLATNER, E. D., QUARLES, C. A., and WILLEN, E. H. (1967). *Phys. Rev. Lett.* **19,** 397.

47. ANKENBRANDT, C. M., CLYDE, A. R., CORK, B., KEEFE, D., KERTH, L. T., LAYSON, W. M., and WENZEL, W. A. (1965). *Nuovo Cim.* **35,** 1052.

48. SACLAY–ORSAY–BARI–BOLOGNA COLLABORATION. (1966). *Nuovo Cim.* A **42,** 606.

49. DECK, R. T. (1964). *Phys. Rev. Lett.* **13,** 169.

50. GOOD, M., WALKER, W. D. (1960). *Phys. Rev.* **120,** 1857; FEINBERG, E. L. and POMERANCHUK, I. YA. (1956). *Nuovo Cim. Sup.* **3,** 652.

51. MAGLIC, B. (1969). *Proceedings of the Lund Conference.*

52. ZYZEWSKI, O., ESCOOBES, B., GOLDSCHMIDT-CLARMONT, Y., GUINEA, M., MOORHEAD, T., HOFMOKL, T., LEWISCH, R., MORRISON, D. R. O., SCHNEEBERGER, N., and UNAMUNO, B. (1963). *Proceedings of the International Conference on Elementary Particles, Sienna.* **1,** 271. Societa Italia di Fisica, Bologna.

53. FERBEL, T., FIRESTONE, A., SANDWEISS, J., TAFT, H. D., GAILLOUD, M., MORRIS, T. W., WILLIS, W. J., BACHMAN, A. H., BAUMEL, P., and LEA, R. M. (1965). *Phys. Rev.* B **138,** 1528; FERBEL, T., FIRESTONE, A., JOHNSON, J., SANDWEISS, J., and TAFT, H. D. (1965). *Nuovo Cim.* **38,** 19.

54. SVENSSON, B. E. Y. (1965). *Nuovo Cim.* **39,** 667.

# 7

# KAONS AND STRANGE BARYONS

## 7.1. The discovery of 'curious' or 'strange' particles

LEPRINCE-RINGUET and Lhertier (1944) [1*a*] observed, in cosmic rays, a charged particle of mass about $1000m_e$, in a cloud chamber exposed to cosmic rays. These and V-type events, which were subsequently reported, were referred to as 'curious' (or strange) particles.

Rochester and Butler [1*b*] presented evidence (one event each) for two types of V events. The first type was a neutral V event in which a fork, consisting of the tracks of two charged particles, suddenly originates at a point within the cloud chamber, and looks like an inverted V (see Fig. 7.1(a)). Since these decays occurred in the cloud chamber gas and were not concentrated in the heavy material, such as a 3 cm lead plate extending across the chamber, it was concluded that they were due to spontaneous decay of a neutral particle.

FIG. 7.1.(a) Decay of a neutral V particle into a positive and negative pair of secondary particles. The lead plate was 3 cm thick. (b) Decay of a positive V particle into a positive particle and a neutral.

There was a second kind of V (Fig. 7.1(b)) also observed which gave the appearance of a charged particle undergoing spontaneous decay by emission of one, or more, neutral particles, accompanied by one charged particle. Thus, these events often appeared to represent a sudden and spontaneous scatter in the gas of the charged primary. This latter type of event occurred less frequently than the neutral V.

Subsequently, these particles were found to exhibit lifetimes in the range about $10^{-10}$ to $10^{-8}$ s. Thus, they were too long-lived for their decay to be due to strong, or electromagnetic, interactions and were, eventually, shown to be due to weak interactions.

Further investigations [2*a*, *b*] revealed that these neutral $V^0$ events contained at least two distinct types. The decay products of type I were a proton and a negative pion. This particle is now referred to as a $\Lambda^0$. For type II, the decay products consisted of two light particles (perhaps $\pi^+ + \pi^-$). At present, we call this particle $K^0_{\text{short}}$ (i.e. the short-lived component of the $K^0$).

There was, in the early days, a third type of particle, detected phenomenologically to be the decay at rest of a stopped charged particle, which was referred to as an S particle. This was mostly probably attributable to decay of the $K^+$ meson into various modes.

### 7.2. Hyperons

As we have just seen, the neutral $V^0$ particles were observed to decay into either a $p+\pi^-$ pair or a $\pi^+ + \pi^-$ pair. In the former case, the particle is a baryon heavier than a proton, and is called a hyperon.

The earliest identified hyperon was the $\Lambda^0 \to p+\pi^-$. Direct identification of the p and $\pi^-$, as well as the demonstration that the proton and $\pi^-$ and the incident $\Lambda^0$ direction are coplanar, established [2*c*], [3] that the $\Lambda^0$ decay is two-body. It was shown that the determination of the disintegration energy gave (except in a few cases) a unique value of $Q$ as approximately $37 \pm 2$ MeV and a corresponding unique mass of the $\Lambda^0$ of about $1115 \pm 3$ MeV. By this time, the lifetime of $\Lambda^0$ had been measured by a number of investigators, and the data was combined by Barker [4] to give

$$\tau = \left(3{\cdot}7\begin{matrix}+0{\cdot}8\\-0{\cdot}6\end{matrix}\right) \times 10^{-10}\ \text{s}.$$

About this time, it was observed [5] that there is a charged hyperon, similar to the $\Lambda^0$, observed in photographic plate exposures (Bonetti *et al.* 1953; Lal 1953). This charged hyperon was denoted $\Sigma$, and the observed decay modes could be classified as

$$\Sigma^+ \to \begin{cases} n+\pi^+ \\ p+\pi^0 \end{cases}, \tag{7.1a}$$

$$\Sigma^- \to n+\pi^-. \tag{7.1b}$$

The $\Sigma^-$ seemed to be observed also (2–3 events) in

$$\pi^- + p \to \Sigma^- + K^+. \tag{7.2}$$

(Fowler 1954 [6]).

The $Q$-value of the two-body $\Sigma$ decay was shown by Fry (1954) [7] to be

$$Q = (117 \pm 2)\ \text{MeV}.$$

The lifetime was deduced to be about the same order as the lifetime of the $\Lambda^0$. At first, it was thought that the $\Sigma^\pm$ and the $\Lambda^0$ might form an isotopic triplet, but the vast difference in $Q$-value shed doubt on this hypothesis. As was discovered later, there is a neutral

$$\Sigma^0 \to \Lambda^0 + \gamma \tag{7.3}$$

which differs in mass by less than 5 MeV from the charged $\Sigma$. This $\Sigma^0$, together with the charged $\Sigma^{\pm}$, form an isotopic triplet ($T = 1$), whereas the $\Lambda^0$ is an isotopic singlet.

A negative hyperon $\Xi^-$ was also observed (Cowan 1954 [8]), which decayed through a two-step cascade process.

$$\begin{array}{c} \Xi^- \rightarrow \Lambda^0 + \pi^- \\ \qquad\downarrow \\ \qquad p + \pi^-. \end{array} \tag{7.4}$$

The rest mass of this $\Xi^-$ was about 1320 MeV.

### 7.3. K-mesons

Some of the neutral $V^0$ particles were shown to decay into two light charged particles. It was later shown by Thompson and co-workers [9] that the great majority of these events contain two pions and a unique $Q$ disintegration value. They were designated $\theta^0$.

$$\theta^0 \rightarrow \pi^+ + \pi^- + (214 \pm 5)\ \text{MeV}.$$

Hence the mass of the $\theta^0 = 493 \pm 5$ MeV. We should note that this $\theta^0$ particle is what we now call $K^0_{\text{short}}$ (i.e. the short-lived component of the $K^0$). About this time the lifetime of $\theta^0$ was reported to be about $2 \times 10^{-10}$ s.

A charged 'curious' meson of the 'S'-type was discovered in photographic emulsions (Brown 1949 [10*a*]), and further studied by the same techniques. This meson, denoted $\tau$, was characterized by a singly charged stopping particle of mass about $1000 m_e$ ($\sim 500$ MeV), which decayed into three coplanar charged particles which appeared most likely to be pions. In some cases it was possible to show that they were pions.

The $\tau$-meson was determined from the decay of the S to have a mass of $(967 \pm 1)$ MeV, which is strikingly close to the $\theta^0$ mass.

The ratio of $\pi^+/\pi^-$ mesons in the $\tau$-decay seemed to be about 2:1. Thus the most likely explanation of the process was

$$\tau^+ \rightarrow 2\pi^+ + \pi^-.$$

This is not surprising since, although no evidence was found for a $\tau^-$, even if it existed, we would expect it to be captured, where it came to rest, and, thus, not be observed as an S particle decaying.

### 7.4. Theoretical considerations

The experimental discovery of 'curious' or 'strange' particles was a great surprise to the theoretical community. Unlike the case of the pion, which was predicted by Yukawa, there were no theoretical predictions, or even indirect allusions, to the possibility of such particles. It was early observed in cosmic ray investigations that these particles were produced with a cross-section of about one to a few per cent of the cross-section for pion production.

Thus they could not be formed by weak interactions, and required a strong interaction of cross-section of about the order of a few per cent of the pion–nucleon interaction. The behaviour of the cross-section is proportional to nuclear cross-sections, and not a strong function of $Z$. More-or-less equal production by incident neutrons and protons, and other factors, make it clear that electromagnetic processes are not involved. Furthermore, it was found that nuclei exist containing hyperons. Two stars were observed (Bonetti 1954 [11*a*]) in photographic emulsion, typically separated by a short heavily-ionizing track of about 100 $\mu$m, corresponding to a highly charged heavy particle. Since the track is short, spontaneous decay, rather than production, is indicated. Generally the second star is characterized by a large energy release and often (in about half the cases) involved emission of a $\pi$-meson. The suggested interpretation was that a hyperon is formed in the first star, and becomes bound to some nucleus, forming a hyperon–nuclear fragment. These travel together until the hyperon decays, the decay energy being sufficient to disintegrate the residual nucleus. The calculated binding energy for identified $\Lambda^0$, for example, was about 1·5 to a few MeV. This demonstrated that a hyperon ($\Lambda^0$) was bound in a nucleus, with a binding energy of the same order as a neutron (Primakoff 1954 [11*b*]), thus implying that hyperons are capable of emitting virtual pions, like nucleons, and showing that they exhibit a strong nuclear interaction.

Thus the problem arose as to why these strange particles should be produced in strong interactions and have such a long lifetime ($\sim 10^{-8}$–$10^{-10}$ s). For example, we could conceive that the $\Lambda^0$ could decay by strong interaction via the route shown in Fig. 7.2, which is interpreted as follows.

(1) Emit a virtual pion,

$$\Lambda^0 \rightarrow \Lambda^0 + \pi^0.$$

(2) The virtual pion creates a $p+\bar{p}$ pair. Thus

$$\Lambda^0 \rightarrow \Lambda^0 + p + \bar{p}.$$

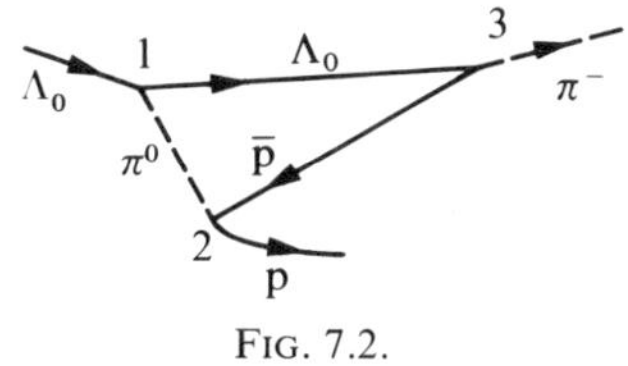


FIG. 7.2.

(3) Finally the $\Lambda^0$ and $\bar{p}$ annihilate giving a $\pi^-$. Thus

$$\Lambda^0 \rightarrow p+\pi^-.$$

The probability of step (1) could be estimated from hyperon–nuclear fragment observations. Step (2) can be estimated from the neutral pion decay, and step (3) can be estimated from the estimated cross-section for

$$n+\bar{p} \rightarrow \pi^-.$$

Taking all three factors into account, we would not expect the lifetime of the $\Lambda^0$ to exceed about $10^{-16}$ s. As we have already noticed, the observed lifetime is about $10^{-10}$ s. Thus, we need to propose an additional selection rule, which explains the suppression of decay rates by an additional factor of about $10^6$. An early attempt was the assumption by Gell-Mann and Pais (1954) [12] that these particles had very large spins† and, thus, angular momentum selection rules slow down the decay. The high spins would also explain the additional observation that these strange particles are produced in pairs. However, it would allow two $\Lambda^0$ particles as easily as a $\Lambda^0+K^0$ pair. The experiments later demonstrated that this was not the case.

Pais (1952) [13] had postulated that strange particles can only be produced (and annihilated) in pairs and not singly. Thus the long lifetime is explained, since a single hyperon cannot be annihilated with an anti-nucleon. Of course, this model does not explain why $2\Lambda^0$ cannot be produced, and does not explain cascade particles.

Gell-Mann and Nishijima [14] introduced a new quantum number called 'strangeness' to explain the data. The values of strangeness assigned to each particle were determined from the experiments, and its conservation in strong, and electromagnetic, interactions was postulated. Its violation, with one unit at a time, was allowed in weak decays. This then explained the long lifetimes. The K-meson was given a strangeness of $+1$, while the $\Lambda^0$ was given a strangeness of $-1$. $\Lambda^0$ and $K^0$ pairs were produceable, but $2\Lambda^0$ pairs were not. In the next section we will investigate this quantum number in detail.

### 7.5. The strangeness quantum number

The properties of the strangeness quantum number, and its relation to other quantum numbers can be deduced phenomenologically by observing the properties of strange particle production and decay. For the sake of clarity, we shall not closely follow the historical development. Furthermore,

† Of course, it is now known that the $\Lambda^0$ spin is $\frac{1}{2}$ and the K spin is zero.

we will restrict ourselves to long- (compared to nuclear interaction) lifetime particles. It is sufficient to note that the following facts were established as a result of both originally cosmic ray and, later, accelerator experiments at the BNL cosmotron (3 GeV), the Berkeley bevatron (6·0 GeV), and other accelerators, where strange particles were copiously produced.

If a strange $K^+$-meson, of mass now known to be $(493{\cdot}82 \pm 0{\cdot}11)$ MeV and with lifetime of $(1{\cdot}235 \pm 0{\cdot}004)\ 10^{-8}$ s, exists, there is an accompanying neutral meson† $K^0$ of mass $(497{\cdot}87 \pm 0{\cdot}16)$ MeV. A $K^+$, or $K^0$, was always found to accompany $\Lambda^0$ production‡ in reactions, such as

$$\pi^- + p \rightarrow \Lambda^0 + K^0, \tag{7.5}$$

$$\pi^+ + n \rightarrow \Lambda^0 + K^+. \tag{7.6}$$

One decay mode of the $K^0$ corresponds to the $\theta$. However, reactions like

$$\pi^- + N \rightarrow \Lambda^0 + \pi, \tag{7.7a}$$

$$\pi^+ + p \rightarrow K^+ + p \tag{7.7b}$$

are strongly forbidden. Although a $K^-$-meson of identical (within error) mass and lifetime as the $K^+$-meson also exists, reactions like

$$\pi^- + n \rightarrow \Lambda^0 + K^- \tag{7.8}$$

are also very strongly forbidden. However, reactions§ like

$$K^- + p \rightarrow \begin{matrix} \Lambda^0 + \pi^0 & (7.9a) \\ \Sigma^\pm + \pi^\mp & (7.9b) \end{matrix}$$

occur with high cross-sections, and a neutral $\Sigma^0$, which decays to a $\Lambda^0 + \gamma$ in about $10^{-16}$ s, occurs in many reactions, including

$$\begin{array}{c} \pi^- + p \rightarrow \Sigma^0 + K^0 \\ \qquad\qquad \downarrow \\ \qquad\qquad \Lambda^0 + \gamma \end{array} \tag{7.10a}$$

and

$$\begin{array}{c} K^- + p \rightarrow \Sigma^0 + \pi^0 \\ \qquad\qquad \downarrow \\ \qquad\qquad \Lambda^0 + \gamma. \end{array} \tag{7.10b}$$

† The neutral meson is composed of two components, $K_{short}$ and $K_{long}$, which have lifetimes of $(0{\cdot}87 \pm 0{\cdot}009) \times 10^{-10}$ s and $(5{\cdot}68 \pm 0{\cdot}26) \times 10^{-8}$ s. The $K^0$ lifetime will be treated in detail in Chapter 12. The spin of the kaon is zero.

‡ The $\Lambda^0$ mass is $(115{\cdot}60 \pm 0{\cdot}08)$ MeV, and its lifetime is $(2{\cdot}51 \pm 0{\cdot}03) \times 10^{-10}$ s. Its spin is $\frac{1}{2}$.

§ $\Sigma^{\pm,0}$ form an isotopic spin-1 multiplet (See Particle Data Group table in the Appendix).

Reactions (7.9) and (7.10) all occur when $K^-$ is stopped in hydrogen as well as $K^- + p$ interactions in flight.†

Thus, the $K^{0,\pm}$ and $\Lambda^0$ seem to carry another quantum number called strangeness which must be conserved. By definition, pions and nucleons have strangeness = 0, then let us define the strangeness of a $\Lambda^0$ equal to $-1$. Thus, in eqns (7.5) and (7.6), if we wish to conserve strangeness, the strangeness of the $K^+$ and $K^0$ must be $+1$. From reaction (7.9) it follows that, in order to conserve strangeness, the $K^-$ must have the same strangeness as the $\Lambda^0$, namely, $-1$. From reactions (7.9) and (7.10), it follows that the $\Sigma^+$, $\Sigma^-$, and $\Sigma^0$ all have the same strangeness as the $K^-$ (namely, $-1$) and, hence, have the same strangeness as the $\Lambda$. The charge exchange reaction

$$K^- + p \rightarrow \overline{K^0} + n$$

has also been observed. We have used a bar over the $K^0$ to distinguish the fact that its strangeness must be $-1$ and, thus, it is a different particle from the $K^0$. In fact it turns out that $K^+$ and $K^-$ are anti-particles (applying the charge conjugation operator to either produces the other), and $K^0$ and $\overline{K^0}$ are anti-particles. No doubly charged K-mesons have ever been discovered.

Now let us assume that isotopic spin is a good quantum number in all strong interactions, even those producing strange particles. Then we would expect, from the above, that $K^+$ and $K^0$ form an isotopic spin doublet ($T = \frac{1}{2}$) of strangeness $+1$, and $K^-$ and $\overline{K^0}$ form an isotopic spin doublet ($T = \frac{1}{2}$) of strangeness $-1$, which contain the anti-particles of the $+1$ strangeness doublet. In this sense, the $\pi^+$ and $\pi^-$ are anti-particles, and the $\pi^0$ is its own anti-particle. No charged partners to the $\Lambda^0$ have ever been discovered. Thus, we can assume that the $\Lambda^0$ is an isotopic singlet ($T = 0$). Since it is a baryon, there would, of course, be an anti-lambda of strangeness $+1$.

Since $\Sigma^+$, $\Sigma^-$, and $\Sigma^0$ seem to form an isotopic triplet and no higher charged partners have ever been discovered, we assume that it is a $T = 1$ isotopic triplet of strangeness $-1$. Obviously there will be an anti-$\Sigma$ isotopic triplet of strangeness $+1$.

Let us continue to assume that both the total isotopic spin, and its third component, were conserved in all strong interactions, and attempt to relate the value of the charge ($Q$) to the third component $T_z$. We recall from earlier discussion that for pions,

$$T_z = Q, \tag{7.11}$$

whereas for nucleons,

$$T_z + \tfrac{1}{2} = Q, \tag{7.12a}$$

† The latest characteristics of the various particles discovered are summarized in the Particle Data Group tables (in the Appendix).

and for anti-nucleons,

$$T_z - \tfrac{1}{2} = Q. \tag{7.12b}$$

Thus

$$Q = T_z + \frac{B}{2}, \tag{7.13}$$

where $B$, the baryon number (i.e. number of nucleons), correctly describes all cases in the pion, nucleon, and anti-nucleon systems. We should note that $B = N - \overline{N}$, where $N$ is the number of nucleons and $\overline{N}$ is the number of anti-nucleons in the system considered.

In the case of the $\mathrm{K}^+$ and $\mathrm{K}^0$, we have

$$Q(\mathrm{K}^+) = 1,$$
$$Q(\mathrm{K}^0) = 0,$$

whereas according to formula (7.13), we would erroneously predict

$$Q(\mathrm{K}^+) = \tfrac{1}{2},$$
$$Q(\mathrm{K}^0) = -\tfrac{1}{2}.$$

Thus, we need to add $\frac{1}{2}$ to formula (7.13). If we assume (as Gell-Mann did) that this displacement is due to the strangeness, we can adjust formula (7.13) by changing it to

$$Q = T_z + \frac{B}{2} + \frac{S}{2}. \tag{7.14}$$

Thus, we find this gives the right relationship for $\mathrm{K}^+$, $\mathrm{K}^0$, $\mathrm{K}^-$ and $\overline{\mathrm{K}^0}$, for $\Lambda^0$, and for $\Sigma^+$, $\Sigma^-$, and $\Sigma^0$. This argument depends on our assumption that $T_z$ is conserved, but all experimental facts to date are consistent with this requirement. Hence we find that the role of the strangeness quantum number is somewhat similar to that of the baryon number. Both of these quantum numbers are conserved in strong interactions, and both shift the relationship between charge and $T_z$. There is one difference: baryon number is absolutely conserved, even in weak interactions, whereas strangeness is violated by weak interactions.

In eqn (7.14) it is the sum of baryon number and strangeness which determines the shift between $Q$ and $T_z$. We usually define this sum ($Y = B + S$) as the hypercharge. $Y$ is conserved in strong interactions.†

† This quantity $Y$ has the same value for each member of a isotopic spin multiplet (just as $B$ and $S$ do).

For the stable particles or metastable particles—$Y = +1$ for N and $\mathrm{K}^{\overset{+}{0}}$, $Y = 0$ for $\Lambda$, $\Sigma$, $\pi$, and $\eta$, $Y = -1$ for $\Xi$ and $\mathrm{K}^{\bar{0}}$, and $Y = -2$ for the $\Omega^-$.

Thus eqn (7.14) becomes

$$Q = T_z + \frac{Y}{2} \tag{7.15a}$$

or

$$Y = 2(Q - T_z). \tag{7.15b}$$

The cascade particle has been found to exist in an isotopic spin doublet $\Xi^-$ and $\Xi^0$, which has been observed in the reactions

$$\pi^- + n \rightarrow \Xi^- + K^0 + K^0$$

and

$$K^- + p \rightarrow \Xi^0 + K^0.$$

We thus conclude that its strangeness is $-2$. A strangeness $-3$ particle, with hypercharge $-2$ (the $\Omega^-$), has also been discovered [15].

### 7.6. Interaction of K-mesons with nucleons

Since the $K^{\pm}$ lifetime ($\sim 1{\cdot}24 \times 10^{-8}$ s) is long enough to allow beams of $K^{\pm}$, produced at high energy accelerators, to interact with protons and nuclei, the direct K–N interaction has been studied.

Figure 7.3(a) shows an earlier compilation of the total cross-section for the scattering of positive K-mesons by protons (for up to several GeV incident energy). Due to the lower production rates, and much shorter decay path, of kaons compared to pions, these data are not as accurate or complete as for the pion case.

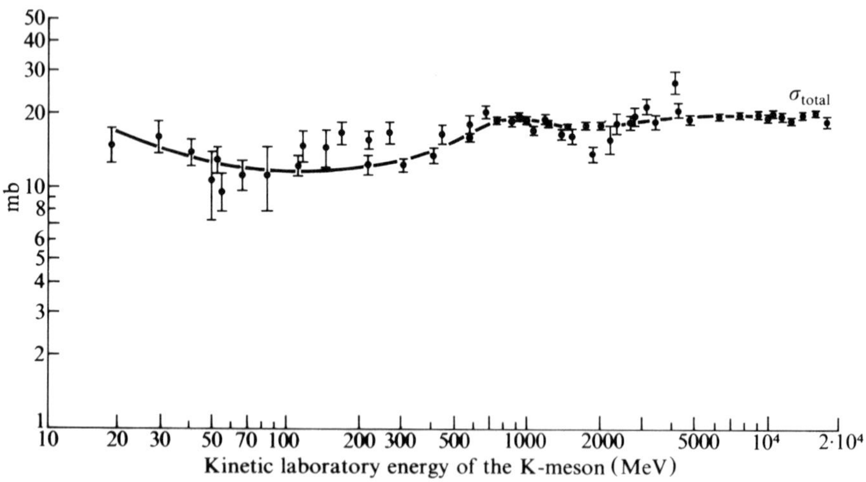


FIG. 7.3.(a) Total $K^+ + p$ cross-section as a function of energy. (Barashenkov and Maltsev. (1961). *Fortr. Phys.* **9**, 549, and Kallen (1964). *Elementary Particle Physics*, Addison–Wesley, Reading, Mass.)

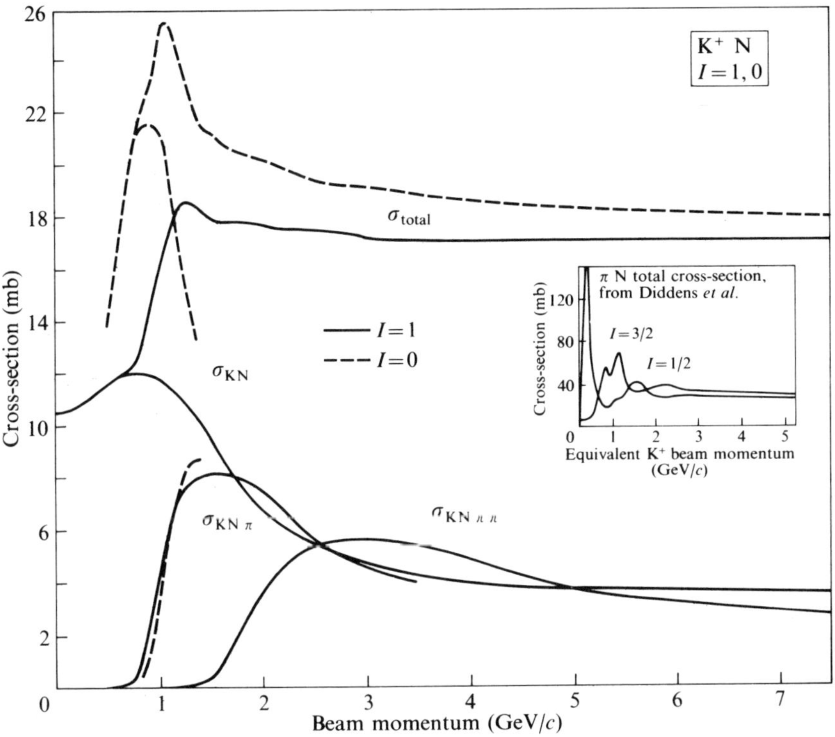


FIG. 7.3.(b) Total $K^+ + N$ cross-sections from Particle Data Group (1970) (UCRL 20 000. September, 1969.)

In contrast to the lower-energy $\pi^\pm + p$ interaction, which exhibited a striking resonance behaviour, we find the $K^+ + p$ total cross-section is smooth and there is little, or no, evidence for peaks. A modern compilation (from Particle Data Group 1970) is shown in Fig. 7.3(b). There is no evidence for peaks of a resonant nature.

The structure in $\pi^\pm$–p total cross-sections was due to the formation of resonant intermediate states by the incoming pion and the proton at certain resonant energies. In the $K^+ + p$ state, namely, a state with baryon number 1, $S = +1$, double charge, but with $T = 1$, the implication is that there are no resonant states. This is (as far as we know from all investigations to date) the case. Nevertheless, recent phase-shift analyses of the $K^+ + p$ elastic scattering between 0·86 and 1·91 GeV/$c$, using polarized targets (combined with other relevant data), have indicated a resonance-like behaviour for the $p_{\frac{3}{2}}$ partial wave, for incident momenta below 1·4 GeV/$c$. However, above 1·5 GeV/$c$, the behaviour of various solutions becomes qualitatively different and rather uneven. Therefore, especially since the uniqueness of phase-shift

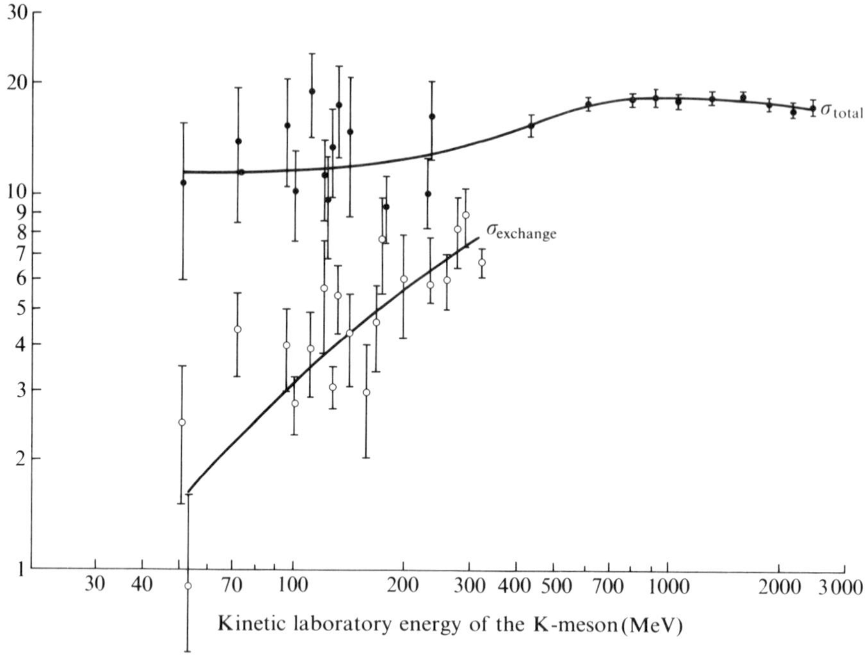


FIG. 7.4. Total $K^+ + n$ cross-section, and the $K^+ + n$ charge-exchange cross-section, as a function of energy. (After Barashenkov and Maltsev, and Kallen (1964). *Elementary Particle Physics*. Addison–Wesley, Reading, Mass.)

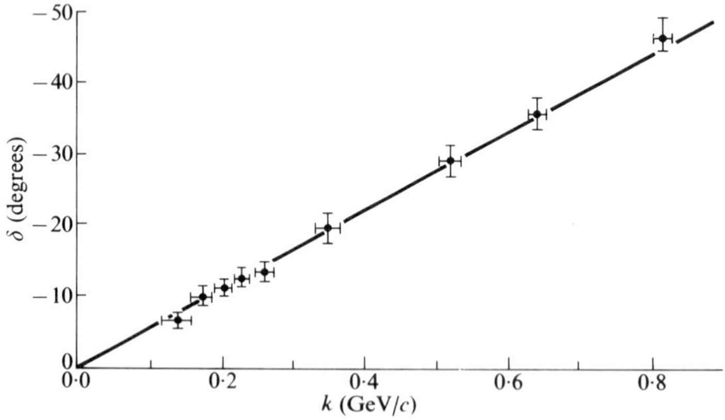


FIG. 7.5. *S*-wave phase shift for $K^+ + p$ scattering protons. (After Goldhaber, Chinowsky, Goldhaber, Lee, O'Halloran, Stubbs, Pjerrou, Stork, and Ticho. (1962). *Phys. Rev. Lett.* **9,** 135.)

analyses are open to question, we cannot yet conclude that a resonance behaviour is indicated. The reader is referred to the paper by Rebka *et al.* [45] for further details.

We can conceive of processes (these exist, as we shall discuss later) where

$$\begin{aligned} K^+ + p \rightarrow\ & K^* + N \\ & K + N^* \\ & K^* + N^*, \end{aligned}$$

where either the K or nucleon or both are excited to higher isobaric states in the collision. However, such threshold effects often do not show up very well in total cross-section measurements. As we observed in p+p collisions, although the onset of single isobar production was clearly indicated by a rise in the cross-section, the onset of double pion production via excitation of two isobars, was hardly noticeable in the total cross-section. Charge exchange is forbidden in $K^+ + p$, since the recoil proton would then have to be doubly charged. However, for $K^+$-mesons incident on neutrons, the charge-exchange reaction

$$K^+ + n \rightarrow K^0 + p$$

is allowed.

Figure 7.4 shows the total $K^+ + n$, and the charge exchange, cross-section as a function of energy. Deuterium targets were used to determine the $K^+ + n$ cross-sections. As we can see, the charge exchange cross-section becomes important as the energy increases. The more-or-less constant cross-section at low energies for $K^+ + p$ implies that *s*-wave scattering dominates and, indeed, this is found to be the case below 1 GeV/*c* [16]. However, *p* and higher waves become important as the energy increases. Figure 7.5 shows the results of a pure *s*-wave phase-shift fit for $K^+ - p$.

At low momentum ($p \lesssim 0{\cdot}6$ GeV) the inelastic cross-section is small enough to be neglected, and a real phase shift was used. Also, the angular distributions were isotropic within errors, thus supporting the *s*-wave dominance. However, at $\gtrsim 1{\cdot}0$ GeV the inelastic cross-section is already important, and these approximations are quite invalid.

The behaviour of the total elastic $K^- + p$ cross-section, and the charge-exchange cross-section $K^- + p \rightarrow K^0 + n$, is given in Fig. 7.6(a). Here we see that there is considerable structure. In retrospect, this is not surprising since, whereas in $K^+ + p$ there are no known resonant states formed by the two incident particles, in $K^- + p$ this is now known not to be the case. The $K^- + p$ system contains a mixture of $T = 0$ and $T = 1$ states.

There are, at present, a number of $T = 0$ and $T = 1$, higher-mass, strangeness $= -1$ baryons, resonances, or isobars, known—some of which were

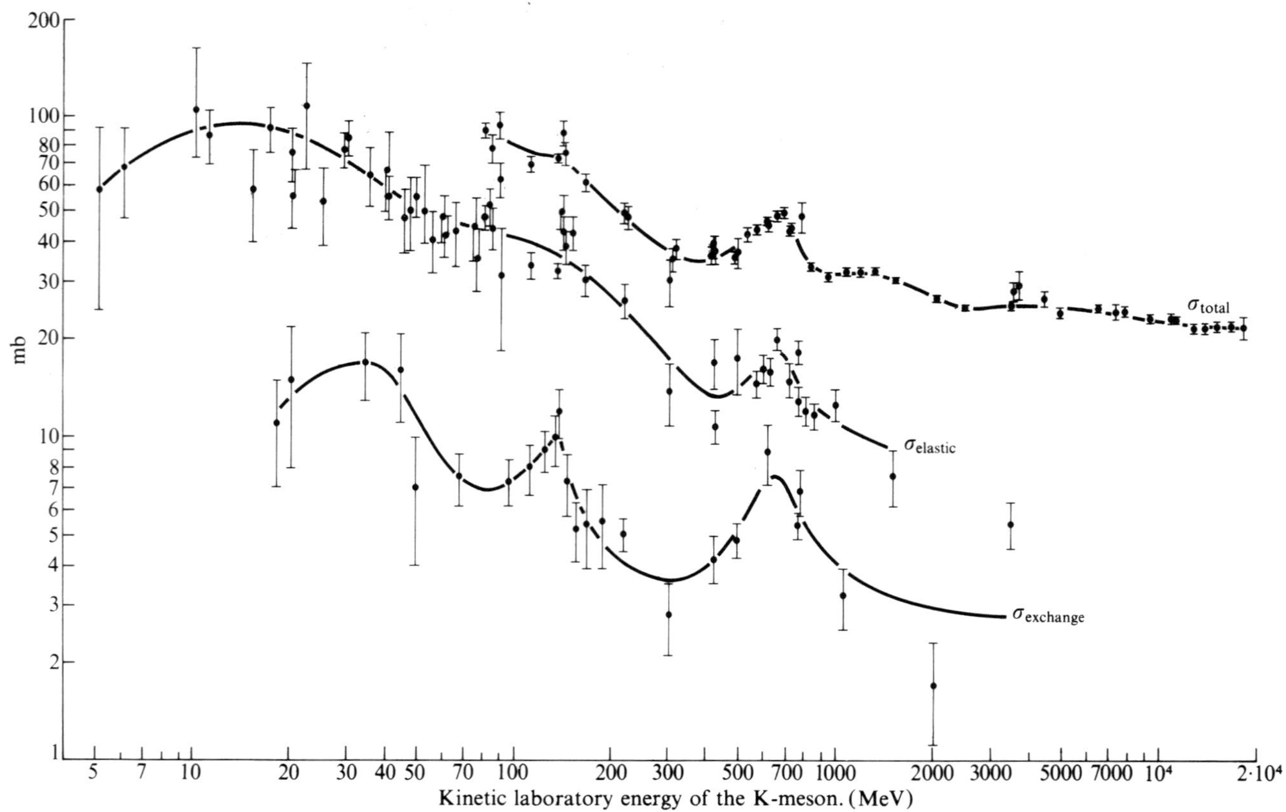


FIG. 7.6.(a) $K^- + p$ total cross-section, elastic cross-section, and charge-exchange cross-section. (After Barashenkov, and Maltsev, and Kallen (1964). *Elementary Particle Physics*. Addison–Wesley, Reading, Mass.)

originally discovered by studying $K^- + p$ interactions. In addition, even if we restrict ourselves to the known metastable particles, $K^- + p$ has a much higher number of inelastic possibilities, which include

$$K^- + p \rightarrow \Lambda^0 + \pi^0,$$

$$K^- + p \rightarrow \Sigma^{\pm,0} + \pi^{\mp,0},$$

$$K^- + p \rightarrow \overline{K^0} + n.$$

A modern compilation of $K^-N$ cross-sections is shown in Fig. 7.6(b).

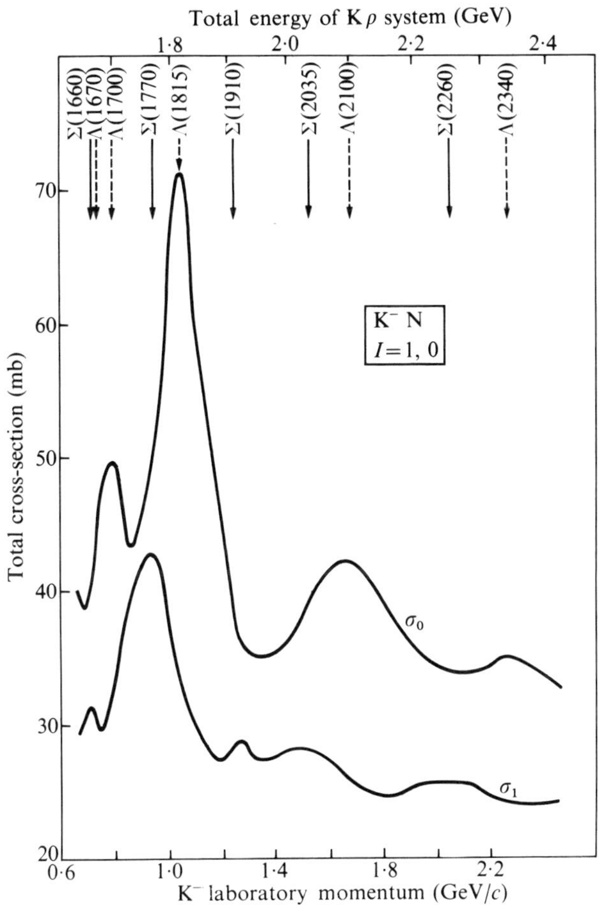


FIG. 7.6.(b) $K^-N$ from Particle Data Group (August 1970) (1970). *Phys. Lett.* **B 33,** 1. The arrows indicate various known resonances.

### 7.7. Spin of the K-meson

The method used for determining the pion spin, by studying a production reaction

$$\mathrm{p}+\mathrm{p} \rightarrow \mathrm{D}+\pi^{+} \tag{7.16}$$

and its inverse

$$\pi^{+}+\mathrm{D} \rightarrow \mathrm{p}+\mathrm{p} \tag{7.17}$$

and using detailed-balance to determine the spin, cannot be used in the K-meson case, since production of a K-meson by itself is forbidden by strangeness conservation. Thus, processes like

$$\mathrm{p}+\mathrm{p} \rightleftarrows \mathrm{D}+\mathrm{K}^{+} \tag{7.18}$$

and

$$\mathrm{K}^{\pm}+\mathrm{p} \rightarrow \pi^{\pm}+\mathrm{p} \tag{7.19}$$

do not occur.

In principle, reactions like

$$\mathrm{K}^{-}+\mathrm{p} \rightleftarrows \pi^{-}+\Sigma^{+} \tag{7.20}$$

$$\mathrm{K}^{-}+\mathrm{n} \rightleftarrows \pi^{-}+\Lambda^{0} \tag{7.21}$$

could be used. However, the lifetime of the $\Lambda$ and $\Sigma$ are too small to allow observation of the inverse processes to sufficient precision to be of any practical value. Fortunately, however, the spin of the K-meson could be determined from its decay characteristics. This is possible because spin conservation is an absolute conservation rule, which is obeyed even by weak interactions. However, since parity is violated by weak interactions (as we shall see in Chapter 12), the parity of the K-meson could not be determined this way. In fact, Dalitz had originally tried to determine the parity of the $\tau$ (i.e. $3\pi$) and $\theta$ (i.e. $2\pi$) decay modes of the K-meson by assuming parity was conserved. This led to the very surprising conclusion that the $\tau$ and $\theta$ had different parities. However, since their other characteristics, such as mass and lifetime, seemed the same within small errors, it appeared likely that they were both the same particle and, therefore, that parity was apparently violated. This was referred to, at the time, as the '$\tau$–$\theta$' puzzle, and led Lee and Yang to investigate the experimental status of parity conservation in weak interactions—this subject will be treated in the chapter on weak interactions. The $\mathrm{K}^0$ can decay into the two pion modes

$$\begin{aligned} \mathrm{K}^{0} &\rightarrow \pi^{+}+\pi^{-} \quad \text{(69 per cent)} \\ &\rightarrow 2\pi^{0} \qquad\quad \text{(31 per cent).} \end{aligned} \tag{7.22}$$

Thus the neutral decay mode has a final state consisting of two identical bosons and, in order to satisfy Bose statistics, must have a wave function which is symmetric under the interchange of the two $\pi^0$. Since, therefore, the orbital angular momentum must be even and both pions are pseudoscalar, the spin–parity combinations allowed in the final state are even–even (i.e. $0^+, 2^+, 4^+, 6^+, \ldots$).

The study of the three-pion decay mode (i.e. the $\tau$ mode),

$$K^+ \to \pi^+ + \pi^+ + \pi^-, \tag{7.23}$$

allows a determination of the spin. There are only two independent pion momenta of the three. We have previously mentioned that a Dalitz plot, with each axis labelled by one of these momenta, is a good way of studying the behaviour of matrix elements, since the density of points in any region is proportional to the local matrix element. Thus, for a spin zero K, we would expect uniform density of points in a Dalitz plot and predictable non-uniformity for other spins. Uniformity was discovered, demonstrating that the $K^+$ spin is zero. However, the parity of the final state is obviously odd, since we have three pions all in an $S$-state. This led to the $\tau$–$\theta$ (opposite parity) puzzle.

In the special case which we are considering, all three final state particles have the same mass and, thus, it is useful to plot the points inside an equilateral triangle (i.e. a symmetric Dalitz plot) as shown in Fig. 7.7. The individual kinetic energies are measured as a distance inward from the sides as shown. $T_1 + T_2 + T_3 = M_K - 3m_\pi =$ constant. The boundary curve corresponds to the case where two pions move parallel, or anti-parallel, to each other. The lines AA′, BB′, CC′ correspond to events for which two of the three pions have equal energy.

The assumption of different spin values for the $K^+$ will lead to the prediction of different regions, in which we would expect depletion of events. A quantitative general analysis of this problem has been made by Zemach [17]. We shall consider here the qualitative features of the problem. We wish to take advantage of the symmetry property imposed on the two identical $\pi^+$ mesons, namely, that their orbital angular momentum about

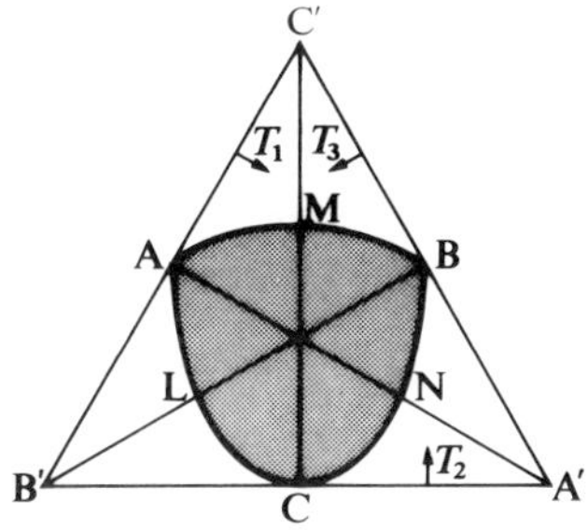


FIG. 7.7.

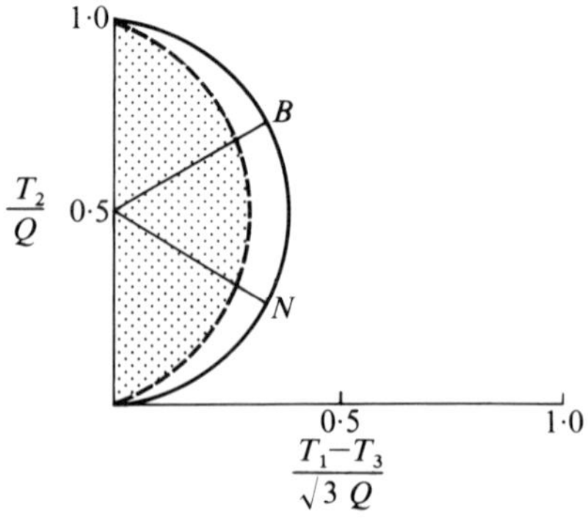


FIG. 7.8. Dalitz plot for 200 $K^+ \rightarrow 2\pi^+ + \pi^-$ events. The events are folded into a half plot, taking advantage of the symmetry between the two $\pi^+$. The axes are labelled similarly to Fig. 7.7, except for the fact that the kinetic energies are expressed in units of the $Q$ value. The dashed boundary is for non-relativistic kinematics (after Orear *et al.* [18].)

their centre of mass must be even, to satisfy Bose statistics. Let us denote this angular momentum by $l_{++}$. The angular momentum of the negative pion relative to the centre of mass of the two positive pions is then denoted by $l_-$. Thus,

$$|l_{++} - l_-| \leq |J| \leq l_{++} + l_- \tag{7.24}$$

and, if $J > 0$, then one of either $l_{++}$ or $l_-$ is greater than or equal to 1. Let us consider the two possible cases

$$l_{++} = 0, \quad \text{or} \quad l_{++} = 2, 4, \ldots \tag{7.25}$$

If $l_{++} \geq 2$, there would be a considerable centrifugal barrier operating when the two positive pions have near zero relative kinetic energy. Kinematically, this corresponds to the case when the negative pion† has its maximum energy. Hence, for $l_{++} = 2, 4, \ldots$, there should be a considerable depletion of events near M on the Dalitz plot. The extent and shape of the depleted region depends on the detailed dynamics. However, we would intuitively expect the depletion to extend over a region of about $kR$, where $R$ is the average radius corresponding to the interaction and $k$ is the relative momentum between the two pions.

By similar reasoning, if $l_- = 1, 2, \ldots$, the centrifugal barrier should deplete points where the $\pi^-$ is near rest (i.e. near C). The appearance of the data is shown in Fig. 7.8.

There is no evidence for any non-uniformity. Thus, it is assumed that the spin of the K-meson is zero. Subsequently, this has been confirmed in many experiments. That the $K^-$ which has the same mass and lifetime, but opposite

† $T_2$ is assigned to the negative pion.

strangeness, is the anti-particle of the $K^+$ has been demonstrated in a number of $K^+K^-$ pair-production experiments and other experiments. That the $K^-$ is accompanied by a neutral anti-K-meson ($\overline{K^0}$) is shown, for example, by the charge exchange reaction

$$K^- + p \rightarrow \overline{K^0} + n. \tag{7.26}$$

To date, the vast body of accumulated experimental data indicates clearly that the spin and parity of all four K-mesons is the same.† That all four ($K^+$, $K^0$, $K^-$, $\overline{K^0}$) K-mesons have the same spin, and parity, is consistent with the general theoretical prediction that anti-bosons have the same spin and parity as the original boson, whereas anti-fermions have the same spin but the opposite parity compared to the original particle. We should note that, for bosons with strangeness $= 0$, like the pion, the anti-particles are contained in the same isospin multiplet as the particles. However, the anti-particles are in a different isospin multiplet for strange boson particles. This is necessarily so, since the anti-particles have the opposite sign of strangeness.

## 7.8. Determination of the spins of the strange baryons

The $\Lambda^0$, $\Sigma$, and some $Y^*$ spins can be determined from the angular distribution of the decays. The lambda is produced in the reaction

$$\pi^- + p \rightarrow \Lambda^0 + K^0, \tag{7.27}$$

and $Y^*$ are produced in reactions initiated by $K^-$ incident on protons. The unknown spin must be aligned along the normal to the production plane, in order to preserve parity conservation in the strong production interaction. That this is so can be seen, since, if the expectation values $\langle \mathbf{J}_\Lambda \cdot \mathbf{p}_K \rangle$ and $\langle \mathbf{J}_\Lambda \cdot \mathbf{p}_\pi \rangle$ are not zero, then they are odd under a parity transformation. Thus, parity will not be conserved. The spin can then be determined by analysis of the angular distribution in terms of a variable assumed spin, which is then deduced from the data.

The determination of spin and/or parity of elementary particles is usually accomplished in two kinds of experiments. The method being pursued here, namely a study of a production process, is generally referred to as the 'production experiment' method. In our study of the first-discovered isobar, namely, the $T = J = \frac{3}{2}$ nucleon isobar, we used the 'formation experiment'

† It often occurs that the accumulated consistency with an assignment throughout many future experiments is as, or even more, important in demonstrating its correctness than the original experiment.

method. In this method the two incident channel particles resonate at a particular energy to form one resonant intermediate state.

7.8.1. *Decay angular distribution in forward direction*

If we could select the $\Lambda^0$ particles produced exactly along the incident beam direction, a particularly simple relationship [19] between the spin and the angular distribution of the pion, which comes from the lambda decay $\Lambda^0 \to p+\pi^-$, would follow. For these forward-produced $\Lambda^0$, the projection of the angular momentum along the beam direction is zero. Furthermore, the spin flip term must be zero for forward produced $\Lambda^0$. Hence, we have a situation where only the $f(\theta)$ term in the angular distribution (i.e. $g(\theta) = 0$) must be considered. If the $\Lambda^0$ particle (or any hyperon), with its spin ($J$) aligned in a given direction, subsequently disintegrates by emitting two other particles of known spin, the angular distribution in this decay process is completely calculable. The state corresponding to the decay of a strange baryon into a pion and a baryon, which conserves parity, can be expressed as†

$$|\Psi_f\rangle = \text{constant } \Sigma\zeta_{m_s} Y_l^{m_l}(\theta)\langle l,m_l;\, J_1 M_{J_1}\,|JM_J\rangle.$$

$M_J$ is the value of the projection of $J$ (the angular momentum of the decaying baryon) along the incoming pion direction, hence it must equal the value for the original target proton, namely $\pm\frac{1}{2}$, with equal probability for each. $J_1$ refers to the baryon decay product. Only one of the two values of $l = J\pm\frac{1}{2}$ will occur due to the requirement of parity conservation. $\theta$ is the direction of motion of the decay pion relative to the incident beam. We can then average over the two values of $M_J$, and use the appropriate Clebsch–Gordan coefficients to obtain the decay angular distribution $D(\theta)$ as a function of $J$ and $l$. It turns out that the angular distributions depend only on $J$, and are independent of parity or $l$. They are given in Table 7.1 below. Hence this treatment and the deduced angular distributions can be applied to the

TABLE 7.1

| $J$ | $D(\theta)$ |
|---|---|
| $\frac{1}{2}$ | $1$ |
| $\frac{3}{2}$ | $1+3\cos^2\theta$ |
| $\frac{5}{2}$ | $1-2\cos^2\theta+5\cos^4\theta$ |

† $J$ is the total angular momentum of the decaying hyperon and $M_J$ is its projection along the incident particle direction of motion. $J_1$ refers to the spin of the decay product accompanying the pion when the hyperon decays.

$\Lambda$ decay which violates parity. Thus, we see that, for spin $\frac{1}{2}$, we expect an isotropic decay distribution. Figure 7.8 (Eisler *et al.* 1958) [20] shows that only spin $\frac{1}{2}$ for the $\Lambda$ fits the data. Eisler *et al.* also determined the spin of the $\Sigma^-$ from the reaction $\pi^- + p \rightarrow \Sigma^- + K^+$, using this method, and obtained $J = \frac{1}{2}$. Of course, there is one limitation of the above method, namely that it is not possible to pick forward-going $\Lambda^0$ only and have sufficient statistics, in a practical case. Therefore, we are forced to make some estimate of how far away from the forward direction we can include statistics and still expect the method to work.

It is usually argued that $\theta \leq \frac{1}{L}$ radians, where $L$ is the maximum angular momentum involved, is acceptable for use with this method. Since the range of interaction is $\sim$ 1 fm then one would expect $L \sim k$, where $k$ is the outgoing wave number in units of $\text{fm}^{-1}$. We can also estimate $L$ from the analysis of the observed angular distribution (i.e. the observed complexity reflects the value of $L$). As previously mentioned, Eisler *et al.* used this method to obtain the $\Lambda^0$ and $\Sigma^0$ spins. Subsequent detailed investigation [20*b*] of the production angular distribution and polarization of the $\Lambda$ indicated that the originally accepted interval, $\theta < 60°$ (for which $k \approx 1{\cdot}3$), chosen to allow sufficient statistics, was too large to be justified. The unforeseen complexities subsequently observed were thought to arise as a result of the dominance of N* resonances of high spin in the incident channel at these energies. The difficulties with this method were surprisingly emphasized when it often failed to give the accepted values when applied to well-known cases, such as N*(1238) and Y*(1385). Stodolsky and Sakurai [21] pointed out that meson–proton collisions leading to the production of fermion resonances often involve vector meson exchange to a considerable, even dominant, degree. Thus we would expect a situation quite different from that assumed in the above method and, as a result, the decay distribution for a $J = \frac{3}{2}$ resonance, produced at small angles from the beam, would be very different from the expected $1 + 3\cos^2\theta$. However, the method does seem to work well when spinless one-meson exchange is involved. For example, when one-pion exchange appears to dominate a reaction, the method seems to work reasonably well. A more general test method [22] was subsequently used to exclude spins $\frac{3}{2}$ and $\frac{5}{2}$ for the lambda [23] and, thus, demonstrate consistency with spin $\frac{1}{2}$. We discuss this method in the next section.

### 7.8.2. *Test-function method*

A method of testing spin hypothesis for metastable fermions was proposed by Lee and Yang [22]. The method requires only a measurement of the polar angular distribution of the decay. Neither the use of the azimuthal decay distribution, nor knowledge of polarization, is required. The method works well only in those cases where there is a large decay asymmetry, such as in hyperon weak interaction decay. As a matter of fact, the large reported decay asymmetry for $\Lambda$ particles in associated production was the major motive for the proposal of the method.

Lee and Yang considered a sample of fermions of spin ($J$) which are an incoherent mixture of $2J+1$ magnetic substates with arbitrary populations.

In order to maximize polarization effects, the axis of quantization is usually taken along the production normal. The method can be illustrated as follows. In general, the decay angular distribution will contain even Legendre polynomials up to order $L \leq 2J-1$.

If the decay violates parity, it will contain odd Legendre polynomials up to an order $L \leq 2J$. The coefficients of the odd $P_L$ terms will be proportional to the decay asymmetry parameter $\alpha$ (where $|\alpha| \leq 1$). The values of $\alpha$ different from zero arise from the interference between two final states of different parities. In this method, for each spin assumed we define $2J+1$ test functions $T_{J,M}$, which are constructed from linear combinations of the Legendre polynomials, such that when each $T_{J,M}$ is averaged over the decay distribution necessarily $|T_{J,M}| \leq 1$. Thus if we find $|T_{J,M}| > 1$ for any of the test functions, the value of $J$ is excluded. Some examples of these test functions are shown in the Table 7.2. This method was used by Crawford *et al.* [23] on 614$\Lambda$ decays from the reaction $\pi^- + p \rightarrow \Lambda^0 + K^0$, for incoming

TABLE 7.2

| $J = \frac{1}{2}$ | $J = \frac{3}{2}$ |
|---|---|
| $T_{\frac{1}{2},\pm\frac{1}{2}} = \pm 3P_1$ | $T_{\frac{3}{2},\pm\frac{3}{2}} = \pm 9P_1 + 5P_2 \mp \frac{1}{3}(P_3)$ |
| | $T_{\frac{3}{2},\pm\frac{1}{2}} = \pm 3P_1 - 5P_2 \pm 7P_3$ |

$\pi^-$ momenta from 0·95–1·23 GeV/$c$. Events from all production angles were used. They concluded that the decay distribution was consistent with a linear dependence on $\cos\theta$. Thus, no need was found for Legendre polynomials higher than $P_1$. Secondly, $\alpha\bar{P} = \langle T_{\frac{1}{2},\frac{1}{2}} \rangle = 0{\cdot}57 + 0{\cdot}066$. $\bar{P}$ is the average over all angles of the $\Lambda$ polarization. Since $\alpha \approx \frac{2}{3}$, it is clear that the lambdas are produced highly polarized. Examining the general expressions for the test functions given by Lee and Yang, we find that if we believe the spin of a particle is $\frac{1}{2}$, in order to rule out a higher spin $J$, $|\alpha\bar{P}|$ must be larger than $\dfrac{1}{2J}$. Thus, if $|\alpha\bar{P}| > \frac{1}{3}$, spin $J \geq \frac{3}{2}$ is ruled out. Since we have just seen that experimentally $\alpha\bar{P} = 0{\cdot}57 \pm 0{\cdot}066$, it is clear that $J \geq \frac{3}{2}$ is ruled out. Evaluation of various test functions for $J = \frac{3}{2}$ and $J = \frac{5}{2}$ led to the following violations of the found conditions

$$\langle T_{\frac{3}{2},\frac{3}{2}} \rangle = 1{\cdot}77 \pm 0{\cdot}244,$$
$$\langle T_{\frac{5}{2},\frac{5}{2}} \rangle = 2{\cdot}99 \pm 0{\cdot}408,$$

thus ruling out $J = \frac{3}{2}$ and $J = \frac{5}{2}$ for $\Lambda$. The physical reason for the exclusion is the large asymmetry in the $\Lambda$ decay. It is impossible to construct, for higher spins, a suitable mixture of magnetic substates which is highly polarized and will also have a linear decay asymmetry.

The only limitation on the method is that the particles must decay in an isolated system—a condition admirably met for weak decays.

### 7.8.3. *S-state capture*

The capture (or production) of particles in an $S$-state leads to selection rules which place severe restrictions on the $J^P$ values. We already have made use of this in the case of the pion (i.e. parity determination), and will shortly find that $S$-state capture is of interest in determining the spin of the produced $\Sigma^+$ in the reaction $K^- + p \rightarrow \Sigma^{\pm} + \pi^{\mp}$. Furthermore, the parity of the $K^-$ meson was deduced from its capture in helium.

It had been shown by calculations that capture for a negative pion occurred almost exclusively from the $S$-state. This is because the $P$-state capture rate is very small compared to the $P \rightarrow S$ radiation transition rate [24]. This is not the case for $\bar{p}$ and $K^-$ capture and, thus, we might have expected that $P$-state capture would be appreciable. However, it appears that Stark effect mixing [25] induced by the electric fields of protons causes the atom to pass near enough to rapidly cause transitions from all $l \neq 0$ states (i.e. $P, D, \ldots$) down to the $S$-state, from which nuclear absorption then occurred quickly. A confirmation of these predictions was obtained in two experiments [26], [27], which searched for $\tau$-decays from $K^-$ coming to rest in liquid hydrogen. The observation of only one $\tau$-decay in more than $10^5$ $K^-$ captures allows us (using the $K^-$ lifetime and $\tau$ branching-ratio) to estimate the $K^-$ capture time to be about $3 \times 10^{-12}$ s. This is in reasonable agreement with the theoretically predicted value of about $2{\cdot}4 \times 10^{-12}$ s, when Stark mixing is taken into account. Without Stark mixing, the predicted capture time would be an order of magnitude slower.

Thus we can conclude [28] that more than 99 per cent of $K^-$ captures in hydrogen occur from an $S$-state. The same situation should occur for $\bar{p}$–p capture processes.

### 7.8.4. *The $\Sigma^{\pm}$ spin*

Very slow $K^-$ mesons captured at, or near, rest in $H_2$ will be captured almost exclusively from $S$-states. Thus in the capture reaction at, or near, rest

$$K^- + p \rightarrow \Sigma^{\pm} + \pi^{\mp}.$$

We can safely conclude [28] that, since the kaon has zero spin, the initial state spin is $J_i = \frac{1}{2}$ and, for any arbitrary spin $J$ of the $\Sigma$, the only possible magnetic substates for the $\Sigma$ along a $z$-axis taken parallel to its line of flight are $J_z = \pm\frac{1}{2}$. This results in a unique angular distribution for $\Sigma$ decay for each $J$-value assumed. For $J = \frac{1}{2}$ the angular distribution is isotropic, and for $J = \frac{3}{2}$ the angular distribution is $1+3\cos^2\theta$. An application of this method (Leitner *et al.*) [29] to $145\Sigma^+$, produced by stopped $K^-$ in a hydrogen bubble chamber gave a distribution consistent with isotropy, but was about 20 standard deviations different from the $J = \frac{3}{2}$ distribution. A much larger sample of about $2000\Sigma^{\pm}$ was subsequently analysed [30], and the same conclusions reached. Thus the spin of $\Sigma^+$, and $\Sigma^-$, is $\frac{1}{2}$.

### 7.8.5. *Moments method*

The moments method (Byers and Fenster) [31] can be applied to a sample of fermions of arbitrary spin, produced in a reaction, provided the fermions decay into a spin-$\frac{1}{2}$ fermion and a spin-0 meson. Whenever the production mechanism is such that the magnetic substates become unequally populated, there can be anisotropies introduced into the decay angular distribution, and the polarization of the decay fermion. The method then often allows the spin and parity (provided the decay preserves parity) to be deduced. The method has been used in the following cases

$$M + N \rightarrow Y \text{ (or } Y^*) + M, \tag{7.28a}$$

$$Y^* \rightarrow Y + M, \tag{7.28b}$$

$$Y \rightarrow N + M, \tag{7.28c}$$

where M is a meson ($\pi$ or K) which conserves strangeness in eqn (7.28a), Y is a spin $\frac{1}{2}$ hyperon, $Y^*$ is an excited hyperon state, and N is a nucleon.

The initial strong interaction is parity conserving. Thus any polarization of the Y (or $Y^*$) must be normal to the production plane. The complexity of the angular distribution, in the second decay interaction, allows us to estimate the minimum spin of the $Y^*$ or Y. If the second step ($Y^*$ decay) is still parity-conserving,† the third step should be parity-violating to be

† This is required if we wish to determine parity, as well as spin.

useful. This is necessary since a parity-violating decay in eqn (7.28c) allows us to determine the polarization state of the Y from the asymmetry. We shall develop this method more generally than we require for the present, as it will also be useful in spin–parity determination of hyperon resonances.

The above analysis assumes that the amplitude of the resonant intermediate state Y* is large enough for other background amplitudes to be neglected. Furthermore, it assumes that the Y* lives long enough to escape the interaction region and, thus, decay without being influenced by final state interactions with the recoiling meson. There are no restrictions limiting the angular region involved, so that almost all events can be used.

Consider production of a Y* of arbitrary $J$, and represent its angular distribution by $I(\theta, \varphi)$ and its polarization by $P(\theta, \varphi)$. The polar axis ($z$-axis) is selected as the normal to the production plane. The incoming beam direction is along the $x$-axis. We now wish to define the moments of the intensity and polarization in terms of the spherical harmonics $Y_L^M$, as follows

$$\langle Y_L^M \rangle = \int I Y_L^M \, \mathrm{d}\Omega, \tag{7.29a}$$

$$\langle P Y_L^M \rangle = \int I P Y_L^M \, \mathrm{d}\Omega. \tag{7.29b}$$

Equation (7.29b) can be evaluated bv obtaining experimental knowledge of the decay distribution of the parity-violating reaction $\mathrm{Y} \rightarrow \mathrm{N} + \mathrm{M}$. If parity is conserved in the Y* production reaction, only even-$M$ moment terms can be different from zero (in order to satisfy the required parity-reflection invariance). If it should be the case that parity is conserved also in the Y* decay, only even values of $L$ can contribute to the intensity distribution (eqn (7.29a)) and only odd values of $L$ can contribute to the polarization distribution (eqn (7.29b)). For a state of spin $J$, the complexity of the angular distribution, and polarization, is limited to $L \leq 2J$. These experimentally calculated moments are then related to the density matrix $\rho$ describing the spin state of the Y*, and we then obtain a limit relation bracketing the spin value within an interval and, also, a means of determining the parity.

The moments method can be applied also to two-body decay sequences in which the Y* decay violates parity. However, when parity is violated, the intensity and polarization moments contain both even and odd $L$ terms. Carmony *et al*. [32] applied the method to the parity violating $\Xi$ decay, and excluded spin $\frac{3}{2}$ by 2·8 standard deviations. In this case, since the polarization of the $\Lambda$ from $\Xi$ decay is determinable from the decay asymmetry of the $\Lambda$, the method is more sensitive than the test-function method, which does not use the information in the polarization.

The applications of this method have included [33]

$$\begin{aligned} K^- + p &\to \Xi^*(1530) + K \\ \Xi^*(1530) &\to \Xi\pi \\ \Xi &\to \Lambda\pi. \end{aligned} \tag{7.30}$$

$J = \frac{1}{2}$ was eliminated and $J = \frac{3}{2}$ preferred, but $J = \frac{5}{2}$ could not be excluded. The favoured angular momentum state was a $P_{\frac{3}{2}}$ state.

In the $Y_1^*(1385)$ produced by 1·1–1·75 GeV/$c$, $K^- + p$, ([34]–[36]) $J = \frac{3}{2}$, and a decay state of $P_{\frac{3}{2}}$ were strongly favoured. The method has also been extended to handle $Y^{**} \to (0, \frac{3}{2})$ and $B^* \to (0, 1)$. The details of the method are given in Byers and Fenster [31].†

The latest situation in regard to which particles have been observed, and the quantum numbers that have been established, is periodically reviewed in the Particle Data Group articles and tables (see Appendix). The instances covered in this chapter are partly for illustrative purposes. It is often the case that the many verifications which follow the assumption, and use, of the original classification are as, or even more, convincingly accurate as the original work on the assignment.

## 7.9. Relative parities of strange particles

All the experimental evidence to date is highly consistent with the assumption that parity is conserved in strong interactions. Thus we can assign a relative parity to strange particles by observing their strong interactions. It is obvious that, since parity is violated in weak interactions and only pairs of strange particles like $K\Lambda$ and $K\Sigma$ are produced in the strong interactions (where parity is conserved), that only the relative parity $P_{\Lambda KN}$ of the $\Lambda$–K system, considered with respect to the nucleon, is well-defined. Similarly, $P_{\Sigma KN}$, $P_{\Lambda\Sigma}$, and $P_{\Xi N}$ are well defined and of interest. Any two of the three quantities $P_{\Lambda\Sigma}$, $P_{\Lambda KN}$, $P_{\Sigma KN}$ are independent, and the third is deducible from the known values of the other two.

### 7.9.1. *Lambda parity*

It is conventional to define the parity of the $\Lambda$ as the same as that of the nucleon, namely, positive. Once the $\Lambda$ parity is thus defined, we can now speak of the K-parity, the $\Sigma$-parity, or the $\Xi$-parity. The latter ($P_{\Xi N}$) can, in principle, be directly defined by reactions like $K^- + p \to \Xi^- + K^+$.

† Also Tripp. *A. Rev. Nucl. Sci.* **15**, p. 325, to which the reader is referred for a general treatment of spins and parities of particles.

7.9.2. K-*meson parity*

All the particles in a group describable as an isotopic spin multiplet should have the same space–time properties and, thus, the same parity, as well as the same spin. The parity of the $K^-$ meson was first deduced, indirectly, based on observations in a helium bubble chamber [37]. It was found that, when low-energy K-mesons are allowed to be absorbed by a helium nucleus, two types of hyperfragments were found containing a $\Lambda$ particle,

$$K^- + He^4 \rightarrow He^4_\Lambda + \pi^-, \tag{7.31}$$

$$K^- + He^4 \rightarrow H^4_\Lambda + \pi^0. \tag{7.32}$$

In reaction (7.31), the $He^4_\Lambda$ hyperfragment was composed of two protons, one neutron, and a $\Lambda$-particle (i.e. a neutron in the original $He^4$ was replaced by a $\Lambda$). In reaction (7.32) a proton in the original $He^4$ was replaced by a $\Lambda$-particle, thus resulting in the $H^4_\Lambda$ hyperfragment. These reactions occurred with good probability for incident low-energy $K^-$.

If we can demonstrate that either $He^4_\Lambda$, or $H^4_\Lambda$, have zero spin (at the time of the strong production interaction) then it is very easy to determine the parity of the $K^-$, and show that it is odd, i.e. the same as the pion parity. This is simply demonstrated, since in reactions (7.31) or (7.32), the $K^-$, $He^4$, and $\pi^{0,-}$ all have zero spin and if the hyperfragments also have zero spin then, in order to conserve angular momentum, the orbital angular momentum of the final state must equal the orbital angular momentum of the initial state, thus its effect on parity is the same for both sides. Since the intrinsic parity of a $\Lambda$ is defined to be positive, the same as that of a nucleon, there is no intrinsic parity change in the hyperfragments relative to the $He^4$ nucleus. Thus conservation of the parity requires that

$$P_K = P_\pi = -1.$$

The most questionable point in the above argument is the assumption that the spin of the hyperfragments is zero. If both hyperfragments should have a spin of 1, for example, the orbital angular momentum in the initial and final states may differ by one unit and, thus, vitiate the conclusion on the K-meson parity (i.e. make it positive instead of negative). The question of the spin of the hyperfragment was considered by Dalitz and Downs [38]. They found that they could not exclude the existence of an excited ground state of $He^4_\Lambda$ or $H^4_\Lambda$. However, if such a state existed, its binding energy should be less than 0·1 MeV. The probability that the $\Lambda$ which is produced remains bound in $He^4_\Lambda$ was calculated to be an increasing function of the $\Lambda$ binding energy and would be about 5 per cent for $B_\Lambda = 0{\cdot}1$ MeV and about four times as large for $B_\Lambda \approx 2{\cdot}0$ MeV. The experimental value ((20 ± 7) per cent) obtained thus favoured the conclusion that $He^4_\Lambda$ is produced in its ground state.

Conservation of isotopic spin can be imposed. For $He^4$ the value of $T = 0$. Thus both $He^4_\Lambda$ and $H^4_\Lambda$ must have $T = \frac{1}{2}$, since a nucleon, $T = \frac{1}{2}$, was replaced with a $\Lambda$, $T = 0$. Since the initial state (in eqns (7.31) and (7.32)) has $T = \frac{1}{2}$, $T_z = -\frac{1}{2}$, in order to conserve isotopic spin $He^4_\Lambda$ must have $T = \frac{1}{2}$, $T_z = +\frac{1}{2}$, and $H^4_\Lambda$ must have $T = \frac{1}{2}$, $T_z = -\frac{1}{2}$. Thus it seems reasonable to assume those two hyperfragments represent an isotopic spin doublet. Hence the final state for $K^- + He^4$ can be represented as

$$|\tfrac{1}{2}, -\tfrac{1}{2}\rangle = \sqrt{\tfrac{1}{3}}\,|\pi^0, H^4_\Lambda\rangle - \sqrt{\tfrac{2}{3}}\,|\pi^- He^4_\Lambda\rangle.$$

The isotopic spin conservation would predict, for the $K^- + He^4$ hyperfragment production, that the ratio

$$\left(\frac{He^4_\Lambda + \pi^-}{H^4_\Lambda + \pi^0}\right) = 2.$$

In the earliest observation a ratio $\approx 3 \pm 1{\cdot}5$ was observed. Thus it was concluded that the two hyperfragments belong to an isotopic spin doublet, and that their ground states are the same.

The angular momentum of the hyperfragments can be investigated by studying their decay. It was found that the decay process [39],

$$H^4_\Lambda \rightarrow He^4 + \pi^-,$$

represents $(0{\cdot}67 \pm 0{\cdot}06)$ of all the decays of $H^4_\Lambda$. This branching ratio was estimated theoretically [40] to have a value in agreement with the experiment for a zero spin $H^4_\Lambda$. However, for spin 1, a value of this branching ratio four times lower was estimated, in sharp disagreement with the experiment. We can estimate also the spin of the hyperfragment by making the extremely plausible assumption that the K-meson, in reactions (7.31) and (7.32), is captured from an *S*-state. Hence, there is zero orbital angular momentum in the initial state.

If the hyperfragment has spin zero, then there must be zero orbital angular momentum in the final state. Therefore, the negative pion, emitted in the decay of the hyperfragment, should have an isotropic angular distribution in the rest system of the hyperfragment. On the other hand, if the hyperfragment has a spin of one, the orbital angular momentum in the final state must necessarily be one to conserve angular momentum. Furthermore, the spin of the hyperfragment must be perpendicular in a direction perpendicular to the direction of relative motion in the production process. Thus it appears that the decay angular distribution is proportional to $\cos^2\theta$, where $\theta$ is the angle between the direction of motion of the decay $\pi^-$ meson and the $H^4_\Lambda$ hyperfragment. The experimental results indicated that the angular distribution was isotropic, not dependent on $\cos^2\theta$, thus indicating that the hyperfragment has spin zero, and strongly supporting the conclusion that the $K^-$ parity is odd. As indicated in the Particle Data Group tables, the parity of all kaons is now considered established as negative (relative to the $\Lambda^0$).

### 7.9.3. $\Sigma^0$ *parity*

The electromagnetic decay of the $\Sigma^0$ conserves parity and can, in principle, be used to determine the $\Sigma^0$ parity. The electromagnetic decay $\Sigma^0 \to \Lambda^0 + \gamma$ is replaced about 0·5 per cent of the time by the process $\Sigma^0 \to \Lambda^0 + e^+ + e^-$, which is analogous to internal pair-production by the $\gamma$, already observed in $\pi^0$ decay. When the $\Sigma^0$ are produced, by capture of slow $K^-$ in hydrogen, they are unpolarized. It can be shown [41] that the $e^+e^-$ invariant mass spectrum depends on the relative parity of $\Sigma^0$ and, thus, we can use a measurement of this invariant mass spectrum to determine the relative parity of $\Sigma^0$.

In the decay process $\Sigma^0 \to \Lambda + \gamma$, in order to conserve parity the transition operator must be a scalar for even relative $\Lambda\Sigma$ parity, or a pseudoscalar for odd $\Lambda\Sigma$ parity. The dependence on the polarization vector of the electric field $\mathbf{P}$ must be linear and can contain $\boldsymbol{\sigma}$ and $\mathbf{k}$ (the photon propagation vector), with $\mathbf{k}.\mathbf{P} = 0$ required. Hence, the only form which will satisfy these requirements, for even parity, is $T \propto (\boldsymbol{\sigma}.\mathbf{k} \times \mathbf{P})$, or $T \propto (\boldsymbol{\sigma}.\mathbf{P})$ for odd parity. The general symmetry arguments which we have given for the form of the transition operator still allow us to multiply either of the above transition operators with a scalar function. This scalar must contain the electromagnetic propagator of the virtual photon as its most important element. This propagator will, therefore, strongly favour low invariant mass. The electric and magnetic form factor effects can be estimated, and are expected to be small compared to the propagator effect, since the electric form factor can be taken as zero for a neutral particle without electromagnetic structure.

The major difference between the even parity, and odd parity, transition operator is that the even parity term has an extra linear $k$ dependence. Thus high $k$ is emphasized but, since

$$M^2 = |E_+ - E_-|^2 - k^2,$$

high $k$ means low $M$ (invariant mass). Thus we can expect that high

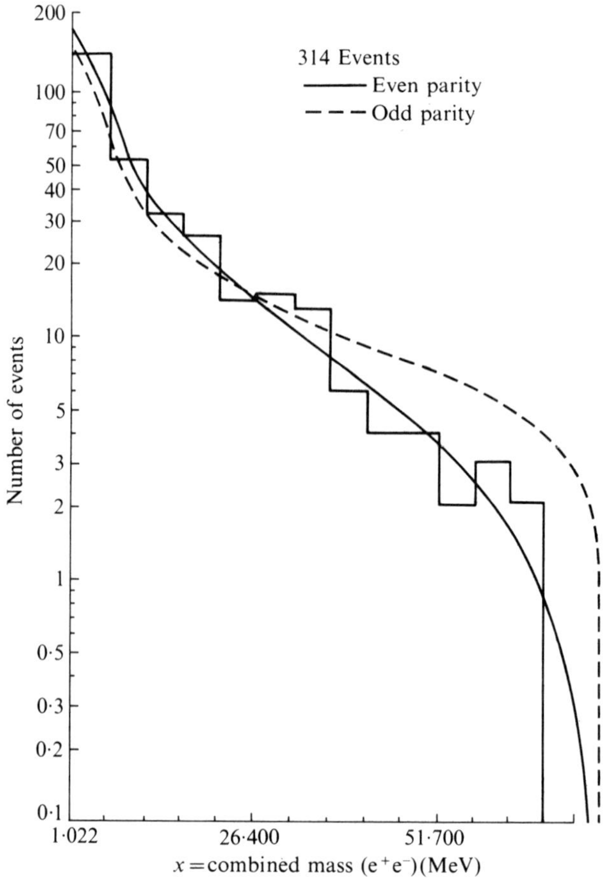


FIG. 7.9. A histogram of the combined mass spectrum of the Dalitz pair, together with the theoretical spectra for odd and even parity. To facilitate comparison, the theoretical spectra have been integrated over the corresponding histogram intervals, and the smoother functions thus obtained are shown. The Dalitz pairs are from the events $\Sigma^\circ \rightarrow \Lambda^\circ + e^+ + e^-$. (From Alff *et al.* (1965). *Phys. Rev.* B **137,** 1105.)

invariant-mass pairs will be relatively suppressed, in a calculable way, for the even parity case.

Courant *et al.* [42] and Alff *et al.* [43] investigated this process, using about a million $K^-$ captures in hydrogen for each investigation. There were somewhat in excess of 300 events obtained for each experiment. Figure 7.9 shows the experimental $e^+e^-$ invariant-mass spectrum [43] compared to the theoretical calculations, for the even and odd parity cases, normalized to the same number of events. The suppression of high mass pairs is clearly indicated (3·5 standard deviations).

Thus the conclusion is that the $\Sigma^0\Lambda^0$ relative parity is even (positive). In order to avoid this conclusion we would need an unreasonably large electric form factor which, if sufficiently large, could make the slopes of both theoretical spectra nearly the same.

The over-all internal conversion rates are estimated to be about $\frac{1}{161}$ for odd parity and $\frac{1}{182}$ for even parity, but the 13 per cent difference is too small to be detected with the statistics available.

### 7.9.4. $\Sigma^{\pm}$ *parity*

The $\Sigma^+$ relative parity was determined to be positive, by analysis of the reactions $K^- + p \rightarrow Y_0^*(1520) \rightarrow \Sigma^+ + \pi^-$. We will treat this in the next chapter. We have also mentioned other determinations of the charged $\Sigma$ parities. Thus, it is reasonable to conclude that the $\Sigma^{\pm,0}$ all have even relative parity.

### 7.9.5. $\Xi$ *parity*

The parity of the $\Xi$ has not been established experimentally. One suggested [44] method is to study the capture of slow $\Xi^-$ in the reaction

$$\Xi^- + p \rightarrow 2\Lambda^0.$$

As we have previously argued, for $K^-$ and $\bar{p}$ capture, we would expect the capture to take place from an $S$-state. The experimental evidence for the spin of the $\Xi$ strongly implies $\frac{1}{2}$. Therefore the initial state must have $J = 0$ or 1 and, hence, be either a ${}^1S_0$ or a ${}^3S_1$ state. Thus the final state must have $J = 0$ or 1. Since both $\Lambda^0$ have spin $\frac{1}{2}$ and, owing to the Pauli principle, the over-all wave function of the final state must be anti-symmetric, ${}^1S_0$, ${}^3P_0$, and ${}^3P_1$ are the only possibilities.

Reference [44] pointed out that there is a different correlation between the polarizations of the two $\Lambda^0$ particles for each of these final states.

The parity of the $\Xi^0$ is assigned as positive on the basis of SU(3). The SU(3) multiplet groupings will be discussed in Chapter 10.

## REFERENCES

1*a*. Leprince-Ringuet, L. and Lheritier, M. (1944). *C.r. hebd. Séanc. Acad. Sci.*, Paris. **219,** 618; *b* Rochester, G. D. and Butler, C. C. (1947). *Nature*. **160,** 855.

2*a*. Armenteros, R., Barker, K. H., Butler, C. C., Cachon, A., and Chapman, A. H. (1951). *Nature*. **167,** 501; *b* Armenteros, R., Barker, K. H., Butler, C. C., Cachon, A., and York, C. M. (1952). *Phil. Mag*. **43,** 597; *c* Leighton, R. B., Wanlass, S. D., and Alford, W. L. (1951). *Phys. Rev*. **83,** 843L.

3. Leprince-Ringuet, L. (1953). Mesons and heavy unstable particles in cosmic rays, *A. Rev. nucl. Sci*. **3,** 39.

4. BARKER, R. (1953). *Proceedings of the Duke University Cosmic Ray Conference, Durham*. p. 2–43.
5. BONETTI, A., LEVI-SETTI, R., and PANETTI, M. (1953). *Nuovo Cim*. **10,** 1736; LAL, D., PAL, Y., and PETERS, B. (1953). *Phys. Rev*. **92,** 438.
6. FOWLER, W. B., SHUTT, R. P., THORNDIKE, A. M., and WHITTEMORE, W. L. (1954). *Phys. Rev*. **93,** 861.
7. FRY, W. F. and SWAMI, M. S. (1954). *Phys. Rev*. **96,** 235L.
8. COWAN, E. W. (1954). *Phys. Rev*. **94,** 161.
9. THOMPSON, R. W., BUSKIRK, A. V., ETTER, L. R., KARZMARK, C. J., and REDIKER, R. H. (1953). *Phys. Rev*. A **91,** 445; THOMPSON, R. W., BURWELL, J. R., COHEN, H. R., HUGGETT, R. W., and KARZMARK, C. J. (1954). *Phys. Rev*. A **95,** 661.
10*a*. BROWN, R. H., CAMERINI, U., FOWLER, P. H., MUIRHEAD, H., POWELL, C. F., and RITSON, D. M. (1949). *Nature*. **163,** 82; (b) Padova conference (1954). Report on the conference on new particles. Supplement to *Nuovo Cim*. **12,** Series IX.
11*a*. BONETTI, A., LEVI-SETTI, R., and PANETTI, M. (1954). *Nuovo Cim*. **11,** 210.
11*b*. PRIMAKOFF, H. and CHESTON, W. (1954). *Phys. Rev*. **93,** 908L.
12. GELL-MANN, M. and PAIS, A. (1954). *Proceedings of the International Conference on Nuclear Physics, Glasgow*.
13. PAIS, A. (1952). *Phys. Rev*. **86,** 663.
14. GELL-MANN, M. (1953). *Phys. Rev*. **92,** 833L; NAKANO, T. and NISHIJIMA, K. (1953). *Prog. theor. phys*. **10,** 581; NISHIJIMA, K. (1954). *Prog. theor. Phys*. **12,** 107.
15. See Chapter 10 for a discussion, and Reference [10] of that chapter.
16. GOLDHABER, S., CHINOWSKY, W., GOLDHABER, G., LEE, W., O'HALLORAN, T., STUBBS, T. F., PJERROU, G. K., STORK, D. H., and TICHO, H. K. (1962). *Phys. Rev. Lett*. **9,** 135.
17. ZEMACH, C. (1964). *Phys. Rev*. B **133,** 1201.
18. OREAR, J., HARRIS, T., and TAYLOR, S. (1956). *Phys. Rev*. **102,** 1676.
19. ADAIR, R. K. (1955). *Phys. Rev*. **100,** 1540.
20*a*. See reference [18].
20*b*. EISLER, F., PLANO, R., PRODELL, A., SAMIOS, N., SCHWARTZ, M., STEINBERGER, J., BASSI, P., BORELLI, V., PUPPI, G., TANAKA, H., WALOSCHEK, P., ZOBOLI, V., CONVERSI, M., FRANZINI, P., MANELLI, I., SANTANGELO, R., SILVESTRINI, V., BROWN, G. L., GLASER, D. A., and GRAVES, C. (1958). *Nuovo Cim*. **7,** 222.
21. STODOLSKY, L. and SAKURAI, J. J. (1963). *Phys. Rev. Lett*. **11,** 90.
22. LEE, T. D. and YANG, C. N. (1958). *Phys. Rev*. **109,** 1755.
23. CRAWFORD, F. S., CRESTI, M., GOOD, M. L., STEVENSON, M. L., and TICHO, H. K. (1959). *Phys. Rev. Lett*. **2,** 114.
24. BRUECKNER, K. A., SERBER, R., and WATSON, K. M. (1951). *Phys. Rev*. **81,** 575.
25. DAY, T. B., SNOW, G. A., and SUCHER, J. (1959). *Phys. Rev. Lett*. **3,** 61.
26. LEON, M. and BETHE, H. (1962). *Phys. Rev*. **127,** 636.
27. DESAI, B. R. (1960). *Phys. Rev* **119,** 1385.
28. TREIMAN, S. B. (1950). *Phys. Rev*. **101,** 1216.
29. LEITNER, J., NORDIN, P., Jr., ROSENFELD, A. H., SOLMITZ, F. T., and TRIPP, R. D. (1959). *Phys. Rev. Lett*. **3,** 238.

30. Humphrey, W. E. and Ross, R. R. (1962). *Phys. Rev.* **127,** 1305.
31. Byers, N. and Fenster, S. (1963). *Phys. Rev. Lett.* **11,** 52.
32. Carmony, D. D., Pjerrou, G. M., Schlein, P. E., Slater, W. E., Stork, D. H., and Ticho, H. K. (1964). *Phys. Rev. Lett.* **12,** 482.
33. Schlein, P. E., Carmony, D. D., Pjerrou, G. M., Slater, W. E., Stork, D. H., and Ticho, H. K. (1963). *Phys. Rev. Lett.* **11,** 167.
34. Shafer, J. B. and Huwe, D. O. (1964). *Phys. Rev.* B **134,** 1372.
35. Ely, R. P., Fung, S. Y., Gidal, G., Pan, Y. L., Powell, W. M., and White, H. S. (1961). *Phys. Rev. Lett.* **7,** 461.
36. Tripp, R. D. (1965). *A. Rev. nucl. Sci.* **15,** 325.
37. Block, M. M., Brucker, E. B., Hughes, I. S., Kikuchi, T., Meltzer, C., Anderson, F., Pevsner, A., Harth, E. M., Leitner, J., and Cohn, H. O. (1959). *Phys. Rev. Lett.* **3**, 291.
38. Dalitz, R. H. and Downs, B. W. (1958). *Phys. Rev.* **111,** 967.
39. Ammar, R. G., Levi-Setti, R., Slater, W. E., Limentani, S., Schlein, P. E., and Steinberg, P. H. (1961). *Nuovo Cim.* **19,** 20.
40. Dalitz, R. H. and Liu, L. (1959). *Phys. Rev.* **116,** 1312.
41. Feinberg, G. (1958). *Phys. Rev.* **109,** 1019.
42. Courant, H., Filthuth, H., Franzini, P., Glasser, R. G., Minguzzi-Ranzi, A., Segar, A., Willis, W., Burnstein, R. A., Day, T. B., Kehoe, B., Herz, A. J., Sakitt, M., Sechi-Zorn, B., Seeman, N., and Snow, G. A. (1963). *Phys. Rev. Lett.* **10,** 409.
43. Alff, C., Gelfand, N., Nauenberg, U., Nussbaum, M., Schultz, J., Steinberger, J., Brugger, H., Kirsch, L., Plano, R., Berley, D., and Prodell, A. (1965). *Phys. Rev.* B **137,** 1105.
44. Treiman, S. B. (1959). *Phys. Rev.* **113,** 355.
45. Rebka, G., Rothberg, J. A., Etkin, A., Glodis, P., Greenberg, J., Hughes, V. W., Kondo, K., Lu, D. C., Mori, S., and Thompson, P. A. (1970). *Phys. Rev. Lett.* **24,** 160.

# 8

# STRANGE BARYONIC ISOBARS (OR RESONANCES)

### 8.1. Y–$\pi$ Resonances

As mentioned earlier, the pion–nucleon interaction was dominated by formation of pion–nucleon resonant states. Thus we might suspect that the $\pi$–Y interaction would exhibit similar characteristics as, for example, was assumed in the generalized isobar model.

### 8.2. $\Lambda$–$\pi$ resonances ($Y_1^*$ (1385))

An early study [1*a*,*b*] of the reaction $K^- + p \rightarrow Y + \pi^+ + \pi^-$ revealed a sharp peak in the Y–$\pi$ invariant mass.

The preliminary data for incident $K^-$ meson momenta of [1*a*] 1·15 GeV/*c*, and [1*b*] 760 MeV/*c*, are shown in Fig. 8.1, in the form of a Dalitz plot with the axes $T_{\pi^+}$ and $T_{\pi^-}$. There is clearly a sharp maximum in both $T_{\pi^+}$ and $T_{\pi^-}$ at 280 MeV. Thus similar to the arguments made previously in observing peaks in recoils from a nucleon isobar, these events in the peak were attributed to the reaction

$$K^- + p \rightarrow Y^{*\pm} + \pi^{\mp}, \tag{8.1}$$

with the subsequent strong interaction decay,

$$Y^{*\pm} \rightarrow \Lambda^0 + \pi^{\pm}. \tag{8.2}$$

being the dominant mode (about 130 events out of an observed total of 170). The calculated average mass for the Y* was 1380 MeV, and its half-width (Breit–Wigner) was 32 MeV.

Since this $Y^{*\pm}(1380) \rightarrow \Lambda^0 + \pi^{\pm}$, with a width characteristic of strong interactions, isotopic spin must be conserved in this decay. Thus, since $\Lambda^0$ has $T = 0$ and $\pi$ has $T = 1$, their sum (the isotopic spin of the Y*) can be only one. We refer to this state as $Y_1^*(1380)$. As can be seen in the Particle Data Group table, this $Y_1^*$ is now known to have a mass of $(1383 \pm 1)$ MeV and a width of $(36 \pm 3)$ MeV. The spin of the $Y_1^*(1385)$ was determined (Shaefer *et al.*) [2] by using the moments method. Both spin $\frac{1}{2}$ and spin $\frac{5}{2}$ were eliminated, and the spin was demonstrated to be $\frac{3}{2}$. A similar result (Malamud and Schlein) [3] was obtained for $Y_1^*$ produced by 1·455 GeV/*c* $K^-$. Only positive parity (i.e. a $\frac{3}{2}^+$ state) yields a satisfactory fit for all momenta studied from 1·1 to 1·75 GeV/*c* incident $K^-$. We should observe that the total mass of a $K^-$ plus a proton, both at rest, is 1436 MeV. Hence

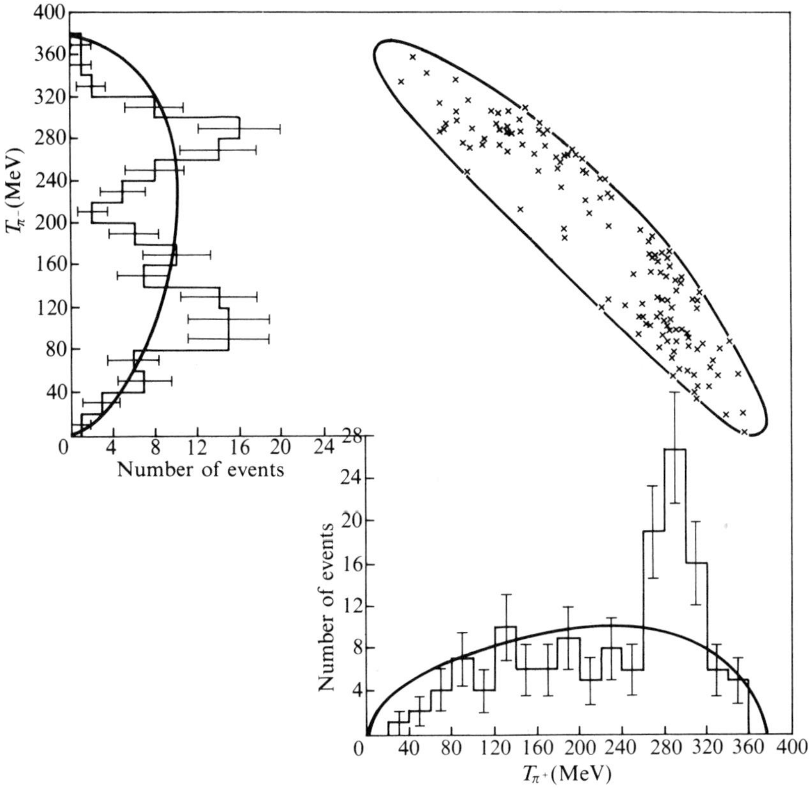


FIG. 8.1. The $\pi - Y$ resonance is deduced by plotting the kinetic energy of the $\pi^+$ and the kinetic energy of the $\pi^-$ (from the reaction $K^- + p \rightarrow Y + \pi^+ + \pi^-$) as a point (in cartesian coordinates formed by $T^{\pi^-}$ and $T^{\pi^+}$) in the central diagram. Then the central diagram is projected on to the $T^{\pi^+}$ and $T^{\pi^-}$ axes. (From Alston *et al.* (1960). *Phys. Rev. Lett.* **5**, 520, or Morpugo. (1960). *A. Rev. nucl. Sci.* p. 81.)

the Y*(1385) occurs below threshold in the $K^- + p$ channel and, thus, cannot be observed in a formation experiment.

### 8.3. $\Sigma - \pi$ resonances ($Y_0^*$ (1405) and $Y_0^*$ (1520))

Observation of the reaction

$$K^- + p \rightarrow (\Sigma^\pm + \pi^\mp) + \pi^0 \tag{8.3}$$

allowed the possibility of observing $Y^* \rightarrow \Sigma^\pm + \pi^\mp$ resonances with $T = 0, 1, 2$.

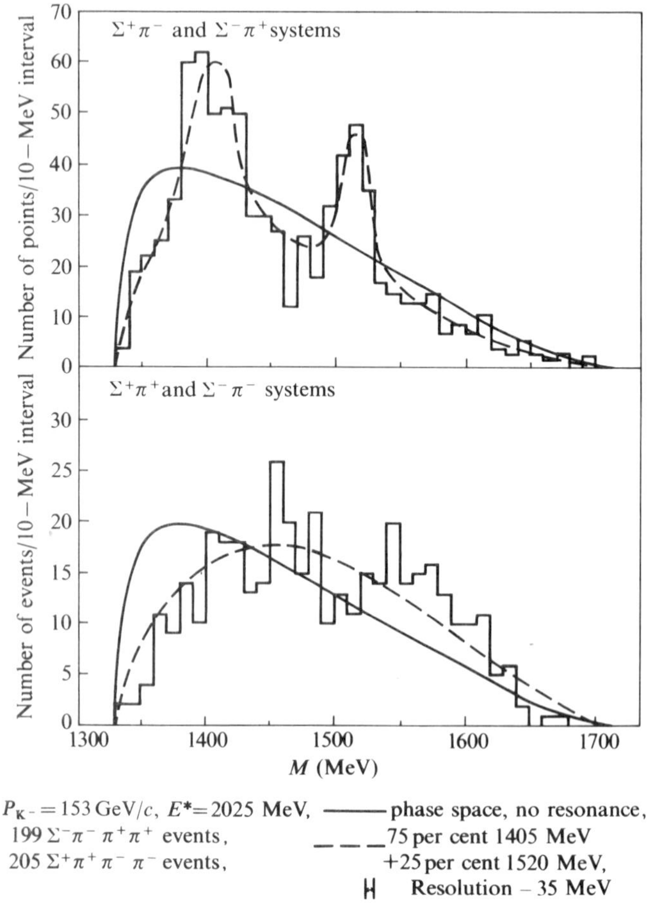


FIG. 8.2. Mass histograms for the $\Sigma\pi$ systems in the reaction $K^- + p \rightarrow \Sigma^{\pm} + \pi^{\mp} + \pi^+ + \pi^-$ at $P_{K^-} = 1{\cdot}51$ GeV/$c$ ($E^* = 2{\cdot}025$ GeV). (From Alston *et al.* (1962). *Proceedings of the 11th International Conference on High Energy Physics*, CERN. p. 314.)

Figure 8.2 shows the early observed [4] distribution in the Dalitz diagram projected onto the axis corresponding to the mass of the $(\Sigma^{\pm} + \pi^{\mp})$. The incoming kaon had a momentum of 1·22 GeV/$c$. There are clearly peaks in the $(\Sigma^{\pm} + \pi^{\mp})$ invariant-mass-squared distribution at values corresponding to masses of 1·5 GeV and 1·405 GeV. We have been considering the $\Sigma$–$\pi$ charge combination which has zero charge. Similar plots for the non-zero charge combinations showed no peaks, thus implying that we are observing $T = 0$ resonant states.

Thus we refer† to these as $Y_0^*$ resonance states. The resonance at 1405 MeV was first observed this way [5]. It is below threshold (1436 MeV) in the $K^- + p$ channel and, thus, could not be observed in a formation experiment. On the other hand, the $Y_0^*$ at 1520 MeV is above threshold in the $K^- + p$ channel, and was previously observed in a formation experiment. Observation and analysis of low-energy $K^- + p$ scattering allows us to determine the quantum numbers of the $Y_0^*(1405)$.

The study of low-energy $K^- + p$ reactions has led to the conclusion that $s$-wave amplitudes are sufficient to explain the dominant features of the interaction, and that these $s$-wave amplitudes are describable by a constant scattering length. Precise studies [6], [7] of the low-energy region predict an $I = 0$ resonance at 1411 MeV, with the width $\Gamma = 37$ MeV. This solution also agrees with the higher-energy data. The difference between 1411 MeV and 1405 MeV, the directly observed mass of the $Y_0^*$, is not significant. Thus, if the $Y_0^*(1405)$ is an $s$-wave resonance (below threshold), $J = \frac{1}{2}$, and the parity is odd (−). Since the parity operator applied to a proton gives a positive sign, the $s$-wave orbital angular momentum gives a positive sign, and the $K^-$ intrinsic parity gives a negative sign, the product is negative (i.e. the parity of the $Y_0^*(1405)$ is odd). The $Y_0^*(1520)$ was observed in a formation experiment [5] [8].

Figure 8.3 shows that the charge exchange cross-section $K^- + p \rightarrow K^0 + n$ shows a peak at about 395 MeV/$c$ in the laboratory, for the incoming $K^-$ meson, which corresponds to a total mass of $1520 \pm 3$ MeV. No corresponding peak was observed in the $K^- + p$ elastic scattering (see Fig. 7.6(a)). However, the cross-section for the inelastic channel $K^- + p \rightarrow \Lambda + \pi^+ + \pi^-$ clearly showed a corresponding peak.

The quantum numbers of the resonances formed were deducible from the behaviour of the elastic scattering angular distribution. The angular distribution of $K^- + p$ elastic scattering (or charge exchange) cross-sections were well-expressed by

$$\frac{d\sigma}{d\Omega} = \lambda^2(A + B\cos\theta + C\cos^2\theta). \tag{8.4}$$

Figure 8.3(b) shows the behaviour of the coefficients $A$, $B$, and $C$ for elastic scattering, or charge exchange, in the neighbourhood of 250 MeV/$c$ to 500–600 MeV/$c$. The most striking feature of the above is that, in both the elastic scattering and the charge exchange scattering, the coefficient $C$ is near

† Our notation (historical) is such that the subscript gives the isotopic spin for the $S = -1$ hyperon. The Particle Data Group tables use $\Sigma$ for the $T = 1$, and $\Lambda$ for the $T = 0$, hyperon families of strangeness $-1$.

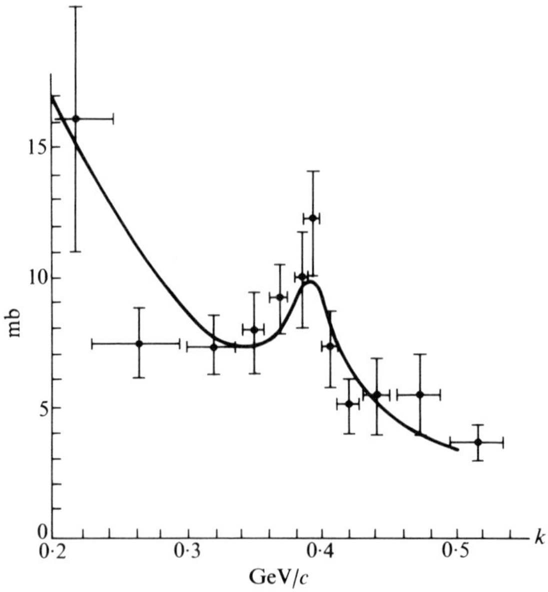


FIG. 8.3.(a) Total cross-section for the charge exchange scattering by protons of negative K-mesons for a laboratory momentum near $\approx 0{\cdot}4$ GeV/$c$. (After Ferro-Luzzi, Tripp, and Watson. (1962). *Phys. Rev. Lett.* **8**, 28.)

zero on both sides of about 400 MeV/$c$ but shows behaviour suggestive of a resonance, in the neighbourhood of 400 MeV/$c$. Particularly in the case of charge exchange scattering, $A$ dips to zero near 400 MeV/$c$ and $B = 0$ (within the errors). For elastic scattering, $B = 0$ (within errors) at 400 MeV/$c$ and $A$ dips to $\lesssim \frac{1}{3}$ of the value of $C$. Thus both the elastic scattering and charge exchange scattering are practically isotropic below, and above, 400 MeV/$c$ but at 400 MeV/$c$, a large amount of a $\cos^2\theta$ term appears in both† [9] accompanied by a dip in the isotropic term. We are evidently (except for the resonance) dealing predominantly with $K^- + p$ $s$-wave scattering. Since the $\cos\theta$ term is very small, this clearly indicates that the parity of the resonance is the same as the parity of the $s$-wave background. For, if the resonance had a parity opposite to the $s$-wave background, there would be a large $\cos\theta$ term. The parity of the $K^- + p$ system in an $s$-wave is obviously negative, thus we attribute negative parity to the $Y_0^*(1520)$. Since $\cos^2\theta$ is the highest term in the angular distribution and there is no evidence for a $\cos^4\theta$ term, the spin $\frac{3}{2}$ is most clearly favoured. The requirement of unitarity also favours $\frac{3}{2}$, and

† This also happens in $K^- + p \rightarrow \Sigma + \pi$.

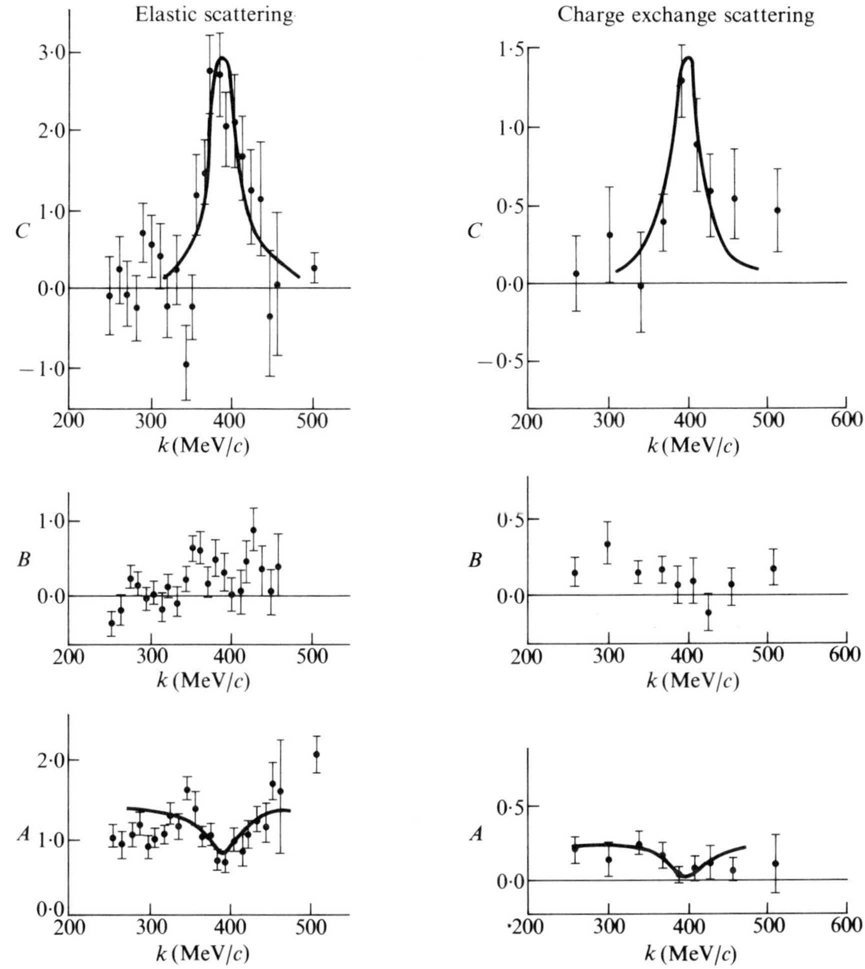


FIG. 8.3.(b) The coefficients $A$, $B$, $C$ for the angular distribution for $K^- + p$ elastic and charge exchange scattering. (After Ferro-Luzzi *et al.* (1962). *Phys. Rev. Lett.* **8,** 28.)

it has been concluded [8] that $J^P = \frac{3}{2}^-$ for the Y*(1520). However, some doubt could be shed on this spin assignment, since we could consider the possibility of non-resonant background through $D_{\frac{3}{2}}$. Thus the probability of a $J = \frac{5}{2}$ resonance fitting all the data was nearly 1 per cent [10]. The $D_{\frac{3}{2}}$ assignment has been subsequently established. Assigning the resonance as $\frac{3}{2}^-$, with a dominant background in $\frac{1}{2}^-$ and with both amplitudes contributing strongly to $\Sigma\pi$ production, we have an ideal situation for measuring the

$\Sigma$ parity. Positive $\Sigma$ parity gives $S_{\frac{1}{2}}$–$D_{\frac{3}{2}}$ interference in the $\Sigma\pi$ channel. Negative $\Sigma$ parity leads to $P_{\frac{1}{2}}$–$P_{\frac{3}{2}}$ interference. Thus, the two parity cases give predictions which differ only by opposite signs of the $\Sigma$ polarization. The relative phase of the non-resonant amplitude, and the resonant circle, can be determined by observation and analysis of the angular distributions of $\Sigma^+\pi^-$ production. The shape of the cross-section gives the magnitudes of the two amplitudes, thus determining all parameters. Charge independence can then be used to relate these to the $\Sigma^-\pi^+$ and $\Sigma^0\pi^0$ channels, and the correctness of the predictions for those angular distributions provide stringent consistency checks on the relative phase between the resonant and non-resonant amplitudes.

The asymmetry in the weak decay, $\Sigma^+ \rightarrow p+\pi^0$, is an excellent analyser of $\Sigma^+$ polarization. Based on a total of 610 decays of $\Sigma^+ \rightarrow p+\pi^0$, clear evidence for positive $\Sigma^+$ parity was obtained [8].

As can be seen in the Particle Data Group tables, there are now six (and possibly seven) additional known $Y_0^*$ resonances. There are also a total of at least six (and probably nine) $Y_1^*$ resonances known.

### 8.4. $\Xi$–$\pi$ resonances ($\Xi_{\frac{1}{2}}^*$ (1532))

Since we have already found that $\Lambda^0$ and $\pi$ form $Y_1^*$ resonances while $\Sigma\pi$ form $Y_0^*$ resonances (both of strangeness $-1$) it would not be unexpected that the $\Xi$–$\pi$ system forms strangeness $-2$ resonances. This was indeed found to be the case [11] in the following observations, by techniques similar to those described above,

$$\begin{aligned} &K^- + p \rightarrow \Xi^{*0} + K^0 \\ &\Xi^{*0} \rightarrow \begin{cases} \Xi^- + \pi^+ \\ \Xi^0 + \pi^0 \end{cases} \end{aligned} \tag{8.5}$$

$$\begin{aligned} &K^- + p \rightarrow \Xi^{*-} + K^+ \\ &\Xi^{*-} \rightarrow \begin{cases} \Xi^- + \pi^0 \\ \Xi^0 + \pi^- \end{cases} \end{aligned} \tag{8.6}$$

The Dalitz diagram [11*b*] is shown in Fig. 8.4. The $s_1$ is the $s$ value for the $\Xi$-particle-plus-one pion. The $s_2$ is the $s$ value (c.m.s. total energy squared) for the K-plus-one pion. The average mass value obtained for the $\Xi^*$ peak, observed at about $s = 2{\cdot}4\ \mathrm{GeV}/c^2$, was

$$M_{\Xi^*} = 1532 \pm 5\ \mathrm{MeV}.$$

The width $\Gamma$ was estimated to be about 20 MeV.

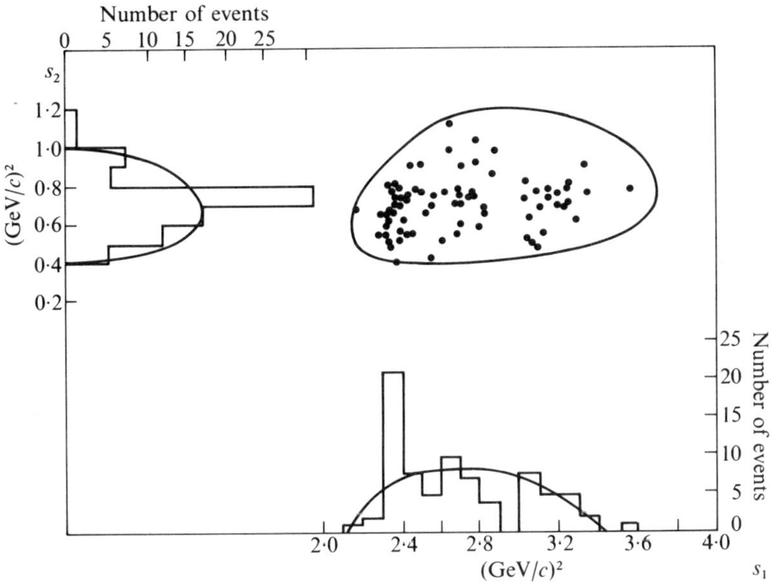


FIG. 8.4. Dalitz plot for $\Xi^*$ production. $s_1$ and $s_2$ are the total c.m.s. energy (of one pion and the $\Xi$ particle) squared. (After Bertanza. (1962). *Phys. Rev. Lett.* **9**, 180.)

The isotopic spin can be determined by the following consideration. Assume that the isotopic spin is $\frac{1}{2}$. Since isotopic spin is conserved in the decay, we have that the final state of the $\Xi^{*0}$, in the reaction (8.5), must be

$$\frac{1}{\sqrt{3}}\left(\sqrt{2}\,|\Xi^-,\pi^+\rangle - |\Xi^0,\pi^0\rangle\right). \tag{8.7}$$

Similarly, for the $\Xi^{*-}$ in reaction (8.6), the final state is

$$\frac{1}{\sqrt{3}}\left(|\Xi^-,\pi^0\rangle - \sqrt{2}\,|\Xi^0,\pi^-\rangle\right). \tag{8.8}$$

Hence we can predict the two branching ratios†

$$r_1 = \frac{\Xi^{*0} \to \Xi^- + \pi^+}{\Xi^{*0} \to \Xi^0 + \pi^0} = 2, \tag{8.9a}$$

† A $\Xi^*$ with $T = \frac{1}{2}$ can be produced in $T = 1$ and $T = 0$ incident channels and, thus, the complete treatment is more complicated than that given, but the conclusion on the ratios remains the same.

and

$$r_2 = \frac{\Xi^{*-} \rightarrow \Xi^0 + \pi^-}{\Xi^{*-} \rightarrow \Xi^- + \pi^0} = 2. \tag{8.9b}$$

The experimental values observed were $r_1 \approx 2{\cdot}5$ [1] and $r_2 \approx 1{\cdot}5$. Although these values differ somewhat from the expected value of 2 for $T = \frac{1}{2}$, they are very different from the values of $\frac{1}{2}$ expected for $T = \frac{3}{2}$. Thus $T = \frac{3}{2}$ is strongly selected. The difference may partially be attributed to the decay products still sensing some interaction with the recoiling K-meson.

The fact that only $\Xi^{*0}$ and $\Xi^{*-}$ charge states of the $\Xi^*$ have been discovered demonstrates that $T = \frac{1}{2}$. However, we can also deduce this by pursuing the present argument. The initial $K^- + p$ state is a mixture of $T = 0$ and $T = 1$ states. Using conservation of isotopic spin, if the isotopic spin of the $\Xi^*$ is $\frac{1}{2}$, the ratio of the production states is

$$r_3 = \frac{K^- + p \rightarrow \Xi^- + \pi^0 + K^+}{K^- + p \rightarrow \Xi^- + \pi^+ + K^0} = \frac{1}{2}\left|\frac{T^{(1)} - T^{(0)}}{T^{(1)} + T^{(0)}}\right|^2, \tag{8.10}$$

where $T^{(0)}$ and $T^{(1)}$ are the matrix elements of the transition matrix $T$ in the incident channels with isotopic spin 1 and 0, respectively. Thus, owing to interference between the $T^{(1)}$ and $T^{(0)}$, any observed value can be explained. However, if the $\Xi^*$ had $T = \frac{3}{2}$, it could only be produced in the incident $T = 1$ channel and, thus, $r_3$ would have to exhibit the unique value which, from the Clebsch–Gordan coefficients, is calculated to be 2. The experimental value observed was $r_3 = 0{\cdot}21 \pm 0{\cdot}07$ [11], thus, supporting the $T = \frac{1}{2}$ assignment.

The spin and parity were assigned by using the moments method [12] $J = \frac{3}{2}$ and positive parity were strongly favoured and, as we can see in the Particle Data Group tables, they are the presently accepted values. As we can further see in the Tables, there are now a total of four $\Xi^*$ resonances identified.

## REFERENCES


1*a*. ALSTON, M., ALVAREZ, L., EBERHARD, R., GOOD, M. L., GRAZIANO, TICHO, H., and WOJCICKI, K. (1961). *Phys. Rev. Lett.* **6,** 300.

1*b*. BERGE, J. P., BASTIEN, P., DAHL, O., FERRO-LUZZI, M., KIRZ, J., and MILLER, D. H. (1961). *Phys. Rev. Lett.* **6,** 557.

2. SHAFER, J. B. and HUWE, D. O. (1964). *Phys. Rev.* B **134,** 1372.

3. MALAMUD, E. and SCHLEIN, P. E. (1964). *Phys. Lett.* **10,** 145.

4. ALSTON, M. H. (1962). *Proceedings of the Annual International Conference on High Energy Physics, CERN*. p. 311.

5. FERRO-LUZZI, M., TRIPP, R. D., and WATSON, M. B. (1962). *Phys. Rev. Lett.* **8,** 28.

6. Sakitt, M. (1964). University of Maryland technical report 400 (unpublished).
7. Kim, J. K. (1961). *Phys. Rev. Lett.* **14,** 29.
8. Tripp, R. D., Watson, M. B., and Ferro-Luzzi, M. (1962). *Phys. Rev. Lett.* **8,** 175.
9. Humphrey, W. E. and Ross, R. R. (1962). *Phys. Rev.* **127,** 1305.
10. Watson, M. B., Ferro-Luzzi, M., and Tripp, R. D. (1963). *Phys. Rev.* **131,** 2248.
11*a*. Pjerrou, G. M., Prowse, D. J., Schlein, P., Slater, W. E., Stork, D. H., and Ticho, H. (1962). *Phys. Rev. Lett.* **9,** 114; *b* Bertanza, L., Brisson, V., Connolly, P. L., Hart, E. L., Mittra, I. S., Moneti, G., Rau, R. R., Samios, N. P., Skillicorn, I., Yamamoto, S. S., Goldberg, M., Gray, L., Leitner, J., Lichtman, S., and Westgard, J. (1962). *Phys. Rev. Lett.* **9,** 180.
12. Schlein, P. E., Carmony, D. D., Pjerrou, G. M., Slater, W. E., Stork, D. H., and Ticho, H. K. (1963). *Phys. Rev. Lett.* **11,** 167.

# 9

# STRANGE AND NON-STRANGE MESON ISOBARS (OR RESONANCES)

## 9.1. K–$\pi$ resonances (K* (890))

JUST as two, or more, pions have been observed to form resonant states or pionic isobars (such as $\rho$ and $\omega$) we might anticipate that the K–$\pi$ system might exhibit strange mesonic resonances. This has indeed turned out to be the case.

The first resonance of this type had been observed in several reactions which, for example, include

$$\mathrm{K}^{+}+\mathrm{p} \longrightarrow (\mathrm{K}^{*})^{+}+\mathrm{p}, \tag{9.1}$$

and

$$\mathrm{K}^{-}+\mathrm{p} \longrightarrow \begin{cases} (\overline{\mathrm{K}}^{*})^{-}+\mathrm{p} \\ (\overline{\mathrm{K}}^{*})^{0}+\mathrm{n} \end{cases}, \tag{9.2}$$

where

$$\mathrm{K}^{*+} \longrightarrow \overset{+}{\mathrm{K}^{0}}+\overset{0}{\pi^{+}}, \tag{9.3a}$$

$$\begin{aligned} \mathrm{K}^{*-} &\longrightarrow \mathrm{K}^{-}\ (\text{or } \mathrm{K}^{0})+\pi^{0}\ (\text{or } \pi^{-}), \\ \mathrm{K}^{*0} &\longrightarrow \mathrm{K}^{+}\ (\text{or } \mathrm{K}^{0})+\pi^{-}\ (\text{or } \pi^{0}). \end{aligned} \tag{9.3b}$$

Hence, in reactions (9.1) and (9.2) we have a three-body final state involving a K, p, and $\pi$. Thus, we can use Dalitz plots to identify K* resonances. A projected distribution from a Dalitz plot for (the first case of) reaction (9.2) ($\mathrm{K}^{-}+\mathrm{p} \rightarrow \overline{\mathrm{K}}^{*-}+\mathrm{p} \rightarrow \overline{\mathrm{K}}^{0}+\pi^{-}+\mathrm{p}$) is shown in Fig. 9.1 (Alston *et al.* 1963) [1]. The mass of the resonance was determined (by averaging over the data from several groups [2]) to be $M_{\mathrm{K}^*} = (888 \pm 3)$ MeV, and the width appeared to be $\Gamma \approx 50$ MeV.

In the ($\mathrm{K}^{+}+\mathrm{p}$) reaction (eqn (9.1)), the initial state can only be $T = 1$. Therefore, since the proton has $T = \frac{1}{2}$, the K* must (isotopic spin being conserved) have $T = \frac{3}{2}$, or $\frac{1}{2}$, only. This is the identical situation to that which we had in the $\Xi^*$ case. Thus, the results for $\Xi^*$ can be converted as follows. If, for the K*, $T = \frac{1}{2}$, then $r_1 = r_2 = 2$ and

$$\frac{1}{r_1} = \frac{1}{r_2} = \frac{1}{2}.$$

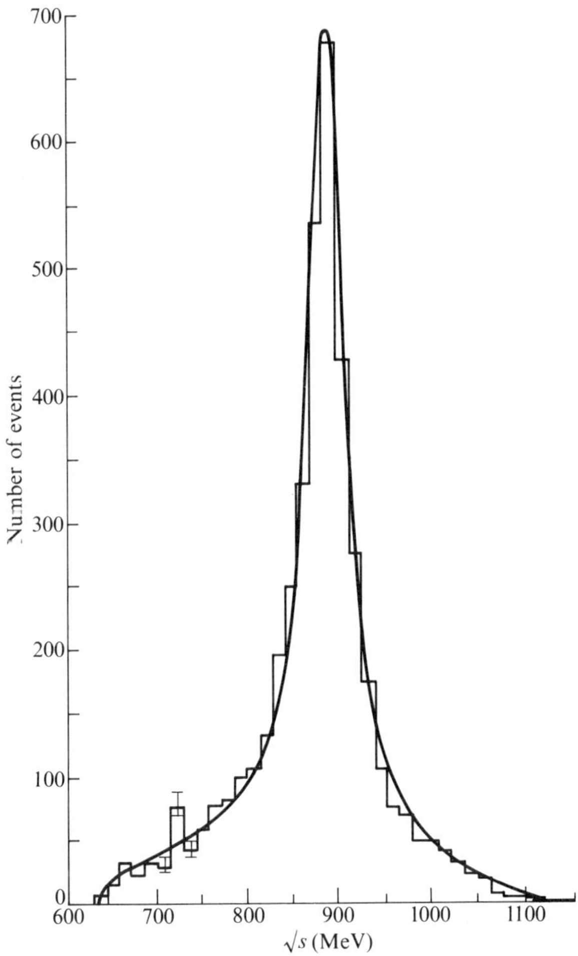


FIG. 9.1. Projection of the Dalitz plot of reaction (9.2) on to the axis of K°π mass. (After Alston *et al*. (1963). *Phys. Lett*. **5**, 283.)

On the other hand, if $T = \frac{3}{2}$, for the K*, then

$$r_1 = r_2 = \tfrac{1}{2} \qquad \text{and} \qquad \frac{1}{r_1} = \frac{1}{r_2} = 2.$$

The experimental conclusion [3] was that $1/r_2 = 0{\cdot}75 \pm 0{\cdot}35$, implying $T = \frac{1}{2}$. No charges higher than unity have ever been found for the K*(890) and, thus, we can conclude that $T = \frac{1}{2}$.

The spin of the K* was determined from the reaction

$$K^+ + p \rightarrow K^{*0} + N^*_{3,3}(1238)$$
$$\downarrow$$
$$(K^+ + \pi^-).$$

An Adair analysis was used to determine the spin [4] of the $K^{*0}$. Since this process is one involving pion exchange, this analysis method is expected to be reasonably reliable.

The angular distribution of the outgoing $K^+$ angle relative to the beam direction in the $K^+$ centre of mass system was found to be

$$w(\theta) = \text{constant}(\approx 0) + \text{constant}\cos^2\theta.$$

Since the decay products of the K* are both spinless, the K* spin is equal to the relative angular momentum, but, if the K* were spinless, the decay state would necessarily be an $S$-state and, thus, the angular distribution would be isotropic. Since we have observed a markedly anisotropic decay angular distribution, the spin of the K* must be $J \geq 1$. The problem is somewhat more subtle than we might imagine because it appears that even if the K* has spin 1, but there is no polarization in the production process, we can still obtain an isotropic distribution. In order to see this we refer to Table 9.1 which shows the decay angular distribution terms as a function of $m_s$ (the spin projection of the incident proton along the incident beam direction).

TABLE 9.1
*Angular distribution for K* (J = 1) decay*

| $m_s$(K*) | $m_s$(N*) | $\lvert\Psi_f\rangle$ | $(\theta)$ |
|---|---|---|---|
| 1 | −1 | $Y_1^1(\theta)$ | $\frac{1}{2}\sin^2\theta$ |
| 0 | 0 | $Y_1^0(\theta)$ | $\cos^2\theta$ |
| −1 | +1 | $Y_1^{-1}(\theta)$ | $\frac{1}{2}\sin^2\theta$ |

We can see that if there is no polarization in the production process, we must take the average of the above three terms which then becomes isotropic. However, if there is any appreciable polarization in the production process, we expect to find the angular distribution anisotropic. Thus, the experimental observation is consistent with the K* having spin one and being polarized perpendicularly to the incident beam. Of course, this argument does not exclude a spin greater than one, but there are other arguments based on angular correlations in the experimental data [5], which indicate that spins of two or larger are not compatible with the data and, as can be seen from the Particle Data Group tables, a spin of one has been established.

Four other K* resonances are listed and described in these tables. Some discussion of these will take place in relation to the later treatment of meson nonets.

## 9.2. Other strangeness zero pionic resonances

We have previously discussed the $\rho$, $\omega$, and $\eta$ pionic resonances.

The Particle Data Group tables† show the status of many other recently discovered pionic resonances. They all have either $T = 0$ (like the $\eta$) or $T = 1$ (like the $\pi$ and $\rho$). The spins include 0, 1, and 2. The relevant details are given in the tables.‡ There is an $\eta'(958)$, or $X^0$, with $T = 0$, $J^P = 0^-$. The dominant decay modes are $\eta\pi\pi(64 \pm 4)$ per cent or $\rho^0\gamma(30 \pm 3)$ per cent. There is some evidence for a $\delta(962)$ with $I \geq 1$, and a $\phi(1019)\, 0^-(1^-)^-$, of which the dominant decay mode is $K^+K^-(45{\cdot}5 \pm 3{\cdot}3)$ per cent. There is an indication for an $A_1(1070)$. Tentatively, it has been assigned the quantum numbers $1^-(1^+)+$ and it has a $3\pi$ decay mode. The B(1235) meson has $T = 1$ and quantum numbers $1^+(1^+)$- with a dominant decay mode $\omega\pi \approx 80$ per cent. There is an f(1260) with quantum numbers $0^+(2^+)+$ with the $\pi$–$\pi$ decay mode $\approx 100$ per cent. There is also an $A_2$, which appear to be split, the heavier one, $A_2^H(1320)$ and the lighter one, $A_2^L(1280)$—both were assigned the quantum numbers $1^-(2^+)+$. Recent experiments at BNL and CERN disagree with the previous $A_2$ work, and do not find a split in the $A_2$, this will be discussed in Chapter 10. $f'(1514)$ has the dominant decay mode $\bar{K}K(72 \pm 12)$ per cent and is considered established as $0^+(2^+)+$. Many other resonances are listed in Particle Data Group tables. Thus it is very clear that the so-called elementary particles are just a part of a complex set of nucleon, hyperon, pion, and kaon isobaric, or resonance, levels. The large number that have been discovered may possibly be only a fraction of those still yet to be discovered. The pion is a universal strong interaction glue and apparently resonates with any other known strongly-interacting particle, including itself.

For the ordinary (non-strange) mesons, the anti-particles are included in the same isotopic spin multiplet and, hence, have the same spin, and isotopic spin. For strangeness-plus-one mesons (like $K^+$ and $K^0$) the anti-particles are strangeness-minus-one mesons (like $K^-$ and $\overline{K^0}$) with the same parity and spin. For baryons, the anti-particles have the opposite sign baryon number, the opposite parity, and the same spin. In addition to $\bar{p}$ and $\bar{n}$, $\Lambda^0$, $\Sigma^-$, $\Sigma^{\mp}$, and $\bar{N}^*_{\frac{3}{2},\frac{3}{2}}$, among others, have been identified.

† We use their numerical notation in describing quantum number (however, our $T$ equals their $I$).

‡ Particle Data Group tables (January 1970) and Appendix.

## 9.3. Strangeness and electromagnetic interactions

We have already assumed, and verified, that strong interactions conserve strangeness. Now let us consider the electromagnetic interaction case. Here, $T$ is not a good quantum number and, thus, one can legitimately wonder whether $T_z$ is conserved. Obviously, if $T_z$ is not conserved, then from eqn (7.14) $S$ cannot be conserved. There are small (a few tenths of a per cent) mass differences within an isotopic spin multiplet such as $\Sigma^{\pm,0}$ and $K^{+,0}$, which can be attributed to the combined effects of both electromagnetic and strong couplings.

We would expect that, if isospin is not conserved in electromagnetic interactions, which are manifestly charge dependent (and there is much clear evidence for this), the full rotational invariance in isotopic spin-space is broken by electromagnetic interactions. However, the experimental evidence on strange particle decays implies that strangeness is not broken by electromagnetic interactions, but only by weak interactions. For example, the $\Sigma^0$ decays to $\Lambda^0+\gamma$. It is experimentally known that the lifetime $\tau$ is less than $10^{-14}$ s, and thus this cannot be a weak interaction decay. In fact, it represents an electromagnetic decay. However, in this case there is no change of strangeness. Thus, the electromagnetic decay of the $\Sigma^0$ does not involve a strangeness violation and $T_z$ must be conserved for this electromagnetic interaction. An electromagnetic interaction in strange particle decays (or elsewhere) that did not conserve $T_z$ has never been observed. This can only be understood if the electromagnetic coupling is 'minimal'. This happens because the photon field is always coupled to the electric charge constant, which is the sum of charge current densities, which transform as an isoscalar, and charge current densities, which transform as an isovector. Thus, the third component of isotopic spin $T_z$ must commute with the electromagnetic coupling Lagrangian. Pais (1952) and Gell-Mann (1956) have pointed out that the metastability of strange particles depends on the assumption of minimal electromagnetic coupling, and Gell-Mann (1956) proposed an example of a non-minimal electromagnetic interaction that violated strangeness.

REFERENCES

1. ALSTON, M. H., KALBFLEISCH, G. R., and WOJCICKI, S. G. (1963). *Phys. Lett.* **5**, 283.
2. GREGORY, B. P. (1962). *Proceedings of the Annual International Conference on High Energy Particles, CERN*. p. 779.
3. ALSTON, M. H., KALBFLEISCH, G. R., and TICHO, H. K. (1962). *Proceedings of the Annual International Conference on High Energy Physics, CERN*. p. 291.
4. CHINOWSKY, M., GOLDHABER, G., GOLDHABER, S., LEE, W., and O'HALLORAN, T. O. (1962). *Phys. Rev. Lett.* **9**, 330.
5. ALSTON, M., ALVAREZ, L. W., EBERHARD, P., GOOD, M. L., GRAZIANO, W., TICHO, H. K., and WOJCICKI, S. G. (1961). *Phys. Rev. Lett.* **6**, 300.

# 10

# ELEMENTARY PARTICLE CLASSIFICATIONS AND THEORIES

THE rapidly growing number of so-called 'elementary particles' seemed to indicate that perhaps they were not so elementary. This stimulated an intensive effort to represent them in terms of conglomerates of simpler building blocks, and, also, an attempt to uncover some underlying symmetry principle which would explain them.

## 10.1. Sakata model

The Sakata model [1*a*] took note of the fact that when we knew only of nucleons and pions, a combination of nucleons and anti-nucleons was sufficient from which to build all particles [1*b*] but, with the introduction of strangeness, another particle carrying strangeness was also necessary. Thus, the fundamental particles, or fields, were assumed to be a triplet ($pn\Lambda^0$) and their anti-particles. For example, we could suggest the following combinations of fundamental particles.

$$\begin{aligned}
\pi^+ &= p\bar{n},\\
\pi^- &= \bar{p}n,\\
K^+ &= \bar{\Lambda}^0 p,\\
K^- &= \Lambda^0 \bar{p},\\
\Sigma^+ &= \Lambda^0 p\bar{n},\\
\Sigma^- &= \Lambda^0 \bar{p}n,\\
\Xi^- &= \Lambda^0\Lambda^0\bar{p}.
\end{aligned}$$

Here, strange baryons are made up of triplets of particles, and mesons (both strange and non-strange) are made up of doublets of particles.

## 10.2. Symmetry groups

### 10.2.1. SU(2)

The early studies of the groups of transformations among the proton and neutron led to the group of transformations SU(2). In this notation U is for a unitary group, S stands for a unimodular group, and the number in parenthesis represents the fact that this group's transformations form all unitary unimodular $2 \times 2$ matrix transformations of the fundamental proton–neutron doublets. It is the assumed invariance of the theory under the SU(2) group which leads to isotopic spin invariance. Thus, we could perhaps hope to find unitary unimodular groups of transformations of the basic triplet pnΛ, under which the theory is invariant, and which lead to a higher symmetry (than isotopic spin) which could, hopefully, provide a grouping of the variety of observed particle states in the same way as isotopic spin invariance explains the observed characteristics of isotopic spin multiplets.

The group SU(2) has rank 1. This means that there is only one Casimir operator (an invariance operator which commutes with all the generators). This Casimir operator is $T^2$. The group, thus, conserves one additive quantum number $T_z$. In other words, only the $T_z$ matrix can be diagonalized. In general, SU($n$) is a group of rank $(n-1)$, has $(n-1)$ Casimir operators, and has only $(n-1)$ quantum numbers that have simultaneously diagonalizable matrices. SU(3) will have a rank of 2 and two Casimir operators to which there correspond two additive quantum numbers $Y$ and $T_z$. The higher symmetry corresponding to invariance under SU(3) is now known as 'unitary symmetry'.

A group† can be represented by a set of matrices if, for each element $\alpha$ of a group, there corresponds a matrix $M(\alpha)$ which obeys the group laws. These matrices (and the complex vectors they act on) are then said to form a representation of the group. This representation is said to be completely reducible if, for every $M$,

$$M = \begin{bmatrix} A & 0 \\ 0 & B \end{bmatrix}, \tag{10.1}$$

since $A$ (and $B$) also form a representation. A most significant characteristic of a group is the dimensionality of its irreducible representation. $U(n)$ is the group of $n \times n$ unitary matrix transformations of an $n$-component complex spinor

$$\zeta_a \rightarrow U_a^b \zeta_b, \tag{10.2}$$

where $U^+ = U^{-1}$ is implied by the unitarity. SU($n$) is the corresponding unimodular unitary group.

Further discussion of the general characteristics of groups, and general reference material for this chapter, can be found in references [2] and [3].

† A group is a set of elements which obey a set of composition laws (usually matrix algebra).

### 10.2.2. SU(3)

During 1959–1960 Ikeda, Ugona, Oknuki and Thirring, Weiss, and Yamaguchi found that the group SU(3), with an eight parameter algebra, had representations which would correspond to the structures allowed in the Sakata model. It was assumed that the process of building particle combinations was invariant with respect to unitary, unimodular transformations of a triplet of three fundamental complex fields identified with Sakata's (pnΛ) set. The difference in the masses was attributed to a propagation of the Λ–N mass difference. Thus, the group SU(3) represents the group of all unitary unimodular $3 \times 3$ matrix transformations of the Sakata fundamental triplet (pnΛ). SU(3) must include SU(2) as a subgroup, since strong interactions are invariant with respect to SU(2). Thus, the Sakata triplet could be represented by the 3 representation, and the anti-particles by a conjugate, 3*, representation. A meson (ordinary or strange) could be represented by a product of the 3 and the 3* representations. For example, $\pi^+ = p\bar{n}$, or $K^+ = \bar{\Lambda}^0 p$, and the natural representation† for the mesons was

$$3 \otimes 3^* = 8 \oplus 1,$$

where the usual rules of group multiplication were employed. Thus it was found that the mesons could be represented by an octet and a singlet. The octet consisted of a $T = \frac{1}{2}$ doublet with $Y = -1$, a $T = \frac{1}{2}$ doublet with $Y = +1$, a triplet with $Y = 0$, and a singlet with $Y = 0$. Assuming the K and $\pi$ were both $0^-$, the $(K^+, K^0)$ doublet and the $(K^-, \bar{K}^0)$ doublets were assigned to the two $T = \frac{1}{2}$, $Y = \pm 1$ doublets. The $\pi^{\pm,0}$ triplet was assigned to the $T = 1$, $Y = 0$ triplet, and there was one missing $(0^-)$ meson with $T = 0$, the $\eta$, which was predicted and subsequently found. This completed the octet. The remaining isotopic singlet ($\eta'$) was subsequently found. However, in its classification of baryons, the Sakata model was very unsuccessful. Since the model had assigned p, n, and $\Lambda^0$ to the fundamental triplet, it required that the additional baryons (Σ and Ξ) belong to multiplets which are contained in $3 \times 3 \times 3^*$, the next simplest representation, after the fundamental triplet, which could provide a unit baryon number. Thus, a vast (and, subsequently, fatal for the model) distinction‡ was made between the baryons and the mesons. Using standard group methods

$$3 \otimes 3 \otimes 3^* = 15 \oplus 6^* \oplus 3 \oplus 3. \tag{10.3}$$

† The representation is valid only for a particle multiplet, each particle of which has the same spin and parity, so that the space–time properties are the same. Thus it was found that the mesons could be represented by an octet and a singlet.

‡ Of course, we could also treat p, n, $\lambda^0$, the basic building blocks, on the same basis in constructing all the baryons. For example, $p = \bar{\lambda}^0 \lambda_0 p_0$, $n = \bar{\lambda}_0 \lambda_0 n_0$, and $\Lambda^0 = \lambda^0 p\bar{p}$. Then all baryons would belong to the $3 \otimes 3 \otimes 3^*$.

## TABLE 10.1(a)
## SU(3) *multiplets*†

$$\begin{aligned}
\mathbf{3} &= (2, \tfrac{1}{3})+(1, -\tfrac{2}{3})\\
\mathbf{6} &= (3, \tfrac{2}{3})+(2, -\tfrac{1}{3})+(1, -\tfrac{4}{3})\\
\mathbf{8} &= (2, 1)+(3, 0)+(1, 0)+(2, -1)\\
\mathbf{10} &= (4, 1)+(3, 0)+(2, -1)+(1, -2)\\
\mathbf{27} &= (3, 2)+(4, 1)+(2, 1)+(5, 0)+(3, 0)\\
&\quad +(1, 0)+(2, -1)+(4, -1)+(3, -2)\\
\mathbf{28} &= (7, 2)+(6, 1)+(5, 0)+(4, -1)+(3, -2)\\
&\quad +(2, -1)+(1, -3)\\
\mathbf{35} &= (5, 2)+(6, 1)+(4, 1)+(5, 0)+(3, 0)+(4, -1)\\
&\quad +(2, -1)+(3, -2)+(1, -2)+(2, -3)\\
\mathbf{64} &= (4, 3)+(5, 2)+(3, 2)+(6, 1)+(4, 1)+(2, 1)\\
&\quad +(7, 0)+(5, 0)+(3, 0)+(1, 0)+(2, -1)+(4, -1)\\
&\quad +(6, -1)+(3, -2)+(5, -2)+(4, -3)
\end{aligned}$$

† Some simple multiplets of SU(3) giving multiplicity and submultiplets in terms of $I$-spin and hypercharge, $(2i+1, Y)$. The adjoint multiplets are got by changing the sign of $Y$ in each term. For the Clebsch–Gordan coefficients see de Swart (1963).

## SU(3) *products*

$$\begin{aligned}
\mathbf{8}\otimes\mathbf{8} &= \mathbf{1}\oplus\mathbf{8}\oplus\mathbf{8}\oplus\mathbf{10}\oplus\overline{\mathbf{10}}\oplus\mathbf{27}\\
\mathbf{8}\otimes\mathbf{10} &= \mathbf{8}\oplus\mathbf{10}\oplus\mathbf{27}\oplus\mathbf{35}\\
\mathbf{8}\otimes\mathbf{27} &= \mathbf{8}\oplus\overline{\mathbf{10}}\oplus\mathbf{10}\oplus\mathbf{27}\oplus\mathbf{27}\oplus\overline{\mathbf{35}}\oplus\mathbf{35}\oplus\mathbf{64}\\
\mathbf{10}\otimes\overline{\mathbf{10}} &= \mathbf{1}\oplus\mathbf{8}\oplus\mathbf{27}\oplus\mathbf{64}
\end{aligned}$$

## TABLE 10.1(b)
## SU(6) *multiplets and products*

$$\begin{aligned}
\mathbf{20} &= (\mathbf{8}, \mathbf{2})\oplus(\mathbf{1}, \mathbf{4})\\
\mathbf{56} &= (\mathbf{10}, \mathbf{4})\oplus(\mathbf{8}, \mathbf{2})\\
\mathbf{70} &= (\mathbf{10}, \mathbf{2})\oplus(\mathbf{8}, \mathbf{4})\oplus(\mathbf{8}, \mathbf{2})\oplus(\mathbf{1}, \mathbf{2})\\
\mathbf{35} &= (\mathbf{8}, \mathbf{3})\oplus(\mathbf{8}, \mathbf{1})\oplus(\mathbf{1}, \mathbf{3})\\
\mathbf{189} &= (\mathbf{27}, \mathbf{1})\oplus(\mathbf{10}, \mathbf{3})\oplus(\overline{\mathbf{10}}, \mathbf{3})\oplus(\mathbf{8}, \mathbf{5})\oplus 2(\mathbf{8}, \mathbf{3})\oplus(\mathbf{8}, \mathbf{1})\oplus(\mathbf{1}, \mathbf{5})\oplus(\mathbf{1}, \mathbf{1})\\
\mathbf{405} &= (\mathbf{27}, \mathbf{5})\oplus(\mathbf{27}, \mathbf{3})\oplus(\mathbf{27}, \mathbf{1})+(\overline{\mathbf{10}}, \mathbf{3})\oplus(\mathbf{10}, \mathbf{1})\oplus(\mathbf{8}, \mathbf{5})\oplus 2(\mathbf{8}, \mathbf{3})\oplus(\mathbf{8}, \mathbf{1})\\
&\quad \oplus(\mathbf{1}, \mathbf{5})\oplus(\mathbf{1}, \mathbf{1})
\end{aligned}$$

$$\begin{aligned}
\mathbf{20}\times\overline{\mathbf{20}} &= \mathbf{1}\oplus\mathbf{35}\oplus\mathbf{175}\oplus\mathbf{189}\\
\mathbf{56}\times\overline{\mathbf{56}} &= \mathbf{1}\oplus\mathbf{35}\oplus\mathbf{405}\oplus\mathbf{2695}\\
\mathbf{35}\times\mathbf{35} &= \mathbf{1}\oplus 2\times\mathbf{35}\oplus\mathbf{189}\oplus 2\times\mathbf{280}\oplus\mathbf{405}\\
\mathbf{35}\times\mathbf{20} &= \mathbf{20}\oplus\mathbf{70}\oplus\mathbf{70}\oplus\overline{\mathbf{540}}\\
\mathbf{35}\times\mathbf{56} &= \mathbf{56}\oplus\mathbf{70}\oplus\mathbf{700}\oplus\mathbf{1134}\\
\mathbf{35}\times\mathbf{70} &= \mathbf{56}\oplus 2\times\mathbf{70}\oplus\mathbf{540}\oplus\mathbf{580}\oplus\mathbf{1134}
\end{aligned}$$

From Mathews [2]. SU(3) × spin components of the simple SU(6) multiplets, where the first number in parenthesis is the multiplicity, and the second number is $2J+1$, where $J$ is the angular momentum. We, also, give the representations appearing in some simple direct products.

The octet and decuplet which have become necessary to classify baryons are obviously missing.

Table 10.1 gives the $T$ and $Y$ values of some simple multiplets of SU(3) and SU(6). It appears that only the 15 (not shown) contains a $Y = -1$ doublet, which is essential to represent the Ξ.

There seemed to be many missing partners to the known baryons N, Λ, and Σ, if the 15 representation was to be assumed. Furthermore, these baryons contained 8 different charged particles, and appeared to have the same spin, parity, and the $T$ and $Y$ content required of an octet.

10.2.3. *Octets*

Ne'eman [4] and Gell-Mann [5] independently proposed that the baryons are also represented by an octet. This famous assignment of both baryons and mesons to octets became known as the 'eight-fold way' [6]. What Ne'eman and Gell-Mann had proposed was that the symmetry group SU(3), discovered in applying the Sakata model to mesons, was universally applicable to baryons as well, and invariance under SU(3) was the basic important conclusion. Their argument was that, if we observed experimentally that the baryons and mesons were representable by the octet representations (and meson–baryon resonance states representations were generated by $8 \otimes 8$) of SU(3), we should accept this fact and not be reluctant to use it, even if we could not see directly how to construct a baryon octet from the basic Sakata triplet.

10.2.4. *Quark model*

It was pointed out, later, in the quark model (Gell-Mann [5] and Zweig [7]) that we can still construct both baryons and mesons from a fundamental triplet and a conjugate triplet, if the fundamental fields (or particles) carry a hypercharge of value of $\frac{1}{3}$ and $-\frac{2}{3}$ for members of the 3 representation, and $\frac{2}{3}$ and $-\frac{1}{3}$ in the 3* representation. The fundamental particles also carry baryon number $\frac{1}{3}$†, and the anti-fields, or anti-particles, in the anti-triplet carry baryon number $-\frac{1}{3}$. Table 10.2 shows the properties of the fundamental triplet, which still obey the relation $Q = (T_z + Y/2)$. Thus, we can construct the mesons from products $3 \otimes 3^*$ (quark-anti-quark bound state) which

† So far it has not been necessary to associate particles with the fractional hypercharge fields. For example, if the triplets carry no hypercharge, this quantum number could, in principle, be contained in an additional unitary singlet field with spin $\frac{1}{2}$ which, when present in an octet, changes it from a meson to a baryon. However, once the field carries fractional baryon number, we must associate a particle with it, because the field operator on the wave function creates a fractional baryon particle which, if we assume baryon number conservation, means that some fractional charge particles must be stable.

TABLE 10.2
*Quark (or 'Ace') properties*

| 3 | $I$ | $I_Z$ | $B$ | $S$ | $Y$ | $Q$ | $J$ |
|---|---|---|---|---|---|---|---|
| $q_1$ = $p_1$ | $\frac{1}{2}$ | $\frac{1}{2}$ | $\frac{1}{3}$ | 0 | $\frac{1}{3}$ | $\frac{2}{3}$ | $\frac{1}{2}$ |
| $q_2$ = $n_1$ | $\frac{1}{2}$ | $-\frac{1}{2}$ | $\frac{1}{3}$ | 0 | $\frac{1}{3}$ | $-\frac{1}{3}$ | $\frac{1}{2}$ |
| $q_3 = \lambda_0$ | 0 | 0 | $\frac{1}{3}$ | $-1$ | $-\frac{2}{3}$ | $-\frac{1}{3}$ | $\frac{1}{2}$ |
| 3* | | | | | | | |
| $\bar{q}_1$ = $\bar{p}_1$ | $\frac{1}{2}$ | $-\frac{1}{2}$ | $-\frac{1}{3}$ | 0 | $-\frac{1}{3}$ | $-\frac{2}{3}$ | $\frac{1}{2}$ |
| $\bar{q}_2$ = $\bar{n}_1$ | $\frac{1}{2}$ | $+\frac{1}{2}$ | $-\frac{1}{3}$ | 0 | $-\frac{1}{3}$ | $+\frac{1}{3}$ | $\frac{1}{2}$ |
| $\bar{q}_3 = \bar{\lambda}_0$ | 0 | 0 | $-\frac{1}{3}$ | $+1$ | $+\frac{2}{3}$ | $+\frac{1}{3}$ | $\frac{1}{2}$ |

carry no baryon number and are still similar to the Sakata model prescription. However, now the baryons can be constructed from the product†

$$3 \otimes 3 \otimes 3 = 10 \oplus 8 \oplus 8 \oplus 1, \tag{10.4}$$

which is a three-quark bound state.‡ Thus, the only formal, but very essential, difference is that the Sakata model product $3 \otimes 3 \otimes 3^*$ has been replaced by $3 \otimes 3 \otimes 3$, but now we have two octets in the decomposition (one of which we have already made use of) and the decuplet. If the quarks did not have fractional charge and baryon number assigned to them, the product $3 \otimes 3 \otimes 3$ would not yield integral charge and baryon number.

So far, searches for quarks have been unsuccessful [8]. Their mass is expected to be large and is now estimated to exceed 5–10 GeV. The theoretical proposal of the fractionally-charged quarks and the resulting quark model (which had a number of successes) led to an intensive experimental search for quarks. The most distinctive property of quarks which would be recognized experimentally is the fractional charge, therefore, it is not surprising that almost all experimental attempts to discover quarks were based on a search for fractional charge particles. There were three general classes of experiments pursued.

(a) Accelerator experiments—quark production in hadronic collisions up to 30 GeV/$c$.

(b) Searches for fluxes of quarks in cosmic rays, at various altitudes, with $\gamma_{\text{quark}} = E/M \gtrsim 1{\cdot}8$, which are not accompanied by other particles.

† For example, now the Ξ could be represented by $\begin{bmatrix} n_1 \\ p_1 \end{bmatrix}[\lambda_0, \lambda_0]$.

‡ The representation, 27, which has not been necessary to fit the clearly indicated multiplet structures (based on observations) is absent here, but does occur in $8 \otimes 8$.

(c) A study of the charge content of matter on the earth, in meteorites, or on the sun.

Reference [8*a*] summarizes the results of these experiments which, are all negative to date. Table 10.3 summarizes the experimental situation. This table is updated in reference [8*b*]. The most notable additions are the following. The work of references [20]–[21] has been improved to where the limit is number of quarks per nucleon $\leq 0{\cdot}5 \times 10^{-18}$. Morpugo [8*b*] emphasized that the strong limit in reference [22], $10^{-27}$ quarks per nucleon in air, should not be taken too literally, since the authors themselves suggest caution in the interpretation of the figures, as they depend on assumptions about the behaviour of quarks absorbed by metal surfaces. Marshall-Libby and Thomas suggested a method for investigating negative stable quarks, or any massive negative stable particle present in the earth or entering with the cosmic radiation. Their suggestion was that such negative particles should be very effective catalysts of Coulomb fission in stock piles of heavy metals. According to their estimates, assuming a quark mass of 5 GeV, each stable quark with charge $-\frac{1}{3}$, should, on the average, induce $> 6 \times 10^9$ fissions when stopped in a block of $^{235}U$ or $^{238}U$. This would give rise to 70·4 curies of neutrons in a time $\geq 0{\cdot}6$ ms ($10^{-13}$ s per fission). According to these authors, the fact that such pathological radioactivity has not been observed can be used to calculate an upper limit to the number $N^-$ of negative heavy stable quarks or negative heavy stable particles, where

$$N^- \leq 2 \times 10^{-30}$$

TABLE 10.3

| Type of experiment | Quantity directly measured | Limit | References |
|---|---|---|---|
| Accelerator | Quark production cross section in hadronic collisions up to 30 GeV/$c$ primary momentum | $\lesssim 10^{-35}$ cm$^2$ | 1–7 |
| Cosmic rays | Flux of quarks at various altitudes | $\lesssim 10^{-10}$ cm$^{-2}$ srad$^{-1}$ s$^{-1}$ | 8–18 |
| Charge content of matter on the earth, in meteorites, or on the sun | Number of quarks per nucleon | $\leq 10^{-15}$ (helium) | 19 |
| | | $\leq 10^{-17}$ (graphite) | 20, 21 |
| | | $\leq 10^{-17}$ (iron meteorites) | |
| | | $\leq 5 \times 10^{-27}$ (air) | 22 |
| | | $\leq 3 \times 10^{-29}$ (sea water) | 22 |
| | | $\leq 10^{-10}$ (solar atmosphere) | 23, 24 |

From Reference [8*a*].

per nucleon of heavy metal. As pointed out by Morpugo, more careful evaluation of these estimates should be made, including a check of the situation for Coulomb fission induced by $U^-$ in heavy elements, since this process is immediately accessible experimentally.

The unexplained 'saturation' phenomenon indicates that three quarks bind, but more than three do not, since this would lead to so-called 'exotic baryons', which have not been observed. One quark–anti-quark pairs bind (i.e. mesons), but two, or more, bound quark–anti-quark pairs, which would lead to so-called 'exotic mesons', have not been observed.

There are variations on the quark model where the number of fundamental triplets is increased and thus fractional values of hypercharge or baryon number do not have to be assigned to the fundamental fields (or particles) [9]. Whether there is any real basis for expecting quarks to exist as particles, or whether they really are a convenient mathematical representation of the symmetry, remains to be seen. Aside from the many conceptual difficulties in a dynamical quark model, and their lack of observation to date, the concept of a fundamental triplet itself is open to question. If such a fundamental triplet existed, we could ask why it would not (by the 'bootstrap' mechanism and the expected strong interactions) 'bootstrap' itself and lose its special identity (i.e. why should it not become a member of Chew's nuclear democracy?). Any serious dynamical theory based on quarks as the fundamental triplet would have to explain the observations and provide a mechanism for keeping the quark in its fundamental status.

10.2.5. *Octets and octet ⊗ octet*

In the eight-fold way, both baryons and mesons were constructed from octets. Thus, it became clear that all observed baryon–meson resonances could be constructed from $8 \otimes 8$, and would preserve SU(3) invariance.†

Standard group theory gave

$$8 \otimes 8 = 1 \oplus 8 \oplus 8 \oplus 10 \oplus 10^* \oplus 27, \tag{10.5}$$

where $27 = 27^*$, because the dimension is a cube of an integer. The representation 27 would break up (when the mass differences are turned on) into an isotopic singlet, triplet, and quintet with $Y = 0$, a doublet and a quartet with $Y = 1$, a doublet and quartet with $Y = -1$, a triplet with $Y = 2$, and a triplet with $Y = -2$. The 10 representation breaks up under the same conditions into a triplet with $Y = 0$, a doublet with $Y = -1$, a quartet with

† We should note that $8^*$ is equivalent to 8. Thus, in considering $8 \times 8$, we must reduce out the direct product $8 \otimes 8$.

$Y = +1$, and a singlet with $Y = -2$. The conjugate representation, $10^*$, appears similar, but the sign of the values of $Y$ is reversed.

As we have just seen, if we wish to maintain SU(3) invariance†, there are only a few representations that can be constructed by combining two octets. These are the $1 \oplus 10 \oplus 10^* \oplus 8 \oplus 8 \oplus 27$. The $T = J = \frac{3}{2}$ isobar certainly does not fit in the $\frac{1}{2}^+$ nucleon octet and, therefore, we need at least one 10 or 27 representation to accommodate this resonance. It was pointed out (1962 Geneva Conference) that, since there was no evidence for any nearby $I = 1$, $Y = +2$ resonance in K–N scattering, the 27 was excluded. That left only the 10 supermultiplet which can be composed of a quartet with $Y = +1$, a triplet with $Y = 0$, a doublet with $Y = -1$, and a singlet with $Y = -2$.

### 10.2.6. *The baryon decuplet*

It was pointed out by Gell-Mann (1962) that, among the multitude of discovered resonances, the $N^*_{\frac{3}{2},\frac{3}{2}}$ (a quartet with $Y = +1$), the $Y^*_1(1385)$ (a triplet with $Y = 0$), and the $\Xi^*_{\frac{1}{2}}(1532)$ (a doublet with $Y = -1$) needed only a singlet with $Y = -2$ to complete the 10 representation, or decuplet, to which we just referred. The spin parity of all the particles would have to be $\frac{3}{2}^+$, like the $T = J = \frac{3}{2}$ isobar. With $Y = -2$, $B = +1$, the strangeness would have to be $S = -3$ and the charge would then have to be $= T_Z + Y/2 = -1$. The mass of this particle was predicted by the G.M.O. mass formula, which we will discuss shortly. A zealous search was made for this particle with 5·0 GeV/$c$ $K^-$ mesons incident on the 80-inch bubble chamber, and an event interpreted as

$$
\begin{array}{l}
K^- + p \rightarrow \Omega^- + K^+ + K^0 \\
\qquad\quad \Omega^- \rightarrow \Xi^0 + \pi^- \\
\qquad\qquad\quad \Xi^0 \rightarrow \Lambda^0 + \pi^0 \\
\qquad\qquad\qquad \Lambda^0 \rightarrow \pi^- + P, \quad \pi^0 \rightarrow \gamma_1 + \gamma_2 \\
\qquad\qquad\qquad\qquad \gamma_1 \rightarrow e^+ + e^-, \quad \gamma_2 \rightarrow e^+ + e^-,
\end{array}
\tag{10.6}
$$

was observed [10*a*]. A number of subsequent observations of $\Omega^-$ (a singlet)

† The isotopic spin group SU(2) must be a subgroup of SU(3), and in fact, it can be shown that

$$
SU(3) = \left[\begin{array}{c|c} SU(2) & 0 \\ \hline 0 & 1 \end{array}\right].
$$

production were reported [10*b*]. It is clear that the strangeness of the $\Omega^-$ is $-3$ in the above reaction.

### 10.2.7. *The mass formula for baryon octets*

If unitary symmetry were exactly conserved, all the masses of the particles of an octet or decuplet would be exactly identical. Since we are assuming that unitary symmetry (SU(3) invariance) holds for strong interactions, it is clear that the massive differences between $\pi$ and K mesons in the $0^-$ meson octet, and between $\Xi$ and N in the $\frac{1}{2}^+$ baryon octet, imply that there are medium-strong interactions which violate unitary symmetry. These medium-strong interactions (M.S.I.) must not affect SU(2) invariance, the small mass splitting within SU(2) multiplets being explained by the electromagnetic interactions. Also, they must not lead to violations of strangeness in either strong, or electromagnetic, interactions. The strong interactions which do not violate unitary symmetry are referred to as very strong interactions (V.S.I.).

Unitary symmetry invariance is analogous to the degeneracy of the different substates $J_z$ of a definite angular momentum state of an atom in a field-free space. However, if we turn on an external magnetic field, the symmetry will be broken and, to first order, we will find a splitting of the energy levels

$$\Delta E = -(\Psi_{JM}, \mu\Psi_{JM}).\mathbf{H} = \alpha_J H_z M, \tag{10.7}$$

where $\mu$ is the magnetic dipole moment. Hence, we will have equal spacing. If the interaction term contained a quadratic term in $H$, we would, of course, have a more complicated spacing. These spacing relations depend on the general transformation properties of the perturbation term, and the detailed effects due to the system are representable by simple constants. Thus, we can expect that the simple transformation properties of the symmetry-breaking term in the eight-fold way will determine the mass splitting. Denote the mass operator by $M$. Then, for exact unitary symmetry invariance,

$$(\Psi_{r,i} M \Psi_{r,i}) = m_r, \tag{10.8}$$

which represents complete mass degeneracy, where $r$ labels the representation and $i$ denotes the states within it. $M$ will contain an important term which transforms as a unitary singlet, and a number of irreducible tensors. Since there is (ignoring electromagnetic effects) still complete degeneracy for the states of a particular $Y$-value, within a $T$-spin multiplet, it follows that, when we expand $M$ into irreducible tensor operators, only those operators for which $Y = T = 0$ can be included.

Let us now consider the masses contained in an octet,

$$8 \otimes 8 = 1 \oplus 8 \oplus 8 \oplus 10 \oplus 10^* \oplus 27. \qquad (10.9)$$

Since the decuplet does not have any $Y = T = 0$ multiplets, its tensors make no contribution to the mass. The singlet contribution gives the mass (before splitting). Okubo [11] (and Gell-Mann) proposed that the octet provides the dominant contribution to mass splitting operator and that a lowest-order perturbation theory treatment was sufficient. This led to the G.M.O. mass formula.

$$(\Psi_{r,i} M \Psi_{r,i}) = M(\text{mass}) = M_0 + bY + c\{\tfrac{1}{4}Y^2 - T(T+1)\}. \qquad (10.10)$$

Let us now apply this formula to the baryon octet. There are four experimentally known masses and only three unknown constants $M_0$, $b$, and $c$. Thus, we can eliminate the three constants, and obtain a relationship between the four masses, which is

$$\tfrac{1}{2}(M_N + M_\Xi) = \tfrac{3}{4}M_\Lambda + \tfrac{1}{4}M_\Sigma.$$

Taking the average value for each multiplet from the tables, we obtain

$$\tfrac{1}{2}(939 + 1318) = \frac{3(1115{\cdot}6) + 1193{\cdot}2}{4}, \qquad (10.11)$$

$$1128{\cdot}5 = 1135{\cdot}0.$$

Thus, the formula is satisfied to within 6·5 MeV, which is an exceedingly small fraction of the masses, and is still only a few per cent of the splitting. For the decuplet decomposition, the octet appears only once and, thus, the above formulae can be replaced by the simpler form, which can be derived as

$$M = M_0 + bY. \qquad (10.12)$$

Both the equal spacing and the mass of the $\Omega^-$ were predicted by eqn (10.12) in advance of the discovery.

The experimental values of the mass spacing for the $\frac{3}{2}^+$ decuplet are given by the following

| | Mass | $\Delta M$ |
|---|---|---|
| $M_{\Omega^-}$ | $= 1672{\cdot}4 \pm 0{\cdot}6$ | |
| $M_{\Xi^*}(1530)$ | $= 1530{\cdot}5 \pm 1{\cdot}5$ | $142 \pm 2{\cdot}1$ |
| $M_{Y^*}(1385)$ | $= 1385{\cdot}2 \pm 2$ | $145{\cdot}3 \pm 3{\cdot}5$ |
| $M_{N^*}(1230)$ | $= 1236{\cdot}0 \pm 0{\cdot}6$ | $149 \pm 2{\cdot}6.$ (10.13) |

The discovery of the $\Omega^-$ considerably enhanced belief in both SU(3) invariance for strong interactions, and a high degree of dominance of octet symmetry-breaking in the medium-strong interactions, responsible for the mass splitting.

10.2.8. *Meson octet mass formula*

Mesons can be treated in a similar fashion, and mass formulae for the known octets developed. However, there are certain differences from the baryon case. The first is the fact that the mass terms are contained in the Lagrangian in quadratic terms of the form $m^2\phi^2$. Thus, when we estimate the effects of a symmetry-breaking term in the Lagrangian, we obtain estimates of $\delta m^2$, and the mass formulae are in terms of $m^2$. The second difference is that, whereas in the baryon multiplets the decomposition of $8\times 8$ includes two types of octets (the symmetric type (D type) and the anti-symmetric (F type)), in the case of the mesons, the mass term is the result of a pure D-type coupling, since $\phi^*$ is related to $\phi$ via charge conjugation. Thus, we obtain, for the $0^-$ meson octets,

$$\begin{aligned} 3M_{\eta}^2+M_{\pi}^2 &= 4M_{K}^2, \\ 3(548{\cdot}6\pm0{\cdot}4)^2+(138{\cdot}1\pm2)^2 &= 4(495{\cdot}8\pm2) \\ \approx 93\times10^4+1{\cdot}91\times10^4 &\approx 98\times10^4,\text{ or,} \\ 95\times10^4 &\approx 98\times10^4, \end{aligned} \tag{10.13}$$

which again is reasonable agreement.

The unitary singlet meson, with the same quantum numbers which accompanies the $0^-$ pseudoscalar meson octet, is presumed to be the $\eta'(958)$ (or previously-labelled $X^0$) which has the dominant decay modes $\eta\pi\pi(66\pm4)$ per cent and $\rho^0\gamma(30\pm3)$ per cent. Although the unitary singlet which accompanies the octet is in a different representation, the fact that the $I = 0$ member of the octet has the same quantum numbers as the unitary singlet allows the symmetry-breaking interaction to mix these two $I = 0$ states, such that the physically observed states are a linear combination of each. We shall discuss this in more detail in the section on vector mesons, where the effect was important for an early understanding of the $1^-$ vector meson nonet (octet + singlet = nonet). In the $0^-$ nonet the mixing effects are estimated to be small, and the mass formula for the octet fits well.

One problem has arisen with regard to the $0^-$ nonet [12]. The E(1420) meson, which has $T = 0$ and positive $G$ parity, and decays into $K^*\overline{K}+K\overline{K}^*(50\pm10)$ per cent and $\pi_N(1016)\pi \sim (50\pm10)$ per cent, has been (after careful analysis by the CERN–College De France group) assigned a $J^P$ of $0^-$. If this should be the case, we would have ten $0^-$ mesons, and the E could be the $\eta'$ (singlet member of the nonet) instead of the $X^0$. The mass formula could be satisfied with a consistent mixing angle. This would release $X^0$ to allow its association with the $\delta$-meson, which it resembles very much. This is just one of many indications that the classification of mesons is not as unique, or satisfactory, as for the baryon case.

### 10.2.9. *The vector meson* ($1^-$) *octet and singlet* (*or nonet*)

The $1^-$ vector mesons† $K^*_{\frac{1}{2}}(890)$, $\rho_1(765)$, and $\omega_0(784)$, for example, have the proper multiplet content to form a vector $1^-$ octet. Thus the $\phi_0(1019)$‡ could be the $1^-$ unitary singlet accompanying this $1^-$ octet, or perhaps the $K^*_{\frac{1}{2}}(890)$, the $\rho_1(765)$, and the $\phi_0(1019)$ could form the vector octet, and the $\omega_0(783)$ could be the accompanying unitary singlet. We could appeal to the mass formula to distinguish between these two cases. However, it was found that the conventional mass formula was badly contradicted for either of these choices. Okubo [14] realized that, in this vector meson case, the existence of an $I = 0$ SU(3) singlet, of the same quantum numbers as the $I = 0$ member of the SU(3) octet, caused interference effects between them. Even though they both belong to different representations, the nature of the symmetry-breaking term is such that they can be mixed by the symmetry-breaking interaction. Thus, we could expect that the physically observed $\omega$ and $\phi$ states are not pure, but that each is a linear superposition of two $I = 0$ $\omega$-mesons, which we can denote as $\omega_1$ and $\omega_8$, where the subscript denotes the unitary singlet or octet.

If we now label the pure octet and singlet wave functions as $\Psi^{\beta}_{\alpha}(8)$ and $\Psi(1)$, respectively, we obtain for the mass term in the vector meson

† Our notation here is that the subscript represents the value of $T$ and the number in parenthesis is the mass in MeV.

‡ In studying the reactions

$$K^- + p \rightarrow \begin{cases} \Lambda^0 + K^+ + K^- \\ \Lambda^0 + K^0 + \overline{K}_0 \end{cases},$$

a peak in the K–$\overline{K}$ mass distribution was observed at 1020 MeV, with a width $\Gamma = 3{\cdot}1 \pm 0{\cdot}6$ MeV [13]. The smallness of the width is due to the extremely low $Q$ of the reaction reducing the phase-space. The isotopic spin of the $\phi$ is deduced to be $T = 0$, since for the reaction

$$K^- + p \rightarrow \begin{matrix} \Sigma^0 + K^+ + K^- \\ \Sigma^+ + K^- + K^0 \end{matrix},$$

only the $K^+K^-$ pair, for which $T = 0$ or 1, shows a peaking, while the $K^- + K^0$ pair, for which $T = 1$, shows no peak. The $\phi$ spin has been deduced to be unity. Thus, the parity can be seen to be negative in the decay $\phi \rightarrow K^+ + K^-$. Hence, $\phi$ is a $1^-$ meson.

Lagrangian

$$\mathscr{L}_{\text{mass}} \sim C_1\Psi^{\beta}_{\alpha}(8)\Psi^{\alpha}_{\beta}(8)+C_2\Psi^{3}_{\alpha}(8)\Psi^{\alpha}_{3}(8)+C_3\Psi(1)\Psi(1)+C_4\Psi^{3}_{3}(8)\Psi(1), \tag{10.15}$$

where the $C_i$ are real parameters. The above equation can also be written in the form

$$\mathscr{L}_{\text{mass}} \sim C_1\rho^2+\left(\frac{2C_1+C_2}{2}\right)\overline{\text{K}}^*\text{K}^*+\tilde{v}^T\tilde{M}\tilde{v}, \tag{10.16}$$

where $\tilde{v} = \begin{pmatrix}\omega_8\\ \omega_1\end{pmatrix}$ and (10.17a)

$$\tilde{M} = \begin{bmatrix} \dfrac{3C_1+2C_2}{3} & \dfrac{C_4}{\sqrt{6}} \\ \dfrac{C_4}{\sqrt{6}} & C_3 \end{bmatrix} \tag{10.17b}$$

is the matrix containing the mass term contributions of the $\omega$ and the $\phi$. This matrix can now be diagonalized to give the physical states $\phi$ and $\omega$, and the parameters, adjusted to agreement with the experimentally observed mass relations, which can be represented as

$$m_\rho^2 \approx m_\omega^2$$

and

$$m_\phi^2 - m_{\text{K}^*}^2 \approx m_{\text{K}^*}^2 - m_\rho^2. \tag{10.18}$$

The choices of the parameters,

$$C_1+C_2 \approx C_3$$

and

$$C_2 \approx 2C_4, \tag{10.19}$$

allow agreement of the theoretical mass relation with the experimentally observed masses. This behaviour led Okubo to propose the following form of the effective mass terms for the vector meson nonet

$$\Psi^{\beta}_{\alpha}(9) = \Psi^{\beta}_{\alpha}(8)+\frac{1}{\sqrt{3}}\,\delta^{\beta}_{\alpha}\Psi(1), \tag{10.20a}$$

$$\phi = \Psi^{3}_{3}, \tag{10.20b}$$

$$\omega = \frac{1}{\sqrt{2}}\,(\Psi^{1}_{1}+\Psi^{2}_{2}). \tag{10.20c}$$

(See Sakita [3], p. 219 for further details.)

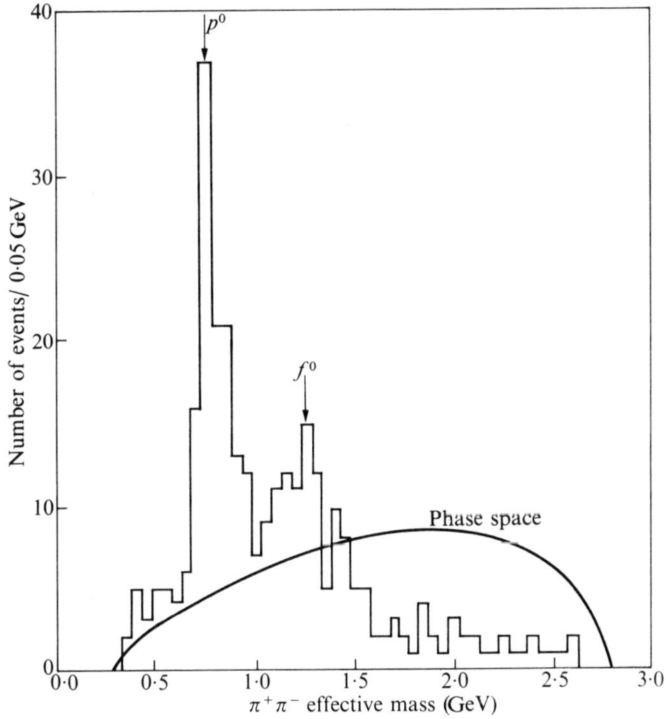


FIG. 10.1. The $\pi^+\pi^-$ effective mass distribution observed in the reaction $\pi^+ + p \to \pi^+ + \pi^- + (N^*)^{++}$ produced by 8 GeV/$c$ incident $\pi^+$. (From the Aachen–Berlin–CERN collaboration. (1964). *Phys. Lett.* **12,** 356.

### 10.2.10. *The* $2^+$ *nonet*

The evidence for the $2^+$ nonet was summarized by Goldhaber (Berkeley conference, 1966) and Butterworth [12] (Heidelberg Conference, 1968).

*The* $f^0$. This meson resonance was first observed [15] as a peak in the $\pi^+\pi^-$ invariant mass at $\approx 1250$ MeV, with a $\Gamma \approx 100$ MeV, in the pion production reaction

$$\pi^- + p \to \pi^+ + \pi^- + n. \tag{10.21}$$

The reaction

$$\pi^- + p \to \pi^- + \pi^0 + p \tag{10.22}$$

does not show this peak, thus it is concluded that $T = 0$. Figure 10.1 [16] shows the observation of the $f^0$, and its relation to the previously-known $\rho^0$ peak. In order to meet the Bose statistics requirement for a symmetric state, for $f^0 \to \pi^+ + \pi^-$, it is obvious that the $f^0$ spin must be even. The angular distribution [17] in the decay mode $f^0 \longrightarrow 2\pi^0$ requires that a $\cos^4\theta$ term be used in the fit. Thus we conclude that the $J^P$ of the $f^0$ are $2^+$.

*The* $A_2$ *meson.* Goldhaber *et al.* investigated [18] the reactions

$$\pi^+ + p \longrightarrow \pi^+ + \pi^- + \pi^+ + p, \tag{10.23}$$

at an incident energy of 3·65 GeV/$c$. It was observed that $N^*_{\frac{3}{2}}(1238)$ production dominated these interactions, and it was often accompanied by a $\rho_0$. For example, in reaction (10.23), $\rho^0 N^{*++}$ occurs about 30 per cent of the time, $\rho^0[\pi^+ p]$ (non-resonant) occurs about 25 per cent and $\pi^+\pi^- N^{*++}$ occurs about 30 per cent of the time. The double resonance production $\rho^0 N^{*++}$ appears to have the peripheral characteristics (i.e. dominance of small momentum transfers) which we would expect from the diagram

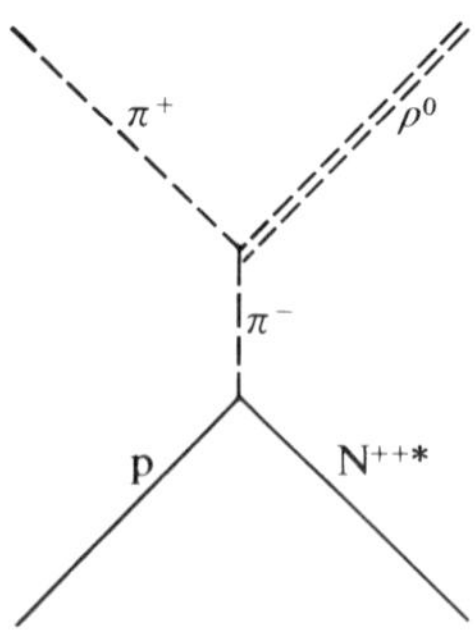


However, the $\rho^0$ accompanied by non-resonant $\pi^+ + p$ do not seem to exhibit peripheral production characteristics. If we select these events, the $\pi^+\rho^0$ invariant mass exhibits a broad peak in the 1·0 to 1·4 GeV mass region.

A study of the reaction

$$\pi^- + p \longrightarrow \pi^- + \pi^+ + \pi^- + p \tag{10.24}$$

at 3·2 GeV/$c$ was also undertaken. It was observed that, if the invariant mass of the three pions was plotted, the broad peak previously observed was resolved into one at mass 1090 MeV, referred to as the $A_1$ meson, and a second, higher-mass peak at 1320 MeV, called the $A_2$ meson. The $\Gamma$ of the $A_2$ was 80 MeV, and the observed decay of $A_2 \longrightarrow \rho + \pi$ (the dominant decay mode), by strong interactions, limits the isotopic spin $T$ to 0, 1, or 2. The decay [19]

$$A_2 \rightarrow K^{\pm} + K^0 \tag{10.25}$$

has also been observed with about a 10 per cent branching ratio. Thus, $T$ is not 0 or 2, and therefore $T = 1$ is selected. The decay of the $A_2$ was observed in the reaction

$$
\begin{array}{l}
\pi^- + \mathrm{p} \to \mathrm{n} + \mathrm{A}_2^0 \\
\qquad\qquad\quad \downarrow \\
\qquad\qquad\quad \mathrm{K}_S^0 \quad + \quad \mathrm{K}_S^0 \\
\qquad\qquad\quad \downarrow \qquad\qquad \downarrow \\
\qquad\qquad \pi^+ + \pi^- \qquad \pi^+ + \pi^-
\end{array}
\tag{10.26}
$$

Thus, the observation of the $A_2$ decay into two neutral kaons, and the requirements of Bose–Einstein statistics limit the $J^P$ to even–even, hence, $0^+$, $2^+$, $4^+$... are selected. As we have already pointed out, a $0^+$ particle cannot decay into three pions, thus, $A_2$ cannot have $0^+$, and the lowest possible assignment was $2^+$, which was supported by the Dalitz plot analysis.

As indicated in the Particle Data Group tables (1970), several recent experiments [20]–[23] have indicated a split $A_2$. The tables listed both an $A_2^L$ and an $A_2^H$ with the assignment of quantum numbers as follows.

$A_2^L(1280)$ has quantum numbers $1^-(2^+)+$, with mass $(1280 \pm 4)$, and $\Gamma = 22 \pm 4$ MeV.

$A_2^H(1320)$ has quantum numbers $1^-(2^+)+$, with mass $(1220 \pm 5)$, and $\Gamma = 21 \pm 4$ MeV.

Thus, the quantum numbers and characteristics of $A_2^L$ and $A_2^H$ were essentially identical, except for a 40 MeV mass shift (the $A_2^H$ is 40 MeV heavier than the $A_2^L$). The evidence for the two split $A_2$ levels is summarized below. Prior to this evidence for a split $A_2$, most experiments had concluded that there was a single wide $A_2$, which was listed in the tables as

$$I^G(J^P)C_n = 1^-(2^+)+,$$

with $M = 1297 \pm 10$ MeV and $\Gamma = 91 \pm 10$ MeV.

*The evidence for a split* $A_2$. That the $A_2$ peak was split into two peaks was first concluded by a CERN group, using a missing-mass spectrometer [20]. They studied the reaction

$$\pi^- + \mathrm{p} \to \mathrm{X} + \mathrm{P}. \tag{10.27}$$

By determining the momentum of the recoil proton (in the $|t|$ range 0·2–0·3), they determined the missing mass, $X$, needed to conserve four-momentum in the reaction.

Figure 10.2 shows a summary of the data for the split $A_2$. This splitting of the $A_2$ was confirmed by a bubble chamber group [21*a*] in the same $t$ range as the CERN group, but the splitting tended to disappear at low $t$. This may

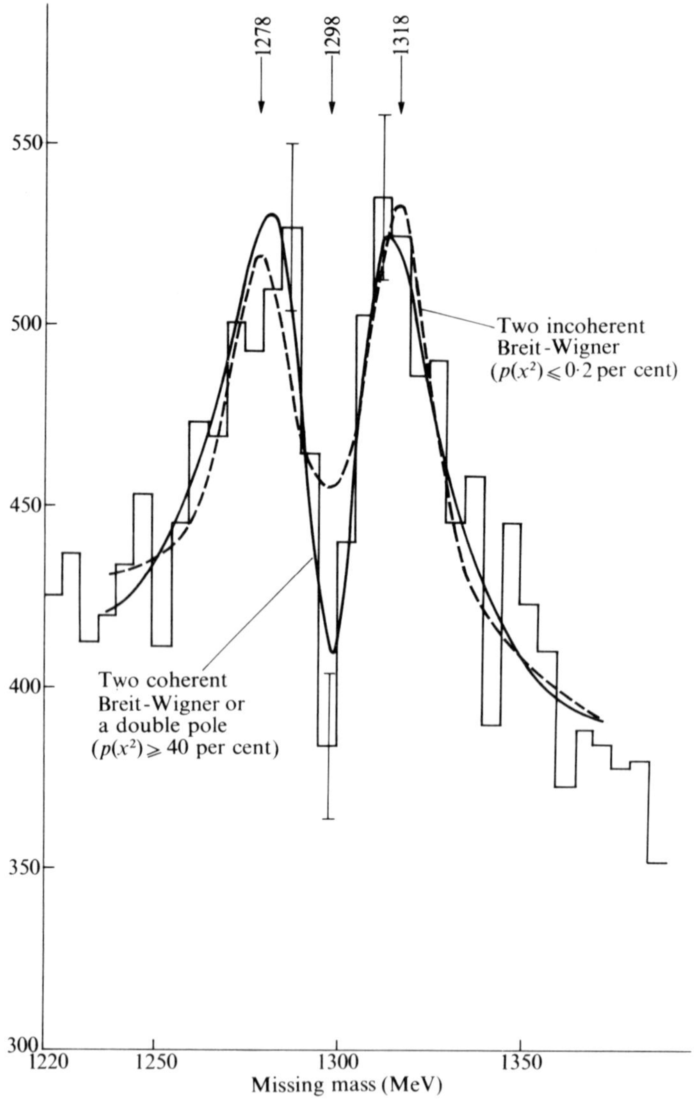


FIG. 10.2. $A_2$-splitting in the missing-mass spectrum using combined data of various spectrometer runs. (From Maglic, Lund Conference.)

be due to the worsening resolution at low $|t|$ which occurs in both experiments. The two $A_2$ states were then labelled as $A_2^L(1269 \pm 5)$ ($\Gamma_{A_2^L} = 24 \pm 10$ MeV), and $A_2^H(1315 \pm 5)$ ($\Gamma_{A_{2H}} = 12 \pm 10$ MeV). The dominant mode of decay is $\rho\pi$($\sim$ 86 per cent), but $\eta\pi$ is about $(11 \pm 2)$ per cent, and this implies that

$$J^P = \begin{cases} 0^+, 2^+ \ldots \\ 1^-, 3^- \ldots \end{cases}$$

The $A_2^H$ shows a $K\bar{K}$ decay mode ($K_S K_S$) with a sharp narrow mass peak at $(1311 \pm 5)$ MeV, and a $\Gamma = 21 \left.\begin{matrix} +10 \\ -6 \end{matrix}\right\}$ MeV and a sharp change in the $K_S^0, K_S^0$ angular distribution at 1311 MeV, where it changes from an isotopic distribution to a forward–backward peaking distribution. Thus, the $A_2^H$ corresponds to the previously-listed quantum number assignment. This experiment led to speculation that $I^G = 1^-$ for both $A_2^L$ and $A_2^H$, but the spin and parities may be different. This would be possible since, for the $K^0\bar{K}^0$ system, $1^-$ is forbidden, but it is possible in the $\rho\pi$ and $\eta\pi$ decays. Thus, $A_2^H$ could have $J^P = 2^+$, and $A_2^L$, $J^P = 1^-$.

A subsequent experiment (CERN–College de France–Liverpool collaboration) [21*b*] observed both peaks $A_2^L$ and $A_2^H$ in the $K^0K^\pm$ decay states produced in the annihilation reaction

$$\bar{p} + p \rightarrow K^0 K^\pm \pi^\mp,$$

at incident $\bar{p}$ momenta 0·7 and 1·2 GeV/$c$. Furthermore, Dalitz plots of various $\rho\pi$ decay states of the $A_2$ excluded $1^-$ for either half, and it was generally concluded [22] that $J^P = 2^+$ for either half of the $A_2$ data if it were split†. In a subsequent experiment [23] the split was observed when the reaction

$$\begin{aligned} \pi^- + p \rightarrow {} & A_2^- + p \\ & \downarrow \\ & K^0 + K^- \end{aligned}$$

was selected. This latter data, together with the previous data for a split, was then interpreted as most consistent with a so-called 'double pole' resonance.

There where various speculations, ranging from two close-by resonances, with the same quantum numbers, to the 'exotic' dipole hypothesis. A split

† Or even in those experiments in which the resolution was too poor to resolve a split.

$A_2$ would be a puzzling phenomenon. The dipole resonance† explanation, which was favoured, would cause difficulties (as we shall see shortly) in understanding the $2^+$ nonet, and why the other partners of the $A_2$ are not split also. It would not be understandable either on the basis of the quark model.

*Recent evidence against an* $A_2$ *split*. A group [24] at the Lawrence Radiation Laboratory studied $\pi^+ + p$ interactions at 7 GeV/*c* in an $H_2$ bubble chamber, which produced an $A_2^+$ meson by

$$\pi^+ + p \rightarrow A_2^+ + p.$$

Their results clearly disagreed with the prior conclusion of a split in the $A_2^+$. Of course, there were some differences between their work and the CERN missing-mass spectrometer work. First, the incident channel was $\pi^+ + p$, instead of $\pi^- + p$. Secondly, an $A_2^+$ was produced, instead of an $A_2^-$, and thirdly, their statistically significant data were obtained by grouping together the data from all $|t|$, and not just the narrow $|t| = 0{\cdot}2$–$0{\cdot}3$ range used in the CERN Missing-mass spectrometer work. Nevertheless, if we wish to reconcile the two experiments, we will have to assume that the production mechanism of the $A_2$ is a strong function of $|t|$ or charge state—or both.

Two recent higher energy [25], [26] experiments, by a BNL and a CERN–Munich group, studied the reaction

$$\begin{array}{l} \pi^- + p \rightarrow A_2^- + p \\ \qquad\qquad \hookrightarrow K^0 + K^- \\ \qquad\qquad\qquad \hookrightarrow \pi^- + \pi^+ \end{array} \tag{10.28}$$

for 20·3 and 17·2 GeV/*c* negative pions, incident on a liquid $H_2$ target. This reaction is particularly suited to studying whether or not the $A_2$ is split, since the background in the $K^0K^-$ channel, under the $A_2$ peak, is very small ($\lesssim$10 per cent) compared to other decay channels, or the background in

† We have described already the characteristics of a resonance, in terms of the Argand diagram behaviour. In a single pole (Breit–Wigner resonance) in an elastic resonance, the $A_{l,j}$ of the Argand diagram moves around the circle once (from 0 to $2\pi$). As the resonance is approached, and traversed, in a double pole resonance, the circle is traversed twice. Analogies in behaviour in resonances which are partly inelastic also occur between single and dipole resonances. The shape for a Breit–Wigner dipole resonance would be

$$\left\{\frac{(m-m_0)\Gamma}{(m-m_0)^2+(\Gamma/2)^2}\right\}.$$

missing-mass experiments. Both of these experiments used wire spark chambers, before and after a large magnet, in order to detect forward-going $A_2^-$ particles, and measure the momenta of all three decay particles. Reference [25] had a somewhat better mass resolution ($\approx$ 25 per cent) of $\approx$ 4·7 MeV (one standard deviation), but reference [26] had about 25 per cent less statistical errors. The data from both experiments is shown in Fig. 10.3. Both groups concluded that when the data at all $|t|$ was grouped together, the probability against a split of the dipole type was very large (i.e. $<1$ part in $10^{-6}$ in [25]), whereas a single ($l = 2$) Breit–Wigner resonance curve fits very well. Both experiments were much better fitted by a single Breit–Wigner resonance even in the $|t| > 0{\cdot}2$ region, thus disagreeing strongly with the previous conclusions on a split in the $A_2$. Both experiments demonstrated, from an analysis of the angular distribution of the decay products and statistical requirements, that $J^P = 2^+$ (and this was true for both halves of the $A_2$, separately), It was deducible also that the $G$-parity was negative.

Reference [25] obtained, for the mass of the $A_2$,

$$m_{A_2} = 1{\cdot}313 \pm 0{\cdot}004 \text{ GeV},$$

and the width

$$\Gamma_{A_2} = 0{\cdot}114 \pm 0{\cdot}01 \text{ GeV}.$$

Reference [26] obtained

$$M_{A_2} = 1{\cdot}321 \pm 0{\cdot}003 \text{ GeV},$$

$$\Gamma_{A_2} = 0{\cdot}123 \pm 0{\cdot}01 \text{ GeV}.$$

A third experimental group [27] attempted to reproduce the original CERN missing-mass spectrometer experiments. They studied the reaction

$$\pi^{\pm} + p \rightarrow X^{\pm} + p, \tag{10.29}$$

at the $\pi^-$ incident momenta of 5 and 7 GeV/$c$ and the $\pi^+$ incident momentum of 5 GeV/$c$. Their calculated resolution was $\Gamma = 16 \pm 1$ MeV (full width at half maximum), which should have been adequate to observe any split effect of the type originally reported by the CERN missing-mass group. The result of their investigation (see Fig. 10.3(c)) clearly contradicted the original work of the CERN missing-mass group. They estimate

$$m_{A_2} \approx 1{\cdot}305 \pm 0{\cdot}004 \text{ GeV},$$

$$\Gamma_{A_2} \approx 0{\cdot}110 \pm 0{\cdot}006 \text{ GeV}.$$

Thus, when we consider the totality of recent experiments, including the one which clearly contradicts the original CERN missing-mass spectrometer work, it appears very likely that the $A_2$ is unsplit. If there should be more than one particle in the $A_2$ mass region, it is likely that only a subtle interference

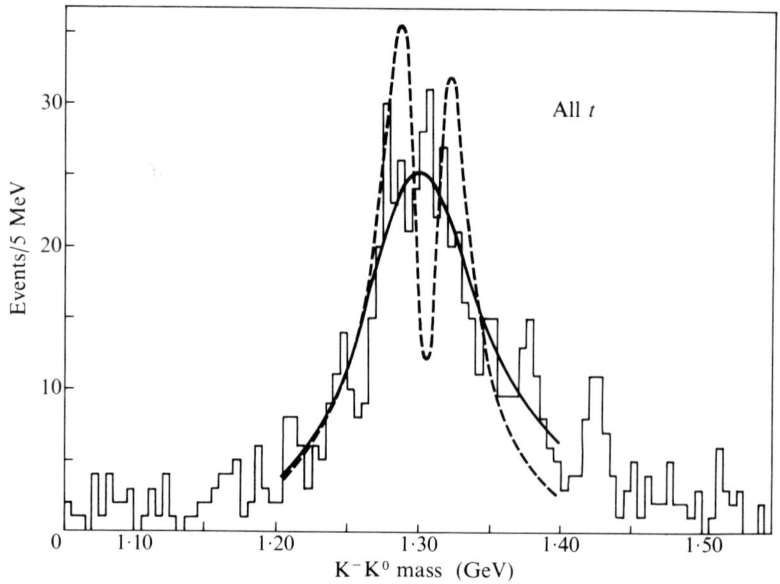


FIG. 10.3.(a) The $K\text{–}K_s^\circ$ effective (i.e. invariant) mass spectrum observed in the reaction $\pi^- + p \to K^\circ + K^- + p$ for 20·3 GeV/$c$ incident $\pi^-$ on hydrogen. The solid line is a Breit–Wigner-type fit corresponding to $l = 2$. The dashed line is a 'dipole fit' which is unacceptable. (From Foley *et al.* (1971). *Phys. Rev. Lett.* **26,** 413.)

$$K^\circ \downarrow K_s^\circ$$

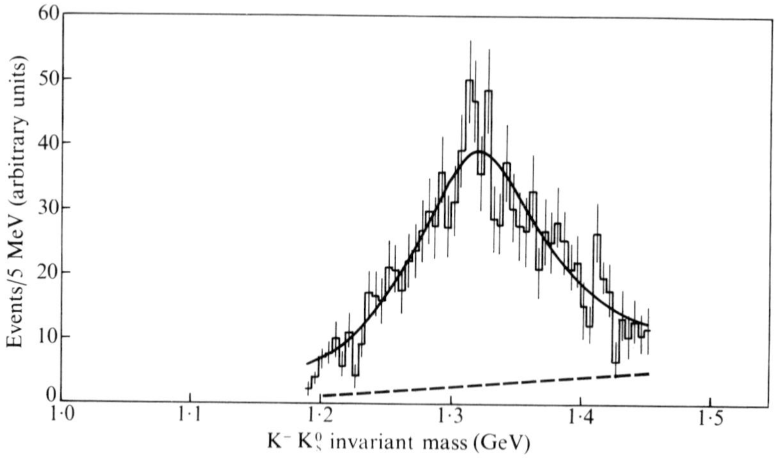


FIG. 10.3.(b) The $K^-K_s^\circ$ invariant mass (GeV) observed in the reaction $\pi^- + p \to K^- + K^\circ + p$, for 17·2 GeV/$c$ $\pi^-$ incident on hydrogen. The solid line is a Breit–Wigner-type fit corresponding to $l = 2$. The dashed line is a linear fit to the background. (From Grayer *et al.* (1971). *Phys. Lett.* B **34,** 333.)

$$K^\circ \downarrow K_s$$

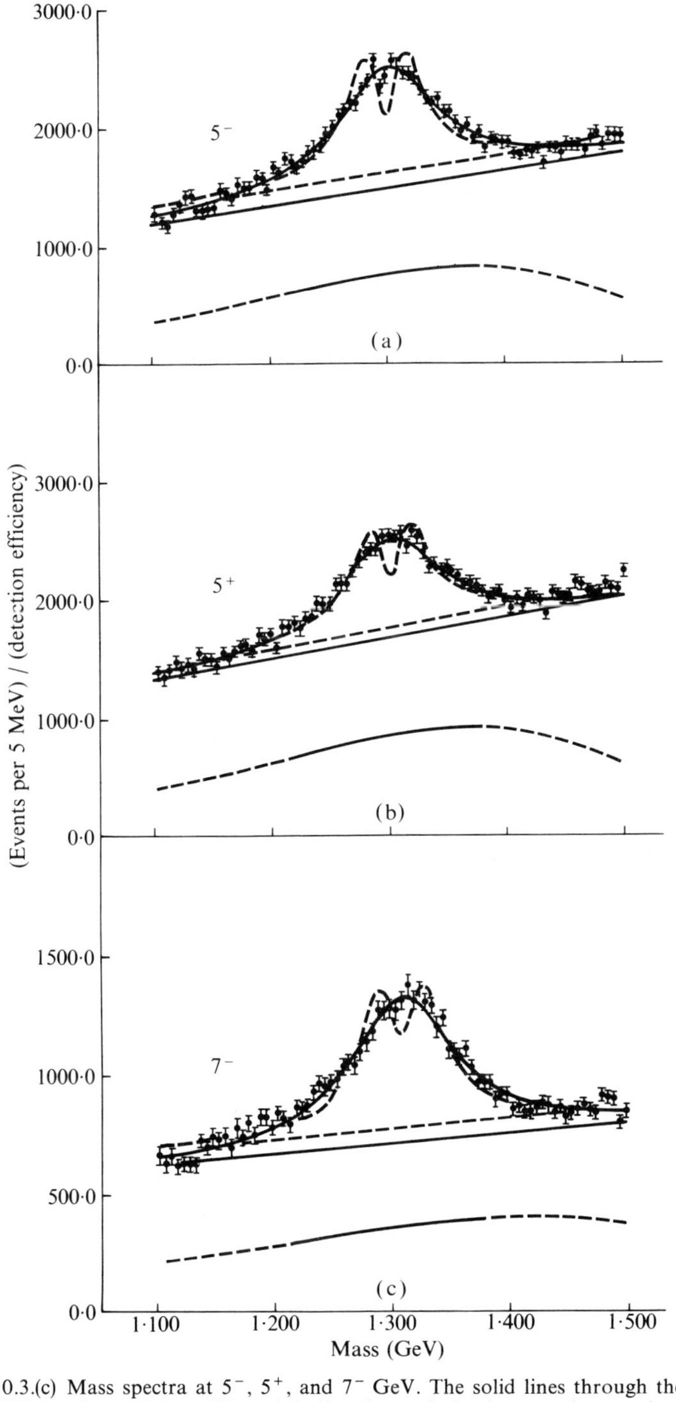


FIG. 10.3.(c) Mass spectra at $5^-$, $5^+$, and $7^-$ GeV. The solid lines through the data are the Breit–Wigner fits, and the solid straight lines beneath the data are the associated fitted linear backgrounds. The dashed lines in the region of the data are the dipole fits ($\Gamma$ dipole $= 28$ MeV, fixed) and their associated backgrounds. The calculated detection efficiencies plotted against mass are shown (arbitrary units) as dashed lines ($1{\cdot}10 \leq M \leq 1{\cdot}22$ GeV, and $1{\cdot}38 \leq M \leq 1{\cdot}50$ GeV) and as solid lines ($1{\cdot}22 \leq M \leq 1{\cdot}38$ GeV, 'resonance' region). The detection efficiencies have been normalized so that, at $M = 1{\cdot}300$ GeV, the ordinates on the graphs indicate the actual number of events detected in the experiment per 5 MeV bin. (From Bowen *et al.* [27].)

effect between different-width different-quantum-number resonances, possibly with different $s$ and $t$ and incident charge dependences, could occur.

*The* K*(*1410*) *meson*. Evidence for a K* resonance [28] in the K–$\pi$ system, with a mass of about 1410 MeV, was observed in the following reactions

$$K^- + p \rightarrow (\bar{K}^0 + \pi^-) + p \qquad (3{\cdot}5\ \mathrm{GeV}/c), \tag{10.29a}$$
$$(K^- + \pi^0) + p$$
$$\pi^- + p \rightarrow \Sigma^{\pm}\pi^{\mp}K^0 \qquad (4{\cdot}10\ \mathrm{GeV}/c)$$
$$\Sigma^-\pi^0K^+$$
$$\Lambda^0\pi^0K^0$$
$$\Lambda^0\pi^-K^+, \tag{10.29b}$$

An analysis of the angular distribution favours the $2^+$ assignment. $T = \frac{1}{2}$ is deduced [29] from the branching ratios. This meson is referred to in the Particle Data Group tables as $K_N(1420)$, with $T = \frac{1}{2}$ established, but the $J^P = 2^+$ assignment is still not considered established. The mass is $(1409 \pm 4)$ MeV, and $\Gamma = 96 \pm 7$ MeV.

*The* $f^*$ *(or* $f'$*) meson*. An investigation of 4·6–5·0 GeV/$c$ $K^-$ mesons incident in a Hydrogen bubble chamber [30] led to a study of the following reactions

$$K^- + p \rightarrow (\Lambda^0, \Sigma^0)K^0K^0$$
$$\Sigma^{\pm}K^{\mp}\begin{pmatrix} K^0 \\ \overline{K^0} \end{pmatrix}$$
$$\Lambda^0K^0\bar{K}^0\pi^0$$
$$\Lambda^0\bar{K}_0K^+\pi^-$$
$$\Lambda^0K_0K^-\pi^+$$
$$\Sigma^{\pm}K^0K^0\pi^{\mp} \tag{10.30}$$

An enhancement was observed in the $K_sK_s$ system (i.e. both K-mesons decay into $\pi^+ + \pi^-$ pairs), with $M \approx 1500$ MeV, and width $\Gamma \approx 85$ MeV. The existence of this decay mode requires that the $f^*$ mesons have even spin and parity. However, $f^* \rightarrow \bar{K}K^*$ and $KK^*$ was observed also and this excludes a $0^+$ spin–parity assignment.

Figure 10.4 shows the $Y^0K^0\bar{K}^0$ Dalitz plot, and exhibits the previously-known $\phi$ peak and the new f* peak. The decay angular distribution is anisotropic, also excluding the $0^+$, and favouring the $2^+$, assignment. There is no evidence for $f^* \rightarrow K^{\pm}K^0$, which implies $T = 0$. As can be seen in the table it has, subsequently, been concluded that the f′(1514) has $I_G = 0^+$, $J^P = 2^+$, and $C_N = +$.

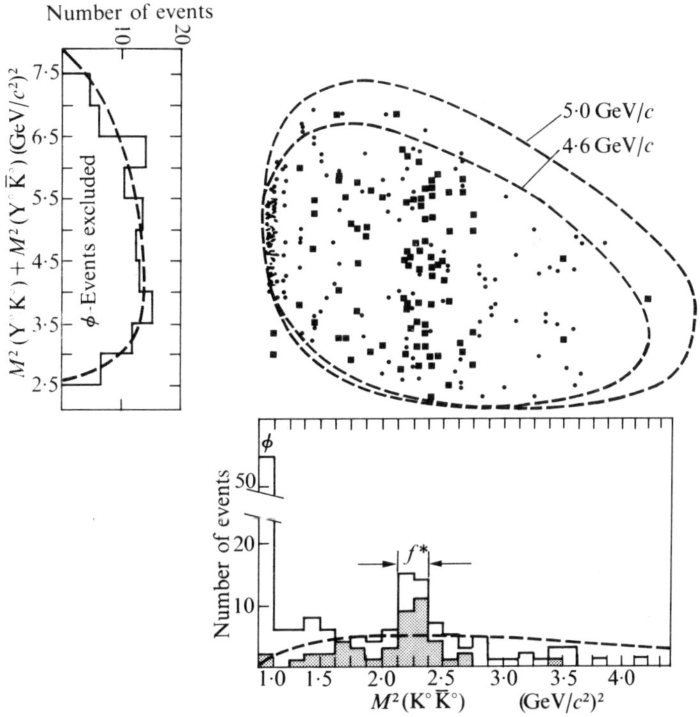


FIG. 10.4. $Y^\circ K^\circ \bar{K}^\circ$ Dalitz plot for 162 events plotted twice. The dots are for events with only one K decay visible, i.e. $K_1 \to \pi^+ + \pi^-$. The crosses are for events with two visible $K_1$ decays. In the $M^2(\bar{K}^\circ K^\circ)$ projection each event is plotted once. The shaded area is for those events with two visible $K_1$ decays. In the $M^2(Y^\circ K^\circ)$ projection, outside the $\varphi$ region, each combination is plotted as half an event. (From Barnes *et al.* (1965). *Phys. Rev. Lett.* **15,** 322.)

Thus, we have the following $2^+$ mesons, $f^0$, $f'$, $A_2$, and $K_N$ and $\bar{K}_N$, which together have the proper multiplet structure to form an SU(3) nonet. The mass formula for a meson octet is

$$M^2 = M_0^2 + \beta\{T(T+1) - \tfrac{1}{4}Y^2\}.$$

Thus, if we assume the $K_N(1420)$ and the $A_2$ are members of an octet, we can solve for $M_0^2$ and $\beta$. We use about 1300 MeV as the $A_2$ mass. Thus, we have, for the $A_2$,

$$(1300)^2 \approx M_0^2 + 2\beta, \tag{10.31}$$

and, for the $K_N(1420)$, we obtain

$$(1420)^2 \approx M_0^2 + \beta\{(\tfrac{1}{2})(\tfrac{3}{2}) - \tfrac{1}{4}\} \approx M_0^2 + \beta/2. \tag{10.32}$$

Thus,

$$\tfrac{2}{3}\beta = (169-201)\times 10^4;\ \beta \approx -21{\cdot}7\times 10^4\,(\text{MeV})^2 \tag{10.33}$$

and

$$M_0^2 = 169\times 10^4 + 43{\cdot}4\times 10^4 = 212{\cdot}4\times 10^4, \tag{10.34a}$$

$$M_0 \approx 1450\ \text{MeV}. \tag{10.34b}$$

Since, for the missing $I$-spin singlet to complete the octet, $T = Y = 0$, its predicted mass is equal to $M_0$ and, thus, 1450 MeV. Now neither f(1260) nor f′(1515), have the required mass, implying that there is considerable interference, and resultant mixing, if we wish to complete the nonet with f and f′. However, it was shown [31] that, with a mixing angle of the order obtained for the vector mesons, it was possible to complete the octet with f(1260), and use the f′(1515) for the unitary singlet to complete the nonet, and satisfy the mass formula.

10.2.11. *The* $0^+$ *nonet*

The $0^-$, $1^-$, and $2^+$ nonets are considered well-established, although we still have the possibility that the E-meson may replace the $X^0$ (or $\eta'$) in the $0^-$ nonet. There are some additional mesons, some of which appear to form part of a $0^+$ nonet. These include $\eta_{0^+}(700)$ or $\epsilon$, $0^+(0^+)+$, with the dominant decay mode to $\pi\pi$, and the $\pi_N(1016) \to K\bar{K}$, with the tentative quantum numbers $1^-(0^+)+$.

We can tentatively consider constructing a $1^+$ nonet from the following possible list of $1^+$ mesons. The D, E, $A_1$, $K_A(1320)$, or $K_A(1780)$, also called L. The $J^P = 1^+$ assignment may apply to all of these but unfortunately the assignment has not been definitely established for any of the above.

The *D(1285) meson* has been assigned the quantum numbers $0^+(A)+$ and decays to $K\bar{K}\pi$ with the $K\bar{K}$ pair, mainly forming a $\pi_N(1016)$. The $J^P$ assignment is $0^-, 1^+, 2^-, \ldots$, with $1^+$ favoured by the analysis.

The *E(1420) meson* has the preferred quantum numbers $0^+(0^-)+$. It decays via $K^*\bar{K}+\bar{K}^*K \sim (50\pm 10)$ per cent, and $\pi_N(1016)\pi \sim (50\pm 10)$ per cent. However, $J^P = 1^+$ is the second choice, and is not yet excluded. If, indeed, $J^P = 0^-$, for the E(1420), we have the previously-mentioned problem of 10 pseudoscalar $(0^-)$ mesons.

The *$A_1$(1070) meson* is observed in various reactions. Its tentative quantum numbers are $1^-(1^+)+$, where $T = 1$, $G = -1$, is considered confirmed. The dominant decay mode is $3\pi$.

The isotopic spin of the $K_A(1320)$ is considered established as $\frac{1}{2}$, but for the $K_A(1240)$ (or C-meson), $T = \frac{1}{2}$ appears likely but is not definitely established. $J^P = 1^+$ is implied for the $K_A(1320)$, but $2^-$ is not completely excluded. The $K_A(1240)$ has an A (abnormal) $J^P = 1^+, 2^- \ldots$, with $1^+$ favoured.

Some investigators see a broad enhancement in mass of the $K\pi\pi$ system extending from 1200 MeV to 1350 MeV. Other investigators see structure. Another peak at 1280 MeV, with $\Gamma = 80$ MeV, has been suggested. In the light of this confusion, the masses, widths, quantum numbers, and branching ratios, listed in the tables, are tentative and unreliable. For the mass region 1200–1350 MeV, the decay rate to the K*(890) and a $\pi$ is large and the $K\pi\pi$ decay is seen. *The* $K_A(1775)$ (*or L*) has the quantum numbers $\frac{1}{2}$(A), with the dominant decay modes $K\pi\pi$, $K_N(1420)\pi$, and $K\rho$ seen also.

Thus, one of the $K_A$, the $A_1$, and the D and E would provide two $Y = \pm 1$ isotopic doublets, an isotopic triplet with $Y = 0$, and two isotopic singlets with $Y = 0$, which is just sufficient to form a $1^+$ nonet. In view of the uncertainties just discussed, it is premature to consider applying the mass formula to this tentative nonet.

The ever-increasing number of identified particles does not allow us to be too satisfied with the partial classification we have made under symmetry schemes. There are still a number of outstanding problems. For example, the $\delta^-$-meson, and its possible relation to the $\eta'$ (or $X^0$), is one such puzzling case. The $\delta^-$ was originally observed in a missing-mass experiment,

$$\pi^- + p \rightarrow p + X^-. \tag{10.35}$$

There followed a number of experiments which alternately confirmed the existence, and then did not find evidence for its existence. Nevertheless, considering all the available information, it appears that the $\delta(962)$ probably exists, has a mass of $(962 \pm 5)$ MeV, and a $\Gamma < 5$ MeV. Since $\delta(962)$ has been deduced to have unit negative charge, $T \geq 1$ is required. It could be related, possibly, to the $\pi_N(1016) \rightarrow K\bar{K}$, which has a mass of $(1016 \pm 10)$ MeV, and $\Gamma \approx 25$ MeV. Its quantum numbers are $1^-(0^+)+$. However, an even more suspicious coincidence is that the $\delta(962)$ has the same mass, within the errors, as the $\eta'(958)$ (with mass $(957{\cdot}7 \pm 0{\cdot}8)$ MeV) or $X^0$. This has led to some theoretical speculation as to their relationship, which is too preliminary to warrant discussion here at this time.

It is an interesting commentary on the state of flux of so-called elementary particle knowledge that the Particle Data Group tables often show great changes from one edition to another.

### 10.2.12. *Higher unitary symmetry* (SU(6))

We started with the observation of isotopic spin multiplets, and discovered that they are understandable as a result of invariance under the group SU(2). We found, subsequently, that many of these multiplets could be grouped into various supermultiplets, including octets and decuplets (and of course, with singlets into nonets in the case of mesons) resulting from our postulating invariance of the strong interactions under SU(3). Each of these SU(3) supermultiplets has the same space–time external properties (i.e. a common spin and parity). Thus, SU(3) invariance applies to charge, isotopic spin, and hypercharge, and the invariant transformations under SU(3) allow these internal quantities to vary and transform among themselves, but there is no connection made between SU(3) invariance and invariance with respect to the rotations and reflections in space–time, which are directly related to intrinsic particle spins and parities. Thus, we have discovered $J^P$ supermultiplets, including octets with $0^-$, $\frac{1}{2}^+$, $1^-$, and $2^+$, a decuplet of $\frac{3}{2}^+$, and unitary singlets of $0^-$, $1^-$, and $2^+$, which are close to degenerate (especially in the $1^-$ case) with the singlet in the octet. Furthermore, we have no clues why these particular $J^P$ combinations occurred. Also, just as SU(3) combined various isotopic spin multiplets which exhibited SU(2) symmetry, into supermultiplets exhibiting SU(3) symmetry, we can perhaps anticipate that space–time and SU(3) symmetries might be combined in a higher symmetry. We have just discussed the fact that SU(3) symmetry can be provided by a basic triplet $(q_1q_2q_3)$ of quarks. Each of these quarks can have a spin which is 'up' or 'down'. Thus, we might expect that we have six basic (or fundamental) states, also, we might suspect that there is a higher SU(6) symmetry under which the strong interactions are invariant. This SU(6) symmetry would rotate any one of the six states into any other. Sakita (1964) [32*a*] and Gursey and Radicati (1964) [32*b*] first introduced this SU(6) symmetry group.

The infinitesimal generators of the SU(6) group in the basic representation are formed by the following cross products,

$$\boldsymbol{\sigma} \otimes \lambda_0,\ 1 \otimes \lambda_i,\ \boldsymbol{\sigma} \otimes \lambda_i, \tag{10.36}$$

where the $\lambda_i$ are the eight $(3 \times 3)$ unitary unimodular matrices† representing

† $$\lambda_1 = \begin{bmatrix} 0 & 1 & 0 \\ 1 & 0 & 0 \\ 0 & 0 & 0 \end{bmatrix}, \quad \lambda_4 = \begin{bmatrix} 0 & 0 & 1 \\ 0 & 0 & 0 \\ 1 & 0 & 0 \end{bmatrix}, \quad \lambda_7 = \begin{bmatrix} 0 & 0 & 0 \\ 0 & 0 & -i \\ 0 & i & 0 \end{bmatrix},$$

$$\lambda_2 = \begin{bmatrix} 0 & -i & 0 \\ i & 0 & 0 \\ 0 & 0 & 0 \end{bmatrix}, \quad \lambda_5 = \begin{bmatrix} 0 & 0 & -i \\ 0 & 0 & 0 \\ i & 0 & 0 \end{bmatrix}, \quad \lambda_8 = \frac{1}{\sqrt{3}}\begin{bmatrix} 1 & 0 & 0 \\ 0 & 1 & 0 \\ 0 & 0 & -2 \end{bmatrix},$$

$$\lambda_3 = \begin{bmatrix} 1 & 0 & 0 \\ 0 & -1 & 0 \\ 0 & 0 & 0 \end{bmatrix}, \quad \lambda_6 = \begin{bmatrix} 0 & 0 & 0 \\ 0 & 0 & 1 \\ 0 & 1 & 0 \end{bmatrix}.$$

SU(3), and

$$\lambda_0 = \frac{\sqrt{2}}{3}\begin{bmatrix} 1 & 0 & 0 \\ 0 & 1 & 0 \\ 0 & 0 & 1 \end{bmatrix} \tag{10.37}$$

is for normalization. The above yields 35 ($= 6^2 - 1$) Hermitian traceless matrices, which generate transformations of SU(6). Our basic spinor is now a six-component quark, the sextet

$$\begin{matrix} q_{j,\alpha} = \\ j = 1, 2, 3 \\ \alpha = 1, 2 \end{matrix} \begin{cases} q_{1,1}\uparrow \\ q_{1,2}\downarrow \\ q_{2,1}\uparrow, \\ q_{2,2}\downarrow \\ q_{3,1}\uparrow \\ q_{3,2}\downarrow \end{cases} \tag{10.38}$$

and the anti-sextet corresponding to the anti-particles $q^*_{j,\alpha}$. We can assume the sextet and the anti-sextet are the fundamental building blocks, and that the interactions (binding effects) are, as far as the strong interactions are concerned, independent of both spin and SU(3). Then we classify the particle states according to the irreducible representations of SU(6).

### 10.2.13. SU(6) *particle multiplets*

The group SU(6) contains SU(2) × SU(3). Thus, we can plainly exhibit the spin and SU(3) content of an SU(6) supermultiplet, by its reduction under the subgroup SU(2) × SU(3). Hence, the mesons can be formed from the $6^* \times 6$ quark–anti-quark pairs ($B = 0$), which is the simplest case, or possibly from $6^* \otimes 6^* \otimes 6 \otimes 6$. The $6^* \otimes 6$ contains 36 states of the type $q^{j',\alpha'}q_{j,\alpha}$. Transformations under SU(6) lead to 35 of these states transforming among themselves (a 35 representation), while one (a singlet) remains invariant. Thus, the traceless (matrix) combination of quark and anti-quark is the regular representation which can be written as [2], [3], [33]

$$M^{j',\alpha'}_{j,\alpha} = (\bar{q}^{j',\alpha'}q_{j,\alpha} - \tfrac{1}{6}\delta^{j',\alpha'}_{j,\alpha}\langle q\bar{q}\rangle). \tag{10.39}$$

Thus

$$6^* \otimes 6 = 35 + 1. \tag{10.40}$$

The meson multiplet content in $6^* \otimes 6$ is determined using standard methods. Clearly, SU(2) and SU(3) must be subgroups of SU(6). $\frac{1}{2} \otimes \frac{1}{2} = 0 + 1$, thus, the SU(6) multiplets are (8, 3), (8, 1), (1, 3), and (1, 1), where our notation is $(\mu, \nu)$, and where $\mu$ is the SU(3) representation (dimensionality), and $\nu$ is

$2J+1$, $J$ being the spin (i.e., $v$ is the spin multiplicity, or SU(2) dimensionality). Thus, the irreducible representations 35 and 1 contain

$$\begin{aligned} 35 &= (8, 3) \oplus (8, 1) \oplus (1, 3), \\ 1 &= (1, 1). \end{aligned} \tag{10.41}$$

We can associate all the mesons in the $0^-$ octet, the $1^-$ octet, and $1^-$ singlet into a 35 SU(6) supermultiplet as follows†

$$\{(8, 1), (8, 3), (1, 3)\} = \{(\pi, \mathrm{K}, \bar{\mathrm{K}}, \eta); (\boldsymbol{\rho}, \mathrm{K}^*, \bar{\mathrm{K}}^*, \boldsymbol{\omega}_8); (\boldsymbol{\omega}_1)\} \tag{10.42}$$

The previously-mentioned mixing of the $\boldsymbol{\omega}_8$ and $\boldsymbol{\omega}_1$ are provided for, and their mixtures make up the physical $\omega$ and $\phi$.

The additional singlet (1, 1) can be identified as the $\eta'(958)$ or $\mathrm{X}^0$. It is interesting to note that the intrinsic parity of a quark–anti-quark pair bound in an $S$-state is negative, since a fermion and anti-fermion have opposite parities. The baryons must, in the simplest representation, be constructed from

$$\mathrm{q}_{j,\alpha} \otimes \mathrm{q}_{j',\alpha'} \otimes \mathrm{q}_{j''\alpha''} = 6 \otimes 6 \otimes 6, \tag{10.43}$$

in order to obtain a baryon number of $3 \times \frac{1}{3} = 1$. By standard techniques,

$$6 \otimes 6 \otimes 6 = 56 \oplus 70 \oplus 70 \oplus 20. \tag{10.44}$$

When the 56 representation is expressed in terms of its SU(3) and SU(2) subgroups, we obtain

$$56 = (10, 4) \oplus (8, 2). \tag{10.45}$$

Thus, the 56 representation is composed of a decuplet of spin $\frac{3}{2}$, and an octet of spin $\frac{1}{2}$. Hence, it can represent the $\frac{3}{2}^+$ (decuplet) and the $\frac{1}{2}^+$ (octet) of baryons. It is interesting to note that if a baryon is a bound $S$-state of three quarks, and, if we assume that the quarks obey Fermi statistics, the Pauli principle requires that the wave functions have an over-all anti-symmetry, which is the case only for the 20-dimensional representation. Thus, since the 56-dimensional representation is the one that fits the experimental observations, the quark model must be either considered inadequate, except as a basis for generating SU(6), which is then to be regarded as the phenomenologically important symmetry, or we are forced to assume that quarks obey different statistics (called parastatistics [34]).

† The (8, 1) contains the $0^-$ octet, or eight meson states. The (8, 3) contains the $1^-$ octet of mesons, each of which can have three spin states, or 24 mesons, and the (1, 3) contains one $1^-$ meson with three spin states, or three mesons. Thus,

$$8 \times 1 + 8 \times 3 + 1 \times 3 = 35.$$

We have explained the meson (octet+singlet = nonet) multiplets with $J^P = 0^-, 1^-$. However, in order to explain the $2^+$ nonet, we apparently need higher-order representations than $6^* \otimes 6$ since a $6^* \otimes 6$ does not contain $\{(8, 5)+(1, 5)\}$, which is required for the $2^+$ nonet. The next simplest structure for mesons would be q*q*qq. Thus, we must investigate the representation $6^* \otimes 6^* \otimes 6 \otimes 6$.

$$6^* \otimes 6^* \otimes 6 \otimes 6 = 1296 = 2\times 1 \oplus 4\times 35 \oplus 189 \oplus 2\times 280 \oplus 405. \tag{10.46}$$

To represent an octet with $J = 2$, we need a representation that contains (8, 5), and for a singlet with $J = 2$, we need (1, 5). The representations that contain these are 189 and 405. This combination is contained in $20^* \otimes 20$, $56^* \otimes 56$, as well as $35^* \otimes 35$. Hence, it appears that, for the $2^+$ nonet, the higher-order representations, such as $6^* \otimes 6^* \otimes 6 \otimes 6$, would be indicated. It is interesting to note that $6^* \otimes 6^* \otimes 6 \otimes 6$ would have positive parity for *s*-wave binding. Nevertheless, these higher-order representations would contain 'exotic' partners, which have never been observed. Exotic mesons are mesons which cannot, in principle, be built from a quark–anti-quark pair. Exotic baryons are baryons which cannot, in principle, be built from three quarks. Neither exotic mesons nor exotic baryons have ever been demonstrated to exist to date.

Since the $2^+$ nonet can, in principle, be built from a quark–anti-quark combination (if we allow for higher orbital angular momentum in the quark model), it is generally considered to be a normal ('non-exotic') multiplet, in spite of the foregoing simple argument.

Reference [7] of Chapter 11 (Feld's book) gives a detailed review of the quark model.

### 10.2.14. *Mass splitting relations in* SU(6)

Just as we were able to obtain mass relations for a SU(3) supermultiplet, we could similarly expect to obtain mass relations for an SU(6) supermultiplet.

The SU(6) symmetry-breaking mass term can be separated into two principle terms, (1) the SU(3) symmetry and spin dependent part, (2) the part which breaks SU(3) symmetry. We can assume that this term is a generalization of the G.M.O. mass term, and tentatively assign it to the (8, 1) component of the adjoint representation of SU(6).

Let us now consider the mesons which belong to the 35 representation. In terms of the meson wave function $\Phi_{j,\alpha}^{j',\alpha'}$, we can write the mass formula as

$$M_0^2\langle\Phi\Phi\rangle - M_1^2\Phi_{j,\alpha}^{j',\alpha}\Phi_{j',\alpha'}^{j,\alpha'} + M_2^2\Phi_{j,\alpha}^{3\alpha}\Phi_{3,\alpha}^{j\alpha}. \tag{10.47}$$

The first term is SU(6) invariant. The second term is spin dependent, and the third term breaks SU(3) symmetry. The 35 representation can be decomposed (as we have done previously) in SU(2) $\times$ SU(3) irreducible representations. Thus, we now write the above in terms of the vector nonet $V$, and the pseudoscalar octet $P$, as follows

$$M_0^2(\langle V.V\rangle + \langle PP\rangle) - m_1^2\langle PP\rangle + m_2^2(V_j^3.V_3^j + P_j^3 P_3^j). \qquad (10.48)$$

The above also yields a relation between the pseudoscalar meson octet masses, and the vector meson masses.

$$m_{\mathrm{K}^*}^2 - m_\rho^2 = m_{\mathrm{K}}^2 - m_\pi^2. \qquad (10.49)$$

The experimental values lead to agreement with the prediction within $\approx 15$ per cent.

Let us now consider the baryons which, as we have seen, appear to belong to the 56-dimensional representation. The mass term can be written as

$$M_0(\bar{\Psi}\Psi) + M_1\bar{\Psi}^{j,\alpha,j',\alpha',j'',\alpha''}\Psi_{j,\alpha',j',\alpha j'',\alpha''} + M_2\bar{\Psi}^{3,\alpha,j',\alpha',j'',\alpha''}\Psi_{3,\alpha,j',\alpha',j'',\alpha''}. \qquad (10.50)$$

we can then evaluate this in terms of the baryonic wave functions of the 56 representation, expressed in terms of the $\frac{3}{2}^+$ decuplet, and the $\frac{1}{2}^+$ octet. When this is done, we obtain, analogous to the meson case, a relation between decuplet and octet masses

$$\begin{aligned} M_\Xi - M_\mathrm{N} &= M_{\Xi^*} - M_{\mathrm{N}^*}, \\ 384 &\approx 294 \text{ MeV}, \end{aligned} \qquad (10.51)$$

which is in poor agreement. This discrepancy is understandable since the simple symmetry-breaking term employed does not take account of the $\Sigma - \Lambda$ mass difference. In order to take this into account, higher-dimensional representations, such as the 405 must be taken into account. This was done by Beg and Singh [35] who then obtained the following formula

$$M(J, T, Y) = \alpha + \beta J(J+1) + \gamma Y + \delta\{T(T+1) - \tfrac{1}{4}Y^2\}, \qquad (10.52)$$

where $\alpha$, $\beta$, and $\gamma$ are constants. Using this mass formula we obtain the following relation between the baryon octet and decuplet masses

$$M_\Xi - M_\mathrm{N} - \tfrac{3}{2}(M_\Sigma - M_\Lambda) = M_{\Xi^*} - M_{\mathrm{N}^*}. \qquad (10.53)$$

We obtain experimentally 268 MeV $\approx$ 292 MeV, which is in reasonable agreement.

### 10.2.15. *Magnetic moments*

In the static limit, we can describe the electromagnetic interaction of matter by the interaction Hamiltonian density

$$\mathscr{H} = \rho\phi - \mathbf{M}.\mathbf{H}, \tag{10.54}$$

where $\rho$ represents the charge density and $\mathbf{M}$ the magnetization density. $\phi$ is the external electric field electrostatic potential and $\mathbf{H}$ is the external magnetic field.

For the quark model, we can express $\rho$ by

$$\psi^* Q \psi, \tag{10.55a}$$

where

$$Q = \frac{e}{3}\begin{bmatrix} 2 & 0 & 0 \\ 0 & -1 & 0 \\ 0 & 0 & -1 \end{bmatrix} \tag{10.55b}$$

is the charge matrix. Furthermore,

$$\mathbf{M} \propto \psi^* Q \boldsymbol{\sigma} \psi. \tag{10.56}$$

Now let us consider the transformation properties of $\rho$ and $Q$ under SU(6). Thus, it follows that

$$\rho = \rho^{j,1}_{j,1}. \tag{10.57}$$

Therefore $\rho = (8, 1)$ component of the 35-dimensional representation, whereas

$$\mathbf{M} = \sigma^{j'}_{j} M^{j,1}_{j',1} = (8, 3)$$

component of the 35-dimensional representation. $\rho$ must transform like a scalar under the spin group, and have octet transformation properties under SU(3). $\mathbf{M}$ must transform like a vector under the spin group, and have octet transformation properties under SU(3). Thus, the above SU(6) transformation properties obtained by employing the quark model are the most likely choices, even from the general point of view.

Hence, the effective electromagnetic interaction of the baryons is of the form

$$\mathscr{H}_{\text{eff}} = 3eJ^1_1\phi - 3\mu \mathbf{J}^1_1.\mathbf{H}, \tag{10.58}$$

where $\mu$ is the magnetic moment of the proton, and where $J^{\alpha'}_{\alpha} = J^{j,\alpha'}_{j,\alpha}$.

$$\mathbf{J}^{\alpha'}_{\alpha} = \sigma^{j'}_{j}(J^{j\beta}_{j',\alpha} - \tfrac{1}{3}\delta^{\alpha'}_{\alpha} J^{j,\gamma}_{j',\gamma}). \tag{10.59}$$

Substituting the explicit expression for $\psi$ in the current, and evaluating the resultant expression, we obtain relationships among the magnetic moments of the baryons and, in particular, the relationship

$$\frac{\mu_p}{\mu_n} = -\tfrac{3}{2}. \tag{10.60}$$

The experimental value is $-1{\cdot}459$, strikingly confirming the SU(6) prediction. The magnetic moments of the other baryons can now be obtained via the Coleman–Glashow relation [36] previously obtained by assuming that the electromagnetic current can be treated as the $j_1^1$ component of the octet tensor current. This is implied for the Dirac quarks interacting via the minimal electromagnetic interaction.

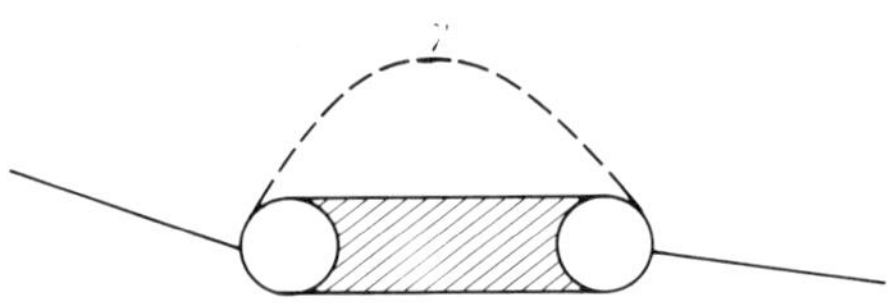

FIG. 10.5.

The following relations were obtained in terms of the values of $\mu_p$ and $\mu_n$ as the two parameters

$$\begin{aligned}
\mu_{\Sigma^+} &= \mu_p,\\
\mu_{\Sigma^0} &= -\tfrac{1}{2}\mu_n,\\
\mu_{\Sigma^-} &= -\mu_p-\mu_n,\\
\mu_{\Lambda} &= \tfrac{1}{2}\mu_n,\\
\mu_{\Xi^-} &= -\mu_p-\mu_n,\\
\mu_{\Xi^\cdot} &= \mu_n,
\end{aligned} \tag{10.61}$$

and, also,

$$\mu_{\Sigma\Lambda} = \frac{\sqrt{3}}{2}\,\mu_n,$$

where $\mu_{\Sigma\Lambda}$ is equal to the transition magnetic moment of $\Sigma^\circ \to \Lambda+\gamma$.

The electromagnetic mass of a particle results from the second order diagram (Fig. 10.5).

### 10.2.16. *Electromagnetic mass shifts*

The assumption that the electromagnetic current transforms under SU(3) as $T_1^1$ allows us to obtain expressions relating the electromagnetic mass shift of the various SU(2) isospin multiplets, in an SU(3) multiplet. We obtain

$$\Xi^- - \Xi^0 = \Sigma^- - \Sigma^+ - (n-p), \qquad (10.62a)$$

$$N^*_- - N^*_0 = Y^*_- - Y^*_0 = \Xi^*_- - \Xi^{**}_0, \qquad (10.62b)$$

$$N^*_{++} - N^*_- = 3(N^*_+ - N^*_0). \qquad (10.26c)$$

Equation (10.62a) represents the most precise check, since all the six masses are well known. We obtain for eqn (10.62a),

$$6{\cdot}6 \pm 0{\cdot}25 = 7{\cdot}9 \pm 0{\cdot}3 - 1{\cdot}3 \pm 0{\cdot}01 \qquad (10.63)$$

or,

$$6{\cdot}6 \pm 0{\cdot}25 = 6{\cdot}6 \pm 0{\cdot}3.$$

The agreement is excellent.

However, the 56-dimensional representation of SU(6) contains both the baryon octet and decuplet. Therefore, we can expect to obtain mass shift relations between them. Since we assume that the transformation properties of the electromagnetic mass operator, for the second order electromagnetic interaction, are contained in

$$35 \otimes 35 = 1 \oplus 35 \oplus 35' \oplus 189 \oplus 280 \oplus 280^* \oplus 405,$$

only matrix elements between baryon states will contribute. Thus, we are interested only in the terms in $56^* \times 56 = 1 + 35 + 405 + 2695$, which are contained also in $35 \otimes 35$. Thus, only 1 (which just gives a constant mass shift), 35, and 405 have non-vanishing matrix elements. Following this line of reasoning, we, finally, obtain

$$n - p = N^*_0 - N^*_+, \qquad (10.64)$$

$$N^*_- - N^*_0 = 2(\Sigma^+ - \Sigma^0) - (\Xi^- - \Xi^0). \qquad (10.65)$$

Thus, combining the above relations, we can obtain

$$N^*_{++} - N^*_0 = 2(p-n) + \Sigma^- - 2\Sigma^0 + \Sigma^+. \qquad (10.66)$$

Evaluating the above, we get

$$-0{\cdot}45 \pm 0{\cdot}85 = 2(-1{\cdot}3) + 1{\cdot}7 \pm 0{\cdot}5 - 0{\cdot}45 \pm 0{\cdot}85 = -0{\cdot}9 \pm 0{\cdot}1 \text{ MeV}, \qquad (10.67)$$

which is excellent confirmation of the SU(6) prediction.

### 10.2.17. *Relativistic SU(6)*

Our definition of SU(6) was clearly a non-relativistic one, since we ascribed two spin states to each quark and, thus, used a six dimensional quark (or sextet) as the fundamental representation. It is known that a relativistically invariant treatment of a particular quark ($q_1$, $q_2$, or $q_3$), according to the

Dirac equation, would result in a four-component Dirac spinor-free particle state (two for positive energy, and two for negative energy). Thus, if we wished to combine the spin properties of such a quark, with the SU(3) internal invariance properties, we would expect that a twelve-component quark, $q_{j,\alpha}$, is required as the basic representation, where $j = 1, 2, 3$ is the SU(3) index and $\alpha = 1, 2, 3, 4$ is the spinor index.

Thus, the meson representation would be $12^* \otimes 12$, or 144 dimensional, and the baryon representation would be $12 \otimes 12 \otimes 12 = 1728$ dimensional. So far, we have applied SU(6) to either static, or near static, problems like multiplet structure, mass differences, and magnetic moments, where it has worked well. However, since it has SU(2) as a subgroup, which corresponds to two spin states, it is not at all clear that we can find a relativistic version of SU(6). However, we might hope that, corresponding to the twelve-component quark there exists a relativistically invariant group which in the static limit reduces to SU(6).

As we recall, SU(6) was obtained by wedding SU(3) and SU(2), thus combining non-relativistic spin with internal symmetry. Similarly, a number of attempts were made to wed the SU(3) group with the Poincaré group (inhomogenous Lorentz group), which expresses the Lorentz invariance symmetry conditions on space–time. Thus, it was hoped that a relativistic generalization of SU(6) might be obtained.

A simple covariant generalization of SU(6) was obtained by Ruhl (1965) [37] Charap, and Mathews (1965) [38] and Nguyen, Van Hien, and Smordinsky (1965) [39]. These authors used a reduced Dirac representation for the proper Lorentz transformations (without reflections) which, in the Weyl representation, select the top two, or bottom two, components of the four-component Dirac spinors, which is thus reduced to the two-component Dirac spinors, usually referred to as the undotted and dotted spinors. The reduced Dirac representation generators operate on the two-component spinors, to produce the group SL(2, C) (the homogeneous Lorentz group), which is the complex generalization of SU(2) (i.e. there are three complex parameters). The direct product of the generators of the reduced Dirac representation and the generators of SU(3) was found to form the extended generators, which, when operating on the six component dotted and undotted quarks, generate the group SL(6, C). In this case, invariance with respect to space reflection has to be additionally required.

### 10.2.18. *I$\tilde{U}$(12) or $\tilde{U}$(6, 6)*

A relativistic (covariant) generalization of SU(6), which is clearly related to the Dirac theory, can be obtained in an analogous way to that by which

SU(6) was generated. The group SU(6) was obtained from the infinitesimal generators, which were obtained as a direct product of the Pauli $\boldsymbol{\sigma}$ matrices and the nine $(3 \times 3)$ $\lambda_i$ matrices of SU(3). Thus, we could perhaps expect to obtain a relativistic version of SU(6), by replacing the Pauli $\boldsymbol{\sigma}$ matrices by their relativistic equivalents, the $(4 \times 4)$ sixteen Dirac matrices, and then extending these by taking the direct product with the nine $(3 \times 3)$ SU(3) generators, $\lambda_i$—thus representing a basic (or fundamental) relativistic quark with twelve components. This is the inhomogenous $\mathrm{I\tilde{U}}(12)$ (or $\tilde{\mathrm{U}}(6, 6)$) proposed by Salam *et al.* (1965) [40] and Sakita and Wali (1965) [41]. The physical unitary representations are, in this case, constructed from the finite-non-unitary represention of $\tilde{\mathrm{U}}(4)$ (or $\tilde{\mathrm{U}}(2, 2)$), and are then extended by forming products with the SU(3) generators, to obtain the twelve-component basic quark. The homogenous group U(6, 6) has 144 infinitesimal generators in the basic $12 \times 12$ representation, which are constructed by taking the (outer) product of the nine basic generators, $\lambda_i$, of U(3), with the sixteen basic generators $\frac{1}{2}\Gamma_R$ of U(2, 2). Thus,

$$F_x = \sqrt{\tfrac{3}{2}}\,\lambda_i \otimes \Gamma_R, \qquad (10.68)$$

where $x = 1 \ldots 144$. The transformation to the inhomogenous group IU(6, 6), or $\mathrm{I\tilde{U}}(12)$, is then accomplished† by taking the semi-direct product with an appropriate 144 parameter Abelian group. At the decisive stage of the introduction of the inhomogenous Lorentz group, the reduction of $\tilde{\mathrm{U}}(12)$ symmetry to $\mathrm{U}(3) \times (\mathrm{I}\mathscr{L}_4)$ correctly reproduces the final physical symmetry situation and assures that the representations we finally work with are unitary. However, the full symmetry of U(12) no longer survives. The kinetic energy is non-invariant and, thus, intrinsically breaks the symmetry.

IU(6, 6) contains a momentum variable, $\tilde{p}$, with one hundred and forty-four components. Obviously, a physical momentum variable only has four components, demonstrating that the symmetry is intrinsically broken in the physical situation. To obtain predictions for physical multiplets and various amplitudes, using the intrinsically broken IU(6, 6) symmetry theory, the method used is the setting of all but the four vector components of $\tilde{p} = \not{p}$ equal to zero. Thus, we can now, within this approximation, construct expressions for the particle multiplets. The simplest multiplet is the twelve-component quark, but is so far not observable. The next simplest multiplet which is observable is that for the mesons

$$12 \otimes 12^* = 143 \oplus 1. \qquad (10.69)$$

† $\tilde{\mathrm{U}}(12)$ is non-compact like the Lorentz group, this is indicated by the tilde ($\sim$) symbol over the U.

The SU(6) content of 143 is

$$143 \rightarrow 1 \oplus 35. \tag{10.70}$$

Thus IU(6, 6) naturally combines the $0^-$ $\eta(960)$ pseudoscalar with the 35 representation.

In order to construct the baryons and conserve baryon number we require

$$12 \otimes 12 \otimes 12 = 220 \oplus 2 \times 572 \oplus 364. \tag{10.71}$$

The SU(6) content of the above IU(6, 6) multiplets is, according to Table IX of reference [2],

$$\begin{aligned} 220 &\rightarrow 20, \\ 572 &\rightarrow 70, \\ 364 &\rightarrow 56. \end{aligned} \tag{10.72}$$

Thus, the SU(6) content of the IU(6, 6) multiplets for the baryons is the same set of multiplets as in SU(6), which we saw can fit the experimental results quite well.

The next-simplest meson multiplets can be obtained from the reduction of

$$143 \otimes 143 \rightarrow 1 \oplus 143_F \oplus 143_D \oplus 4212 \oplus 5005 \oplus 5005^* \oplus 5940. \tag{10.73}$$

Thus, the next-simplest meson multiplets and their SU(6) contents are

$$4212 \rightarrow 1 \oplus 35 \oplus 189, \tag{10.74a}$$

$$5940 \rightarrow 1 \oplus 35 \oplus 405. \tag{10.74b}$$

The parity is positive for these multiplets. Thus, we can accommodate the $2^+$ nonet, just as we can do in the SU(6) case.

### 10.2.19. *Form factors*

The photon transformation property is like the $1^-$ member of a 143. Employing this, we can extend the proton and neutron magnetic moment relationships, to obtain the corresponding form factors as follows

$$G_E^n(q^2) = 0, \tag{10.75}$$

i.e. the neutron charge form factor is zero.

$$G_M^p(q^2) = -\tfrac{3}{2}G_M^n(q^2). \tag{10.76}$$

If we assume that the 143 meson pole term dominates the dispersion relations for the nuclear charge and magnetic form factor (i.e. Clementi–Villi model), we obtain

$$\mu_p = 2m/\mu, \tag{10.77}$$

where $\mu$ is an effective meson mass for the $1^-$ meson multiplet, which is not precisely defined, but for which it seems reasonable to take the $\rho$ mass. The agreement with experiment is then reasonable, considering the crudeness of the estimate.

We have just considered single particle states that are transformed by the so-called 'little group' which leaves invariant a single physical momentum spinor (which can be expressed as $\not{p} = m\gamma_0$).

If we now wish to consider 'linear' processes, such as forward scattering, or a vertex, it is obvious that these are dependent on two physical momentum spinors (which can both be expressed as $\not{p} = p_0\gamma^0 + p_3\gamma^3$). The group which leaves both momenta (and consequently the amplitude for a linear process) invariant is the so-called 'lesser group', and has as its generators

$$\frac{1}{2}\frac{\sqrt{3}}{\sqrt{2}}\lambda_i \otimes (1, \gamma_5\gamma_1, \gamma_5\gamma_2, \sigma_{12}). \tag{10.78}$$

These generators then generate the group $\mathrm{U}(1)\times\mathrm{SU}(6)_W$ i.e. the product of baryon number and the so-called $W$-spin (Meshkov) [42].

A two-particle scattering process is 'coplanar' in the sense that it is a function only of three independent momenta, which can be expressed in the form

$$\not{p} = p_0\gamma^0 + p_1\gamma^1 + p_2\gamma^2. \tag{10.79}$$

Thus, the generators which maintain the invariance of these three momenta are

$$\frac{1}{2}\frac{\sqrt{3}}{\sqrt{2}}\lambda_i \otimes \tfrac{1}{2}(1 \pm \gamma_5\gamma_3), \tag{10.80}$$

but these are just the generators that generate the 'least group'

$$\mathrm{U}(3) \otimes \mathrm{U}(3). \tag{10.81}$$

If the situation is any more complicated, we cannot find any commuting generators and, thus, the restrictions placed on amplitudes are merely those of SU(3), and Poincaré group invariance.

### 10.2.20. *Johnson–Treiman relations*

An important prediction of the 'lesser group' is the relation between meson–baryon scattering amplitudes originally obtained by applying SU(6). Meson–baryon scattering can be described by

$$143 \otimes 364 = 364 \oplus 572 \oplus 16016 \oplus 35100. \tag{10.82}$$

For finite-angle scattering there are a large number of IU(6, 6) invariant amplitudes, containing derivative couplings, resulting from the external momenta ($\vec{p}_{A}^{B}$) which are restricted by the 'least group'.

If we now restrict ourselves to forward scattering only, the 'lesser group' restricts the amplitude to just four amplitudes.

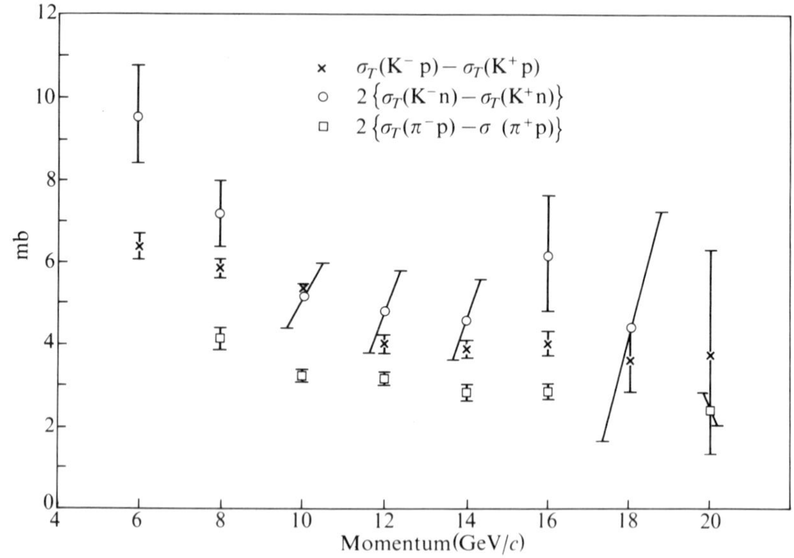


FIG. 10.6. According to the Johnson–Treiman relations, the three experimental quantities which are plotted should be the same (from [50]). See [50] for the experimental data.

The two relations obtained between six amplitudes, which are unrelated by SU(3), are the following [43],

$$F(\pi^{+}+\mathrm{p})-F(\pi^{-}+\mathrm{p}) = \tfrac{1}{2}\{F(\mathrm{K}^{+}+\mathrm{p})-F(\mathrm{K}^{-}+\mathrm{p})\}$$
$$= F(\mathrm{K}^{+}+\mathrm{p})-F(\mathrm{K}^{-}+\mathrm{n}). \qquad (10.83)$$

The comparison of these predictions to the experimental situation is shown in Fig. 10.6. The agreement is, at first glance, quite poor. This is not wholly unexpected due to the large K/$\pi$ mass ratio, since, in the approximation for which these relations were generated,

$$\frac{m_{\mathrm{k}}}{m_{\pi}} \approx 1.$$

Thus, it is understandable that ratios involving both $\pi$ and K mesons disagree more than those involving K mesons only. Therefore at higher incident momenta one might expect to obtain better agreement.

10.2.21. *The intrinsically broken symmetries, $I\tilde{U}(12)$, etc.*

The fact that it was found that $\mathrm{I\tilde{U}}(12)$ is an intrinsically broken symmetry, in the physical states, and all attempts to marry the Poincaré group and SU(3) have led to broken symmetry groups, has been clarified by the establishment of several theorems (by Michel (1965), [44] O'Rafertaigh (1965) [45], and Feldman and Mathews (1966) [46]).

(1) It was demonstrated that any direct extension of the Poincaré group, by combining it with an internal symmetry group (provided the combination is non-trivial), requires that the energy–momentum vector be extended to more than four components. Thus, the symmetry must be broken, when we make contact with the physical case.

(2) If the extension is made, as in U(6, 6) theories, to the index transformations of the Lorentz group, unitarity is violated, and in order to avoid this, we necessarily must give up the well-known features of relativistic quantum theory associated with anti-particles, spin, and statistics, CPT invariance, and the substitution law. Mathews and Sakita have reviewed this work in detail.

There have been attempts to understand the well-established SU(6) multiplet structure of hadrons, by a non-relativistic bound-state quark model [47]. The results for mesons have been reasonable, but treating the baryons necessarily involves the three-body problem with its usual difficulties. A second approach to understanding dynamical hadron symmetries problems is to use the algebra of current components [48]. The objective is that we may be able to explain both the hadron supermultiplets, as well as the dynamics of their interactions. If we assume the existence of a certain SU(6) multiplet of hadrons, and that the commutation relations of the current components are saturated by the multiplet, then it has been shown [48*c*] that we obtain the same relations for hadron interactions, as were previously obtained from employing index symmetry. Since the most successful features of current algebra applications have been in obtaining relations for hadron interactions, we can regard them as a dynamical substitute for phenomenological index symmetry, which can even provide corrections to SU(6) such as the Alder–Weissberger relations [49] which account for the discrepancy between the SU(6) predictions of $g_A/g_V$ and the measured value. In essence, additional states other than those in the SU(6) multiplets can also be taken into account. The Alder–Weissberger relations suggest that more than one SU(6) multiplet is necessary in order to satisfy the commutation relations of the current components.

## REFERENCES

1*a*. SAKATA, S. (1956). *Prog. theor. Phys.* **16,** 686.

1*b*. FERMI, E. and YANG, C. N. (1949). *Phys. Rev.* **76,** 1739.

2. MATTHEWS, P. T. (1967). *High energy physics*, Vol. 25-1. (edited by Burhop). Academic Press, New York. p. 392.

3. SAKITA, B. (1967). *Advances in particle physics*. Interscience Publishers, New York. p. 219.

4. NE'EMAN, Y. (1961). *Nucl. Phys.* **26,** 222.

5. GELL-MANN, M. (1964). *Phys. Lett.* **8,** 214.

6. GELL-MANN, M. and NE'EMAN, Y. (1964). *The eightfold way*. W. A. Benjamin, New York.

7. ZWEIG, G. (1964). *CERN reports, TH401 and 412.*

8. For general reviews see (*a*) LIPKIN, H. J. (p. 253) and ZICHICHI, A. (p. 266) (1967). *Proceedings of the Heidelberg International Conference on Elementary Particles.* North Holland Publishing Co., Amsterdam. (*b*) MORPUGO, G. (1968). *XIVth International Conference on High Energy Physics, Vienna*. p. 266.

9. GURSEY, F., LEE, T. D., and NAUENBERG, M. (1964). *Phys. Rev.* B **135,** 467.

10(a). BARNES, V. E., CONNOLLY, P. L., CRENNELL, D. J., CULWICK, B. B., DELANEY, W. C., FOWLER, W. B., HAGERTY, P. E., HART, E. L., HORWITZ, N., HOUGH, P. V. C., JENSEN, J. E., KOPP, J. K., LAI, K. W., LEITNER, J., LLOYD, J. L., LONDON, G. W., MORRIS, T. W., OREN, Y., PALMER, R. B., PRODELL, A. G., RADOJICIC, D., RAHM, D. C., RICHARDSON, C. R., SAMIOS, N. P., SANFORD, J. R., SHUTT, R. P., SMITH, J. R., STONEHILL, D. L., STRAND, R. C., THORNDIKE, A. M., WEBSTER, M. S., WILLIS, W. J., and YAMOMOTO, S. S. (1964). *Phys. Rev. Lett.* **12,** 204.

11. OKUBO, S. (1962). *Prog. theor. Phys.* **27,** 949.

12. See Ian Butterworth Rapporteur's talk, Heidelberg International Conference on Elementary Particles (1968).

13. BERTANZA, L., *et al.* (1962). *Phys. Rev Lett.* **9,** 229; CONNOLLY, *et al.* (1963). *Proceedings of the Sienna Conference on Elementary Particles*. p. 163.

14. OKUBO, S. (1963). *Phys. Rev. Lett.* **5,** 165.

15. SELOVE, W., HAGOPIAN, V., BRODY, H., BAKER, A., and LEBOY, E. (1962). *Phys. Rev. Lett.* **9,** 272.

16. A.B.C. COLLABORATION. (1964). *Phys. Rev. Lett.* **12,** 356.

17. SODICKSON, L., WHALIG, M., MANNELLI, I., FRISCH, D., and FACKLER, O. (1964). *Phys. Rev. Lett.* **12,** 485.

18. GOLDHABER, G., BROWN, J., GOLDHABER, S., KADYK, J. A., SHEN, B. C., and TRILLING, G. (1964). *Phys. Rev. Lett.* **12,** 336.

19. CHUNG, S. U., DAHL, O. I., HARDY, L. M., HESS, R. I., KALBFLEISCH, G. R., KIRZ, J., MILLER, D. H., and SMITH, G. (1964). *Phys. Rev.* **12,** 621.

20. FOCACCI, M. N., KIENZLE, W., LEVRAD, B., MAGLIC, B., and MARTIN, M. (1966). *Phys. Lett.* **17,** 890; CHICKOVANI, G. E., FOCACCI, M. N., KIENZLE, W., LECHANOINE, C., LEVRAT, B. L., MAGLIC, B., MARTIN, M., SCHÜBELIN, P., DUBAL, L., FISCHER, M., GRIEDER, P., NEAL, H. A., and NEF, C. (1967). *Phys. Lett.* B **25,** 44; BENZ, H.,

CHIKOVANI, G. E., DAMGAARD, G., FOCACCI, M. N., KIENZLE, W., LECHANOINE, C., MARTIN, M., NEF, C., SCHÜBELIN, P., BAUD, R., BOSNJAKOVIC, B., COTTERON, J., KLANNER, R., and WEITSCH, A. (1968). *Phys. Lett.* B **28,** 233.

21*a*. CRENNELL, D. J., KARSHON, U., LAI, K. W., SCARR, J. M., and SKILLICORN, I. O. (1968). *Phys. Rev. Lett.* **20,** 1318.

21*b*. AGUILAR-BENITEZ, M., BARLOW, J., JACOBS, L. D., MALECKI, P., MONTANET, L., TOMAS, M., DELLA NEGRA, M., COHEN-GANOUNA, J., WEST, N., and LÖRSTAD, B. (1969). *Phys. Lett.* B **29,** 62.

22. MAGLIC, B. (1969). *Proceedings of the Lund International Conference on Elementary Particles, Berlingska Boktryckeriet, Lund, Sweden.*

23. CHICKOVANI, G. E., FOCACCI, M. N., KIENZLE, W., MARTIN, M., KRUSE, W., LECHANOINE, C., and SCHÜBELIN, P. (1969). *Phys. Lett.* B **28,** 526; BAUD, R., BENZ, H., BOSNJAKOVIC, B., BOTTERILL, D. R., DAMGAARD, G., FOCACCI, M. N., KIENZLE, W., KLANNER, R., LECHANOINE, C., MARTIN, M., NEF, C., SCHÜBELIN, P., WEITSCH, A., BLUMENFELD, H., JOSTLEIN, H., ROINISHVILI, V., and LECOMTE, P. (1970). *Phys. Lett.* B **31,** 397.

24. ALSTON-GARNJOST, M., BARBARO-GALTIERI, A., BUHL, W. F., DERENZO, S. E., EPPERSON, L. D., FLATTE, S. M., FRIEDMAN, J. H., LYNCH, G. R., OTT, R. L., PROTOPOPESCU, S. D., RABIN, M. S., and SOLMITZ, F. T. (1970). *Phys. Lett.* B **33,** 607.

25. FOLEY, K. J., LOVE, W. A., OZAKI, S., PLATNER, E. D., LINDENBAUM, S. J., SAULYS, A., and WILLEN, E. H. (1971). *Phys. Rev. Lett.* **26,** 413.

26. GRAYER, G., HYAMS, B., JONES, C., SCHLEIN, P., BLUM, W., DIETL, H., KOCH, W., LIPPMANN, H., LORENZ, E., LÜTJENS, G., MÄNNER, W., MEISSBURGER, J., STIERLIN, U., and WEILHAMMER, P. (1971). *Phys. Lett.* B **34,** 333.

27. BOWEN, D., EARLES, D., FAISSLER, W., GARELICK, D., GETTNER, M., GLAUBMAN, M., GOTTSCHALK, B., LUTZ, G., MOROMISOTO, J., SHIBATA, E. I., TANG, Y. W., VON GOELER, E., BLEIDEN, H. R., FINOCCHIARO, G., KIRZ, J., and THUN, R. (1971). *Phys. Rev. Lett.* **26,** 1663.

28*a*. ALSTON, M., ALVAREZ, L., EBERHARD, P., GOOD, M. L., GRAZIANO, W., TICHO, H., and WOJCICKI, S. (1961). *Phys. Rev. Lett.* **6,** 300.

28*b*. HAQUE, N., SCOTTER, D., MUSGRAVE, B., BLAIR, W. M., GRANT, A. L., HUGHES, I. S., NEGUS, P. J., TURNBULL, R. H., AHMAD, A. A. Z., BAKER, S., CELNIKIER, L., MISBAHUDDIN, S., SHERMAN, H. J., SKILLICORN, I. O., ATHERTON, A. R., CHADWICK, G. B., DAVIES, W. T., FIELD, J. H., GRAY, P., LAWRENCE, D. E., LOKEN, J. G., LYONS, L., MULVEY, J. H., OXLEY, A., WILKINSON, C. A., FISHER, C. M., PICKUP, E., RANGAN, L. K., SCARR, J. M., and SEGAR, A. M. (1965). *Phys. Lett.* **14,** 338.

29. HARDY, L. M., CHUNG, S. U., DAHL, O. I., HESS, R. I., KIRZ, J., and MILLER, D. H. (1964). *Phys. Rev. Lett.* **14,** 401.

30. BARNES, V. E., CULWICK, B. B., GUIDONI, P., KALBFLEISCH, G. R., LONDON, G. W., PALMER, R. B., RADOJICIC, D., RAHM, D. C., RAU, R. R., RICHARDSON, C. R., SAMIOS, N. P., SMITH, J. R., GOZ, B., HORWITZ, N., KIKUCHI, T., LEITNER, J., and WOLFE, R. (1965). *Phys. Rev. Lett.* **15,** 322.

31. GLASHOW, S., and SOCOLOW, R. (1965). *Phys. Rev. Lett.* **15,** 329.
32*a*. SAKITA, B. (1964). *Phys. Rev.* B **136,** 1756.
32*b*. GURSEY, F. and RADICATI, L. A. (1964). *Phys. Rev. Lett.* **13,** 299.
33. ZWEIG, G. (1965). *Symmetries in elementary particle physics.* Academic Press, New York.
34. GREENBERG, O. W. (1964). *Phys. Rev. Lett.* **13,** 598.
35. BEG, M. A. and SINGH, V. (1964). *Phys. Rev. Lett.* **13,** 418.
36. COLEMAN, S. and GLASHOW, S. L. (1961). *Phys. Rev. Lett.* **6,** 423.
37. RUHL, W. (1965). *Nuovo Cim.* **37,** 301, 319.
38. CHARAP, J. M. and MATTHEWS, P. T. (1965). *Proc. R. Soc.* **A284,** 146.
39. VAN HIEN, N. and SIMORDINSKI, Y. A. (1965). *J. nucl. Phys. USSR.* **2,** 543.
40. SALAM, A., DELBOURGO, R., and STRATHDEE, J. (1965). *Proc. R. Soc.* **A284,** 146.
41. SAKITA, B. and WALI, K. C. (1965). *Phys. Rev. Lett.* **14,** 404.
42. MESHKOV, S., LEVINSON, C. A., and LIPKIN, H. J. (1963). *Phys. Rev. Lett.* **10,** 361.
43. JOHNSON, K. and TREIMAN, S. B. (1965). *Phys. Rev. Lett.* **14,** 189.
44. MICHEL, L. (1965). *Phys. Rev.* **B137,** 405.
45. O'RAFERTAIGH, L. (1965). *Phys. Rev.* **B139,** 1052.
46. FELDMAN, G. and MATTHEWS, P. T. (1966). *Phys. Rev.* **151,** 1176.
47. DALITZ, R. H. (1965). *Proceedings of the Oxford International Conference on Elementary Particles.*
48*a*. GELL-MANN, M. (1964). *Phys. Lett.* **1,** 63.
48*b*. DASHEN, R. F. and GELL-MANN, M. (1965). *Phys. Lett.* **17,** 142, 145.
48*c*. LEE, B. W. (1965). *Phys. Rev. Lett.* **14,** 676.
49. ADLER, A. (1965). *Phys. Rev. Lett.* **14,** 1051; WEISSBERGER, W. I. (1965). *Phys. Rev. Lett.* **14,** 1047.
50. LINDENBAUM, S. J. (1965). *Proceedings of the Oxford International Conference on Elementary Particles.*

# 11

# SCATTERING OF ELECTRONS FROM NUCLEONS AND FORM FACTORS

## 11.1. Point electron–point proton elastic electromagnetic scattering

THE interaction of a point electron with a point proton (or a heavy positron) can be treated by the conventional electrodynamics methods (primarily one-photon exchange).

The resultant elastic differential scattering formula, which is generally expressed in the laboratory frame, is referred to as the Mott scattering formula

$$\left(\frac{\mathrm{d}\sigma}{\mathrm{d}\Omega}\right)_{\mathrm{Mott}} = \frac{\alpha^2}{4E^2\sin^4\theta/2}\left(\frac{\cos^2\theta/2}{1+2(E/M)\sin^2\theta/2}\right). \tag{11.1}$$

This is the formula which is applicable to scattering of a point electron from any point-charge spin-$\frac{1}{2}$ target particle (which obeys the Dirac equation). If we consider a physical proton, which has an extended charge distribution, the elastic scattering of fast electrons $\lambda\!\!\!^{-}_{\mathrm{e}} \ll r_{\mathrm{p}}$ should be modified from that expected for a point proton, as soon as the invariant four-momentum transfer squared ($q^2 = -t$) becomes different from zero (i.e. for any finite positive value of $q^2$).

## 11.2. The Rosenbluth Factor—form factor effects

The description of the scattering of fast electrons from physical protons can be formulated as follows

$$\frac{\mathrm{d}\sigma}{\mathrm{d}\Omega} = \left(\frac{\mathrm{d}\sigma}{\mathrm{d}\Omega}\right)_{\mathrm{Mott}} R(q^2, E) = \left(\frac{\mathrm{d}\sigma}{\mathrm{d}\Omega}\right) R(q^2, \theta) \tag{11.2}$$

or, in Lorentz invariant form,

$$\frac{\mathrm{d}\sigma}{\mathrm{d}t} = \left(\frac{\mathrm{d}\sigma}{\mathrm{d}t}\right)_{\mathrm{Mott}} R(t, s), \tag{11.3}$$

where $R$ is the so-called Rosenbluth factor. That $R$ was different from 1 was discovered by Hofstadter *et al.* [1], who utilized electron beams of up to 900 MeV, produced in electron linacs (linear accelerators) at Stanford.

Figure 11.1 shows a plot of the angular distribution for elastic scattering of 400 MeV electrons by protons against increasing laboratory angle ($\theta$ in degrees), which corresponds to increasing $q^2$. The solid points (with flags)

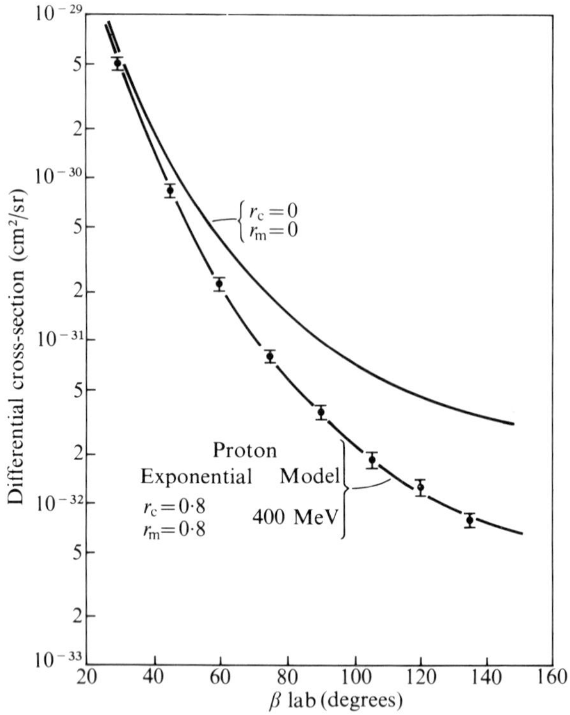


FIG. 11.1. Typical angular distribution of elastically scattered 400 MeV electrons incident upon protons. The solid line is a theoretical estimate with an exponential model with r.m.s. radius = $0{\cdot}80 \times 10^{-13}$ cm, and agrees quite well with experiment. (From Hofstadter. (1957). *Rev. mod. Phys.* **28,** 214.)

represent the data. The upper solid curve (labelled $r_c$ (= charge radius) = 0, $r_m$ (= magnetic radius) = 0) represents the Mott scattering prediction. The lower solid curve represents a particular extended model fit to the data. $r_c$ and $r_m$ are measured in fm (= $10^{-13}$ cm) units.

It is clear from Fig. 11.1 that

$$\frac{\left(\dfrac{d\sigma}{d\Omega}\right)}{\left(\dfrac{d\sigma}{d\Omega}\right)_{\text{Mott}}} \equiv R(q^2, E) \tag{11.4}$$

decreases as $q^2$ (i.e. $\theta_{\text{lab}}$) increases, which is what we would expect for an extended charge (and magnetization) distribution for the proton. Thus, Hofstadter *et al.* in their classic experiment demonstrated that the proton charge distribution was extended. These effects are called form factor effects. In a classical treatment we define the (suitably normalized) Fourier transform of a spherically symmetric charge density $\rho(r)$ as an electric form factor, and the Fourier transform of a density of magnetization $M(r)$ as a magnetic form factor.

Thus

$$F_{\text{electric}}(q^2) = \int_0^\infty \frac{\rho(r)\sin qr 4\pi r^2 \, dr}{qr}, \tag{11.5}$$

where $q$ is the magnitude of the three-dimensional momentum transfer. The normalization is chosen such that

$$\int_0^\infty 4\pi\rho(r)r^2 \, dr = 1. \tag{11.6}$$

As long as the charge distribution has a finite range $R$, the behaviour of the form factor, for $qR \ll 1$, can be shown to depend on only one parameter $\langle r^2 \rangle$, the mean square charge radius. This can easily be seen, using the expansion

$$\sin qr = qr - \frac{(qr)^3}{6} + \dots. \tag{11.7}$$

For $qR \ll 1$ we need retain only the first two terms. Then

$$F_{\text{electric}}(q^2) = \int_0^\infty \rho(r) 4\pi r^2 \left\{ \frac{qr - (qr)^3/6}{qr} \right\} dr \tag{11.8}$$

$$= \int_0^\infty \rho(r) 4\pi r^2 \, dr - \int_0^\infty \rho(r) \frac{4\pi r^2 (qr)^2}{6} \, dr, \tag{11.9}$$

but, since,

$$\int \rho(r) 4\pi r^2 \, dr = 1,$$

$$F_{\text{electric}}(q^2) = 1 - \tfrac{1}{6} q^2 \langle r^2 \rangle. \tag{11.10}$$

The above treatment, which interprets the Fourier transforms of three-dimensional charge (or magnetization) distributions as form factors, is not relativistically invariant. It is obvious that in order to treat the scattering of relativistic electrons we must have Lorentz invariant quantities.

However, in the case of elastic scattering in the c.m.s. frame, the zero-th component of the four-momentum transfer vanishes and, thus, we can still regard the form factors as functions of three-momentum transfer, when they are referred to the appropriate spatial densities in the nucleon's rest frame. The essential changes in the above three-dimensional non-relativistic treatment are to replace the three-dimensional momentum transfer $\mathbf{q} = \mathbf{k}-\mathbf{k}'$ (where $\mathbf{k}'$ is the momentum of the scattered electron) by the relativistic four-momentum transfer, and then replace the three-dimensional radius vector $\mathbf{r}$ by its four-dimensional relativistic counterpart.

As can be seen from Fig. 11.1, the charge distribution radius

$$\langle r^2 \rangle_{\mathrm{e}}^{\frac{1}{2}} \approx 0{\cdot}80 \text{ fm}. \tag{11.11}$$

The model which yields the curve fitting the data in Fig. 11.1 is an exponential distribution with

$$\rho(r) = \mathrm{e}^{-r/R}. \tag{11.12}$$

From this particular distribution,

$$\langle r^2 \rangle = 12R^2. \tag{11.13}$$

### 11.3. The Dirac form factors

When the Rosenbluth factor is derived, from a treatment utilizing the Dirac equation, the effects of the charge distribution are represented by the so-called Dirac form factor $F_1(q^2)$, with

$$F_1(q^2 = 0) = Q = \tfrac{1}{2}+t_3. \tag{11.14}$$

There are further additional effects due to the anomalous Pauli magnetic form factor, which is generally referred to as $F_2(q^2)$ with†

$$F_2(q^2 = 0) = 1. \tag{11.15}$$

In terms of $F_1(q^2)$ and $F_2(q^2)$, the Rosenbluth factor is

$$R(q^2, \theta) = F_1^2(q^2)+\left(\frac{q^2}{4M^2}\right)\mu_{\mathrm{N}}^2 F_2^2(q^2)$$
$$+2\left(\frac{q^2}{4M^2}\right)\{F_1(q^2)+(\mu_{\mathrm{N}})F_2(q^2)\}^2\tan^2\theta/2, \tag{11.16}$$

where $\mu_{\mathrm{N}}$ is the anomalous moment of the nucleon, in units of nuclear magnetons $\left(\dfrac{e\hbar}{2M_{\mathrm{p}}c}\right)$.

† $q^2 = -t$.

### 11.4. The $G_E$ and $G_M$ form factors

Since $F_2$ specifically takes account of the anomalous magnetic moment effects, then the values of $F_1$ and $F_2$ represent the results of the combination (mixture) of electric and magnetic effects. Furthermore, it is clear that the cross-section contains interference terms between $F_1$ and $F_2$. It is, therefore, often convenient to define a second set of form factors

$$G_E = F_1 - \frac{q^2}{4M^2}\mu_N F_2, \tag{11.17}$$

$$G_M = F_1 + \mu_N F_2. \tag{11.18}$$

This second set of form factors is such that they can be defined in terms of the relativistically invariant four-vector current density $(i\rho(r), \mathbf{j}(r))$, as follows,

$$\mathbf{j}(r) = \frac{ie}{(2\pi)^3}\int d^3q(\boldsymbol{\sigma}\times\mathbf{q})G_M(q^2)e^{i\mathbf{q}\cdot\mathbf{r}}, \tag{11.19a}$$

$$\rho(r) = \frac{e}{(2\pi)^3}\int d^3q G_E(q^2)e^{i\mathbf{q}\cdot\mathbf{r}}. \tag{11.19b}$$

Thus, $G_M$ is directly connected only with current (i.e. magnetization) effects, whereas $G_E$ is directly connected only with charge effects. Substituting $G_E$ and $G_M$ into the Rosenbluth factor, we find

$$R(q^2, \theta) = \frac{G_E^2 + \dfrac{q^2}{4M^2}G_M^2}{1 + \dfrac{q^2}{4M^2}} + \frac{2q^2}{4M^2}G_M^2\frac{\tan^2\theta}{2}, \tag{11.20}$$

or, equivalently,

$$R(t, \theta) = \frac{G_E^2(t) - \dfrac{tG_M^2(t)}{4M^2}}{1 - \dfrac{t}{4M^2}} - \frac{t}{2M^2}G_M^2(t)\tan^2\theta/2. \tag{11.21}$$

Thus,

$$\frac{d\sigma}{d\Omega} = \left(\frac{d\sigma}{d\Omega}\right)_{\text{Mott}}\left\{\frac{G_E^2(t) - \dfrac{tG_M^2(t)}{4M^2}}{1 - \dfrac{t}{4M^2}} - \frac{t}{2M^2}G_M^2(t)\tan^2\theta/2\right\}. \tag{11.22}$$

Hence, if we measure $\dfrac{d\sigma}{d\Omega}$ for various incident energies, at a fixed value of $t$, and plot

$$\frac{d\sigma}{d\Omega}\bigg/\left(\frac{d\sigma}{d\Omega}\right)_{\text{Mott}} \text{ against } \tan^2\theta/2,$$

a straight line should result, the slope of which yields $G_M^2(t)$, and the value of the intercept can be used to determine $G_E^2(t)$. This straight-line behaviour comes about because of the assumption of one-photon exchange as the

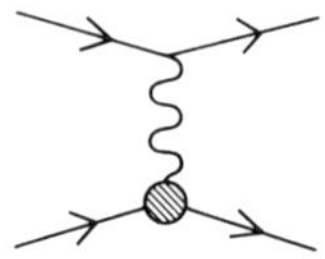

mechanism responsible for the scattering (as illustrated above). This rectilinear behaviour has been found to hold well [2] in the region $0 < -t < 1{\cdot}8$ $(\text{GeV}/c)^2$. This demonstrates that two photon exchanges (see diagram) lead to only small effects.

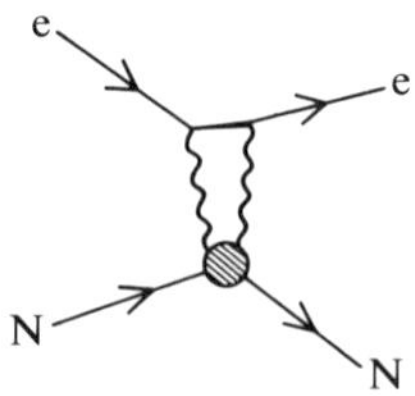


(Two-photon exchange)

The proton form factors are determined by e–p scattering. In order to determine the neutron form factors, e–n scattering should, in principle, be performed. In practice, since there are no free neutron targets, $e+d \rightarrow n+p+e$ scattering experiments are performed, and the e–n scattering cross-section extracted from this data, using the already measured e–p cross-section, and the theoretical corrections.

It follows from the previous definitions that

$$G_E^p(0) = 1 \tag{11.23a}$$

$$G_M^p(0) = 1+\mu_p \approx +2{\cdot}79, \tag{11.23b}$$

$$G_E^n(0) = 0, \tag{11.23c}$$

$$G_M^n(0) = -1{\cdot}91. \tag{11.23d}$$

The proton and neutron form factors as a function of $-t$ are shown in Figs. 11.2 and 11.3 (from Hughes *et al.* [3]).

Attempts to explain the early observations on the nuclear form factors by means of a pion cloud surrounding a nuclear core were unsuccessful.

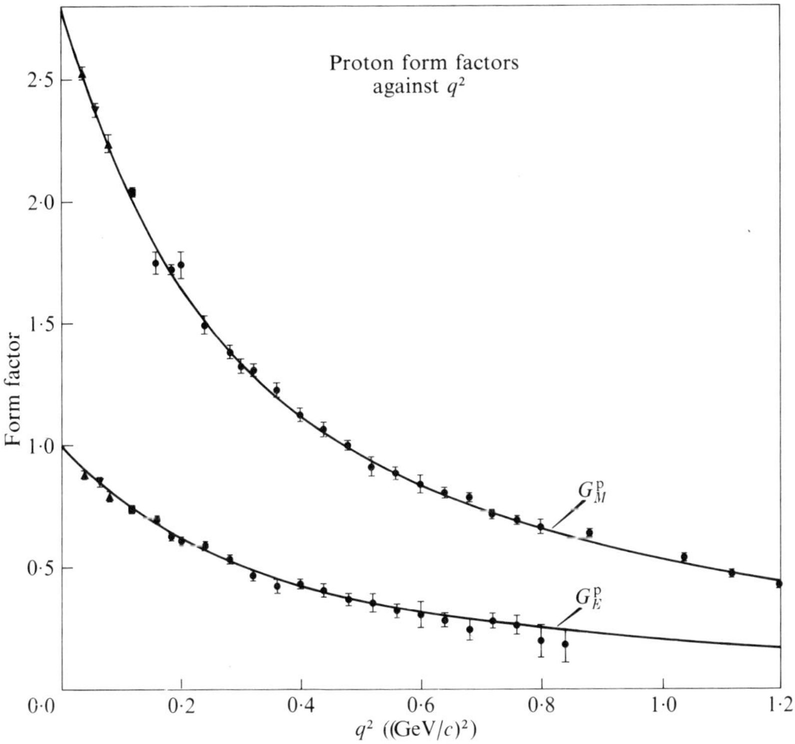


FIG. 11.2. The proton form factors as a function of $q^2 = -t$ (GeV/$c$)$^2$. (Hughes *et al.* (1965). *Phys. Rev.* **B139,** 458).

In fact, Frazer and Fulco [4] predicted that a vector meson, heavier than the pion, with $J^P = 1^-$, which corresponds to a mass about equal to the $\rho$ mass, would be required to explain the data. Since neutral $1^-$ vector mesons have identical quantum numbers to the $\gamma$ (with, of course, the exception of the mass), we can conceive that the $\gamma$, being exchanged, transforms to a $1^-$ vector meson, which is coupled to the nucleons and, thus, the corresponding Feynman diagram will be

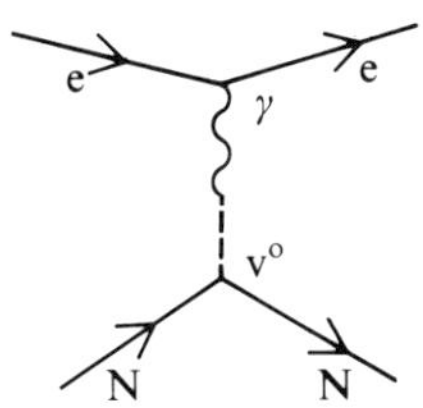

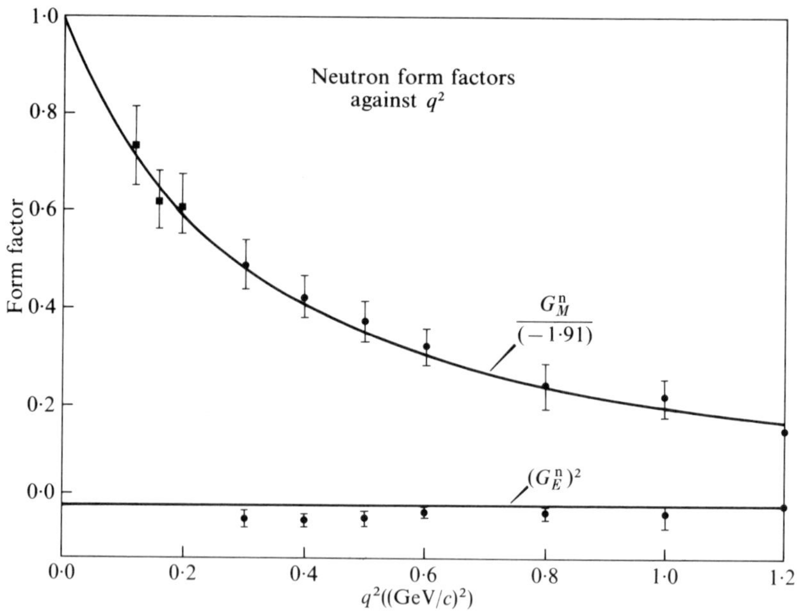


FIG. 11.3. The neutron form factors as a function of $q^2 = -t\ (\mathrm{GeV}/c)^2$. (Hughes *et al.* (1965). *Phys. Rev.* **B139**, 458).

The result of such a coupling would lead to the following form factor

$$G = \frac{1}{1+\dfrac{q^2}{m_v^2}}. \tag{11.24}$$

## 11.5. The isoscalar and isovector form factors for the nucleons

The form of the electric current operator,

$$j_\mu(x) = e\overline{\Psi}(x)\gamma_\mu\left(\frac{1+\tau_3}{2}\right)\Psi(x)+\ldots, \tag{11.25}$$

will exhibit mixed transformation properties when rotations in isotopic spin-space are considered. We can conclude that part of the current will behave as a scalar under the transformation, and part as the third component of an isovector. Thus we can express the nucleon currents in terms of a sum of isoscalar and isovector currents, which in turn will lead to corresponding isoscalar and isovector form factors for the nucleons, as follows

$$G^{\mathrm{N}}_{\mathrm{E,M}} = G^{(\mathrm{s})}_{\mathrm{E,M}}+2\tau_3 G^{(\mathrm{v})}_{\mathrm{E,M}}, \tag{11.26}$$

but, since $2\tau_3 = 1$ for proton (p), and 0 for neutron (n), we obtain

$$G^{\mathrm{p}}_{\mathrm{E,M}} = G^{(\mathrm{S})}_{\mathrm{E,M}} + G^{(\mathrm{V})}_{\mathrm{E,M}}, \tag{11.27}$$

$$G^{\mathrm{n}}_{\mathrm{E,M}} = G^{(\mathrm{S})}_{\mathrm{E,M}} - G^{(\mathrm{v})}_{\mathrm{E,M}}. \tag{11.28}$$

If we assume that the vector meson coupling dominates, then the $\rho$-meson (isovector) would be the important one, coupled to $G^{(\mathrm{v})}$. The $\omega$- and $\phi$-mesons would be the important (isoscalar) mesons, coupled to $G^{(\mathrm{s})}$. From the definitions of the $G$s and their values at $t = q^2 = 0$, we can demonstrate that

$$G^{(\mathrm{v})}_{\mathrm{E}}(0) = G^{(\mathrm{S})}_{\mathrm{E}}(0) = \tfrac{1}{2}, \tag{11.29a}$$

$$G^{(\mathrm{S})}_{\mathrm{M}}(0) = 0{\cdot}44, \tag{11.29b}$$

$$G^{(\mathrm{v})}_{\mathrm{M}}(0) = 2{\cdot}35. \tag{11.29c}$$

The experimental measurements on all e–N form factors up to† $q^2 = 600f^{-2} \approx 25\ (\mathrm{GeV}/c)^2$ are expressible to a quite good degree of accuracy‡ by the so-called 'universal dipole fit' [5].

$$G^{\mathrm{p}}_{\mathrm{E}} \approx \frac{G^{\mathrm{p}}_{\mathrm{M}}}{\mu_{\mathrm{p}}} \approx \frac{G^{\mathrm{n}}_{\mathrm{M}}}{\mu_{\mathrm{n}}} \approx \frac{G^{\mathrm{n}}_{\mathrm{E}}}{(q^2/4M^2)\mu_{\mathrm{n}}} \approx \{1 + q^2/0{\cdot}71\ (\mathrm{GeV}/c)^2\}^{-2}.$$

This form would correspond to nucleons which have a continuing exponential distribution of charge and magnetization (i.e. no charged core). If this $\frac{1}{q^4}$ behaviour of the form factors were to continue to infinite $q^2$ or $-t$, then we could not fit the data by a superposition of simple poles (which contribute terms of order $1/q^2$), but would necessarily require a dipole form.

Considerable future work at the higher energies should do much to resolve these questions. The reader is referred to the books by Gasiorowicz [6] and Feld [7] for treatments of this subject.

## REFERENCES

1. HOFSTADTER, R. (1957). *Rev. mod. Phys.* **28,** 214; HOFSTADTER, R. (1957). *A. Rev. nucl. Sci.* **7,** 231; HOFSTADTER, R. (1963). (editor). *Electron scattering and nuclear and nucleon structure*. W. A. Benjamin, New York.

† It should be noted that a common unit in use for $q^2$ in e–N scattering is $(\mathrm{fm})^{-2}$, and this corresponds to $\approx 0{\cdot}04\ (\mathrm{GeV}/c)^2$.

‡ Coward *et al.* (*Phys. Rev. Lett* (1968) **20,** 292) report a small deviation which is considered significant. There also is some indication that the form factor may become relatively constant for deep inelastic scattering (i.e. large excitations corresponding to several GeV). This would imply a point nucleon interaction (at least for part of the interaction), or imply point-like clusters of charge (see Panofsky, *Proc. XV International Conference on High Energy Physics, Vienna.* (1968)). This had led to the Bjorken scaling hypothosis.

2. Yennie, D. R. (1963). *Electron scattering and nuclear and nucleon structure*. W. A. Benjamin, New York.
3. Hughes, E. B., Griffy, T. A., Yearian, M. R., and Hofstadter, R. (1965). *Phys. Rev.* B**139,** 458.
4. Frazer, W. R. and Fulco, J. R. (1960). *Phys. Rev.* **117,** 1609.
5. Dunning, J. R., Jr., Chen, K. W., Cone, A. A., Hartwig, G., Ramsey, N. F., Walker, J. K., and Wilson, R. (1966). *Phys. Rev.* **141,** 1286.
6. Gasiorowicz, S. (1967). *Elementary particle physics*. John Wiley and Sons, Inc., New York.
7. Feld, B. (1969). *Models of elementary particles*. Blaisdell Publishing Company, Waltham, Massachusetts.

# 12

# WEAK INTERACTIONS

## 12.1. $\beta$-decay

THE observation of $\beta$-decay in nuclei was the first observation of weak interactions. The basic process for the decay of the free neutron

$$\mathrm{n} \to \mathrm{p}+\mathrm{e}^-+\bar{\nu}_\mathrm{e} \tag{12.1}$$

has been observed to have a lifetime of about 17 minutes. Since the neutron is slightly (1·3 MeV) heavier than the proton, the beta decay of the free neutron is energetically possible, while the beta decay of a free proton is not observable. However, in nuclei, since the magnitude of the binding energy of the product nucleus can be greater than the binding energy of the initial nucleus by an amount larger than the neutron–proton mass difference, plus the rest mass of a positron, we can observe the beta decay process

$$\mathrm{p} \to \mathrm{n}+\mathrm{e}^++\nu_\mathrm{e}. \tag{12.2}$$

We have anticipated the results of later experiments, and their theoretical interpretation, and have labelled the neutrino accompanying the positron in $\beta$-decay as $\nu_\mathrm{e}$ (i.e. the particle), and the neutrino accompanying the electron as $\bar{\nu}_\mathrm{e}$ (the anti-particle). In our notation, $\mathrm{e}^-$ is considered a particle, and $\mathrm{e}^+$ an anti-particle. Thus, our notation preserves the conservation of leptons.† An intimately-related process is K-capture, where one of the K-shell electrons of the atom is absorbed by one of the protons inside the nucleus, converting this e–p pair into an n–$\nu$ pair.

The 'neutrino' was postulated by Pauli (1931) to explain the puzzling observation of a continuous $\beta$-decay energy spectra, and still allow conservation of energy and momentum, but the first theory of $\beta$-decay was proposed by Fermi in 1934 [4]. He postulated, in analogy with quantum electrodynamics, that the interaction Hamiltonians involve a Lorentz invariant vector coupling of the form

$$\mathscr{H}_\mathrm{I} = C_\mathrm{V}\int \mathrm{d}^3x\, \overline{\Psi}_\mathrm{p}(x)\gamma^\mu\Psi_\mathrm{n}(x)\overline{\Psi}_\mathrm{e}(x)\gamma_\mu\Psi_\nu(x). \tag{12.3}$$

The vector coupling constant $C_\mathrm{V}$ has dimensions of inverse mass squared. We can, thus, justify the use of a perturbative theory treatment if $C_\mathrm{V}$ is small

† References [1]–[3] are good general texts on this topic, in particular, the recent book by Marshak, Riazuddin, and Ryan [3] is the most advanced and comprehensive treatment of weak interactions.

enough and, as we shall see, this is the case for the weak interactions, at all presently available energies, since $C_V E_{max}^2 \ll 1$. The Fermi coupling was only one of the possible non-derivative couplings which satisfy proper Lorentz invariance. Thus, the following is the most general form of the simplest (non-derivative) Hamiltonian which is invariant under proper Lorentz transformations,† and contains the product of the four fields necessary for the creation and annihilation of the particles involved in beta decay

$$\mathscr{H}_{\mathrm{I}} = \frac{1}{\sqrt{2}} \int \mathrm{d}^3x \left\{ \left( \sum_i C_i \overline{\Psi}_{\mathrm{p}} \Gamma_i \Psi_{\mathrm{n}}(x) \overline{\Psi}_{\mathrm{e}} \Gamma_i \Psi_{\nu} \right) + \text{Hermitian conjugate} + \right.$$

$$\left. + \frac{1}{\sqrt{2}} \left( \sum_i C_i' \overline{\Psi}_{\mathrm{p}} \Gamma_{\mathrm{i}} \Psi_{\mathrm{n}}(x) \overline{\Psi}_{\mathrm{e}} \Gamma_i \gamma_5 \Psi_{\nu} \right) + \text{Hermitian conjugate} \right\}. \quad (12.4)$$

The $\Gamma_i$ are the complete set of Dirac matrices ($1, \gamma_\mu, \sigma_{\mu\nu}, \gamma_5$, and $\gamma_\mu\gamma_5$), where

$$\sigma_{\mu\nu} = -\frac{\mathrm{i}}{2}(\gamma_\alpha\gamma_\beta - \gamma_\beta\gamma_\alpha).$$

By straightforward application of the operations involved, it can be shown that the conditions for invariance under the operations C, P, T are as shown in Table 11.1.

TABLE 11.1

| Operation | Requirement for invariance |
|---|---|
| P | $C_i' = 0$ |
| C | Real $C_i$, Imaginary $C_i'$ |
| T | Real $C_i$, and Real $C_i'$ |
| CP | Real $C_i$, and Real $C_i'$ |

In the early (pre-1957) work on $\beta$-decay, it was generally believed that time reversal (T), and space inversion (P), invariance were valid for weak interactions. In this case it is easily seen that only the first (i.e. the $C_i$ terms) set of

† Invariance under the improper Lorentz transformations, P and T, (space and time reflections), and under C (charge conjugation) are not required. Of course, the Hamiltonian is invariant under CPT (and its permutations). The CPT theorem states that every possible state of a system of particles is also a possible state of the corresponding system of anti-particles with space and time inverted. This theorem can be derived from Lorentz-invariance, the assumed existence of a vacuum state and local commutativity.

terms are necessary and that these constants, which are in general complex, are all real. Thus,

$$\begin{aligned}\mathscr{H}_{\mathrm{I}} = \int \mathrm{d}^3x\{ & C_{\mathrm{V}}(\overline{\Psi}_{\mathrm{p}}(x)\gamma^{\alpha}\Psi_{\mathrm{n}}(x))(\overline{\Psi}_{\mathrm{e}}(x)\gamma_{\alpha}\Psi_{\nu}(x)) + \\ & + C_{\mathrm{A}}(\overline{\Psi}_{\mathrm{p}}(\alpha)\gamma^{\alpha}\gamma_5\Psi_{\mathrm{n}}(x))(\overline{\Psi}_{\mathrm{e}}(x)\gamma_{\alpha}\gamma_5\Psi_{\nu}(x)) + \\ & + C_{\mathrm{S}}(\overline{\Psi}_{\mathrm{p}}(x)\Psi_{\mathrm{n}}(x))(\overline{\Psi}_{\mathrm{e}}(x)\Psi_{\nu}(x)) + \\ & + C_{\mathrm{T}}(\overline{\Psi}_{\mathrm{p}}(x)\sigma^{\alpha\beta}\Psi_{\mathrm{n}}(x))(\overline{\Psi}_{\mathrm{e}}(x)\sigma_{\alpha\beta}\Psi_{\nu}(x)) + \\ & + C_{\mathrm{P}}(\overline{\Psi}_{\mathrm{p}}(x)\mathrm{i}\gamma_5\Psi_{\mathrm{n}}(x))(\overline{\Psi}_{\mathrm{e}}(x)\mathrm{i}\gamma_5\Psi_{\nu}(x))\}. \end{aligned} \tag{12.5}$$

V, A, S, T, and P refer, respectively, to the vector, axial vector, scalar, tensor, and pseudoscalar interactions. Since $\beta$-decay always involves non-relativistic nucleons, the P term is not expected to contribute, and no information can be obtained on this term. The S and V terms only contribute when $|\langle 1\rangle|^2 \neq 0$, and $\Delta P = \Delta J = 0$, where we are referring to changes from initial to final nucleus (or nucleon) state. These are the so-called 'Fermi transitions'. Since they include the V interactions, which was the one Fermi originally proposed, they obviously involve zero angular momentum change, and no parity changes. The T and A terms only contribute when $|\langle\sigma\rangle|^2 \neq 0$. Furthermore, $\Delta J = 0 \pm 1$, but $0 \to 0$ forbidden since $\langle J = 0\,|\sigma|\,J = 0\rangle = 0$, and $\Delta P = 0$. These are Gamow–Teller transitions.

Both the Fermi and the Gamow–Teller transitions are the allowed transitions which, in analogy to the optical spectroscopy, correspond to states with zero orbital angular momentum of the light particle. The 'forbidden' transitions correspond to those with non-zero orbital angular momentum for the light particles. The treatment of these transitions is very dependent on the theory of nuclear structure and will not be considered here.

Experimentally, it was found that both pure Fermi ($O^{14}$, $Cl^{34}$, etc.) and pure Gamow–Teller decays ($He^6$ and $Ne^{23}$, etc.) occurred in nature. The free neutron decay satisfies both the Fermi and Gamow–Teller selection rules.

Thus, $\mathscr{H}_{\mathrm{I}}$ must contain at least two terms, one corresponding to Fermi (S and/or V), and another (T and/or A) corresponding to Gamow–Teller. The interference terms between (S and V), or those between T and A, have a characteristic, and marked, energy dependence, involving an additional factor $\left(1 + b\,\dfrac{m_{\mathrm{e}}}{E_{\mathrm{e}}}\right)$ which was never observed in the experimental $\beta$-decay spectrum. From the absence of these so-called Fierz [5] interference terms, it was concluded that†

$$\begin{aligned} C_S C_V &\approx 0 \\ C_T C_A &\approx 0. \end{aligned} \tag{12.6}$$

Thus, the pure Fermi interaction is either all S, or all V, and the pure GT interaction is either all T, or all A and, thus, the two terms involved were one of the following combinations, $\begin{bmatrix} \mathrm{S} \\ \mathrm{T} \end{bmatrix}$, $\begin{bmatrix} \mathrm{S} \\ \mathrm{A} \end{bmatrix}$, $\begin{bmatrix} \mathrm{V} \\ \mathrm{T} \end{bmatrix}$, or $\begin{bmatrix} \mathrm{V} \\ \mathrm{A} \end{bmatrix}$. We now know that these 'Fierz interference terms' vanish for the special kind of maximal parity violation which occurs. However, at the earlier time it was assumed (with parity conservation as an axiom) that the absence of these interference terms meant that both S and V, or both T and A, did not occur. The conclusion was correct, even though the reasoning was entirely wrong.

To determine the magnitude of $C_S$ or $C_V$ we select a $0 \to 0$, $\Delta P = 0$ transition, since this can only be a pure Fermi transition. In this case, the S and V $\beta$-decay differ only in the electron–neutrino correlation. In the vector interaction, the neutrino and electron prefer to come out parallel, whereas in the scalar interaction the preference is anti-parallel. To determine the magnitude of A or T, we can select $|\Delta J| = 1$, $\Delta P = 0$ transitions, which can only be Gamow–Teller transitions. These two possibilities can again be decided between by observing the electron–neutrino correlations. These experiments were difficult to perform, and before 1957 the experimental results available led to contradictory conclusions.

The decay

$$\begin{array}{ccc} \mathrm{He}^6 \;\to & \mathrm{Li}^6 & +\mathrm{e}^- + \bar{\nu}_e \\ (J = 0) & (J = 1) & \\ \Delta P = 0 & & \end{array} \tag{12.7}$$

showed a negative correlation in one experiment, which suggested an axial vector interaction. Another experiment showed a positive correlation, suggesting a tensor interaction. The decay

$$\begin{array}{ccc} \mathrm{Ne}^{19} \;\to & \mathrm{F}^{19} & +\mathrm{e}^+ + \nu_e \\ (J = \tfrac{1}{2}) & (J = -\tfrac{1}{2}) & \\ \Delta P = 0 & & \end{array} \tag{12.8}$$

† If the more general treatment were followed, we would have obtained

$$\mathrm{Re}(C_S C_V^* + C_S C_V'^*) = 0,$$
$$\mathrm{Re}(C_T C_A^* + C_T' C_A'^*) = 0.$$

Thus, there are other ways of satisfying these requirements such as, for example, there is a $\pi/2$ phase difference between the two complex coupling constants.

indicated that S and T dominated, but the errors were large. On the other hand, the decay

$$A^{35} \rightarrow Cl^{35} + e^{+} + \nu_{e} \tag{12.9}$$

clearly fit the assumption of vector interaction dominance.

The $A^{35}$ and $He^{6}$ experiments were then repeated [6], and the results demonstrated that V and A were the dominant interactions.† It is interesting to note that the values obtained for the coupling constants from experiments with unpolarized particles were‡ (setting $\hbar = c = m_e = 1$ (natural units))

$$|C_V| \approx (3{\cdot}01 \pm 0{\cdot}01) \times 10^{-12} \text{ natural units} = G_F \tag{12.10a}$$

and,

$$|C_A| = (3{\cdot}55 \pm 0{\cdot}08) \times 10^{-12} \text{ natural units} = G_{GT}. \tag{12.10b}$$

These indicate the near equality of $|C_V|$ and $|C_A|$.

## 12.2. Leptonic decay of $\pi$- and $\mu$-mesons

The leptonic decay of the $\pi$- and $\mu$-mesons observed in high-energy physics experiments exhibited long enough lifetimes ($\sim 10^{-8}$ and $10^{-6}$ s, respectively) to ensure that the interaction responsible was much weaker than strong ($\tau \lesssim 10^{-20}$ s) or electromagnetic ($\tau \lesssim 10^{-16}$ s) interactions. The weak decay interactions of the muon

$$\mu^{+} \rightarrow e^{+} + \nu_{e} + \tilde{\nu}_{\mu}, \tag{12.11a}$$

$$\mu^{-} \rightarrow e^{-} + \bar{\nu}_{e} + \nu_{\mu}, \tag{12.11b}$$

and the nuclear capture reaction

$$\mu^{-} + A_Z \rightarrow A_{Z-1} + \nu_{\mu}, \tag{12.12}$$

were analysed in terms of a four-fermion interaction, similar to that assumed in beta decay, and it was found [7] that the observed characteristics of the

† Sudarshan and Marshak ([6*b*] and [42]) strongly emphasized the V–A interaction, based on chirality invariance and, in fact, stimulated the repeating of the $A^{35}$ and $He^{6}$ experiments, the results of which clearly supported their V–A proposal.

‡ Since parity and charge conjugation are badly violated, a more exact treatment would have given the following

$$|G_F| = \frac{1}{\sqrt{2}}(|C_V|^2 + |C'_V|^2)^{\frac{1}{2}},$$

$$|G_{GT}| = \frac{1}{\sqrt{2}}(|C_A|^2 + |C'_A|^2)^{\frac{1}{2}}.$$

In weak interactions, if we use the nucleon mass to obtain a dimensionless analogue of $\alpha_{EM}$ for the electromagnetic interactions, we obtain, for decay interactions, $\alpha_{weak} \approx 10^{-12}$, whereas for scattering interactions, we obtain, $\alpha_{weak} \approx 10^{-6}$. If we determine the equivalent $\alpha$ for the gravitation interaction, we obtain $\approx 10^{-39}$.

decay spectra and the lifetime could both be explained with a coupling constant, which was approximately the same as that required for beta decay of the neutron. We must recall here that the neutron $\beta$-decay lifetime is about $10^3$ s, while the muon decay lifetime is about $10^{-6}$ s. However, the phase-space factor in the muon decay case, which varies as the fifth power of the maximum electron energy, is about $10^9$ larger in muon decay and, thus, about the same coupling constant is involved in both neutron and muon decay.

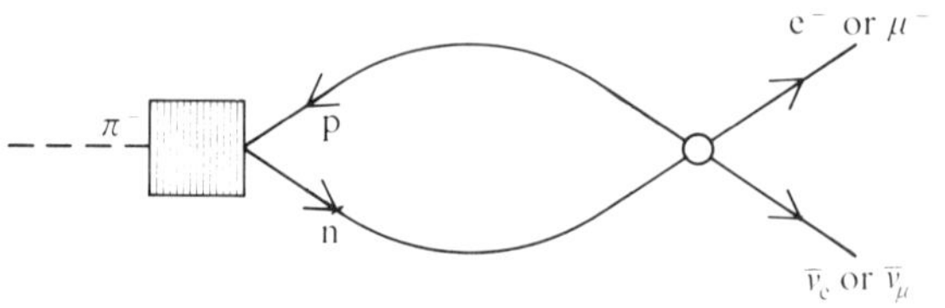


FIG. 12.1.

Thus, it was expected that the leptonic decays of the pion might also be explained in terms of this universal four-fermion interaction. We could, therefore, represent the leptonic decay of the pion by the two-step process shown in Fig. 12.1. The diagram represents the two decay processes

$$\pi^- \to \mu^- + \bar{\nu}_\mu,$$
$$\pi^- \to e^- + \bar{\nu}_e.$$

The $\pi^-$ is transformed into an anti-proton–neutron pair which then annihilates into a negative lepton, plus an anti-neutrino. The amplitudes for the decay via the two step process illustrated are to the lowest order of the form

$$(\Psi_{e^-\bar{\nu}_e}\mathscr{H}_I\Psi_\pi) = \sum_i C_i(\Psi_{e^-\bar{\nu}_e}, \bar{\Psi}_e\Gamma_i\Psi_{\nu_e}\Psi_0)(\Psi_0, \bar{\Psi}_P\Gamma^i\Psi_n\Psi_\pi)$$

or,

$$(\Psi_{\mu^-\nu_\mu}\mathscr{H}_I\Psi_\pi) = \sum_i C_i(\Psi_{\mu^-\nu_\mu}, \bar{\Psi}_\mu\Gamma^i\Psi_{\nu_\mu}\Psi_0)(\Psi_0, \bar{\Psi}_p\Gamma_i\Psi_n\Psi_\pi). \quad (12.13)$$

In order to conserve parity and angular momentum in the second (strong interaction) matrix element factor, the axial vector and pseudoscalar terms are the only non-vanishing terms.

The decay rate calculated by the standard techniques, involving the squared matrix element becomes

$$\Gamma_{\pi\to\mu+\nu} = \frac{m_\pi}{8\pi}\left(1-\frac{m_\mu^2}{m_\pi^2}\right)^2\left(\frac{m_\mu}{m_\pi}f_A+f_P\right)^2, \quad (12.14)$$

$$\Gamma_{\pi\to e+\nu} = \frac{m_\pi}{8\pi}\left(1-\frac{m_e^2}{m_\pi^2}\right)^2\left(\frac{m_e}{m_\pi}f_A+f_P\right)^2, \quad (12.15)$$

where $f_A$ and $f_P$ are, respectively, the axial and pseudoscalar coupling constants. The lifetime of the pion can be used to estimate

$$f_A \approx 2{\cdot}1\times 10^{-7}.$$

From eqns (12.14) and (12.15) we obtain the ratio of the decay rates

$$R = \frac{\Gamma_{\pi\to e+\nu}}{\Gamma_{\pi\to\mu+\nu}} = \frac{\left(1-\dfrac{m_e^2}{m_\pi^2}\right)^2\left(\dfrac{m_e}{m_\pi}f_A+f_P\right)^2}{\left(1-\dfrac{m_\mu^2}{m_\pi^2}\right)^2\left(\dfrac{m_\mu}{m_\pi}f_A+f_P\right)^2}, \tag{12.16}$$

or,

$$R \approx \frac{1}{(1-m_\mu^2/m_\pi^2)}\frac{\left(\dfrac{m_e}{m_\pi}f_A+f_P\right)^2}{\left(\dfrac{m_\mu}{m_\pi}f_A+f_P\right)^2}. \tag{12.17}$$

Owing to the smallness of the factor $\dfrac{m_e}{m_\pi}$ ($\approx 0{\cdot}004$) in the first term of the numerator, the value of this ratio is very sensitive to even small values of the ratio $f_P/f_A$. Experimentally [8], this ratio is $(1{\cdot}25\pm0{\cdot}03)10^{-4}$, which is about what we would expect for $f_p \ll 0{\cdot}004 f_A$, thus implying the absence of pseudoscalar interactions.

### 12.3. Polarization of electrons and positrons in β-decay

At this point it is worth noting that, by observing the polarization of the electron in β-decay, for both pure Fermi and pure Gamow–Teller transitions, [9] it was determined that the electrons are polarized with their spin anti-parallel to their momentum (negative helicity†), while the positrons were polarized along their momentum (positive helicity) as follows,

$$P \approx \begin{cases} -\beta & \text{for electron} \\ +\beta & \text{for positrons,} \end{cases} \tag{12.18}$$

where $\beta = v/c$. If we analyse this in terms of the $C_i$ and $C_i'$, this would imply

$$C_A \approx C_A', \quad C_v \approx C_v', \tag{12.19a}$$

which should imply parity violation, since parity conservation gives

$$C_A' = C_v' = 0, \tag{12.19b}$$

as was previously determined.

† If a particle is completely polarized with its spin parallel (anti-parallel) to the direction of motion, it has, by definition, a helicity of $+1$ $(-1)$. If the polarization along the momentum is $\pm P$, the helicity is $\pm P$.

## 12.4. Parity violation in weak interactions

We have already mentioned the $\theta^+ - \tau^+$ puzzle in the kaon decay process, observed in high-energy experiments. Owing to the long lifetime involved, it was demonstrated by simple calculations that this must be a weak interaction decay. The $\theta^+$ and $\tau^+$ seemed to be identical, within experimental errors, in all respects except that the $\theta^+$ and $\tau^+$ appeared to have opposite parities. Thus, either parity was violated in weak interactions, or the $\theta^+$ and $\tau^+$ were different particles. There were many proposals made, including the existence of parity doublets to attempt to conserve parity, since it had been quite generally assumed prior to this that even weak interactions conserved parity. Lee and Yang [10] then re-analysed the $\beta$-process to determine whether the assumed parity conservations were experimentally verified, or just an assumption. They also investigated parity conservation in strong interactions, by estimating the fraction $|\mathscr{F}|$ of mixing in of the wrong, or forbidden, parity amplitude in atomic spectroscopy. Lee and Yang estimated

$$|\mathscr{F}|^2 \lesssim 10^{-6}.$$

In nuclear reactions a limit on parity violation was set (Tanner 1957) [11] by studying the reaction

$$p + F^{19} \rightarrow Ne^{20*} \rightarrow O^{16} + \alpha, \tag{12.20}$$

in the neighbourhood of the $Ne^{20*}$ resonance state, which is known to be $1^+$. Since both $O^{16}$ and $\alpha$ are known to be spinless, the final state can only be $1^-$. Hence, the reaction will not be seen if parity is conserved, and its non-observation can be used to set a limit on parity conservation. Tanner found no evidence, within his sensitivity of measurement, and concluded that

$$|\mathscr{F}|^2_{\text{nucl}} \lesssim 4 \times 10^{-8}. \tag{12.21}$$

Wilkinson [12] studied the problem further and pointed out that there were three classes of checks.

*Class I*—A search for violations based on absolute selection rules.

*Class II*—A search for longitudinal polarizations of product heavy particles, or circular polarization of $\gamma$-rays. These checks are directly sensitive to $f$.

*Class III*—Searches for odd powers of $\cos\theta$ in the angular distribution, or correlations, of radiation emitted from well-isolated nuclear states and, here again, one is sensitive to $|\mathscr{F}|^2$.

Wilkinson also pointed out that the Tanner check was of Class I, but assumed that the parity non-conserving forces do not change the isotopic spin. Wilkinson's check of the Class I type involved a search for radiative capture of deuterium $He^4(d\gamma)Li^6$, through the $0^+$ state at 3·56 MeV. He obtained $\mathscr{F}^2 \lesssim 10^{-7}$, but this check involved a change of isospin. He also obtained results of similar accuracy for a check of type Class II.

For electromagnetic interactions, the empirical, 'Laporte rule', which states that for dipole radiation even states combine exclusively with odd states, requires parity conservation for its validity. This rule has been demonstrated to hold to quite high precision and, thus, the fraction of parity violating amplitudes is small, and has been estimated to be

$$|\mathscr{F}|^2_{\text{E.M.}} \lesssim 10^{-12}.$$

Subsequently, it was observed [13] that the circular polarization of a $\gamma$ transition in $Ta^{181}$ was of the order of $10^{-4}$. From this we could apparently conclude

$$|\mathscr{F}|^2 \sim 10^{-12}.$$

Lobashov *et al.* [14] have, however, done a similar but improved experiment, from which we can conclude

$$|\mathscr{F}|^2 \sim 10^{-14},$$

which is the order of magnitude violation which we would estimate as contributed by the violation in weak interactions. Hence, it appears that parity is, as far as we know, completely conserved in strong and E.M. interactions.

Our most general $\mathscr{H}_{\mathrm{I}}$ for $\beta$-decay (eqn (12.3)) is composed of two sets of terms, the $C_i$ terms and the $C_i'$ terms. The $C_i$ terms are parity conserving, and the $C_i'$ terms are (due to the extra $\gamma_5$ factor) parity violating. We have already implied this since, when we previously insisted on parity conservation, we found the $C'_i$ must equal zero.

The experiments that had been performed at the time gave results for $\beta$-decay spectra, $\beta$–$\nu$, and $\beta$–$\gamma$ correlations, and the results were expressed in terms of the five coupling constants, $C_i$, with $C_i'$ set equal to zero, to require parity conservation. If the $C'_i$ were not assumed to be zero, the only effect was to replace terms like $C_i^*C_i$ by $C_i^*C_i + C_i'^*C_i'$, thus leading to a renormalization. This simple substitution would not occur if the neutrino rest mass were non-zero. Unfortunately, no interference terms of the form $C_i^*C_i'^*$ occurred. Thus, Lee and Yang pointed out that, in order to test parity conservation, we must observe pseudoscalar quantities, and see if their expectation values vanish, as would be the case for parity conservation. If parity is violated, nature distinguishes left from right, and the pseudoscalar quantities have a finite expectation value. For nuclear beta decay, the pseudoscalar is

$$\mathbf{J}.\mathbf{P}_{\mathrm{e}}.$$

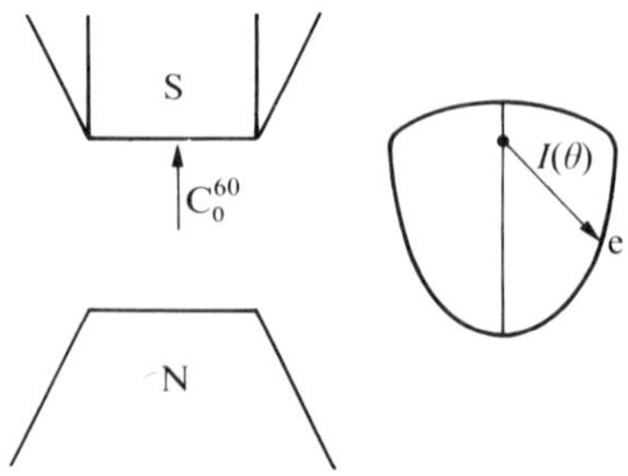


FIG. 12.2.

The corresponding pseudoscalar for decay of a free neutron would be

$$\boldsymbol{\sigma}.\mathbf{P}_e.$$

Thus, we need to observe $\beta$-decay from polarized nuclei.

Wu and Ambler *et al.* [15] used paramagnetic $Co^{60}$ polarized by an external magnetic field, at very low temperatures, to avoid thermal depolarization. A schematic representation of the experiment is given in Fig. 12.2. The predicted normalized angular distribution of the electrons emitted from the $Co^{60}$ is

$$I(\theta) = \frac{1}{4\pi}\left(1+\alpha\frac{\langle J_z\rangle}{J}\beta\cos\theta\right),$$

where $\theta$ is the polar angle relative to the nuclear spin, and $|\alpha| \leq 1$. If parity is conserved, $\alpha = 0$. The results of the experiment gave $\alpha \approx -1$, which demonstrates a maximal (or near-maximal) parity violation. Very shortly thereafter, the violation of parity in meson decay was demonstrated [16]–[17]. Consider the reactions

$$\pi^+ \to \mu^+ + \nu_\mu$$
$$\hookrightarrow e^+ + \nu_e + \bar{\nu}_\mu.$$

Let us assume parity is violated in both decays. Then we can expect that the expectation values of the pseudoscalars $\langle\sigma_\mu . P_\mu\rangle$ and $\langle\sigma_e . P_e\rangle$ are greater than zero. Hence, a $\pi^+$ decaying, at rest, will be expected to give rise to a muon polarized along (or anti-) its direction of motion and, thus, the subsequent $\beta$-decay of the muon should exhibit an angular distribution relative to its direction of emission similar to the $\beta$-decay from cobalt 60. Therefore,

$$I(\theta) = \frac{1}{4\pi}(1+\alpha\cos\theta).$$

It was found that $\alpha \approx (-\frac{1}{3}\pm 10)$ per cent.

It was demonstrated, shortly afterwards, that parity was violated also in the non-leptonic weak decays of strange particles [18]–[19].

### 12.5. Muon or neutrino helicity in $\pi \rightarrow \mu + \nu$ decay

The helicity (of the leptons) is defined as the algebraic value of the polarization along the momentum vector. Negative helicity particles are often referred to as 'left-handed', whereas positive helicity particles are often referred to as 'right-handed'.

In the decay

$$\pi^{\pm} \rightarrow \mu^{\pm} + \nu,$$

the helicity of the muons has been measured directly via observations on high-energy knock-on electrons, produced by the high-energy muons in iron which is magnetized [20], [21].

Backenstoss *et al.* [21] deduced that the helicity of the $\mu^-$ was $+1{\cdot}2 \pm 0{\cdot}3$. Thus, the $\mu^-$ was right-handed, and had a large helicity consistent with $+1$ (i.e. completely polarized along the momentum direction).

Bardon *et al.* (1961) [22] employed muons from 43 MeV decaying negative pions, which then underwent Mott scattering in lead, in order to measure the helicity of negative muons from negative pion decay.

The longitudinally polarized muons in the centre of mass system exhibit a transverse polarization component in the laboratory system. The muons are then scattered in lead plates, and the muons scattered at large angles are then detected in counters, which lie in planes which are intersected by the beam line. Thus, the left–right asymmetry of the scattering can be measured. The energy of the negative muons and the angle of scattering was selected so as to result in an appreciable left–right asymmetry for polarized muons. Nevertheless, the asymmetry obtained is still small, even for completely polarized muons and, thus, considerable effort must be made to assure preserving azimuthal symmetry about the pion beam. The result was that the helicity of $\mu^-$ is $+1{\cdot}1 \pm 0{\cdot}3$, which agrees with the result of Backenstoss *et al.* Alikhanov *et al.* [20] found that the $\mu^+$ emitted in $\pi^+$ decay were left-handed.

Let us now use the above information to deduce the relation between spins and momenta, for the neutrino emitted in pion decay. In the centre of mass system, the momenta and spin (helicity) vectors have the relationships shown in Fig. 12.3. Since the $\mu^-$ is known experimentally to be emitted with right-handed helicity, the accompanying anti-neutrinos must be right-handed also to conserve angular momentum. Since the neutrino quantities are deduced, they are shown as dotted lines. Similarly, since the $\mu^+$ is

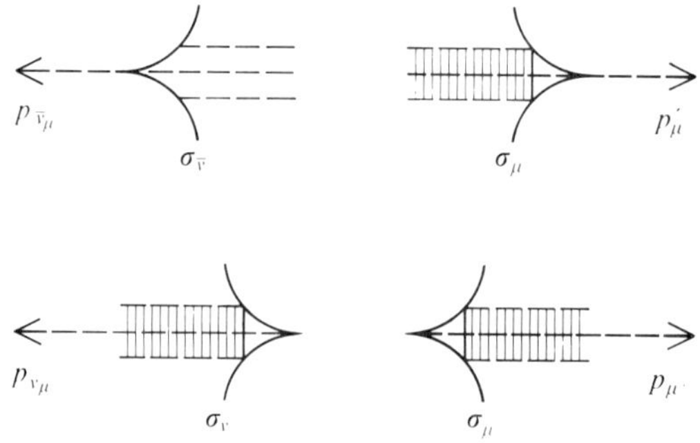


FIG. 12.3.

experimentally known to be left-handed, the neutrino emitted must be left-handed.

The convention generally employed calls $\mu^-$ and $e^-$ particles and denotes as anti-neutrinos the particles accompanying them. On the other hand, $\mu^+$ and $e^+$ are anti-particles, and the particles accompanying them are neutrinos. This convention satisfies the conservation of leptons, which we shall soon show to be the case.

If we apply the parity transformation to Fig. 12.3, it is clear that the momenta will reverse, but the spins will not, thus, this situation represents the maximum violation of parity. Similarly, charge conjugation interchanges $\mu^+$ and $\mu^-$, but does not change the helicities (as required to fit the experimental facts) and, thus, is violated. However, applying both parity and charge conjugation (CP) transforms the two diagrams into each other and, thus, we can expect CP conservation to be consistent with these experimental facts.

## 12.6. Electron and neutrino helicities, and correlation functions in $\beta$-decay

It was previously discovered that the $e^\pm$ in beta decay have a definite sign of helicity (its value is $\approx -\beta$ for electrons, and $\approx +\beta$ for positrons).

Observations of the correlation function between the neutrino and the electron can answer the question of whether the Fermi interaction is pure S, or V, or a mixture. Such observations have been accomplished, indirectly, by measuring the momentum of the recoiling nucleus as well as the electron, and then using conservation of angular momentum to obtain the momentum of the outgoing neutrino.

The angle $\theta$ between the momentum of the electron and the momentum of the neutrino has the distribution (see [1] for more details and definitions of $\xi$, $a$, and $b$).†

$$I(q) = \frac{\xi}{4\pi^3}\, q^2(E_{\max}-E)^2\left(1-a\beta\cos\theta+b\,\frac{2m}{E}\right)\sin\theta\, d\theta, \qquad (12.22)$$

where for a pure Fermi transition

$$a = \frac{|G_v|^2 - |G_S|^2}{|G_v|^2 + |G_S|^2}, \qquad (12.23a)$$

$$|G_i|^2 = \tfrac{1}{2}[|C_i|^2 + |C_i'|^2]. \qquad (12.23b)$$

Such measurements have been performed on $A_r^{35}$, which decays by emitting a positron, to $Cl^{35}$. The transition is, practically, a pure Fermi transition, and gives the result [23]

$$a = 0{\cdot}97 \pm 0{\cdot}14.$$

Thus

$$|G_s|^2 \ll |G_v|^2. \qquad (12.24)$$

The Gamow–Teller interaction can be analysed from a study of the electron–neutrino angular correlation in the $\beta$-decay of $He^6$.

$$He^6 \to Li^6 + e^- + \bar{\nu}_e. \qquad (12.25)$$

$He^6$ has 0 spin and positive parity, whereas $Li^6$ has spin $= 1$ and positive parity. Thus, the $\beta$-decay is a pure Gamow–Teller transition. In this case, the angular coefficient

$$a = \frac{1}{3}\frac{|G_T|^2 - |G_A|^2}{|G_T|^2 + |G_A|^2}. \qquad (12.26)$$

The measured result was [24]

$$a = -0{\cdot}35 \pm 0{\cdot}03,$$

and a measurement of $Ne^{23}$ gave a confirming result, with slightly lower precision [23].

This demonstrates that the A interaction dominates Gamow–Teller transition, and is consistent with the assumption that $C_T = C_T' = 0$. From measurement of the correlation function of the outgoing electron, and a knowledge of the polarization direction of a polarized neutron, we can conclude that the relative phase between $C_V$ and $C_A$ is $\pi$.

† $\xi$, $a\xi$, and $b\xi$ are functions of the absolute value of coupling constants squared, and $|M_F|^2$ and $|M_{GT}|^2$.

To summarize the information obtained, we have, for the various coupling constants in the $\beta$-decay interaction,

V $C_V = C'_V = g = (3{\cdot}01 \pm 0{\cdot}01) \times 10^{-12}$ natural units or $(1{\cdot}418 \pm 0{\cdot}004) \times 10^{-49}$ erg cm$^3$,

A $C_A = C'_A = -(1{\cdot}18 \pm 0{\cdot}03)g$,

P undetermined.

The natural units have $\hbar = c = m_e = 1$. This weak interaction is often referred to as the V–A interaction.

### 12.7. The two-component theory of the neutrino

The above-described interaction can be represented by the following Hamiltonian

$$\mathscr{H}_I = \frac{C_V}{\sqrt{2}} \int d^3x \overline{\Psi}_p \gamma^\alpha \Psi_n(x) \overline{\Psi}_e(x) \gamma_\alpha (1+\gamma_5) \Psi_\nu(x) - $$
$$- \frac{C_A}{\sqrt{2}} \int d^3x \overline{\Psi}_p(x) \gamma^\alpha \gamma_5 \Psi_n(x) \overline{\Psi}_e(x) \gamma_\alpha (1+\gamma_5) \Psi_\nu(x) + \text{Hermitian conjugate.} \tag{12.27}$$

The appearance of the factor $(1+\gamma_5)$ in front of the neutrino field operator, wherever it occurs, determines the helicity of the neutrinos. We can see this in the following way. If $\Psi$ is a solution of the Dirac equation, $i\gamma q \Psi = 0$, for zero mass, then it follows that $\gamma_5 \Psi$ is also a solution, since $\gamma_5$ anti-commutes with all $\gamma_\alpha$. Thus, for each solution ($\Psi$) we can introduce the following two expressions

$$\Psi_\pm = \tfrac{1}{2}(1 \pm \gamma_5)\Psi, \tag{12.28}$$

where it is easy to show that

$$\gamma_5 \Psi_\pm = \pm \Psi_\pm, \tag{12.29a}$$

with

$$\Psi = \Psi_+ + \Psi_-. \tag{12.29b}$$

Thus, all solutions to the Dirac equation with zero mass have the eigenvalue of either $+1$ or $-1$ for the matrix $\gamma_5$, and, hence, this new quantum number can characterize the solution. $(1 \pm \gamma_5)$ acts as a projection operator for positive (negative) helicity states of a massless particle. Hence, the form of the V–A interaction strongly suggests that neutrinos of only one definite helicity take part in weak $\beta$-decay interactions.

By working out the V–A Hamiltonian in terms of the two neutrino component theory, we can show that anti-neutrinos have unit positive helicity

$(+1)$, while neutrinos have unit negative helicity $(-1)$. Thus, the helicities are the same as for the neutrinos accompanying the muons in pion decay. Hence, the neutrinos can be represented by a two-component rather than a four-component wave function, normal for a spin $\frac{1}{2}$ particle.

This two-component theory of the neutrino has maximum parity violation naturally built into it and, therefore, was not taken seriously when first introduced [25] but became very appropriate [26] to explain the parity violation in weak interactions. Of course, we can formally treat even a four-component neutrino as the sum of two two-component neutrinos.

Now let us look at the factor representing the creation and annihilation of light particles in the V–A interaction. This factor is

$$\overline{\Psi}_e(x)\gamma_\alpha(1+\gamma_5)\Psi_\nu(x). \tag{12.30}$$

We can also express it as

$$\tfrac{1}{2}\overline{\Psi}_e(x)(1-\gamma_5)\gamma_\alpha(1+\gamma_5)\Psi_\nu(x). \tag{12.31}$$

Now the electron field also is multiplied by $1+\gamma_5$, since†

$$((1+\gamma_5)\Psi(x))^*\gamma_0 = \Psi^*(x)(1+\gamma_5)\gamma_0 = \overline{\Psi}(x)(1-\gamma_5). \tag{12.32}$$

If the electron is relativistic, so that its kinetic energy is large compared to its rest mass, we could treat it, approximately, in terms of the two neutrino component theory. Equation (12.31) shows that the two leptons emitted in a $\beta$-decay have helicities of opposite sign.

Now let us consider a pure Fermi decay, and assume the electron and anti-neutrino are emitted in exactly opposite directions. Since they have opposite helicities, the total spin of the light particles is one which violates angular momentum conservation for a Fermi transition and, thus, cannot occur. Hence, the angular correlation function must vanish for $\theta = \pi$, in a pure Fermi transition. This is borne out by the calculated correlation function which is

$$W(\theta) = 1+\cos\theta \tag{12.33}$$

for a pure Fermi transition. This is experimentally observed to be the case.

If we assume $C_A \approx -C_V$ (i.e. neglect the factor $1{\cdot}18\pm0{\cdot}03$, by which $|C_A|$ is larger than $|C_V|$) the interaction Hamiltonian becomes

$$\mathscr{H}_I \sim g\int d^3x(\overline{\Psi}_p(x)(1-\gamma_5)\gamma^\alpha(1+\gamma_5)\Psi_n(x))(\overline{\Psi}_e(x)(1-\gamma_5)\gamma_\alpha(1+\gamma_5)\Psi_\nu(x)) + \\ +\text{Hermitian conjugate}, \tag{12.34}$$

† We are using $\gamma_0, \gamma_1, \gamma_2, \gamma_3$, as the four components of $\gamma$. Authors often use $\gamma_1, \gamma_2, \gamma_3, \gamma_4$.

where

$$C_V = -C_A = g.$$

Thus, the interaction part of the $\beta$-decay Hamiltonian can be formulated in terms of $(1+\gamma_5)\Psi(x)$ for all fields. This can be expressed by saying that the interaction part of the Hamiltonian is invariant under

$$\Psi(x) \to \gamma_5 \Psi(x), \tag{12.35a}$$

$$\overline{\Psi}(x) \to \overline{\Psi}(x)\gamma_5. \tag{12.35b}$$

It is easy to show that if eqns (12.35a) and (12.35b) hold, or if all fields can be formulated in terms of $(1+\gamma_5)\Psi(x)$, then eqn (12.34) is the only possible form of the interaction Hamiltonian and, thus, V–A is the only possible theory. The experimental data is nearly consistent with this hypothesis.

## 12.8. Conservation of leptons

### 12.8.1. *Comparison of neutrinos from Fermi and Gamow–Teller transitions*

Let us accept the hypothesis that there are only two component neutrinos. However, we should check whether the neutrino which appears in Gamow–Teller transitions has the same helicity as the neutrino which appears in Fermi transitions. In fact, our entire treatment of $\beta$-decay so far has made this assumption.

The obvious way to test the assumption is to observe a transition that contains both Fermi and Gamow–Teller terms, and to see if interference occurs. Only if the two neutrinos have the same helicity can the V (and/or S) amplitude interfere with the A (and/or T) amplitude.

A nucleus which first $\beta$-decays, thus becoming polarized, and then undergoes a $\gamma$-decay, allows us to measure the nuclear polarization via a measurement of the helicity of the photon [27]. This led to the conclusion that Fermi–Gamow–Teller interference was observed.

For the processes considered so far, we can associate a leptonic number (or charge) of $+1$ for particles (electrons and neutrinos), and $-1$ for anti-particles (positrons and anti-neutrinos), and all our experimental results, and conclusions, are consistent with conservation of leptons.

### 12.8.2. *Double beta decay interactions—lepton number conservation*

Consider the $\beta$-decays

$$(Z, A) \to (Z+1, A) + e^- + \bar{\nu}_e, \tag{12.36a}$$

$$(Z', A') \to (Z'+1, A') + e^- + \bar{\nu}_e. \tag{12.36b}$$

We can expect to have a cascade reaction

$$(\mathrm{Z}, \mathrm{A}) \to (\mathrm{Z}+1, \mathrm{A})+\mathrm{e}^- + \bar{\nu}_\mathrm{e}$$
$$\hookrightarrow \bar{\nu}_\mathrm{e} + (\mathrm{Z}'+1, \mathrm{A}') \to (\mathrm{Z}', \mathrm{A}')+\mathrm{e}^+, \qquad (12.36\mathrm{c})$$

where the second nucleus $(\mathrm{Z}'+1, \mathrm{A}')$ serves as a detector for the anti-neutrino, emitted by the original $\beta$-decaying nucleus (Z, A). Such reactions conserve leptonic number, and have been observed.

The intense source of anti-neutrinos, needed for performing such experiments have come from the decay of neutrons generated in a nuclear reactor (pile). Experiments [28] have shown that anti-neutrons from a pile have been captured by protons in the reaction

$$\bar{\nu}_\mathrm{e} + \mathrm{p} \to \mathrm{n} + \mathrm{e}^+, \qquad (12.37\mathrm{a})$$

which is the expected type (12.36c) and conserves lepton number. The observed average cross-section ($\approx 10^{-44}\ \mathrm{cm}^2$) agrees well with the theoretical prediction of the two-component neutrino theory. However, alternatively, if

$$(\mathrm{Z}', \mathrm{A}') \to (\mathrm{Z}'-1, \mathrm{A}')+\mathrm{e}^+ + \nu_\mathrm{e},$$

or, equivalently,

$$(\mathrm{Z}', \mathrm{A}')+\mathrm{e}^- \to (\mathrm{Z}'-1, \mathrm{A}')+\nu_\mathrm{e},$$

and, if leptons are conserved, we can expect that

$$\nu_\mathrm{e} + (\mathrm{Z}'-1, \mathrm{A}') \to (\mathrm{Z}', \mathrm{A}')+\mathrm{e}^-,$$

but we cannot expect the cascade

$$(\mathrm{Z}, \mathrm{A}) \to (\mathrm{Z}+1, \mathrm{A})+\mathrm{e}^- + \bar{\nu}_\mathrm{e}$$
$$\hookrightarrow \bar{\nu}_\mathrm{e} + (\mathrm{Z}'+1, \mathrm{A}') \to (\mathrm{Z}'+2, \mathrm{A}')+\mathrm{e}^-, \qquad (12.37\mathrm{b})$$

since this would violate conservation of leptons in the final states.

So far, searches have not observed reactions of this type. The intense anti-neutrino flux required is produced in a pile. The detector nucleus reaction employed is

$$\mathrm{K\ (electron)} + \mathrm{A}^{37} \to \mathrm{C}_1^{37}. \qquad (12.38)$$

The reaction studied was

$$\bar{\nu}_\mathrm{e} + \mathrm{C}_1^{37} \to \mathrm{A}^{37} + \mathrm{e}^-, \qquad (12.39)$$

and a cross-section consistent with zero (i.e. $(0{\cdot}1 \pm 0{\cdot}6)\ 10^{-45}\ \mathrm{cm}^2$ per atom) was found. The theoretically predicted value for the two-component neutrino theory, with $L_\mathrm{e}$ violated maximally, was $2 \times 10^{-45}\ \mathrm{cm}^2/\mathrm{atom}$, thus implying that $L_e$ is conserved, although the above limit placed on the $L_e$ non-conservation process is not too high.

A well-known set of related processes is the double $\beta$-decay process

$$(Z-2, A) \rightarrow (Z, A)+e^- +e^-, \tag{12.40a}$$

$$(Z-2, A) \rightarrow (Z, A)+e^- +e^- +\bar{\nu}_e+\bar{\nu}_e. \tag{12.40b}$$

Process (12.40a) (neutrinoless double $\beta$-decay) clearly violates lepton conservation, since $L = 0$ in the initial state, and $L = -2$ in the final state. On the other hand, process (12.40b) (double $\beta$-decay with two (anti) neutrinos) is consistent with lepton conservation. The phase-space factor for neutrinoless double beta decay is about $10^5$ larger for the double $\beta$-decay with two neutrinos, than for the neutrinoless double beta decay. An experimentally distinguishable characteristic of the neutrinoless double $\beta$-decay is that the electrons in this reaction share all the available energy. The typical predicted half lifetime range for neutrinoless double $\beta$-decay is about $10^{16\pm2}$ years, whereas for double $\beta$-decay with two neutrinos, the typical prediction is about $10^{22\pm3}$ years.

Reference [29] reviews a number of experiments which place half lifetime limits on double $\beta$-decay, implying neutrinoless double $\beta$-decay does not occur. Recent experiments have detected double $\beta$-decay.

It has been demonstrated [30], from geological evidence, that

$$T_e^{130} \rightarrow X_e^{130}+2e^-,$$

exhibiting a half-life of $10^{21\cdot34\pm0\cdot12}$ years. The theoretically estimated neutrinoless lifetime is about $10^{16\cdot3\pm2}$ years, whereas the estimated lifetime for the case with two neutrinos is about $10^{22\cdot5\pm2\cdot5}$ years and, thus, the experiment agrees with the latter and, hence, strongly supports lepton conservation.

An experiment (Goldhaber, M. *et al.*) [31] in which the neutrino helicity was measured, confirmed the V–A form of the $\beta$ interaction. $Eu^{152}(0^-)$ captured a K-electron, and became an excited state $Sm^{152}(1^-)$ which decayed to the ground state, of zero spin, by gamma emission. It turned out that, in this case, where the photon moves in a direction opposite to the neutrino, resonant absorption, or resonant scattering, can occur. The direction of the recoiling nucleus was measured to determine the neutrino direction. Resonant scattering was found to occur and, thus, allowed a measurement of the photon helicity, which then allowed Goldhaber *et al.* to demonstrate that the neutrino had negative helicity and, thus, that the interaction was axial. If we accept the two neutrino hypothesis, the interaction must be some combination of axial and vector only.

### 12.8.3. *Conservation of leptons in pion and muon decay*

The assignments

$$\pi^+ \to \mu^+ + \nu_\mu, \quad (12.41a)$$

$$\pi^- \to \mu^- + \bar{\nu}_\mu. \quad (12.41b)$$

explicitly satisfy lepton conservation if $\mu^-$ (like $e^-$) is called a particle with lepton number

$$L_{\mu^-} = +1 \qquad (L_{e^-} = +1).$$

Correspondingly, $\mu^+$ (like $e^+$) is called an anti-particle with lepton number

$$L_{\mu^+} = -1 \qquad (L_{e^+} = -1),$$

$\nu_\mu$ and $\nu_e$ have

$$L_{\nu_\mu} = +1 \qquad (L_{\nu_e} = +1),$$

$\bar{\nu}_\mu$ and $\bar{\nu}_e$ have

$$L_{\bar{\nu}_\mu} = -1 \qquad (L_{\bar{\nu}_e} = -1)$$

and,

$$L = 0$$

for all hadrons (p, n, $\bar{p}$, Y, $\pi$, K, $\bar{K}$, etc.) and gamma rays.

### 12.8.4. *The experimentally observed differences between $\nu_\mu$ and $\nu_e$*

High energy muon–neutrino beams have been formed at the Brookhaven AGS and the CERN proton cyclotron, by allowing high energy $\pi^\pm$ to decay in a drift space, and then using massive shielding to attempt to stop as much as possible of all products from leaving the shield, except the neutrinos, which, due to their exceedingly small ($\sigma \lesssim 10^{-38}\,\text{cm}^2$) cross-section, easily penetrate any practical shield. Owing to the low $\dfrac{\pi\text{–e}}{\pi\text{–}\mu}$ decay ratio, we have essentially an almost pure muon–neutrino beam.

A number of experiments have now been performed (Danby *et al.* 1962 [32*a*], Bernardini *et al.* 1964 [32*b*], Burns *et al.* 1965 [33], and Bartley *et al.* 1965 [34]). The experiments all employed neutrinos from high-energy pions decaying in flight, produced by high-energy 15–30 GeV protons [35].

Figure 12.4 shows the layout of the more recent CERN experiment. Some of the more important features to note in the CERN experiment are

(1) a magnetic focusing device which strongly enriches the $\pi^+$, and reduces the $\pi^-$, content of the beam (or vice versa);
(2) the path allowed for the pions to decay, thus forming neutrinos;
(3) a massive shield which stops muons and electrons by electromagnetic energy loss, and effectively removes all hadrons by nuclear interaction;
(4) the detector which is well shielded, and is a massive spark chamber, or heavy liquid chamber.

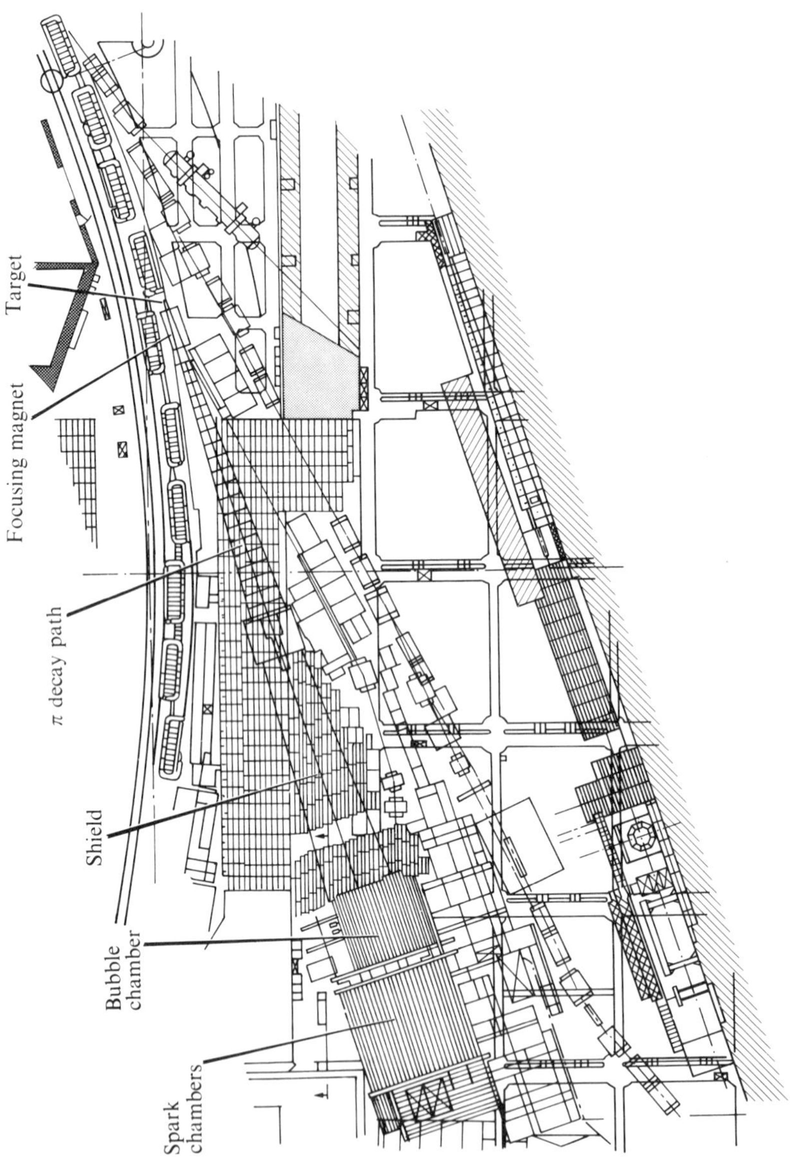


FIG. 12.4. The CERN neutrino experimental arrangement. (From [35].)

If the muon and electron neutrinos were different, and leptons were conserved, separately, for electron-coupled leptons, and muon-coupled leptons, we would expect

$$\pi^+ \rightarrow \mu^+ + \nu_\mu, \tag{12.42a}$$

$$\pi^- \rightarrow \mu^- + \bar{\nu}_\mu, \tag{12.42b}$$

$$\bar{\nu}_\mu + \mathrm{p} \rightarrow \mu^+ + \mathrm{n}, \tag{12.43a}$$

$$\nu_\mu + \mathrm{n} \rightarrow \mu^- + \mathrm{p}. \tag{12.43b}$$

On the other hand, if $\nu_\mu$ and $\nu_e$ were the same, we would also expect the reactions

$$\bar{\nu}_\mu + \mathrm{p} \rightarrow \mathrm{e}^+ + \mathrm{n}, \tag{12.44a}$$

$$\nu_\mu + \mathrm{n} \rightarrow \mathrm{e}^- + \mathrm{p}. \tag{12.44b}$$

If $\nu_\mu$ and $\nu_e$ were the same, we might expect the same order of magnitude for reactions (12.44a) and (12.44b) as for (12.43a) and (12.43b). The first experiments (Danby *et al.*) strikingly showed that reactions (12.44a) and (12.44b) did not occur (or were limited to a small fraction of reactions (12.43a) and (12.43b)). It was further established by the subsequent experiments that the number of electron events induced by muon neutrinos was down by at least two orders of magnitude compared to muon events, thus demonstrating clearly that there were two kinds of neutrinos, $\nu_\mu$ and $\nu_e$. The reader is referred to the review by Willis and Thompson [35] as a general reference.

### 12.8.5. *Conservation of μ lepton number and* e *lepton number separately*

In the heavy-liquid bubble chamber experiments (Bartley 1965) it was observed that, when neutrinos (originating from $\pi^+$ decay) were incident, only $\mu^-$, and not $\mu^+$, was produced. Conversely, when anti-neutrinos (originating from $\pi^-$ decay) were used, only $\mu^+$, and not $\mu^-$, was observed. Thus indicating that the muon lepton number was separately conserved, to within the accuracy set by the contamination of the anti-particles in the neutrino beam due to the lack of a perfect focusing arrangement, etc.

Having two different neutrinos, and demanding conservation of leptons separately for muon and electronic leptons, affects the predictions for muon decay. Thus,

$$\mu^+ \rightarrow \mathrm{e}^+ + \bar{\nu}_\mu + \nu_e, \tag{12.45a}$$

$$\mu^- \rightarrow \mathrm{e}^- + \nu_\mu + \bar{\nu}_e. \tag{12.45b}$$

If $\nu_\mu$ and $\nu_e$ were the same, (i.e. $\nu_e = \nu_\mu = \nu$) we could expect to observe

$$\mu^+ \rightarrow \mathrm{e}^+ + \bar{\nu} + \nu \rightarrow \mathrm{e}^+ + \gamma, \tag{12.46}$$

where the $\bar{\nu}$ and $\nu$ annihilated, and then a gamma is emitted.

Experimentally, (Parker *et al.* 1964) [36] it has been shown that

$$\frac{\mu^+ \to e^+ + \gamma}{\mu^+ \to e^+ + \nu + \bar{\nu}} < 0{\cdot}6 \times 10^{-8}. \tag{12.47}$$

The theoretical estimates cannot be made too precisely. However, it is clear that they must be several orders of magnitude greater than the experimental value, thus implying lepton conservation separately, for electron and muon leptons, to a high accuracy and, thus, the necessity for two different neutrinos.

## 12.9. The helicity rule

### 12.9.1. *Muon decay*

The helicity $h$ of the positrons emitted in the decay

$$\mu^+ \to e^+ + \nu_e + \bar{\nu}_\mu \tag{12.48}$$

was measured by Duclos *et al.* (1964) [37] by utilizing Bhabha scattering on a magnetized iron foil. They obtained $h_{e^+} = 1{\cdot}04 \pm 0{\cdot}18$, demonstrating that the positrons in $\mu^+$ decay obey the so-called 'helicity rule' (i.e. + for anti-particles). The over-all average, including other experiments, (Buhler *et al.* (1963), and Bloom *et al.* (1963)) [38] is

$$h_{e^+} = 1{\cdot}03 \pm 0{\cdot}04. \tag{12.49}$$

The positron spectrum in $\mu^+$-decay can be used to test the two-component theory of the neutrino, with lepton conservation. This theory would, for example, predict as we have shown above that the $\nu_e$ is left-handed, while the $\bar{\nu}_\mu$ is right-handed.

It has been shown experimentally [16], [17] that, in $\mu^+$-decay, the positrons ($e^+$) prefer to be emitted in the opposite direction to the $\mu^+$

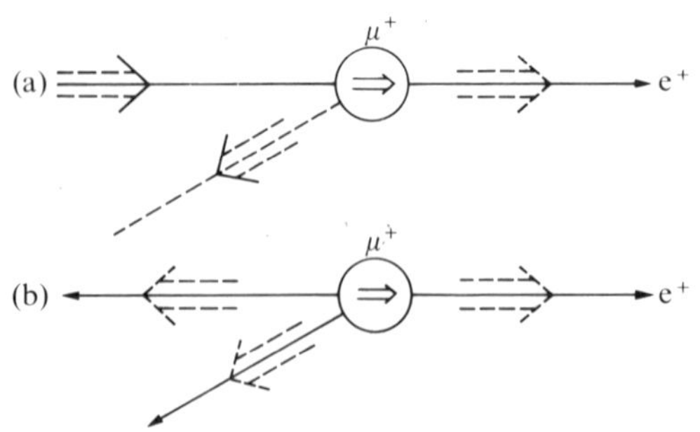


FIG. 12.5.

direction. Thus, taking account of the preceding diagram (on $\pi \to \mu + \nu$ decay), we obtain the following preferential decay possibilities for the $\mu^+$, depending on whether one neutrino is left-handed and one right-handed (a). or both, for example, right-handed (b). These are illustrated in Fig. 12.5 where arrow directions refer to spin.

Case (b) would correspond to Lepton non-conservation, and it is obvious from the schematic, that it is difficult to conserve angular momentum in (b),

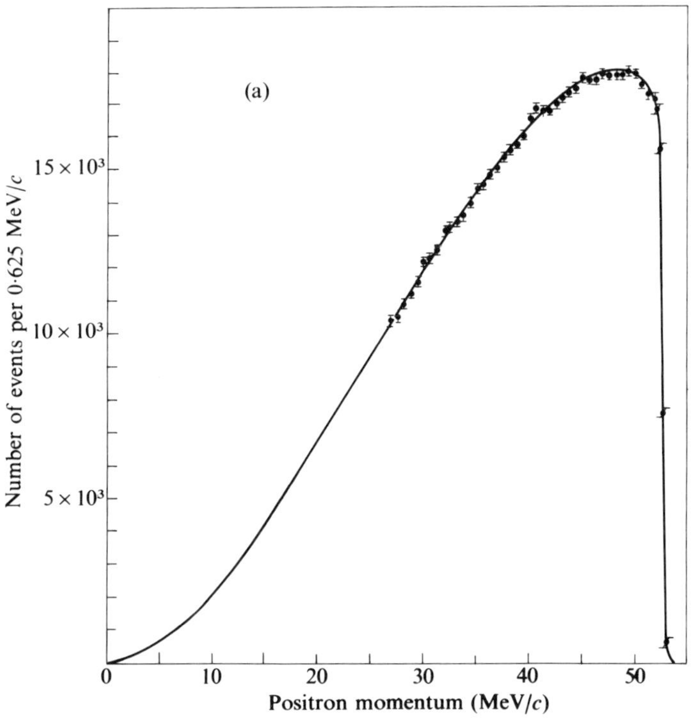


FIG. 12.6. The observed spectrum of $e^+$ in the decay $\mu^+ \to e^+ + \nu + \bar{\nu}$. A theoretical prediction based on the two-component neutrino theory, which is corrected for electromagnetic effects, is shown as a solid line. (From Bardon *et al.* (1965). *Phys. Rev. Lett.* **14,** 449.)

when the electron moves in an opposite direction to the two neutrinos, but this is just the case which corresponds to the maximum energy $e^+$ and, thus, there would (in (b)) be a dip in this spectrum at the high energies.

Figure 12.6 (from Bardon *et al.* 1965) [39] reveals that this dip does not occur and, thus, the combination is $\nu_L + \nu_R$, thus indicating lepton conservation occurs.

### 12.9.2. *Kaon decay*

As we have seen, the helicity rule is a useful tool in predicting certain angular correlations in $\mu$-decay. It is, also, possible to deduce correlations in leptonic decays connecting two spin-zero hadrons, such as, for example,

$$K^+ \to \pi^0 + e^+ + \nu_e \tag{12.50a}$$

and

$$K^0 \to \pi^\pm + e^\mp + \nu_e. \tag{12.50b}$$

The $e^\pm$ emitted in these decays are relativistic enough for us to expect the helicity rule to work. In the $K_{e_3}$ decay, both the $\pi$ and the $K^+$ are spinless. Therefore, for example, in collinear decays the helicity rule, and conservation of angular momentum, enhance some correlations and forbid others. For example, in the $K_{e_3}$ decay, when the $\pi$ energy is near zero, and the $e^+$ and $\nu_e$ have opposite momentum, since by the helicity rule they have opposite helicities (i.e. positron right-handed, neutrino left-handed), their combined spin–angular momentum is unity. Thus, since the orbital angular momentum is zero, angular momentum is not conserved and, hence, this configuration is forbidden.

Another collinear situation occurs when the $\pi$ has maximum energy, and is in a direction opposite to that of the leptons, both of which have parallel momentum. In this case, owing to the helicity rule, the combined angular momentum is zero, and this configuration is favoured.

Analogous arguments can be made for the $K^0 \to \pi^\pm + e^\mp + \nu_e$ decays. However, in the $K_{\mu_3}$ decay, the muon is non-relativistic (i.e. slow) enough so that we cannot expect the helicity rule to apply and, indeed, both helicities occur for the muon.

## 12.10. The nature of the weak interaction

### 12.10.1. *The two-component theory of the neutrino*

The experimental data on muon decay is quite consistent with the form provided by the two-component neutrino theory. The most general form of the latter is†

$$\tilde{C}_s(\overline{\Psi}_\nu(1+\gamma_5)\Psi_\mu)(\overline{\Psi}_e(1-\gamma_5)\Psi_\nu)+\tilde{C}_V(\overline{\Psi}_\nu\gamma^\alpha(1-\gamma_5)\Psi_\mu)(\overline{\Psi}_e\gamma_\alpha(1-\gamma_5)\Psi_\nu). \tag{12.51}$$

Since $\tilde{C}_s$ must be quite small, the interaction reduces to

$$\tilde{C}_V(\overline{\Psi}_\nu\gamma^\alpha(1-\gamma_5)\Psi_\mu)(\overline{\Psi}_e\gamma_\alpha(1-\gamma_5)\Psi_\nu) \tag{12.52}$$

† Where $\tilde{C}_i$ is related to $C_i'$ via a Fierz transformation (see [1] and [2]).

Hence, we can write, for the interaction Hamiltonian,

$$\mathscr{H}_{\mu}(x) = \frac{1}{\sqrt{2}} G_{\mu}(\overline{\Psi}_{\nu}\gamma^{\alpha}(1-\gamma_5)\Psi_{\mu})(\overline{\Psi}_{e}\gamma_{\alpha}(1-\gamma_5)\Psi_{\nu}) \tag{12.53}$$

From our previous discussion of the $\beta$-decay interaction, its interaction Hamiltonian can be written as

$$\mathscr{H}_{\mathrm{I}} = \frac{1}{\sqrt{2}}(C_{\mathrm{V}}\overline{\Psi}_{\mathrm{p}}\gamma^{\alpha}\Psi_{\mathrm{n}} + C_{\mathrm{A}}\overline{\Psi}_{\mathrm{p}}\gamma^{\alpha}\gamma_5\Psi_{\mathrm{n}})(\overline{\Psi}_{e}\gamma_{\alpha}(1-\gamma_5)\Psi_{\nu}), \tag{12.54}$$

with $C_{\mathrm{A}} \approx -1{\cdot}18C_{\mathrm{V}}$. Equation (12.54) becomes

$$\frac{1}{\sqrt{2}}C_{\mathrm{V}}(\overline{\Psi}_{\mathrm{p}}\gamma^{\alpha}(1-1{\cdot}18\gamma_5)\Psi_{\mathrm{n}})(\overline{\Psi}_{e}\gamma_{\alpha}(1-\gamma_5)\Psi_{\nu}) \tag{12.55a}$$

or

$$\approx \frac{1}{\sqrt{2}} C_{\mathrm{V}}(\overline{\Psi}_{\mathrm{p}}\gamma^{\alpha}(1-\gamma_5)\Psi_{\mathrm{n}})(\overline{\Psi}_{e}\gamma_{\alpha}(1-\gamma_5)\Psi_{\nu}) \tag{12.55b}$$

where, in the last approximation, we have set

$$\frac{C_{\mathrm{A}}}{C_{\mathrm{V}}} \approx -1,$$

instead of $-1{\cdot}18 \pm 0{\cdot}02$ which is the calculated value, based on the half-life of the neutron [40]. The vector part of the interaction does not seem to have any form factor value different from unity (i.e. the p and n act as if they were bare, as far as the weak interaction are concerned). However, the axial part of the interaction seems to exhibit a form factor effect.

The lifetime of the muon can be used to calculate the magnitude of $G_{\mu}$, and, as we have discussed previously, $C_{\mathrm{V}}$ and $C_{\mathrm{A}}$ are determined from $\beta$-decay experiments. When the appropriate electromagnetic radiative and Coulomb corrections are taken into account, we obtain [41]

$$\frac{G_{\mu} - C_{\mathrm{V}}}{G_{\mu}} \cong 0{\cdot}025. \tag{12.56}$$

12.10.2. *The universal weak interaction and the* W-*meson*

The implication is that there is a universal weak interaction [42]–[44] which we can express by writing the interaction Hamiltonian in the form

$$\mathscr{H}_{\mathrm{w}} = \frac{G}{\sqrt{2}}\int \mathrm{d}^3x J_{\alpha}(x)J^{\alpha+}(x). \tag{12.57}$$

The above represents the Fermi theory, in the form where all spinors are required to have the form $(1-\gamma_5)\Psi$. The current $J_\alpha(x)$, in the above, would be of the form

$$J_\alpha(x) = \overline{\Psi}_e\gamma_\alpha(1-\gamma_5)\Psi_\nu+\overline{\Psi}_\mu\gamma_\alpha(1-\gamma_5)\Psi_\nu+\overline{\Psi}_n\gamma_\alpha(1-\gamma_5)\Psi_p+$$
+terms corresponding to strange particle decays (which we have not yet considered), etc. (12.58)

The form of the weak Hamiltonian can be cast in a form suggesting a vector meson, coupled by a Yukawa type coupling,

$$\mathscr{H}_w \propto \sqrt{G}\, J_\alpha W^\alpha. \tag{12.59}$$

The observed four-particle interaction would then be explained by analogy to Moller scattering (i.e. a second-order process). These vector mesons, in addition to being heavy, would have to possess 'semi-weak' interaction properties, since, if they did not, $\beta$-decay rates would be faster than muon decay by a factor of $\dfrac{1}{G}$.

A lower limit on the mass of such a vector meson would be set by the K-meson mass, otherwise the weak decay process $K^+ \rightarrow W^+ + \gamma$ would be faster, by a factor $\dfrac{1}{G}$, than the observed process

$$K^+ \rightarrow \mu^+ + \nu. \tag{12.60}$$

Searches have been made for the W-meson, by utilizing high energy neutrino beams from high energy pion beam decays. We would expect to produce the W-meson in a process illustrated by Fig. 12.7, where the effective coupling constant at each vertex is $\sqrt{G}$. The expected $\mu^+e^-$ pairs for such a process were not seen, allowing [33a] a lower limit to be placed on the mass of the W-meson. Thus, we can place a lower limit on the mass of the W-meson, if

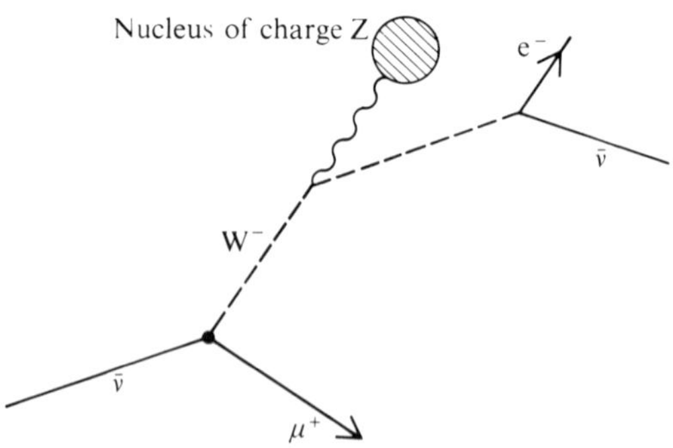


FIG. 12.7.

it exists, and the result is

$$m_W > 2 \text{ GeV}.$$

The intermediate vector boson can also be produced by a combination of strong and semi-weak interactions, which occur in first order in the semi-weak. See Marshak *et al.* [3] for a discussion of this.

## 12.11. Current analysis of the weak interactions

### 12.11.1. *Conserved or nearly (i.e. partially) conserved currents*

The fact that the weak interaction can be represented by the current–current form takes on particular significance when we consider the possibility that some components of this current may be conserved or nearly (i.e. partially) conserved. This suggests itself clearly for the vector component of the current, because of the observed equality of the vector coupling constants in weak interactions of nuclear $\beta$-decay and in the decay of the muon.

In the case of muon decay, even the axial vector coupling constant is the same. However, in ordinary $\beta$-decay the axial vector coupling constant, as we have seen, is about 18 per cent larger in magnitude. We might attribute this latter slight discrepancy to renormalization effects, since in $\beta$-decay, the fact that the neutron and proton can interact strongly with the pion could be expected to lead to renormalization effects on $C_A$ of even more than 18 per cent. Of course, the same renormalization effects could also be expected to drastically change the $C_V$, in $\beta$-decay, compared to muon decay. It was pointed out [45] that a straightforward way to obtain an additive conserved quantity, which is not subject to renormalization, is the imposition of the appropriate gauge-invariance requirement on the Lagrangian density. This naturally provides a definition for a conserved current.

The conservation laws for rigorously conserved quantities, like charge and baryon number, are obtained as follows. The integral over space of the zero-th component of the conserved quantity is equivalent to the conserved quantum number. As we shall see later, it appears that the hypothesis of a conserved vector current (CVC) has led to experimental predictions which are, in a number of cases, in agreement with the data. The axial current does not appear to be quite conserved since $C_A \approx -1{\cdot}18C_V$. Furthermore, it appears that, if the axial current were precisely conserved, the decay

$$\pi \to \mu + \nu$$

would be strictly forbidden for finite pion mass. This has led to the proposed partially conserved axial current hypothesis (PCAC).

For the present, let us proceed with the current analysis. So far we have considered leptonic currents, which we can denote as

$$J_l^\alpha(x) = \overline{\Psi}_e\gamma^\alpha(1-\gamma_5)\Psi_{\nu_e}(x)+\overline{\Psi}_\mu\gamma^\alpha(1-\gamma_5)\Psi_\nu(x), \tag{12.61}$$

and we have considered currents of strongly interacting particles, which we can refer to as

$$J_{st}^\alpha(x),$$

$J_{st}^\alpha(x)$ is, in general, composed of two components

$$J_{st,\,\Delta S=0}^\alpha(x) \quad \text{and} \quad J_{st,\Delta S\neq 0}^\alpha(x),$$

where, up till now, we have only considered $J_{st,\,\Delta S=0}^\alpha(x)$. Since it is observed experimentally, that strangeness either does not change, or changes at most by $\pm 1$ unit, in a weak decay, it is obvious that $J_{st,\,\Delta S=0}^\alpha(x)$, and $J_{st,\,\Delta S=\pm 1}^\alpha(x)$ are the only possibilities. The vector and axial currents which we analysed in beta and muon decay are $\Delta S = 0$ currents, and we can represent them by

$$J_{V,\,\Delta S=0}^\alpha \quad \text{and} \quad J_{A,\,\Delta S=0}^\alpha,$$

respectively. The sum of the vector and axial $\Delta S = 0$ currents (V–A) were represented by

$$J_{V,\,\Delta S=0}^\alpha + J_{A,\,\Delta S=0}^\alpha = \overline{\Psi}_n(x)\gamma^\alpha(1-\gamma_5)\Psi_p(x). \tag{12.62}$$

A leptonic weak interaction involves only leptons, for example,

$$\mu^+ = e^+ + \nu_e + \bar{\nu}_\mu. \tag{12.63}$$

A semi-leptonic interaction involves hadrons plus leptons, and is of the form

$$H \to H' + \begin{Bmatrix} e^\pm \\ \mu^\pm \end{Bmatrix} + \nu. \tag{12.64a}$$

for example,

$$n \to p + e^- + \bar{\nu}_e, \tag{12.64b}$$

$$K^+ \to \pi^0 + e^+ + \nu_e. \tag{12.64c}$$

When we consider what happened in the semi-leptonic interaction characterized by these currents, and look at the effect on the hadronic charge in going from the neutron ($\Psi_n$ on the left-hand side) to the transformed hadron, the proton ($\Psi_p$ on the right-hand side), we see that the hadronic charge† has been changed by $+1$ units. Thus the often-seen statement that $\Delta Q = +1$ for this current† refers to just this change of hadronic charge by the currents. If we wish to represent currents which change the charge

† These decays involve leptons, however, we ignore them in calculating $\Delta Q$, considering only the hadrons.

by $-1$ (i.e. $\Delta Q = -1$) these will be $J^{\alpha+}_{V,\Delta S=0}$ and $J^{\alpha+}_{A,\Delta S=0}$. The form of the (current–current) Hamiltonian which we have assumed for the weak interactions,

$$\frac{G}{\sqrt{2}}\int J_{\alpha}(x)J^{\alpha+}(x)\,d^3x, \tag{12.65}$$

and the necessity of rigorously conserving charge, limits our choice of terms in $J^{\alpha}_{st}$ to only those terms which have $\Delta Q = 1$ and, accordingly, $J^{\alpha+}_{st}$ terms also must have $\Delta Q = -1$.

In order to fit the experimental observations, we shall, just as in the $\Delta S = 0$ case, need both vector and axial terms in the $|\Delta S| = 1$ currents. This can be seen from the following examples.

In order to explain the commonly-observed decay of the $K^{\pm}$-mesons,

$$K^{\pm} \to \mu^{\pm} + \nu, \tag{12.66a}$$

we need

$$(\Psi_0 J^{\alpha}_{A,\,\Delta S=1}\Psi_K) \neq 0. \tag{12.66b}$$

The observation of the alternate, and less frequent, decay mode of the charged K-mesons,

$$K^{\pm} \to \pi^0 + e^{\pm} + \nu, \tag{12.67a}$$

requires that

$$(\Psi_{\pi^0} J^{\alpha}_{V,\,\Delta S=1}\Psi_K) \neq 0, \tag{12.67b}$$

and the observed decay rates in the above processes determine the values of their matrix elements.

### 12.11.2. *The $\Delta S = \Delta Q$ rule*

The experimental observation that, in all known cases, $|\Delta S| \leq 1$ (i.e. $|\Delta S| \neq 2$) was employed [46] to limit the possibilities of the structure of $|\Delta S| = 1$ leptonic currents. These considerations led to the $\Delta S = \Delta Q$ rule, which the experimental evidence confirms [47]. We can see from the following diagram that, if we allow $\Delta S = -\Delta Q$ currents such as $\Sigma^+ \to n$ to occur, how we can expect to generate, via the four-fermion interaction, the unobserved reaction $\Xi^- \to n + \pi^-$, which has $\Delta S = 2$. In Fig. 12.8 at step (1) there is a $\Xi^- \to K^- + \Sigma^0$, which is a strong interaction, at step (2) $\Sigma^0 \to \Sigma^+ + \pi^-$ (strong interaction), and at step (3) $K^- \to \bar{n} + \Sigma^-$ (forward). At step (4), if we allow $\Sigma^+ \to n$, which is a $\Delta S = -\Delta Q$ weak interaction current, as part of the four-fermion weak interaction, we predict the process $\Xi^- \to n + \pi^-$ will be generated via the four-fermion interaction. This $\Delta S = 2$ interaction has never been observed.

So far, the $\Delta S = \Delta Q$ rule has been confirmed by a number of experiments. For example,

$$\Sigma^- \to n + e^- + \bar{\nu}_e$$

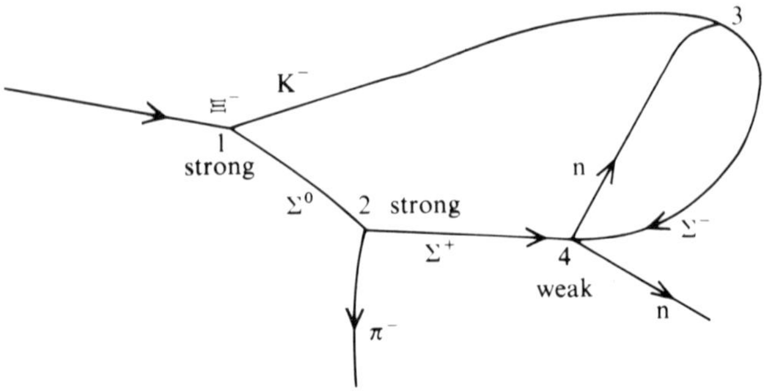


FIG. 12.8.

is observed to occur with a $(1{\cdot}06 \pm 0{\cdot}05) \times 10^{-3}$ branching ratio compared to $\approx 1$ for the highly dominant mode $\Sigma^- \to n + \pi^-$. For this decay, $\Delta S$ is $+1$, and $\Delta Q$ (applied to the hadrons on both sides) is also $+1$. On the other hand, the reaction

$$\Sigma^+ \to n + e^+ + \nu_e$$

is not observed. The branching ratio is $< 0{\cdot}7 \times 10^{-5}$ here, and $\Delta S = +1$, but $\Delta Q = -1$. Thus, $\Delta Q = -\Delta S$. The following $K^+$ decay reaction is not observed,

$$K^+ \to 2\pi^+ + e^- + \bar{\nu}_e.$$

The branching ratio is $< 7 \times 10^{-7}$, here $\Delta S = -1$, and $\Delta Q = +1$. Thus $\Delta S = -\Delta Q$. On the other hand, the very similar reaction,

$$K^+ \to (\pi^+ + \pi^-) + e^+ + \nu_e,$$

is observed. The branching ratio is $(3{\cdot}3 \pm 0{\cdot}3) \times 10^{-5}$. Here, $\Delta S = -1$ and $\Delta Q = -1$ (i.e., $\Delta S = \Delta Q$).

12.11.3. *The* $\Delta T = \frac{1}{2}$ *rule*

The $\Delta S = \Delta Q$ rule is a specialized form of another rule, the so-called $\Delta T = \frac{1}{2}$ rule. That such a rule might exist was strikingly implied by the enormous difference (about a factor of 500) in lifetimes between the K-meson decays.

$$K^0 \to \pi^+ + \pi^- (\tau \sim 10^{-10}\ \text{s}), \tag{12.68a}$$

$$K^+ \to \pi^+ + \pi^0 (\tau \sim 6 \times 10^{-8}\ \text{s}). \tag{12.68b}$$

Superficially, we would not expect much difference between the two rates. The clue as to where the difference lies was discovered in the change in isotopic spin between the initial and final states. For the $K^+$, since the final state must have spin zero, the only possibility which satisfies the required wave function symmetry is $T = 2$, $T_z = 1$ (i.e. the $T = 0$ state which is also possible on symmetry grounds cannot provide a charged final state). In the case of a $K_S^0$ decay into two pions, both $T = 0$ and $T = 2$ states are allowed. Therefore, in the $K^0$ pion decay, the possibilities are $\Delta T = \frac{1}{2}, \frac{3}{2}, \frac{5}{2}$†. In the $K^+$ decay, only $\frac{3}{2}$ and $\frac{5}{2}$ are possible. Thus, by requiring that $\Delta T = \frac{1}{2}$, we can explain the large difference in rate for the $K_S^0$ decay component, compared to the $K^+$ two-pion decay. This rule follows from the assumption that the weak interaction Hamiltonian transforms as an isospinor. The observed violation of the rule by the $K^+$ two-pion decay rate can be used to estimate the admixture of $\Delta T = \frac{3}{2}$ matrix elements. Electromagnetic radiative corrections would be expected to introduce some $\Delta T = \frac{3}{2}$ transitions, even if the original weak interaction Hamiltonian did not contain any of this component, but this would account for only a few per cent of the observed rate. The $\Delta T = \frac{1}{2}$ rule results in the prediction that in the $K_S^0$ decay into two $\pi$-mesons, the final predominant state involved is‡

$$|T = 0, T_z = 0\rangle = \frac{1}{\sqrt{3}}(\sqrt{2}\,|\pi^- + \pi^+\rangle - |\pi^0 + \pi^0\rangle), \qquad (12.69)$$

thus predicting that, in the $K_S^0$ two-pion decay, two-thirds of the time there is a $\pi^+ + \pi^-$ pair and only one-third of the time a $\pi^0 + \pi^0$ pair. Since the $K^+$ can also decay into two pions, there is a correction to the above prediction that allows it to vary by about $\pm 10$ per cent. The weighted experimental result is that the $\pi^+ + \pi^-$ decay occurs $(68{\cdot}7 \pm 0{\cdot}6)$ per cent while the $\pi^0 + \pi^0$ decay occurs $(31{\cdot}3 \pm 0{\cdot}6)$ per cent of the time (from the Particle Data Group), thus confirming the prediction. So far we have considered the $\Delta T = \frac{1}{2}$ rule in the case of non-leptonic interactions where the change in $\Delta T$ is clearly definable. In the case of semi-leptonic interactions, such as the decays

$$K^+ \rightarrow \pi^0 + \mu^+ + \nu_\mu, \qquad (12.70a)$$

$$K^0 \rightarrow \pi^- + \mu^+ + \nu_\mu, \qquad (12.70b)$$

$$K^0 \rightarrow \pi^+ + \mu^- + \bar{\nu}_\mu, \qquad (12.70c)$$

we can only associate isotopic spin with the pion, thus the extended meaning of the $\Delta T = \frac{1}{2}$ rule will be in comparing only the $T$ of the $K^{+}_{0}$ and the $T$ of the $\pi^{0\pm}$. Thus reactions (12.70a) and (12.70b) can proceed and satisfy the $\Delta T = \frac{1}{2}$ rule, but reaction (12.70c) cannot. We notice that the $\Delta T = \frac{1}{2}$ rule

† The $\Delta T$ is taken in the vector sense, and we are referring to the magnitude (i.e. both signs are allowed).

‡ The other possibility $|T = 2, T_z = 0\rangle$ was ruled out since $\Delta T = \frac{3}{2}$ would be required.

gives similar predictions to the $\Delta S = \Delta Q$ rule. The relationship is that the $\Delta T = \frac{1}{2}$ rule is stronger than the $\Delta S = \Delta Q$ rule, since the latter follows from the former. However, the $\Delta T = \frac{1}{2}$ rule is capable of predicting ratios of rates between the various allowed processes, while the $\Delta S = \Delta Q$ rule has no such rate-predictive power.

A simple method of calculating predictions of the $\Delta T = \frac{1}{2}$ rule is to introduce the so-called 'spurion' field, which transforms as an isospinor and represents the $\Delta T = \frac{1}{2}$ change, and nothing more. Then the decay can be treated as if isotopic spin were conserved, and various branching ratios worked out, using the standard techniques. The experimental observations generally agree with the $\Delta T = \frac{1}{2}$ rule, as in the just-treated case of the ratio of the two-pion decay rates of $K^+$ and $K^0$. Good examples of agreement with the $\Delta T = \frac{1}{2}$ rule are in non-leptonic $\Lambda^0$ and $\Sigma$ decays. Let us first consider the $\Lambda^0$ decays

$$\Lambda^0 \to \pi^- + p \tag{12.71a}$$

or

$$\Lambda^0 \to \pi^0 + n. \tag{12.71b}$$

As we can recall from the treatment of $\pi$–N scattering earlier in the book, the final states contain a mixture of only $T = \frac{3}{2}$ and $T = \frac{1}{2}$ states. Since the left-hand side contains only a lambda, for which $T = 0$, the $\Delta T = \frac{1}{2}$ rule would predict that the right-hand side contains only the $T = \frac{1}{2}$ state. From the decomposition of the $\pi$–N state with $T = \frac{1}{2}$, $T_z = -\frac{1}{2}$, it is easy to show that the ratios of

$$\frac{\Lambda^0 \to n + \pi^0}{\Lambda^0 \to p + \pi^-} = \tfrac{1}{2}. \tag{12.72}$$

The foregoing follows from

$$A(\Lambda^0 \to \pi^0 + n) = \sqrt{\tfrac{2}{3}}\, A_{\frac{3}{2}} - \sqrt{\tfrac{1}{3}}\, A_{\frac{1}{2}} \tag{12.73a}$$

$$A(\Lambda^0 \to \pi^- + p) = \sqrt{\tfrac{1}{3}}\, A_{\frac{3}{2}} + \sqrt{\tfrac{2}{3}}\, A_{\frac{1}{2}}. \tag{12.73b}$$

Thus, if we set $A_{\frac{3}{2}} = 0$,

$$\frac{A(\Lambda^0 \to \pi^0 + n)}{A(\Lambda^0 \to \pi^- + p)} = -\sqrt{\tfrac{1}{2}}, \tag{12.73c}$$

or the ratio of the decay rates which are proportional to the absolute square of the ratio of the amplitudes becomes

$$\left|\frac{A(\Lambda^0 \to \pi^0 + n)}{A(\Lambda^0 \to \pi^- + p)}\right|^2 = \tfrac{1}{2}. \tag{12.74a}$$

Let us define

$$\frac{\Lambda^0 \to n+\pi^0}{\Lambda^0 \to p+\pi^-} = R_\Lambda. \tag{12.74b}$$

Then, if the predominant amplitude were $A_{\frac{3}{2}}$ ($A_{\frac{1}{2}} = 0$), the prediction for the ratio $R_\Lambda$ would be $\frac{1}{2}$. Experimentally, we find from the Particle Data Group table, $R_\Lambda = \dfrac{34{\cdot}7 \pm 1{\cdot}3}{65{\cdot}3 \pm 1{\cdot}3} = 0{\cdot}53 \pm 0{\cdot}02$ in excellent agreement with the $\Delta T = \frac{1}{2}$ rule.

A second check of the $\Delta T = \frac{1}{2}$ rule is obtained from the fact that the general form of the decay amplitudes are [48]

$$A(\Lambda \to \pi^0+n) = (S_0+P\boldsymbol{\sigma}.\hat{\mathbf{n}}),$$
$$A(\Lambda^0 \to \pi^-+p) = (S_-+P\boldsymbol{\sigma}.\hat{\mathbf{n}}),$$

where $\boldsymbol{\sigma}$ is the polarization, and $\hat{\mathbf{n}}$ is the decay proton, or neutron, direction in the $\Lambda^0$ rest system.

For the $\Delta T = \frac{1}{2}$ rule, only the $T = \frac{1}{2}$ amplitude contributes, and the prediction is $S_0/S_- = P_0/P_-$, which agrees with the experimental observations. Let us now apply similar considerations to the $\Sigma$ non-leptonic decays. Let the decay amplitudes for

$$\Sigma^+ \to n+\pi^+ \quad \text{be denoted by } \Sigma_+^+, \tag{12.75a}$$
$$\Sigma^+ \to p+\pi^0 \quad \text{be denoted by } \Sigma_0^+/\sqrt{2}, \tag{12.75b}$$
$$\Sigma^- \to n+\pi^- \quad \text{be denoted by } \Sigma_-^-. \tag{12.75c}$$

Since the $\Sigma$ has $T = 1$, the $\Delta T = \frac{1}{2}$ rule allows both $T = \frac{1}{2}$ and $T = \frac{3}{2}$ final states. When the $\Delta T = \frac{1}{2}$ rule is applied to these interactions, and the above three amplitudes are plotted as vectors in the $S$, $P$ cartesian frame, where $S$ is the spin-independent, and $P$ the spin-dependent, coefficient (as in the case of $\Lambda^0$ decay) in the amplitudes, the $\Delta T = \frac{1}{2}$ rule predicts that we obtain a closed triangle. Whereas earlier experiments did not agree with the $\Delta T = \frac{1}{2}$ rule, as we can see in Fig. 12.9, the more recent experimental data closes the triangle very well and, thus, confirm the $\Delta T = \frac{1}{2}$ rule prediction.

The data on non-leptonic decay modes of $\Xi^0$ also agree with the $\Delta T = \frac{1}{2}$ rule prediction, which is

$$\Gamma\frac{(\Xi^0 \to \Lambda^0+\pi^0)}{(\Xi^- \to \Lambda^0+\pi^-)} = \frac{1}{2}. \tag{12.76}$$

The weighted averages of measurements of the $\Xi^0$ and $\Xi^-$ lifetime can be used to determine this ratio, even though there are some discrepancies between individual measurements, and we obtain (from the Particle Data

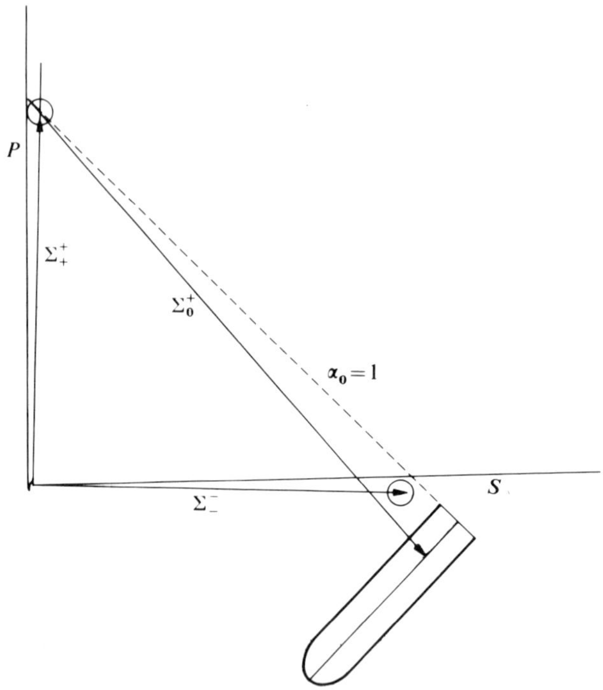


FIG. 12.9. Check on the $\Delta T = \frac{1}{2}$ rule. The three vector quantities shown should form a closed triangle according to the prediction of the $\Delta T = \frac{1}{2}$ rule. (From [48].)

Group table)

$$\tau_{\Xi^-} = (1{\cdot}66 \pm 0{\cdot}04) \times 10^{-10} \text{ s}, \tag{12.77a}$$

$$\tau_{\Xi^0} = (3{\cdot}03 \pm 0{\cdot}18) \times 10^{-10} \text{ s}. \tag{12.77b}$$

Thus

$$\frac{\Gamma(\Xi^0 \to \Lambda^0 + \pi^0)}{\Gamma(\Xi^- \to \Lambda^0 + \pi^-)} = \frac{\tau_-}{\tau_0} = \frac{1{\cdot}66 \pm 0{\cdot}04}{3{\cdot}03 \pm 0{\cdot}18} \approx 0{\cdot}55 \pm 0{\cdot}02, \tag{12.78}$$

which represents good agreement. However, an admixture of $\text{Re}\left(\frac{A_{\frac{3}{2}}}{A_{\frac{1}{2}}}\right)$ of about 0·02 would be accommodated by the present error limits on the data.

The decay parameters are

$$\alpha_{\Xi^-} = -0{\cdot}41 \pm 0{\cdot}04,$$

$$\alpha_{\Xi^0} = -0{\cdot}35 \pm 0{\cdot}08.$$

The $\Delta T = \frac{1}{2}$ rule predicts

$$\alpha_{\Xi^-} = \alpha_{\Xi^0}.$$

Thus, there is agreement within experimental error.

The $\Delta T = \frac{1}{2}$ rule can be used, also, to predict ratios in the three-pion decay modes of $K^+$ and $K_L$. In two cases the agreement is excellent, whereas in two other cases it is off by about 10–20 per cent, when compared with experiments [49].

The $\Delta T = \frac{1}{2}$ rule is also satisfied in the following leptonic decay cases where the $K^\pm$ meson decays to

$$\mu^\pm + \begin{pmatrix} \nu_\mu \\ \bar{\nu}_\mu \end{pmatrix} \quad \text{or} \quad e^\pm + \begin{pmatrix} \nu_e \\ \bar{\nu}_e \end{pmatrix}.$$

The $\Delta T = \frac{1}{2}$ rule (including a small correction due to phase-space) predicts that

$$\frac{2\Gamma(K^+ \to \pi^0 l^+ \nu)}{\Gamma(K_L \to \pi^+ l^- \bar{\nu}) + \Gamma(K_L \to \pi^- l^+ \nu)} = 0{\cdot}96.$$

A compilation of the relevant rates [50] led to an experimental value for the above ratio of $1{\cdot}09 \pm 0{\cdot}04$, which differs by 3·3 standard deviations. This is possibly an indication of, but hardly a basis for, concluding a small violation of the $\Delta T = \frac{1}{2}$ rule.

### 12.12. Decay of the $K^0$ meson

In a strong interaction such as

$$\pi^- + p \to K^0 + \Lambda, \tag{12.79}$$

the conservation of strangeness means that the $K^0$ (at the time of creation) has a definite strangeness eigenvalue $S = +1$.

Let us now consider the following reaction

$$K^- + p \to \bar{K}^0 + n. \tag{12.80}$$

Owing to conservation of strangeness, the $\bar{K}^0$ has a definite strangeness eigenvalue of $S = -1$. However, the weak interactions violate strangeness and, in fact, that with a lifetime of about $8{\cdot}6 \times 10^{-10}$ s, the $K^0$ (and $\bar{K}^0$) decays to a $\pi^+ + \pi^-$ pair. Thus, we can anticipate that, under the influence of the weak interactions,

$$K^0 \rightleftarrows \pi^+ + \pi^- \rightleftarrows \bar{K}^0. \tag{12.81}$$

There is a small probability of these transformations occurring between the $K^0$ and $\bar{K}^0$, due to the expected close similarity of the mass and lifetimes expected under CPT, for the $K^0$ and $\bar{K}^0$. This will occur very slowly since we are talking about a $\Delta S = \pm 2$ process.

It was pointed out [50] that we could thus expect that, after a time characteristic of the weak interactions, there would result a state which although originally a pure $K^0$ state formed in the strong interaction would at a later time, be a mixture of $K^0$ and $\bar{K}^0$ (a linear combination) and, similarly, for a state which was originally a $\bar{K}^0$. Thus, we might expect that two linear combinations might exist.

Since the $|K^0\rangle$ and $|\bar{K}^0\rangle$ are strong interaction eigenstates, of which the phases can be arbitrarily chosen. We can choose the phases so that

$$C\,|K^0\rangle = |\bar{K}^0\rangle, \tag{12.82a}$$

$$C\,|\bar{K}^0\rangle = |K^0\rangle. \tag{12.82b}$$

From the above, it follows that

$$CP\,|K^0\rangle = -|\bar{K}^0\rangle, \tag{12.83a}$$

since, as we have seen previously, applying the $P$ operator to a pseudoscalar particle like the $|K^0\rangle$ results in a minus sign. Similarly,

$$CP\,|\bar{K}^0\rangle = -|K^0\rangle. \tag{12.83b}$$

Thus the $|K^0\rangle$ and $|\bar{K}^0\rangle$ are definitely not eigenstates of the $CP$ operator.

Since the $|K^0\rangle$ and $|\bar{K}^0\rangle$ have zero spin, the two-pion decay state must be an $S$-state (and this is verified by experimental observations). Thus, in order to satisfy Bose statistics, the isotopic spin wave function must be even (i.e. $T = 0$ or $T = 2$). Both of these isotopic spin wave functions for two pions are even under charge conjugation as can be seen by writing out their wave functions and applying the $C$ operation.

The parity operator, applied to a two-spin $S$-state, leads to a factor of $+1$. Thus a two-pion $J = 0$ state is an eigenstate of $CP$, with eigenvalue $= +1$. However, in the appropriate eigenstates for the weak decays, charge conjugation × parity (CP) is, to a high degree of approximation, conserved.† Thus, the $|K^0\rangle$ and $|\bar{K}^0\rangle$ cannot (under the assumption of CP invariance) be the states that are decaying into pion pairs.

Hence, we might expect that, under the weak interaction, the $K^0$ and $\bar{K}^0$ states transform into the two linear combinations which essentially represent

† Until recent experiments demonstrated a violation of about $10^{-3}$, it was assumed that CP conservation was exact in weak interactions. This would be expected, for example, for the current–current form of the weak interaction Hamiltonian. With both parity and charge conjugation violation, natural laws would be different in a left- or right-handed coordinate system in a universe. They would also be different in an anti-universe, from those in a universe. However, if CP invariance is valid, then natural laws would be identical in an anti-universe, which is a mirror image of the universe.

sums and differences of these original strong interaction states,†

$$|K_L^0\rangle = \frac{p\,|K^0\rangle + q\,|\bar{K}^0\rangle}{(|p|^2 + |q|^2)^{\frac{1}{2}}}, \tag{12.84}$$

$$|K_S^0\rangle = \frac{r\,|K^0\rangle + s\,|\bar{K}^0\rangle}{(|r|^2 + |s|^2)^{\frac{1}{2}}}. \tag{12.85}$$

Three of the four phases of $|K_S^0\rangle$, $|K_L^0\rangle$, $|K^0\rangle$, and $|\bar{K}^0\rangle$ can be arbitrarily selected. Thus, we can select $p$, $q$, and $r$ all as real and positive.‡ Let us first require that CP is conserved. Then $|K_S^0\rangle$ and $|K_L^0\rangle$ will be eigenstates of CP. For this case, denote $|K_S^0\rangle \equiv |K_1\rangle$, and $|K_L^0\rangle \equiv |K_2^0\rangle$. Thus, we obtain

$$|K_1^0\rangle = \frac{1}{\sqrt{2}}(|K^0\rangle - |\bar{K}^0\rangle), \tag{12.86}$$

$$|K_2^0\rangle = \frac{1}{\sqrt{2}}(|K^0\rangle + |\bar{K}^0\rangle). \tag{12.87}$$

If we now apply CP to $|K_1^0\rangle$ and $|K_2^0\rangle$, we obtain the following

$$CP\,|K_1^0\rangle = \frac{1}{\sqrt{2}}(-|\bar{K}^0\rangle + |K^0\rangle) = |K_1^0\rangle, \tag{12.88}$$

$$CP\,|K_2^0\rangle = \frac{1}{\sqrt{2}}(-|\bar{K}^0\rangle - |K^0\rangle) = -|K_2^0\rangle. \tag{12.89}$$

† We are assuming CPT invariance, but not CP invariance. We do not, at present, have a consistent theoretical framework to discuss interactions within which we do not require CPT invariance. All experimental evidence to date is consistent with this hypothesis.

‡ This will preserve our previous results

$$CP\,|K^0\rangle = -|\bar{K}^0\rangle,$$

and

$$CP\,|\bar{K}^0\rangle = -|K^0\rangle.$$

Furthermore, we can normalize, so that

$$|p|^2 + |q|^2 = 1,$$
$$|r|^2 + |s|^2 = 1.$$

We are at present, using the phase conventions of Bell and Steinberger (1965) [51*a*]. As they point out, a common phase convention is obtained from the present one, by the substitution

$$|K\rangle_{\text{present}} = i\,|K\rangle_{\text{others}},$$
$$|\bar{K}\rangle_{\text{present}} = -i\,|\bar{K}\rangle_{\text{others}}.$$

This field has been characterized by an unusual variety of notations (almost as many as authors), and the reader is referred to the article by Okun and Rubbia [51*b*] for some correlation between the various notations employed.

Hence, the $|K_1^0\rangle$ is the only state which can decay into a two-pion $J = 0$ state, which is even under CP (i.e. has an eigenvalue $= +1$).

In the case of the three-pion final state, the $P$ operator acting on a three-pion (pseudoscalar particle) state, with all relative angular momentum zero (i.e. a relative $S$-state), produces the same final state multiplied by $-1$. The state is even under the $C$ operator, which exchanges $\pi^+$ and $\pi^-$, and leaves $\pi^0$ unchanged.

Thus $CP = -1$ for the three-pion $S$-state decay mode and, hence, $K_1^0$ cannot decay into this state, but the $K_2^0$ can. Both $|K_1^0\rangle$ and $|K_2^0\rangle$ can, in principle, decay into the $\pi^\pm + e^\mp + \nu$ state. However, in the case of $|K_1^0\rangle$, the phase-space, and probability, for the two-pion decay is much larger and, thus, this is the only decay effectively observed.

Thus we see that the $|K_1^0\rangle$, and $|K_2^0\rangle$, have entirely different properties for their decay states and, hence, we should expect considerably different lifetimes, with the $|K_1^0\rangle$ exhibiting a much shorter lifetime. Experimentally, this was observed to be the case. In reactions which produce a $|K^0\rangle$ such as, for example,

$$\pi^- + p \rightarrow |K^0\rangle + \Lambda, \tag{12.90}$$

we can, using eqns (12.86) and (12.87), express the $|K^0\rangle$ in terms of $|K_1^0\rangle$ and $|K_2^0\rangle$ as follows,

$$|K^0\rangle = \frac{1}{\sqrt{2}}(|K_1^0\rangle + |K_2^0\rangle). \tag{12.91}$$

Thus, we would expect to find the decay products of the $|K^0\rangle$ appear for about half the time as pion pairs with a short lifetime, phenomenologically referred to as,†

$$|K_S^0\rangle \approx |K_1^0\rangle, \tag{12.92a}$$

and this has been observed to be the case, the $K_S^0$ being a $T(J^P) = \frac{1}{2}(0^-)$ particle, with an observed lifetime of $(0{\cdot}862 \pm 0{\cdot}006) \times 10^{-10}$ s. The decay products are $\pi^+ + \pi^-$ $(68{\cdot}7 \pm 0{\cdot}6)$ per cent, and a $\pi^0 + \pi^0$ pair $(31{\cdot}3 \pm 0{\cdot}6)$ per cent of the time, consistent with the prediction of the $\Delta T = \frac{1}{2}$ rule.

It has also been observed that there is an associated much longer-lived component of about equal intensity‡

$$|K_L^0\rangle \approx |K_2^0\rangle, \tag{12.92b}$$

† $|K_S^0\rangle$ would exactly equal $|K_1^0\rangle$ if CP were rigorously conserved in weak interactions. However, there is a slight difference due to the small observed CP violation effects.

‡ The $|K_L^0\rangle$ would exactly equal the $|K_2^0\rangle$ if CP were rigorously conserved in weak interactions. However, there is a slight difference due to the small observed CP violation effects (See Particle Data Group tables for its characteristics).

with a lifetime of $(5{\cdot}38 \pm 0{\cdot}19) \times 10^{-8}$ s. The most important decay modes of this are

$$\pi^0\pi^0\pi^0 \ (21{\cdot}5 \pm 0{\cdot}7 \text{ per cent}), \tag{12.93a}$$

$$\pi^+\pi^-\pi^0 \ (12{\cdot}6 \pm 0{\cdot}3 \text{ per cent}), \tag{12.93b}$$

$$\pi\mu\nu \ (26{\cdot}8 \pm 0{\cdot}7 \text{ per cent}), \tag{12.93c}$$

$$\pi e\nu \ (38{\cdot}8 \pm 0{\cdot}8 \text{ per cent}), \tag{12.93d}$$

and the recently-observed CP violating mode

$$\pi^+\pi^- \ (0{\cdot}157 \pm 0{\cdot}005 \text{ per cent}).$$

On the basis of the experiments, it has been verified that the initial $|K^0\rangle$, or $|\bar{K}^0\rangle$, state is represented by approximately equal parts of $|K_S^0\rangle$ and $|K_L^0\rangle$, and that the equality is checked experimentally to within a few per cent.

## 12.13. C, P, and CP violations in weak interactions

### 12.13.1. *C and P violations*

As we have already seen, charge conjugation (C) and parity (P) are violated in weak interactions. However, in our studies of the helicities of $\mu$ and $\nu$ (in $\pi \rightarrow \mu + \nu$), we discovered that the helicity rule observed conserved CP. This is a general expectation for the two-component neutrino theory or the V–A interaction and, until recently, all experimental evidence was consistent with CP conservation, which we assumed in our treatment of the $K^0$ meson decay.

### 12.13.2. *CP violation*

In 1964, an experiment (Christenson *et al.*) [52] revealed that the decay

$$|K_L^0\rangle \rightarrow \pi^+ + \pi^- \tag{12.94}$$

occurred about once in each 500 $|K_L^0\rangle$ decays.

Figure 12.10 shows the experimental arrangement. A $|K_L^0\rangle$ beam was formed at the Brookhaven AGS, by allowing so long a decay path in a neutral beam that, effectively, all $|K_S^0\rangle$ had decayed. Two magnetic spectrometers with spark chamber detectors were arranged symmetrically around the beam line as shown in Fig. 12.10, and were used to measure both the direction, and momenta, of the two pions. The trigger for the spark chambers required a coincidence between the Cerenkov and scintillation counters, behind the spark chambers. The water Cerenkov counters selected candidates for pions, or lighter particles. Helium was used to fill the spectrometers, to minimize the possibility of interactions. In order to check the performance

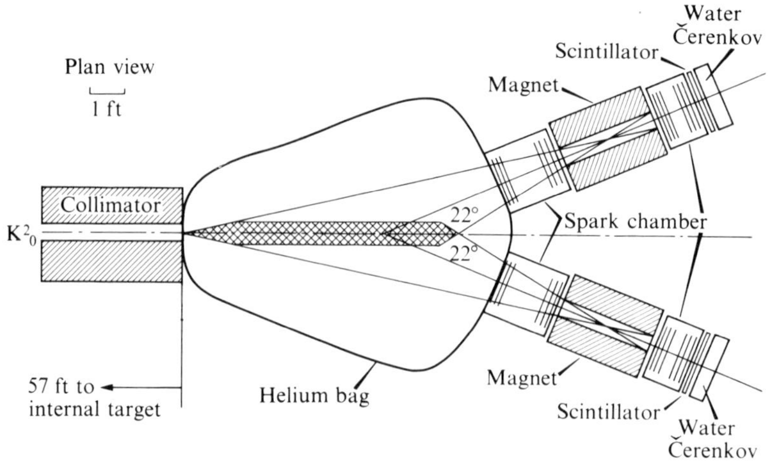


FIG. 12.10. Plan view of the detector arrangement. (From Christenson *et al.* [52].)

of the apparatus and the analysis, a $K_S^0$ beam was obtained by using a thick slab of copper to regenerate† some $K_S^0$. The pion-pair decays were identified by assuming that there were only two pions in the decay, and constructing the $\pi$–$\pi$ invariant mass and, also, determining the angle to the beam of the resultant vector-momentum of the two particles. Observing the two charged particles from the much more abundant decays,

$$K_L^0 \to \pi + e + \nu \tag{12.95a}$$
$$\to \pi + \mu + \nu \tag{12.95b}$$
$$\to \pi^+ \pi^- + \pi^0,$$

would not give the $K^0$ rest mass and, in general, would give a $\cos\theta$ quite different from 1, whereas for the two-pion decay of the $K_L^0$, $\cos\theta$ would be expected to be very close to 1, since the original $K_L^0$ was well-defined in beam direction. It was found that, near the forward direction (events for which $\cos\theta > 0{\cdot}9995$), there were about $(45 \pm 9)$ events out of an estimated number of 22 700 $K_L$ decays. Thus,

$$\frac{K_L \to \pi^+ + \pi^-}{K_L \to \text{all modes involving charged particles}} = (2{\cdot}0 \pm 0{\cdot}4) \times 10^{-3}. \tag{12.95c}$$

† This regeneration effect will be discussed later.

Abashian *et al.* (1964) [53] also came to similar conclusions. Various experiments [54]–[58] have, since, measured the ratio

$$\frac{K_L \to \pi^+ + \pi^-}{K_L \to \text{all modes involving charged particles}}. \tag{12.96}$$

A weighted average, from the Particle Data Group tables, is

$$\frac{K_L \to \pi^+ + \pi^-}{\text{all charged modes}} = \frac{(0{\cdot}157 \pm 0{\cdot}004) \text{ per cent}}{(88{\cdot}5 \pm 0{\cdot}7) \text{ per cent}} = (1{\cdot}79 \pm 0{\cdot}05) \times 10^{-3}.$$

12.13.3. *Phenomenological analysis of CP violating experiments*

A commonly-defined useful quantity is the ratio of decay amplitudes

$$\eta_{+-} = \frac{\langle \pi^+\pi^- |T| K_L^0\rangle}{\langle \pi^+\pi^- |T| K_S^0\rangle}, \tag{12.97}$$

which can be computed from the decay ratios and the decay lifetimes, obtaining

$$|\eta_{+-}| = (1{\cdot}92 \pm 0{\cdot}05). \tag{12.98}$$

Owing to the small observed failure in the CP conservation prediction that $\eta_{+-} = 0$, we must modify our treatment somewhat, restricting ourselves to the assumption of only CPT invariance. We can then express the relationships between the various $K^0$-meson states as follows [48], [51]

$$|K_S^0\rangle = \frac{|K_1^0\rangle + \epsilon |K_2^0\rangle}{(1+|\epsilon|^2)^{\frac{1}{2}}}, \tag{12.99a}$$

$$|K_L^0\rangle = \frac{|K_2^0\rangle + \epsilon |K_1^0\rangle}{(1+|\epsilon|^2)^{\frac{1}{2}}}, \tag{12.99b}$$

where $\epsilon \approx 10^{-3}$. The parameter $\epsilon$ measures the small fractional admixture of the opposite CP eigenvalue state, due to CP violation. The parameter $\epsilon$ is a function of $p$ and $q$, and can be shown to be

$$\epsilon = \frac{p-q}{p+q}. \tag{12.100}$$

The parameter $\eta_{+-}$, which we introduced, can be expressed as

$$\eta_{+-} = |\eta_{+-}|\, e^{i\phi_{+-}} \equiv \frac{\langle \pi^+\pi^- |T| K_L^0\rangle}{\langle \pi^+\pi^- |T| K_S^0\rangle}. \tag{12.101a}$$

Similarly, let us introduce the parameter

$$\eta_{00} = \frac{\langle \pi^0\pi^0 \,|T|\, \mathrm{K_L^0}\rangle}{\langle \pi^0\pi^0 \,|T|\, \mathrm{K_S^0}\rangle} = |\eta_{00}|\, \mathrm{e}^{\mathrm{i}\phi_{00}}. \tag{12.101b}$$

Let us consider the two possible isotopic spin states for the two-pion decay, namely, $T = 0$ and $T = 2$. Then we can define

$$\epsilon_0 = \frac{\langle 2\pi_{T=0} \,|T|\, \mathrm{K_L^0}\rangle}{\langle 2\pi_{T=0} \,|T|\, \mathrm{K_S^0}\rangle}, \tag{12.102a}$$

$$\epsilon_2 = \frac{\langle 2\pi_{T=2} \,|T|\, \mathrm{K_L^0}\rangle}{\sqrt{2}\,\langle 2\pi_{T=0} \,|T|\, \mathrm{K_S^0}\rangle}, \tag{12.102b}$$

and the ratio

$$\omega = \frac{\langle 2\pi_{T=2} \,|T|\, \mathrm{K_S^0}\rangle}{\langle 2\pi_{T=0} \,|T|\, \mathrm{K_S^0}\rangle}. \tag{12.102c}$$

Utilizing the Clebsch–Gordan coefficients for the decomposition of the initial $2\pi$ states, into $T = 0$ and $T = 2$ states, we obtain

$$\langle \pi^+\pi^-| = \sqrt{\tfrac{2}{3}}\,\langle 2\pi_{T=0}| + \sqrt{\tfrac{1}{3}}\,\langle 2\pi_{T=2}|, \tag{12.103a}$$

$$\langle \pi^0\pi^0| = -\sqrt{\tfrac{1}{3}}\,\langle 2\pi_{T=0}| + \sqrt{\tfrac{2}{3}}\,\langle 2\pi_{T=2}|. \tag{12.103b}$$

Substituting eqn (12.103a) into eqn (12.101a), we obtain

$$\eta_{+-} = \frac{\sqrt{\tfrac{2}{3}}\,\langle 2\pi_{T=0} \,|T|\, \mathrm{K_L^0}\rangle + \sqrt{\tfrac{1}{3}}\,\langle 2\pi_{T=2} \,|T|\, \mathrm{K_L^0}\rangle}{\sqrt{\tfrac{2}{3}}\,\langle 2\pi_{T=0} \,|T|\, \mathrm{K_S^0}\rangle + \sqrt{\tfrac{1}{3}}\,\langle 2\pi_{T=0} \,|T|\, \mathrm{K_S^0}\rangle}. \tag{12.104}$$

Dividing numerator and denominator by $\sqrt{\tfrac{2}{3}}\,\langle 2\pi_{T=0} \,|T|\, \mathrm{K_S^0}\rangle$, we obtain

$$\frac{\dfrac{\langle 2\pi_{T=0} \,|T|\, \mathrm{K_L^0}\rangle}{\langle 2\pi_{T=0} \,|T|\, \mathrm{K_S^0}\rangle} + \dfrac{\langle 2\pi_{T=2} \,|T|\, \mathrm{K_L^0}\rangle}{\sqrt{2}\,\langle 2\pi_{T=0} \,|T|\, \mathrm{K_S^0}\rangle}}{1 + \dfrac{1}{\sqrt{2}}\dfrac{\langle 2\pi_{T=2} \,|T|\, \mathrm{K_S^0}\rangle}{\langle 2\pi_{T=0} \,|T|\, \mathrm{K_S^0}\rangle}}. \tag{12.105}$$

Using the definitions in eqns (12.102), we obtain

$$\eta_{+-} = \frac{\epsilon_0 + \epsilon_2}{\left(1 + \dfrac{\omega}{\sqrt{2}}\right)}. \tag{12.106}$$

Similarly, by substituting eqn (12.103b) into eqn (12.101b), we obtain

$$\eta_{00} = \frac{-\sqrt{\tfrac{1}{3}}\,\langle 2\pi_{T=0} \,|T|\, \mathrm{K_L^0}\rangle + \sqrt{\tfrac{2}{3}}\,\langle 2\pi_{T=2} \,|T|\, \mathrm{K_L^0}\rangle}{-\sqrt{\tfrac{1}{3}}\,\langle 2\pi_{T=0} \,|T|\, \mathrm{K_S^0}\rangle + \sqrt{\tfrac{2}{3}}\,\langle 2\pi_{T=2} \,|T|\, \mathrm{K_S^0}\rangle}. \tag{12.107}$$

Dividing the numerator and denominator by $\sqrt{\frac{1}{3}}\,\langle 2\pi_{T=0}\,|T|\,\mathrm{K}_\mathrm{S}^0\rangle$, we obtain

$$\eta_{00} = \frac{\dfrac{\langle 2\pi_{T=0}\,|T|\,\mathrm{K}_\mathrm{L}^0\rangle}{\langle 2\pi_{T=0}\,|T|\,\mathrm{K}_\mathrm{S}^0\rangle} - \dfrac{\sqrt{2}\,\langle 2\pi_{T=2}\,|T|\,\mathrm{K}_\mathrm{L}^0\rangle}{\langle 2\pi_{T=0}\,|T|\,\mathrm{K}_\mathrm{S}^0\rangle}}{1 - \dfrac{\sqrt{2}\,\langle 2\pi_{T=2}\,|T|\,\mathrm{K}_\mathrm{S}^0\rangle}{\langle 2\pi_{T=0}\,|T|\,\mathrm{K}_\mathrm{S}^0\rangle}}. \tag{12.108}$$

Thus,

$$\eta_{00} = \frac{\epsilon_0 - 2\epsilon_2}{(1-\omega\sqrt{2})}. \tag{12.109}$$

Since analysis of experiments measuring the branching ratio

$$\frac{\mathrm{K_S} \to \pi^0\pi^0}{\mathrm{K_S} \to 2\pi}, \tag{12.110}$$

give the result that $|\omega| \ll 1$, we shall approximate by neglecting $\omega$, and obtain

$$\eta_{+-} = \epsilon_0 + \epsilon_2, \tag{12.111}$$

$$\eta_{00} = \epsilon_0 - 2\epsilon_2, \tag{12.112}$$

from which we obtain

$$2\eta_{+-} + \eta_{00} = 3\epsilon_0. \tag{12.113}$$

This relation is referred to as the Wu–Yang triangle (Wu and Yang 1964) [59].

It is convenient to relate the transition matrix element for $\mathrm{K}^0 \to 2\pi$ and $\overline{\mathrm{K}}^0 \to 2\pi$ in terms of the phase shifts $\delta_T$ for $\pi$–$\pi$ scattering at the c.m.s. energy which the two pions have after the $\mathrm{K}^0$ decays. This will allow us to obtain predictions, involving these phase shifts, from weak interaction experiments, and cross-check these predictions with information on $\delta_T$, obtained in strong interactions. Thus, we can define†

$$\langle 2\pi_{T=0}\,|\mathbf{T}|\,\mathrm{K}^0\rangle = A_0 \mathrm{e}^{\mathrm{i}\delta_0}, \tag{12.114}$$

$$\langle 2\pi_{T=2}\,|\mathbf{T}|\,\mathrm{K}^0\rangle = A_2 \mathrm{e}^{\mathrm{i}\delta_2}, \tag{12.115}$$

$$\langle 2\pi_{T=0}\,|\mathbf{T}|\,\overline{\mathrm{K}}^0\rangle = A_0^* \mathrm{e}^{\mathrm{i}\delta_0}, \tag{12.116}$$

$$\langle 2\pi_{T=2}\,|\mathbf{T}|\,\overline{\mathrm{K}}^0\rangle = A_2^* \mathrm{e}^{\mathrm{i}\delta_2}. \tag{12.117}$$

From these definitions, we can obtain

$$\langle 2\pi_{T=0}\,|\mathbf{T}|\,\mathrm{K}_1^0\rangle = \sqrt{2}\,\mathrm{Re}\,A_0 \mathrm{e}^{\mathrm{i}\delta_0}, \tag{12.118}$$

$$\langle 2\pi_{T=2}\,|\mathbf{T}|\,\mathrm{K}_1^0\rangle = \sqrt{2}\,\mathrm{Re}\,A_2 \mathrm{e}^{\mathrm{i}\delta_2}, \tag{12.119}$$

$$\langle 2\pi_{T=0}\,|\mathbf{T}|\,\mathrm{K}_2^0\rangle = \mathrm{i}\sqrt{2}\,\mathrm{Im}\,A_0 \mathrm{e}^{\mathrm{i}\delta_0}, \tag{12.120}$$

$$\langle 2\pi_{T=2}\,|\mathbf{T}|\,\mathrm{K}_2^0\rangle = \mathrm{i}\sqrt{2}\,\mathrm{Im}\,A_2 \mathrm{e}^{\mathrm{i}\delta_2}. \tag{12.121}$$

† The $T = 0$ or 2 subscript refers to isotopic spin, whereas the $\langle|\mathbf{T}|\rangle$ refers to the $T$-matrices, which, to avoid confusion, are printed bold.

Using the foregoing, we can express $\epsilon_0$, $\epsilon_2$, and $\omega$ in terms of $A_0$, $A_2$, $\delta_0$, and $\delta_2$ as follows,

$$\epsilon_0 = \frac{\epsilon \text{ Re } A_0 + \text{i Im } A_0}{\text{Re } A_0 + \text{i } \epsilon \text{ Im } A_0}, \tag{12.122}$$

$$\epsilon_2 = \frac{1}{\sqrt{2}}\left(\frac{\epsilon \text{ Re } A_2 + \text{i Im } A_2}{\text{Re } A_0 + \text{i } \epsilon \text{ Im } A_0}\right)\text{e}^{\text{i}(\delta_2 - \delta_0)}, \tag{12.123}$$

$$\omega = \left(\frac{\text{Re } A_2 + \text{i } \epsilon \text{ Im } A_2}{\text{Re } A_0 + \text{i } \epsilon \text{ Im } A_0}\right)\text{e}^{\text{i}(\delta_2 - \delta_0)}. \tag{12.124}$$

If the phases can also be selected so as to make $A_0$ entirely real, i.e.

$$\text{Im } A_0 = 0, \quad \text{(Wu and Yang 1964)} \tag{12.125}$$

then eqns (12.122–12.124) become

$$\epsilon_0 = \epsilon \tag{12.126a}$$

$$\epsilon_2 = \frac{1}{\sqrt{2}}\left(\frac{\epsilon \text{ Re } A_2 + \text{i Im } A_2}{A_0}\right)\text{e}^{\text{i}(\delta_2 - \delta_0)}, \tag{12.126b}$$

$$\omega = \left(\frac{\text{Re } A_2 + \text{i } \epsilon \text{ Im } A_2}{A_0}\right)\text{e}^{\text{i}(\delta_2 - \delta_0)}. \tag{12.126c}$$

The unitarity requirement has been considered by Wu and Yang (1964), [59], Bell and Steinberger (1965) [51], and Rubbia [48]. It leads to the following relationship which determines arg $\epsilon$, in terms of $m_L - m_S$, and $\Gamma_S$ (i.e. all experimentally determinable quantities),

$$\arg \epsilon = \text{arc tan}\left\{\frac{2(m_L - m_S)}{\Gamma_S}\right\} \equiv \begin{cases}(43 \pm 10)^\circ \\ +(180n)^\circ,\end{cases} \tag{12.127}$$

where $n$ is any integer 0, 1, 2, ....

Let us now consider the determination of $\omega$. It has been shown by Abbud, Lee, and Yang (1967) [60] that

$$\frac{K_S \to \pi^0\pi^0}{K_S \to \text{all } 2\pi} = \frac{|1 - \sqrt{2}\omega|^2 + (5 \pm 5) \times 10^{-3}}{3 + 3\,|\omega|^2}, \tag{12.128}$$

where the last term is due to radiative corrections and c.m.s. mass differences. We have concluded that $|\omega| \ll 1$, and it appears that $|\omega|^2 \lesssim 10^{-3}$ and, hence, we can approximate by omitting terms of order $|\omega|^2$ and, using the

Particle Data Group table value, evaluate the left-hand side. Thus

$$\frac{K_S \to \pi^0\pi^0}{K_S \to \text{all } 2\pi} = 0{\cdot}316 \pm 0{\cdot}011 = \tfrac{1}{3}(1 - 2\sqrt{2}\ \text{Re}\ \omega) + (5 \pm 5) \times 10^{-3} \qquad (12.129a)$$

or

$$\begin{aligned} 0{\cdot}948 \pm 0{\cdot}033 &= 1 - 2{\cdot}82\ \text{Re}\ \omega + (15 \pm 15) \times 10^{-3} + 0{\cdot}052 \pm 0{\cdot}033 \\ &= 2{\cdot}82\ \text{Re}\ \omega - 0{\cdot}015 \pm 0{\cdot}015, \\ \text{Re}\ \omega &\approx (0{\cdot}0185 \pm 0{\cdot}012), \end{aligned} \qquad (12.129b)$$

but, from eqn (12.124),

$$\text{Re}\ \omega \approx \frac{\text{Re}\ A_2}{A_0} \cos(\delta_2 - \delta_0). \qquad (12.129c)$$

Hence,

$$0{\cdot}0185 \pm 0{\cdot}012 = \text{Re}\ \frac{A_2}{A_0} \cos(\delta_2 - \delta_0). \qquad (12.130)$$

An average value from pion production experiments [61], [62] gives

$$(\delta_0 - \delta_2) = 50° \pm 20°.$$

Hence,

$$\frac{\text{Re}\ A_2}{A_0} = \frac{0{\cdot}0185 \pm 0{\cdot}012}{\cos(50^0 \pm 20^0)} \approx 0{\cdot}028 \begin{array}{l} +0{\cdot}025 \\ -0{\cdot}005 \end{array}.$$

The above is within the errors compatible with the result from $K^+$ decay. We can conclude that $A_2/A_0$ is largely real, since even the experiment which gives the highest decay rate $K_L \to \pi^0\pi^0$ gives an $\dfrac{\text{Im}\ A_2}{A_0}$ an order of magnitude smaller that Re $A_2/A_0$ (Cronin 1968 [62]).

## 12.14. Determination of the $K_L$–$K_S$ mass difference

We can expect a slight mass difference between $K_L$ and $K_S$, due to higher-order interaction effects. This mass difference would be expected to exhibit itself most sensitively in interference effects between $K_L$ and $K_S$ in beams where they are coherently superimposed. In particular, we would expect that the phase procession of each state depends upon the mass and, thus, there will develop a slight difference in the relative phase of the two states if there is a mass difference and, thus, a phase procession with time that depends upon the mass difference.

In a strong interaction, owing to strangeness conservation, (depending upon the process) we can produce a $|K^0\rangle$ with $S = +1$, originally at $\tau = 0$, which can be expressed as

$$|K^0\rangle = \frac{1}{2p}(|K_L\rangle + |K_S\rangle) \tag{12.131}$$

or, alternately, we can produce a $|\bar{K}^0\rangle$ with $S = -1$, which can be expressed as

$$|\bar{K}^0\rangle = \frac{1}{2q}(|K_L\rangle - |K_S\rangle). \tag{12.132}$$

As time progresses the $|K_L\rangle$ and $|K_S\rangle$ each change their respective exponential factors and the $|K\rangle$ and $|\bar{K}\rangle$ states develop as follows

$$|K\rangle \rightarrow \frac{1}{2p}\{(e^{-iM_L\tau})\,|K_L^0\rangle + e^{-iM_S\tau}\,|K_S^0\rangle\}$$

$$= \tfrac{1}{2}(e^{-iM_L\tau} + e^{-iM_S\tau})\,|K^0\rangle + \frac{q}{2p}(e^{-iM_L\tau} - e^{-iM_S\tau})\,|\bar{K}^0\rangle. \tag{12.133a}$$

Similarly, we find

$$|\bar{K}\rangle \rightarrow \frac{1}{2q}(e^{-iM_L\tau}\,|K_L^0\rangle - e^{-iM_S\tau}\,|K_S^0\rangle)$$

$$= \tfrac{1}{2}(e^{-iM_L\tau} + e^{-iM_S\tau})\,|\bar{K}^0\rangle + \frac{1}{2}\frac{p}{q}(e^{-iM_L\tau} - e^{-iM_S\tau})\,|K^0\rangle, \tag{12.133b}$$

where

$$\left.\begin{aligned} M_L &= m_L - \tfrac{1}{2}i\Gamma_L \\ M_S &= m_S - \tfrac{1}{2}i\Gamma_S \end{aligned}\right\}. \tag{12.133c}$$

Thus, in addition to decaying there is, during one unit of proper time $\tau$, a rotation in the relative phase between $|K_L\rangle$ and $|K_S\rangle$ equal to $(m_S - m_L)$ and, thus, the resultant angular frequency of the relative phase rotation is

$$\frac{(m_S - m_L)}{2\pi\tau}.$$

Hence, in any common decay channel, we should expect to observe interference effects following this angular frequency (or time-distribution). This change in phase introduces, with time, a $|\bar{K}^0\rangle$ component, into what was originally a pure $|K^0\rangle$ beam (and vice versa). It can be shown from the foregoing, that if, at time $\tau = 0$, we have a pure $K^0$ beam, then after a time

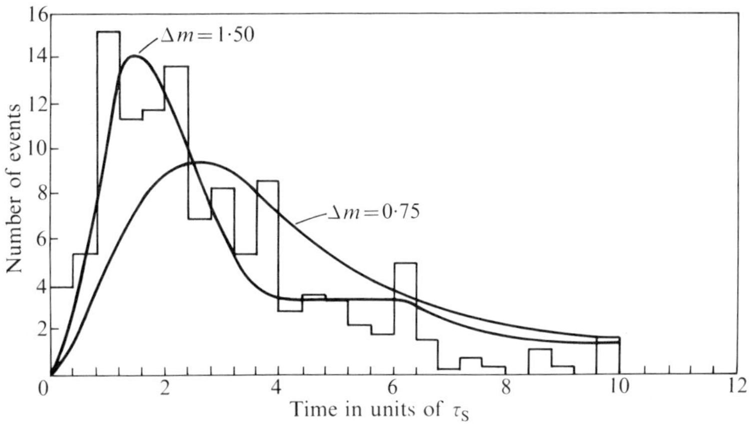


FIG. 12.11. The distribution of $\bar{K}^\circ$ interactions observed as a function of time. The solid curves were the calculated theoretical predictions for the $\Delta m$ values (in units of $\tau_S$) indicated. (From Camerini *et al.* (1962). *Phys. Rev.* **128,** 362.)

$\tau$, there is a probability of observing a $\overline{K}^0$ equal to

$$P(\overline{K}^0, t) \approx \text{constant}\ |(K_S, \tau = 0)|^2\,(1 - e^{-\frac{\Gamma_S \tau}{2}} \cos|m_S - m_L|\,\tau + e^{-\Gamma_S \tau}), \tag{12.134}$$

where $\Gamma_L$ has been set equal to zero, since $\dfrac{\Gamma_L}{\Gamma_S} \sim 500$.

Camerini *et al.* (1962) [63] observed the decay of $|\overline{K}^0\rangle$ measured as a function of time, in an initially pure $|K^0\rangle$ beam. They obtained the result shown in Fig. 12.11, and detected the $|\overline{K}^0\rangle$ by looking at hyperons with strangeness $-1$, produced by the K interaction. As can be seen in Fig. 12.11, the mass difference can be determined from the oscillations. Camerini *et al.* obtained the following result

$$|\Delta m| = (1{\cdot}5 \pm 0{\cdot}2),$$

in units† of $\tau_S$, the mean life of the $|K_S^0\rangle$. Recent experiments by Camerini *et al.* (1966) give $0{\cdot}88 \pm 0{\cdot}2$, and Meissner *et al.* (1966) [64] obtained $0{\cdot}62^{+0{\cdot}33}_{-0{\cdot}27}$.

The time dependence experiments have also been performed by observing leptonic decay channels, and assuming the $\Delta S = \Delta Q$ rule, to distinguish between $K^0$ and $\overline{K}^0$. Thus, the decay state $l^+ + \pi^- + \nu$ is used to identify

† $\Delta m$ is usually given in units of $\tau_S$, thus, we must multiply the numbers given by $\tau_S$ to obtain $\Delta m$.

the $K^0$, while the decay state $l^- + \pi^+ + \bar{\nu}$ is used to identify $\bar{K}^0$, where $l$ is a lepton. For example, various authors [65]–[67] used the reaction $\bar{K}^0 \to \pi^+ + e^- + \bar{\nu}_e$ to detect the $\bar{K}^0$. Fitch *et al.* [65] obtained $1{\cdot}9 \pm 0{\cdot}3$, in units of $\tau_S$, for the mass difference.

Recent experiments by Aubert *et al.* [66] obtained $0{\cdot}47 \pm 0{\cdot}20$, and Baldo-Ceolin *et al.* [67] obtained

$$0{\cdot}15 \begin{matrix} +0{\cdot}35 \\ -0{\cdot}50 \end{matrix}.$$

Hill *et al.* [68] obtained $0{\cdot}62 \pm 0{\cdot}16$. Another way of observing the $\bar{K}^0$ is via charge exchange, to produce $K^-$ mesons [69]. The latest value from the Particle Data Group tables is $m_{K_L} - m_{K_S} = 0{\cdot}469 \pm 0{\cdot}015$. It is interesting to note that, since $\Delta m$ is of the order $\tau_S$, which is of the order $G^2$, it can be demonstrated that there is no $\Delta S = 2$ term in the weak Hamiltonian, since if there was a $\Delta S = 2$ term, the mass difference would be expected to be of the order $G$.

Another class of experiments which measure $|m_L - m_S|$ are the regeneration experiments. The method consists of starting with a beam of almost pure $K_L$, which is obtained by allowing sufficient decay path for the decay of the $K_S$ component, so as to make the remaining $K_S$ negligible. The $K_L$ are then allowed to traverse an absorber—the regenerator. Owing to the differences in interaction for $K^0$ and $\bar{K}^0$, the mixture of these components in the incident $K_L^0$ is changed, and the result is such that a small coherent $K_S$ amplitude is regenerated. Thus, after passing through the regenerator, we can characterize the $|K\rangle$ state as

$$|K\rangle = |K_L\rangle + \rho\, |K_S\rangle, \tag{12.135}$$

$\rho$ is a complex constant defined as the 'regeneration amplitude'. Good (1958) [70] gave a prescription for computing $\rho$,

$$\rho = |\rho|\, e^{i\phi_\rho} = -\frac{\pi N}{m\, \Delta M}(f(0) - \bar{f}(0))(1 - e^{i\Delta M \frac{l m_K}{p_K}}), \tag{12.136}$$

where $N$ is the number of atoms per unit volume, $l$ is the regenerator thickness which the beam passes through, $m_K$ is the kaon mass, and $p_K$ the kaon momentum, and

$$\Delta M = m_L - \frac{i}{2}\Gamma_L - m_s + \frac{i}{2}\Gamma_S.$$

If the regenerator is thin enough, so that $e^{i\Delta M \frac{l m_K}{p_K}} \to 1$, then the regeneration phase $\phi_f$ is given by

$$\phi_f = \arg i(f(0) - \bar{f}(0)). \tag{12.137}$$

In the general case of a thick regenerator

$$\phi_\rho = \phi_f + \arg\left(\frac{1-e^{i\Delta M\frac{lm_K}{P_K}}}{-i\,\Delta M}\right). \tag{12.138}$$

Thus, from measurements of the value of regenerated intensity $|\rho|^2$ as a function of the regeneration thickness, determinations of $|m_L - m_S|$ were made [70], [71] and the results were

$$0{\cdot}84^{+0{\cdot}29}_{-0{\cdot}22} \quad \text{and} \quad 0{\cdot}72^{+0{\cdot}15}_{-0{\cdot}15},$$

respectively.

A variation of this method, which is more sensitive, is to use two sheets of regenerators and vary their separation, measuring the regeneration intensity behind the second one. The constant rate of procession of the relative phase between $K_L$ and $K_S$ in the gap causes interference effects between the amplitudes regenerated from the two individual slabs, and these interferences depend sensitively on the separation. The results obtained were $0{\cdot}55 \pm 0{\cdot}1$ [72] and $0{\cdot}82 \pm 0{\cdot}14$ [73]. The coherent combinations of $|K_L\rangle$ and $|K_S\rangle$, which follow passage of a $|K_L\rangle$ beam through a regenerator, can be observed in the $\pi^+\pi^-$ decay channel as a function of proper time. In the case of $|K_S\rangle$, this would be the CP-conserving decay. In the case of the $|K_L\rangle$, this would be the CP-violating decay mode. Since the final state is the same, we can expect interference effects to occur in the decay rate as a function of proper time; these have been observed, and represent the most accurate determination of $|\Delta m|$. The results obtained were $0{\cdot}455 \pm 0{\cdot}034$ (Alff-Steinberger *et al.* [74]) and $0{\cdot}480 \pm 0{\cdot}024$ (Bott-Bodenhausen *et al.* [75]).

The previously described experiments were sensitive to the magnitude, but not the sign, of $\Delta m$. The reaction

$$K^+ + d \rightarrow |K^0\rangle + p + p$$

was used by Canter *et al.* [76] to produce $|K^0\rangle$, and an investigation of the proper time distribution of the scattering of the neutral kaons on deuterium gave a result which allowed Canter to conclude that $m_L - m_S$ is a positive quantity (i.e. $m_L > m_S$). The sign of the mass difference was also indicated to be such that $m_L > m_S$, by observing the interference effects in regeneration between two different regenerators (carbon and uranium). The sign of the mass difference depends upon the sign of the difference in the real parts of the two regeneration amplitudes, which can be calculated from $K^\pm$ scattering observations and the optical theorem. Melhop *et al.* [77] allowed a $K^0$

beam, originally obtained by charge-exchange in copper, to pass through a copper regenerator, and observed the interferences between the regenerated $K_S$, and the still originally produced surviving $K_S$ amplitudes. The geometry, the sign of $\Delta m$, and the sign of the real part of the regeneration amplitude determine the interference pattern. The sign of the real part of the $\rho$ is calculable, as previously discussed, and, thus, Melhop was able, on the basis of observations, to conclude that $m_L - m_S$ is positive.

### 12.15. Observation of CP violation via charge asymmetries in the long lived leptonic decays

Consider the following four processes,

| | Amplitude denoted by | |
|---|---|---|
| (a) $K_L^0 \to \pi^- + \begin{Bmatrix} \mu^+ \\ e^+ \end{Bmatrix} + \begin{Bmatrix} \nu_\mu \\ \nu_e \end{Bmatrix}$ | $f$ | (12.139a) |
| (b) $\bar{K}_L^0 \to \pi^+ + \begin{Bmatrix} \mu^- \\ e^- \end{Bmatrix} + \begin{Bmatrix} \bar{\nu}_\mu \\ \bar{\nu}_e \end{Bmatrix}$ | $f'$ | (12.139b) |
| (c) $K_L^0 \to \pi^+ + \begin{Bmatrix} \mu^- \\ e^- \end{Bmatrix} + \begin{Bmatrix} \bar{\nu}_\mu \\ \bar{\nu}_e \end{Bmatrix}$ | $g$ | (12.139c) |
| (d) $\bar{K}_L^0 \to \pi^- + \begin{Bmatrix} \mu^+ \\ e^+ \end{Bmatrix} + \begin{Bmatrix} \nu_\mu \\ \nu_e \end{Bmatrix}$ | $g'$ | (12.139d) |

denoted by the amplitudes $f$, $f'$, $g$, $g'$, respectively. Conservation of CPT requires that $f' = f^*$, and that $g' = g^*$. The $f$ and $f'$ amplitudes obey the $\Delta Q = \Delta S$ rule, whereas for the $g$ and $g'$ amplitudes $\Delta Q = -\Delta S$. Thus, let us define $x = g/f$ and, since the $\Delta Q = \Delta S$ amplitude is indicated by experiments to be much larger than the $\Delta Q = -\Delta S$ amplitude, we can expect that $|x| \ll 1$.

If CP were not violated, it is clear that in the $K_L$ state decays, there would be equal numbers of cases where the decay involved a $\mu^+$ lepton or a $\mu^-$ lepton. Similarly, there would be equal numbers of $e^+$ and $e^-$ leptonic decays of $K_L$.

However, owing to CP violation, we expect that

$$\delta = \frac{\mu^+ - \mu^-}{\mu^+ + \mu^-} = \frac{e^+ - e^-}{e^+ + e^-}, \tag{12.140}$$

should be slightly different from zero (i.e. $\sim 10^{-3}$). We shall show that $\delta$ is a function of $\epsilon$ and $x$.

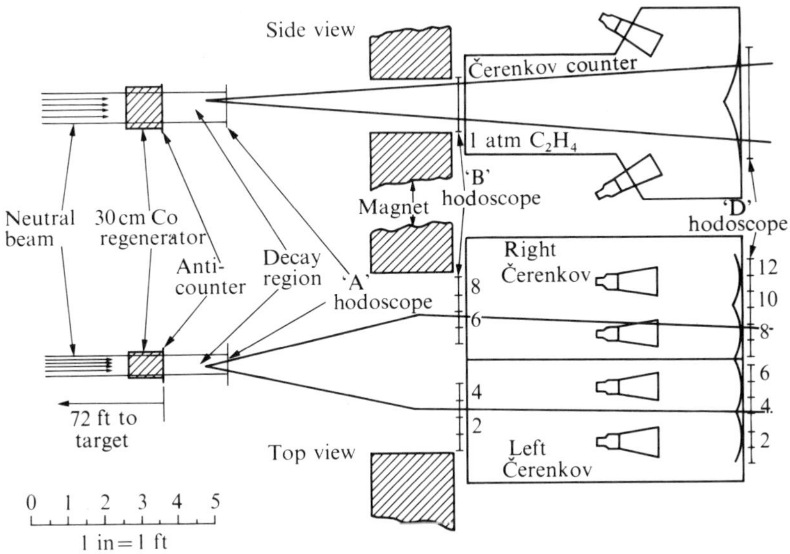


FIG. 12.12. The apparatus used by Bennett *et al.* [78] to determine the electronic charge asymmetry in $K^0_{Le_3}$ decay.

Bennett *et al.* (1967) [78] studied the reactions

$$K^0_{Le_3} \rightarrow \begin{cases} \pi^- e^+ + \nu_e \\ \pi^+ e^- + \bar{\nu}_e \end{cases}. \tag{12.141}$$

Figure 12.12 shows the apparatus used. The position and momentum of both charged particles was determined by counter hodoscopes. The electron (or positron) was identified by gas Cerenkov counters, and its charge sign was determined by the bending in the magnet. The magnetic field was reversed in sign from time to time, to make the result insensitive to geometrical uncertainties. The apparatus and data were monitored by the BNL–OLDF computer. The corrected result obtained was

$$\delta_e = +(2{\cdot}24 \pm 0{\cdot}36) \times 10^{-3}, \tag{12.142}$$

in which about a 40 per cent correction was applied to the raw data. This correction was mainly due to the interaction of pions and leptons with the detector material.

Dorfan *et al.* (1967) [79] has determined the charge asymmetry corresponding to the muon decay $\delta_\mu$. Muons were selected by the requirement that they penetrate a thick enough shield (mostly containing lead) to rule out other particles (i.e. pions and electrons) to a sufficient degree. Magnetic

TABLE 12.1

*Test of $\Delta Q = \Delta S$ in $K^0_{l_3}$ decays*

| Group | Method | No. of $K_{l_3}$ events | $\lvert\Delta m\rvert$ | Re $(x)$ | Im $(x)$ | Ref. |
|---|---|---|---|---|---|---|
| Paris | Freon/Prop.BC $K^+n \quad K^0p$ | 315 | $0{\cdot}47 \pm 0{\cdot}20$ | $0{\cdot}035^{+0{\cdot}11}_{-0{\cdot}30}$ | $-0{\cdot}21^{+0{\cdot}15}_{-0{\cdot}11}$ | a |
| Padua | Freon/Prop.BC $K^+n \quad K^0p$ | 152 | $0{\cdot}15^{+0{\cdot}35}_{-0{\cdot}50}$ | $0{\cdot}06^{+0{\cdot}18}_{-0{\cdot}44}$ | $-0{\cdot}44^{+0{\cdot}32}_{-0{\cdot}19}$ | b |
| Columbia/ Rutgers | HBC $\bar{p}p$ | 109 $\mu$, e | $\equiv 0{\cdot}5$ | $-0{\cdot}08^{+0{\cdot}16}_{-0{\cdot}28}$ | $+0{\cdot}24^{+0{\cdot}40}_{-0{\cdot}30}$ | c |
| Pennsylvania | Spark chamber $\pi^-p \quad \Lambda K^0$ | 116 | ? | $0{\cdot}17^{+0{\cdot}16}_{-0{\cdot}35}$ | $0{\cdot}0 \pm 0{\cdot}25$ | d |
| Brookh./ Carnegie | $D_2$BC $K^+n \quad K^0p$ | 335 | $\equiv 0{\cdot}50$ | $0{\cdot}17 \pm 0{\cdot}10$ | $-0{\cdot}20 \pm 0{\cdot}10$ | e |
| Berkeley | HBC $K^-p \quad \bar{K}^0n$ | 242 $\mu$, e | $\equiv 0{\cdot}47$ | $0{\cdot}22^{+0{\cdot}07}_{-0{\cdot}09}$ | $-0{\cdot}08 \pm 0{\cdot}08$ | f |
| CERN/ Paris | HBC $\bar{p}p$ | 121 $\mu$, e | $\equiv 0{\cdot}47$ | $0{\cdot}09^{+0{\cdot}13}_{-0{\cdot}11}$ | $+0{\cdot}22^{+0{\cdot}29}_{-0{\cdot}37}$ | g |
| Average | | | | $+0{\cdot}14 \pm 0{\cdot}05$ | $-0{\cdot}12 \pm 0{\cdot}05$ | |
| CERN/ Columbia | Counter exp. $K^0_L$ beam | ? | 0·445 | $\frac{1-\lvert x\rvert^2}{\lvert 1-x\rvert^2} = 1{\cdot}06 \pm 0{\cdot}06$ | | h |

(a) B. AUBERT *et al.* (1965). *Phys. Lett.* **17,** 59.
(b) M. BALDO-CEOLIN *et al.* (1965). *Nuovo Cim.* **38,** 684.
(c) P. FRANZINI *et al.* (1965). *Phys. Rev.* B **140,** 127.
(d) L. FELDMAN *et al.* (1967). *Phys. Rev.* **155,** 1611.
(e) D. G. HILL *et al.* (1967). *Phys. Rev. Lett.* **19,** 668
(f) B. R. WEBBER *et al.* (1968). *Phys. Rev. Lett.* **21,** 498.
(g) F. JAMES AND H. BRIAND (1968) *Proc. XIV Int. Cont. High Energy Physics. Vienna.* paper 332.
(h) S. BENNETT *et al.* (1968). *Phys. Lett.* B **27,** 244.

(From Cronin, Vienna conference [62].)

analysis was used to determine the muon sign. The result obtained was

$$\delta_\mu = (4{\cdot}05 \pm 1{\cdot}7) \times 10^{-3}. \tag{12.143}$$

From eqns (12.99) and other general requirements on the K-system, it can be shown that (See Okun and Rubbia [48] and Cronin [62])

$$\delta = \langle K_L \mid K_S \rangle \frac{(1-\lvert x\rvert^2)}{\lvert 1-x\rvert^2} = \frac{2\,\mathrm{Re}\,\epsilon\,(1-\lvert x\rvert^2)}{(1+\lvert\epsilon\rvert^2)^{\frac{1}{2}}\,\lvert 1-x\rvert^2}. \tag{12.144}$$

Averaging the two values of $\delta$, we obtain

$$\delta_1 = (2{\cdot}3 \pm 0{\cdot}4) \times 10^{-3}.$$

The amplitude of oscillation of the charge asymmetry, following a regenerator was used to deduce [80]

$$\frac{(1-|x|^2)}{|1-x|^2} = 1{\cdot}06 \pm 0{\cdot}06.$$

Then, using these numbers, we find

$$\mathrm{Re}\,\epsilon = (1{\cdot}1 \pm 0{\cdot}2) \times 10^{-3}.$$

If we assume $x = 0$, the result would hardly change, i.e.

$$\mathrm{Re}\,\epsilon = (1{\cdot}15 \pm 0{\cdot}2) \times 10^{-3}.$$

Cronin [62] gives a summary of the determinations of Re $x$ and Im $x$, to date, and thus, a summary of the tests of the $\Delta S = \Delta Q$ rule in $K^0_{l_3}$ decays. The measurements, to date, are summarized in Table 12.1 and Fig. 12.13. It appears that there is no clear evidence for $\Delta S = -\Delta Q$ and, thus, we can still assume that the $\Delta S = \Delta Q$ rule is consistent with the measurements of $x$.

As reviewed (Cronin) (Table 12.2), the average value of the phase $\phi_{+-}$, from various experiments, is $59^0 \pm 6^0$. However, there are wide fluctuations in individual experiments. The two methods used were the time distribution of $2\pi$ decay following a regenerator, and the time distribution of $2\pi$ decay following an interaction which produces a nearly pure $K^0$ state.

## 12.16. Measurement of $|\eta_{00}|$ and comparison with $|\eta_{+-}|$

There were a number of results presented at the Vienna Conference (Cronin) (see Table 12.3), where $|\eta_{00}|$ varied from about $3{\cdot}6 \times 10^{-3}$ down to zero. The experiments are difficult to perform owing to the fact that the decay products $2\pi^0 \to 4\gamma$ are difficult to detect and distinguish from the large $3\pi^0$ decay background. The most recent experiments differed. Two (CERN–Ecol. PolyTech, Budagov *et al.* paper 377 and Princeton II, Banner *et al.* paper 911 of the conference papers) give results consistent with $|\eta_{00}| = |\eta_{+-}|$, i.e. $|\eta_{00}| = 2{\cdot}2 \pm 0{\cdot}4$ and $2{\cdot}3 \pm 0{\cdot}3$, respectively. However, two others (Berkeley–Hawaii, Cence *et al.* paper 476, and CERN–RHEL–Aachen II) obtained $3{\cdot}6 \pm 0{\cdot}4$ and $3{\cdot}6 \pm 0{\cdot}6$, respectively, implying that $|\eta_{00}| \neq |\eta_{+-}|$, unless we allow for considerable systematic errors, or statistical fluctuations. Further work is obviously awaited.

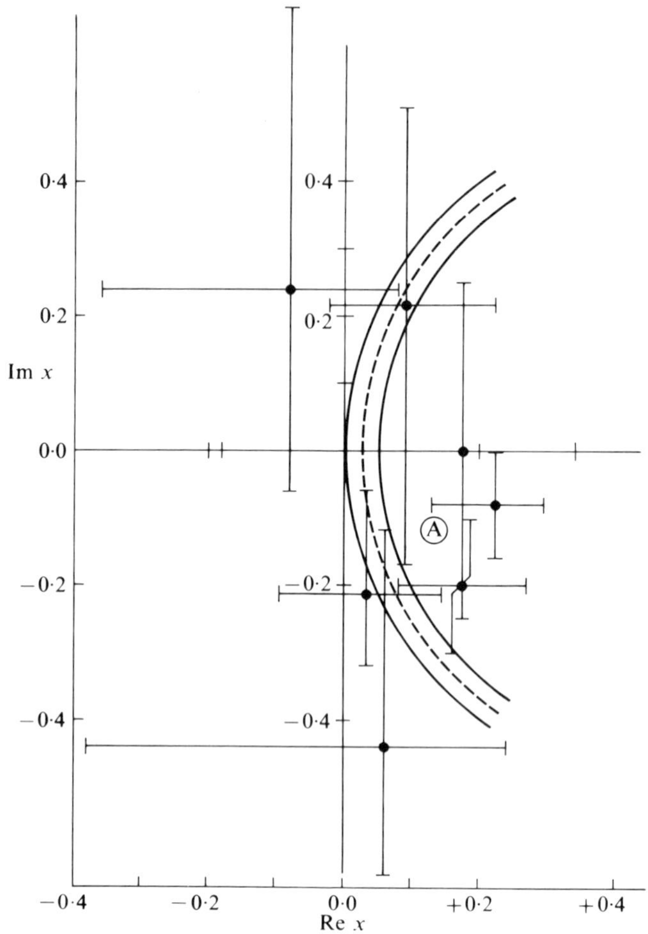


FIG. 12.13. Im $x$ against Re $x$ (Cronin [62]).

If we take the average of these last four measurements, we obtain $|\eta_{00}| \approx 3 \pm 0{\cdot}3$. However, looking at the scatter of the data, we can guess that $|\eta_{00}| \approx 3 \pm 1$ is a more realistic estimate. We could construct a Wu–Yang triangle, using the values of $|\eta_{+-}|$, arg $\eta_{+-}$, Re $\epsilon$, the phase of $\epsilon$, and $|\eta_{00}|$. An example of such a construction is shown in Fig. 12.14, according to eqn (12.113), using a prescription given in Rubbia's review.

In Fig. 12.14 the vector $\eta_{+-}$ is drawn from the origin giving point A. However, there is an uncertainty in both amplitude and phase in $\eta_{+-}$, and this is allowed for by a curve outlining the A region. The limits of the

TABLE 12.2
$\phi_{+-}$

| Group | $P_K$ (GeV/$c$) | Regenerator | $\phi_{+-}-\phi_f$ | $\phi_f$ | $\phi_{+-}$ | Ref. |
|---|---|---|---|---|---|---|
| Princeton | 1·55 | Be | $45°\pm35°$ | $0°\pm15°$ $f_{21}$, $\sigma_{tot}$ Opt. mod. | $45°\pm40°$ | a |
| CERN | 4·8 | C | $90°\pm6°$ | $-20°\pm20°$ $f_{21}$, $\sigma_{tot}$ | $70°\pm21°$ | b |
| CERN–Columbia | 2·7 | Cu | $80·5°\pm10·3°$ | | | c |
| Illinois | 1·1 | Cu | $67°\pm20°$ | $-42°\pm22°$ Opt. mod. | $25°\pm35°$ | d |
| Yale–BNL | 5 | $H_2$ | $120°\pm45°$ | $-90°\pm$? Disp. rel. | $30°\pm45°$ | e |
| CERN | 7 | Vacuum | | | $46°\pm15°$ | f |
| Columbia | 2·5 | Cu | | $-28·8°\pm4·7°$ | $51·2°\pm11°$ combined with (c) | g |
| CERN | 3·4 | Cu | $98°\pm10°$ | | $68°\pm7·5°$ combined with (g) | h |

(a) V. FITCH *et al.* (1965). *Phys. Rev. Lett.* **15,** 73 (1967). *Phys. Rev.* **164,** 1711.
(b) M. BOTT-BODENHAUSEN *et al.* (1966). *Phys. Lett.* **23,** 277, and (1967). *Phys. Lett.* B **24,** 438.
(c) C. ALFF-STEINBERGER *et al.* (1966). *Phys. Lett.* **21,** 595.
(d) R. E. MISCHKE *et al.* (1966). *Phys. Rev. Lett.* **18,** 138.
(e) A. FIRESTONE *et al.* (1966). *Phys. Rev. Lett.* **16,** 556, and (1966). *Phys. Rev. Lett.* **17,** 116.
(f) A. BÖHM *et al.* (1968). *Phys. Lett.* B **27,** 321.
(g) S. BENNETT *et al.* (1968). *Phys. Lett.* B **27,** 239.
(h) V. BISI *et al.* (1968). *Proc. XIV Int. Conf. High Energy Physics. Vienna*, abstract 590.

(From Cronin, Vienna Conference [62].)

Re $\epsilon$ are drawn as two dotted vertical lines, and the limits of the phase range of $\epsilon$ are drawn from the origin. The common region of intersection labelled B is the allowed range of $\epsilon$. Then two parallel dotted circles are drawn to outline the range of uncertainty of $|\eta_{00}|$. We then attempt to draw a straight line through a particular A, B, and C, such that AB $= \frac{1}{2}$BC, and A, B, and C are all in (or near) their required regions. Then the predicted value of $\delta_2-\delta_0$ is as indicated on the diagram and equals ($64°\pm8°$). The value [61] from

TABLE 12.3

$|\eta_{00}|$

| Group | $P_K$ (GeV/$c$) | $K_L \to 2\pi^0$ normalization | $|\eta_{00}|^2 \times 10^6$ | $|\eta_{00}| \times 10^3$ | Remarks | Ref. |
|---|---|---|---|---|---|---|
| CERN–RHEL Aachen I | $1{\cdot}5-2{\cdot}75$ | Regenerated $K_S^0$ | $18{\cdot}5^{+10{\cdot}5}_{-6{\cdot}5}$ | $4{\cdot}3^{+1{\cdot}1}_{-0{\cdot}8}$ | Superseded by II. | a |
| Princeton I | $0{\cdot}25 \pm 0{\cdot}10$ | $3\pi^0$ | $24 \pm 5$ | $4{\cdot}9 \pm 0{\cdot}5$ | Withdrawn | b |
| Princeton | $\sim 0{\cdot}73$ | Regenerated $K_S^0$ | $-2 \pm 7$ | $< 3{\cdot}0$ | 90% confidence level | c |
| Berkeley–Hawaii | $0{\cdot}53 \pm 0{\cdot}05$ | $3\pi^0$ | $13{\cdot}0 \pm 3{\cdot}0$ | $3{\cdot}6 \pm 0{\cdot}4$ | | d |
| CERN–RHEL Aachen II | $1{\cdot}5-2{\cdot}75$ | Regenerated $K_S^0$ | $13 \pm 4$ | $3{\cdot}6 \pm 0{\cdot}6$ | | e |
| CERN Ec.Pol. Orsay | $0{\cdot}4-2{\cdot}0$ | $3\pi^0$ | $4{\cdot}8 \pm 1{\cdot}8$ | $2{\cdot}2 \pm 0{\cdot}4$ | | f |
| Princeton II | $0{\cdot}25 \pm 0{\cdot}10$ | $3\pi^0$ | $5{\cdot}1 \pm 1{\cdot}2$ | $2{\cdot}3 \pm 0{\cdot}3$ | | g |

(a) J. M. GAILLARD *et al.* (1967). *Phys. Rev. Lett.* **18,** 20.
(b) J. CRONIN *et al.* (1967). *Phys. Rev. Lett.* **18,** 25.
(c) D. BARTLETT *et al.* (1968). *Phys. Rev. Lett.* **21,** 558.
(d) R. J. CENCE *et al.*, paper 476 (1968). *Proc. XIV Int. Conf. High Energy Physics, Vienna.*
(e) J. M. GAILLARD *et al.*, paper 139 (1968). *Proc. XIV Int. Conf. High Energy Physics, Vienna.*
(f) I. A. BUDAGOV *et al.*, paper 377 (1968). *Proc. XIV Int. Conf. High Energy Physics, Vienna.*
(g) M. BANNER *et al.*, paper 910 (1968). *Proc. XIV Int. Conf. High Energy Physics, Vienna.*

(From Cronin, Vienna Conference [62].)

pion production experiments is $\delta_0 - \delta_2 = (50° \pm 20°)$. Owing to the uncertainties involved in both the weak interaction quantities, and the phase-shift analysis experiments, in pion production, we cannot yet draw a firm conclusion.

## 12.17. The Cabibbo angle

Cabibbo [81] proposed that the hadronic current $J^\alpha$ have a structure that is related to the SU(3) generating currents as follows

$$J^\alpha = \sin\theta(J^{\alpha 4} + \mathrm{i}J^{\alpha 5}) + \cos\theta(J^{\alpha 1} + \mathrm{i}J^{\alpha 2}). \tag{12.145}$$

The octet of currents in SU(3), which are represented by $J^{\alpha i}$, contains axial and vector current terms, both of the $\Delta S = 0$ and $\Delta S = 1$ variety. The coefficient of $\cos\theta$ contains the sum of the $\Delta S = 0$ currents $J^\alpha_{V,\Delta S=0}$ and $J^\alpha_{A,\Delta S=0}$.

In essence, what is proposed here as universality is that the square of the coupling constant for $\Delta S = 0$ hadronic interactions, plus the square of

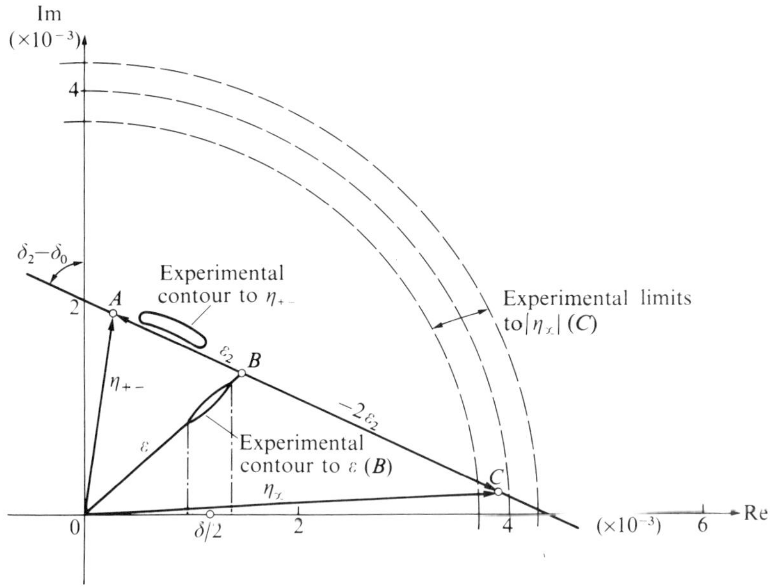


FIG. 12.14. Wu–Yang triangle for the analysis of CP violation in K° decays. The construction is as follows. A line is drawn through points A, B, and C (representing $\eta_{+-}$, $\varepsilon$, $\eta_{00}$) such that 2AB = BC. The angle between the line AC and the imaginary axis is equal to the predicted value of the $\pi\pi$ scattering phase-shift difference $\delta_2 - \delta_0$. (From Rubbia [48].)

the coupling constant for $\Delta S = 1$ hadronic interactions, is an invariant equal to the square of the coupling constant for (pure) leptonic interactions. In principle, $\theta_V$ could be different from $\theta_A$. However, the so-called 'parallelism' (i.e. $\theta_V = \theta_A$) has been employed in eqn (12.145). It can be shown that $|\Delta T| = \frac{1}{2}$ (i.e. the $\Delta T = \frac{1}{2}$ rule) for the $|\Delta S| = 1$ current part (i.e. the strangeness-changing part). Furthermore, $\Delta S = \Delta Q$. Correspondingly for the $\Delta S = 0$ part, $|\Delta T| = 1$. These current forms are modified by SU(3) symmetry-breaking effects. The SU(3) symmetry-breaking corrections to the vector part are smaller than the corrections to the axial part [82] (Ademollo and Gatto 1965). It was shown by Okubo [83] that $J^{\alpha}_{V,\Delta S=1}$ cannot be an exactly conserved current since, in that case, the transition matrix element

$$\int d^3x(\Psi_{\Lambda} J^0_{V,\Delta S=1} \Psi_p) \equiv 0.$$

However, since $J^{\alpha}_{V,\Delta S=1}$ is the current which leads to the observed $\Lambda^0 \to p + \pi^-$ transitions, it cannot be zero and, hence, $J_{V,\Delta S=0}$ cannot be exactly conserved. Since the corrections due to SU(3) symmetry-breaking are small in the vector parts of the current, it appears best to first attempt to determine the Cabibbo angle from vector current interactions. Consider the following two decays

$$\pi^+ \to \pi^0 + e^+ + \nu_e, \tag{12.146a}$$

$$K^+ \to \pi^0 + e^+ + \nu_e. \tag{12.146b}$$

The ratio, using SU(3), becomes

$$\frac{\Gamma(K^+ \to \pi^0 e^+ \nu)}{\Gamma(\pi^+ \to \pi^0 e^+ + \nu)} = \frac{1}{4}\frac{(\sin\theta)^2}{(\cos\theta)^2}\left|\frac{f_+\ |K|}{f_+\ |\pi|}\right|^2. \tag{12.147a}$$

From evaluations of the experimental quantities on the left, and the matrix elements on the right, we can show that

$$\theta_V \approx 0{\cdot}2. \tag{12.147b}$$

Oneda and Sucher (1965) [84] showed how to determine accurately sin $\theta$, from the rates of $K^+_{e_3}$ decay and the $\pi^+ \to \pi^0 + e^+ + \nu_e$ decay. The result obtained is

$$\sin\theta_V = 0{\cdot}202 \pm 0{\cdot}004$$

or, correspondingly,

$$\theta_V \approx 0{\cdot}20 \pm 0{\cdot}004.$$

We can also obtain the Cabibbo angle from the difference between vector coupling constants for $\beta$-decay and muon decay,

$$\frac{G_\mu - C_V}{G_\mu} = 1 - \cos\theta_V. \tag{12.148}$$

The result from this kind of calculation, including corrections, is

$$1 - \cos\theta \approx 0{\cdot}022 \pm 0{\cdot}05 \approx \frac{\theta^2}{2}. \tag{12.149a}$$

Thus,

$$\theta \approx \sqrt{(0{\cdot}044 \pm 0{\cdot}01)} \approx 0{\cdot}21 \pm 0{\cdot}025. \tag{12.149b}$$

For the axial current one can compare the rates for the two reactions

$$\pi^\pm \to \mu^\pm + \nu, \tag{12.150a}$$

$$K^\pm \to \mu^\pm + \nu. \tag{12.150b}$$

The predicted result is

$$\left|\frac{F_K}{F_\pi}\right|^2 = \tan^2\theta, \tag{12.151}$$

where $\left|\frac{F_K}{F_\pi}\right|^2 \approx 0{\cdot}075$ is the square of the absolute value of the ratio of the effective coupling constants. Thus, one obtains $\theta_A \approx 0{\cdot}27$. Considering the experimental and theoretical uncertainties, the angles are equal within the errors.

Cronin [62] has reviewed fits for various results on hyperon decays, using a one-angle Cabibbo fit, and finds good agreement with the data with a single angle $\theta = \theta_V = \theta_A = \sin^{-1} 0{\cdot}227$. It appears that if due allowance is made for the symmetry-breaking effects in the axial vector current terms, we can conclude that the Cabibbo theory fits the data reasonably well, and the isospin ($\Delta T = \frac{1}{2}$) and the $\Delta S = \Delta Q$ requirements are met within the uncertainties involved. It also allows us a framework for understanding the long-observed experimental fact that strangeness changing currents ($\Delta S = \pm 1$) are coupled about an order of magnitude weaker than non-strangeness changing currents.

The Cabibbo angle can be considered to represent some measure of the different ways in which the semi-strong, and weak interactions, violate SU(3) symmetry. We should recall here that the semi-strong violation preserves isotopic spin (SU(2)) as a sub-symmetry, whereas the weak violation does not. Thus, in a sense, the Cabibbo angle is intimately related to the relationship between the SU(3) symmetry-breaking effects for the two interactions.

## 12.18. CPT invariance

We have assumed so far that CPT invariance holds rigorously in all interactions, including the weak. The consequences of failure of CPT invariance would be an enormous complication to all present theories. It is, therefore, worthwhile to review the experimental evidence for CPT invariance.

One of the most important classes of consequences of CPT invariance is the predicted equality of lifetimes, masses, and magnetic moments for particles and their corresponding anti-particles. Owing to the very small difference between the masses of $K^0_S$ and $\overline{K}^0_L$, which can be determined from the experiments previously described, it is concluded that CPT violations are at least $10^{-3}$ times smaller than the second order weak interactions effects (i.e. $\sim 10^{-8}$ eV). Rubbia [48] and Cronin [62] review the results of various experimental checks. The mass difference between $K^+$ and $K^-$ has shown to be consistent with zero within 300 keV [85]. The same experiment also determined $m(\pi^+)$ to be $(139{\cdot}659 \pm 0{\cdot}076)$ MeV. Schaefer *et al.* (1965) [86] determined from $\pi^-$-meson X-ray experiments that $m_{\pi^-} = 139{\cdot}577 \pm 0{\cdot}014$). Thus,

$$m_{\pi^+} - m_{\pi^-} = (82 \pm 80)\ \text{keV}.$$

The ratio $m_{\mu^-}/m_{e^-}$ was determined [87] by a comparison of the $3d$–$2p$ transition in phosphorus with the K absorption edge in lead giving $m_{\mu^-}/m_{e^-} = 206{\cdot}76 \pm 0{\cdot}02$ and, thus, allows a determination of the negative muon mass.

The $g$–2 experiment [88], and a measurement of the precession frequency of muons in a magnetic field, results in

$$\frac{m_{\mu^+}}{m_{e^+}} = 206{\cdot}761 \pm 0{\cdot}005.$$

This allows a determination of the mass of the positive muon. Hence, the difference in mass between a positive and negative muon is, from the above, less than about one part in $10^4$. For baryon particle and anti-particle pairs, the mass differences are, in general, checked to be within a fraction of an MeV.

Particle–anti-particle lifetime differences for $\mu^+$ and $\mu^-$ are zero, to about one part in a thousand [89]. For $K^+$ and $K^-$ Lobkowicz *et al.* obtained [90*a*]

$$\frac{\tau_{K^+}}{\tau_{K^-}} = 1{\cdot}004 \pm 0{\cdot}00097.$$

For $\pi^+$ and $\pi^-$ an accurate result [90*b*] gives

$$\frac{\tau_{\pi^+}}{\tau_{\pi^-}} = 1{\cdot}00064 \pm 0{\cdot}00069,$$

which is somewhat better than one part in a thousand.

## 12.19. T-violation in weak interactions

If we assume CPT invariance in weak interactions, with which all present data are consistent, and accept the experimentally demonstrated violation of CP in the decay $K_L^0 \to 2\pi$, then small T-violations must also occur in the $K_L^0 \to 2\pi$ decays. However, it is interesting to explore the possibility of demonstrating T-violation directly, without using the requirement of CPT invariance to deduce it indirectly.

Casella has shown [91*a*] that if we give up the assumption of CPT invariance, and substitute T invariance instead, while retaining the remainder of the Wu–Yang conditions, we obtain a modified set of triangular relations

$$\eta_{\pm} = \bar{\epsilon} + \epsilon_2,$$
$$\eta_{00} = \bar{\epsilon} - 2\bar{\epsilon}_2,$$

where $\bar{\epsilon}$ and $\bar{\epsilon}_2$ are under the CPT invariance assumption, the previously-defined $\epsilon$ and $\epsilon_2$, respectively, rotated in phase by 90°. Casella shows that, even within the limitations of experimental uncertainties in $|\eta_{00}|$ and its phase, the modified Wu–Yang triangle does not close and, therefore, T-violation is indicated. This method assumes $|\Delta T| = \frac{1}{2}$ dominance, but requires no knowledge of $\phi_{00}$.

A second method of demonstrating T-violation is the failure of the data to satisfy a sum rule derived by Bell and Steinberger [91*b*], which does not require the $|\Delta T| = \frac{1}{2}$ rule, but requires knowledge of $\phi_{00}$. It also indicates [88]–[90*a*] T-violation.

Although the data are consistent with CPT invariance, when formulated in terms of the original Wu–Yang triangle, the T-violation is much more strongly indicated than the confirmation of CPT.

## 12.20. The source of C and CP violations

The observed C and CP (or T) violations, in the weak interactions, can be introduced theoretically, in a variety of ways; there are numerous models [92], and an ever-increasing number being introduced. Their best classification into types of model is in the nature of the interaction which is responsible for the violation. The violation can be placed in the strong, electromagnetic, weak, or superweak interaction and, in general, explains the existing experimental situation.

In the strong interaction violating models, there is assumed a hadron interaction component of strength about $10^{-2}-10^{-3}$ that of the usual strong component, and this small component violates C and T (or PC) but conserves S (strangeness) and P (parity). The electromagnetic models allow a C and T (or PC) violating component, comparable in strength to the normal electromagnetic interaction strength. The electromagnetic and strong interaction violating models both yield T (or PC) violating effects in semi-leptonic and non-leptonic weak interactions to a level of about $10^{-2}$–$10^{-3}$ relative to non-violating effects. The same level of violation occurs for C and T, also in strong reactions. However, the major distinction between these two models occurs in electromagnetic phenomena for which the electromagnetic models predict large violations, comparable with the maximum but, of course, suitably attenuated by the isotopic spin, centrifugal barriers, and other details of the particular interaction. Thus, for example, if the reported $\eta$ decay asymmetry were to be substantiated by more accurate experiments, this would strongly support the concept of the electromagnetic origin of the violation. There would also be an indication that the C-even hadron current behaves like an isovector. This would suggest that, in the $K_L \rightarrow 2\pi$, $|\epsilon|$ and $|\epsilon_2|$ might be comparable.

## 12.21. C and CP violations in electromagnetic interactions

There have been theoretical suggestions that the CP (or assuming CPT invariance, also the T) and C violations may occur in electromagnetic interactions, and a number of experiments have been performed to test these possibilities. For example, Christ and Lee [93] suggested that, if there were a T-violation in electromagnetic interactions, it might be detected in inelastic scattering of electrons. If the electrons have initial momentum **P**, a scattering momentum $\mathbf{P}_S$, after scattering on protons, and a spin orientation $\boldsymbol{\sigma}_P$, the T-violating asymmetry could be detected in the term

$$(\mathbf{P} \times \mathbf{P}_S) . \boldsymbol{\sigma}_p .$$

This subject was reviewed by Panofsky [94]. It was shown that such a term cannot be present in elastic scattering. However, even if such an asymmetry were found, it could be used as a proof of T-violation only if purely one-photon exchange controls the reaction. Thus, if an asymmetry were discovered, the experiment would have to also be repeated with inelastic positron scattering. Also, an inelastic state, with isotopic spin $= \frac{1}{2}$, as for example N*(1512), should be used since the more prominent N*(1238), with isotopic spin $= \frac{3}{2}$, would show no asymmetry if the T-violating interaction were an isotopic scalar.

Experimental results were reported by Appel *et al.* [95] and Chen. The results were negative. However, a similar experiment at higher energy and sensitivity was planned. Following a suggestion by Kosborev *et al.* [96], Prepost *et al.* [97] examined the polarization of recoil deuterons from elastic electron scattering. Since the deuteron has spin 1, the elastic electron scattering can contain T-violating terms which vanish for spin-$\frac{1}{2}$ protons, owing to current conservation. The results of this experiment were also negative. There is, however, an indication of a positive result reported by Lee and collaborators, giving the results shown in Fig. 18 of [62], which exhibits an asymmetry in the $\eta \rightarrow \pi^+ + \pi^- + \pi^0$ decay

$$A = \frac{N(E_{\pi^+} > E_{\pi^-}) - N(E_{\pi^+} < E_{\pi^-})}{N(E_{\pi^+} > E_{\pi^-}) + N(E_{\pi^+} < E_{\pi^-})} = 1{\cdot}5 \pm 0{\cdot}5 \text{ per cent.}$$

Hence, the effect appears to represent three standard deviations, which is about the respectable minimum. We should consider whether the effect is a false asymmetry due to interference with a small background. Treiman [98] treated this process and the various factors involved.

This process is forbidden by the strong interaction selection rules and, thus, it is considered to result from the combined effects of electromagnetic and strong interactions. According to this picture, the $3\pi$ final state must have $C = +1$ and $T = 1$ and, hence, must be symmetric under interchange of $\pi^+$ and $\pi^-$, thus, $A = 0$. $A \neq 0$ can arise only through interference between $C = +1$ and $C = -1$ ($T = 0$ and/or $T = 2$ states).

A detailed analysis of the Dalitz plot distribution indicates that there is a C-violating admixture of which the main component has $T = 2$ states.

Yuta and Okubo (see [98]) pointed out that the three-pion background in the $C = -1$ states could cause interference effects. Even within the narrow width of the $\eta$ resonance, the background contribution is about $10^{-4}$ of the three-pion intensity and, thus, is about $10^{-2}$ of the resonance amplitude (i.e. about 1 per cent) which could generate interference, just about the size of the asymmetry. Lee attempted to determine an upper bound of the $C = -1$ part of the background amplitude by analysing the Dalitz plot in those regions which were slightly outside the $\eta$ resonance region. On the basis of a preliminary analysis, it was concluded that the background interference contributions to the asymmetry are no larger than a few tenths of a per cent. Considering that three standard deviations are about the boundary line for demonstrating a statistically significant effect, and the possibility of systematic errors, and background interference effects, it does not appear that we can conclude that the effect has been demonstrated. The decreasing limits obtained by Ramsey and co-workers on the neutron electric dipole moment tend to cast doubt on the electromagnetic origin of the violations.

*If the CP violation is attributed to the weak interaction*, then we would, in general, expect CP violating effects to occur in other weak interactions than just the kaon decays. Of course, there are many possibilities of having the only large enough effects to be observed occur only in the kaon decays.

The 'superweak' model of Wolfenstein [99] results only in CP impurities in $K_S$ and $K_L$ states and in no other effects where this violation makes itself felt. In other words, all other CP violating amplitudes are zero. This comes about since, in this model, the CP violating term is a $|\Delta Y| = 2$ operator. To give the right value of $\eta_{+-}$, the $H_-$ interaction term must have a dimensionless coupling constant†

$$\frac{G_{H_-} M_P^2}{4\pi} \sim |\eta_{+-}| \left(\frac{G M_P^2}{4\pi}\right)^2.$$

† $H_-$ is the CP violating term in the interaction Hamiltonian.

Since $\frac{GM_P^2}{4\pi}$ is the CP conserving coupling constant ($\approx 10^{-6}$) then

$$\frac{G_{H_-}M_P^2}{4\pi} \sim 10^{-15}.$$

Thus, it is referred to as a 'superweak' interaction.

In the $|K^0\rangle$ or $|\bar{K}^0\rangle$ system, the terms due to $H_-$, which are introduced by the mass operator, are of the order

$$\frac{\frac{G_-M_P^2}{4\pi}}{\left(\frac{GM_P^2}{4\pi}\right)^2} \sim \frac{10^{-15}}{10^{-12}} \sim 10^{-3}.$$

Thus neglecting terms of order $10^{-9}$, we will not observe CP violation in other systems. The major predictions of this theory are $\eta_{+-} = \eta_{00} = \epsilon_0$.

Table 6a, p. 75 from Lee [100] summarizes the coupling constant of $H_-$ (the CP non-conserving part of the Hamiltonian) and the selection rules of $H_-$. In the super-weak theory only $K_L$ and $K_S$ should show the CP violating effects. If the general weak interaction violation case is assumed, we would expect that some other weak interaction would show those effects. So far one has not observed them.

If the CP violation is in the 'non-weak' (i.e. strong or electromagnetic) interactions, we could expect that some strong interaction processes would exhibit small C or T violations. If the CP violation is attributed to the electromagnetic interaction, then we would expect sizeable C, T violation amplitudes to show up in some electromagnetic processes. Of course, it is generally possible for resourceful theorists to arrange the theory to match observations in most models. However, the decreasing neutron electric dipole [101] moments ($(+2\pm 2)\times 10^{-23}$ e cm), for example, are beginning to make life difficult for the electromagnetic model, but on the other hand, the reported asymmetries in $\eta$ decay (if substantiated) support it. See Treiman [98] for a further general discussion of this subject including what one might call 'conglomerate models' (i.e. those that cut across general class lines).

At the 15th international conference on high energy physics (Kiev 1970), the experimental data seemed, in general, to support the super weak model best.

## 12.22. Conserved vector current

Gershtein and Zeldovich, [45], and Feynman and Gell-Mann [46] hypothesized that the strangeness-conserving part of the vector current is conserved. This naturally implies that it be equivalent in structure to the T-spin lowering part of the T-spin current. The scale was set by requiring that it be equal to the T-spin current. Thus, it was assumed that the isovector part of the electromagnetic current and the $J^{\alpha}_{V,\Delta S=0}$, and $J^{\alpha^+}_{V,\Delta S=0}$ transform as if they were members of the same isospin, $T = 1$, triplet. This, and the fact that the electromagnetic current is conserved

$$\left(\text{i.e. } \frac{\partial J^{\alpha}_{\text{el.}}(x)}{\partial x^{\alpha}} = 0\right),$$

leads to the fact that the vector current is also conserved

$$\left(\text{i.e. } \frac{\partial J^{\alpha}_{V,\Delta S=0}(x)}{\partial x^{\alpha}} = 0\right).$$

It was suggested [102] that a sensitive test of the conserved vector current (CVC) hypothesis could be obtained by comparing the decays to the $T = 1$, $J = 1$ triplet of nuclei $B^{12}$, $C^{12*}$, $N^{12}$ to the $C^{12}$, $T = J = 0$ state, as shown in Fig. 12.15(a).

Since the CVC theory predicts a definite magnetic moment form of the vector coupling, if we compare the $\beta^-$ and $\beta^+$ decays, the two spectrum shapes, owing to correction terms, deviate from the allowed transition form in opposite directions, in such a manner that the predicted ratio of the

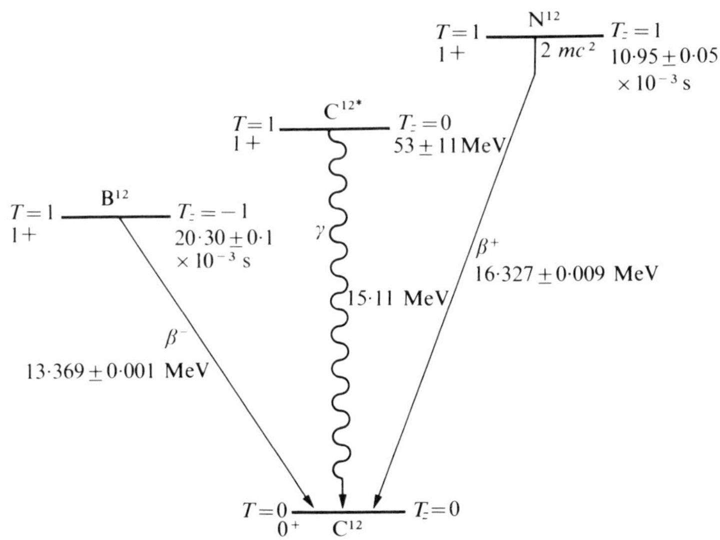


FIG. 12.15.(a) The decay scheme of the $B^{12}$, $C^{12*}$, and $N^{12}$ (Lee *et al.* 1963), [103].

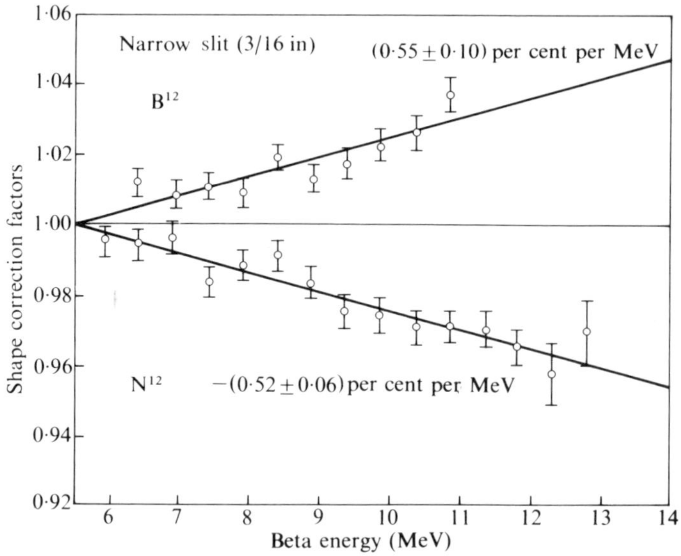


FIG. 12.15.(b) The observations of Lee *et al.* (as re-calculated by Wu 1964), [104], exhibiting the shape correction factors to the allowed spectrum shape due to weak magnetism compared to the predictions of conserved vector current theory.

deviations is quite sensitive to whether the vector current is conserved. The prediction was

$$\frac{C^-(E)}{C^+(E)} \approx 1+(2{\cdot}28\pm 0{\cdot}4)\times 10^{-2}E/m_e.$$

Lee *et al.* (1963) [103] used a refined method and their results recalculated in a review by Wu (1964) [104] are shown in Fig. 12.15(b). They gave, in agreement with the prediction,

$$C^-(E)/C^+(E) = 1+(2{\cdot}08\pm 0{\cdot}24)\times 10^{-2}\times E/m_e.$$

The branching ratio,

$$\frac{\pi^+ \to \pi^0+e^++\nu}{\pi^+ \to \mu^++\nu},$$

can be predicted on the basis of CVC, and the Cabibbo hypothesis (Cabibbo *et al.* 1967 [105]). The predicted result is

$$\frac{\pi^+ \to \pi^0+e+\nu}{\pi^+ \to \mu^++\nu} = (1{\cdot}035\pm 0{\cdot}005)\times 10^{-8}.$$

Based on a number of experiments the Particle Data Group tables gave

$$\frac{\pi^+ \to \pi^0 + e^+ + \nu}{\pi^+ \to \mu^+ + \nu} = (1{\cdot}02 \pm 0{\cdot}07) \times 10^{-8},$$

which is in good agreement with the prediction based on CVC.

The high-energy neutrino experiments can also be used to probe the vector, and axial vector, parts of the weak hadronic currents. The results obtained are consistent with CVC. However, the most useful applications come from starting with CVC as an assumption, and calculating the vector form factors from e–p scattering, or photo-production experiments, and then deducing the axial vector form factors from the neutrino experiments.

### 12.23. Partially-conserved axial current

The axial vector form factor $F_A(q^2 = 0)$ is close to, but significantly different from, unity (i.e. $F_A \approx 1{\cdot}18 \pm 0{\cdot}02$), thus suggesting that the axial current may be partially conserved. It cannot be fully conserved, for then the well-known decay $\pi \to \mu + \nu$ would be forbidden. This pion-decay matrix element has the following form,

$$(\Psi_0 J^\alpha_{A,\Delta S=0} \Psi_\pi) \propto M_p q^\alpha F_\pi(m_\pi^2).$$

If we assume that the axial current is completely conserved, then

$$\partial_\alpha J^\alpha_{A,\Delta S=0} = 0.$$

Hence,

$$(\Psi_0\, \partial_\alpha J^\alpha_{A,\Delta S=0} \Psi_\pi) \propto M_p m_\pi^2 F_\pi(m_\pi^2) = 0.$$

Thus, $F_\pi(m_\pi^2)$, which measures the rate of pion decay, would have to be zero. This suggests that perhaps the axial current would be completely conserved in a theory where the pion rest mass were zero.

The partially-conserved axial current hypothesis (PCAC) [106], [107] uses the assumption that the divergence of the axial current is proportional to $m_\pi^2 \phi(x)$, where $\phi(x)$ is the pseudoscalar meson field operator. With this assumption, we can express $F_A(0)$ in terms of the pion decay rate. The relation so obtained was first derived by Goldberger and Treiman [108] (the Goldberger–Treiman relation) using dispersion theory. The vector currents of the weak and electromagnetic interactions are generating currents of SU(3). Gell-Mann [109] proposed extending the symmetry, so that the pseudoscalar currents, also, are generating currents, and chose their commutation relations to avoid introducing new quantities in the algebra. These commutation relations then determined the scale of the axial currents, just as in the CVC theory the scale of the vector currents is determined.

The commutation relations, and the PCAC hypothesis, were used to determine the values of $F_A(0)$. This investigation gave results which are consistent with the above assumed commutation relations. Weisberger [110*a*] obtained $F_A \approx 1{\cdot}16$, whereas Adler [110*b*] attempted to take into account off-shell effects, and obtained $F_A \approx 1{\cdot}24$. These results agree well, within the uncertainties involved in the estimates and the experimental results. Further application of this method [111] also gave results in agreement with the experiment.

### 12.24. Do the weak interactions become strong at high enough energies?

Consider the following processes,

$$\nu_\mu + e^- \to \nu_e + \mu^-, \tag{12.152a}$$

$$\bar{\nu}_e + e^- \to \bar{\nu}_\mu + \mu^-, \tag{12.152b}$$

which are the minimum which must be present in a theory, since they represent the clearly-observed muon decay.

For $s \gg m_\mu^2$, the total cross-sections (see Marshak *et al.* [3] for a complete treatment and, also, Low [112]) are given by

$$\sigma(\nu_\mu e^-) = 3\sigma(\bar{\nu}_e e^-) = \frac{G^2 s}{\pi}. \tag{12.153}$$

At high enough $s$, eqn (12.153) will eventually lead to a violation of the unitarity limit. Reaction (12.152a) is almost a pure $s$-wave interaction while reaction (12.152b) is mostly $p$-wave, with some $s$-wave mixed in.

If $L_\mu$ is conserved, we should expect the self-current process to occur and, thus, we can expect

$$\nu_l + l^- \to \nu_l + l^-, \tag{12.154a}$$

$$\bar{\nu}_l + l^- \to \bar{\nu}_l + l^-, \tag{12.154b}$$

and for $s \gg m_\mu^2$, eqn (12.153) will hold, with $l^-$ substituted for $e^-$. For reaction (12.154a), which is an $s$-wave interaction, it is estimated that the cross-section, calculated for the first-order allowed process, will become larger than the unitary limit for $E_{\text{cms}}$ (of one particle) ($\gtrsim$350 GeV). Furthermore, if we calculate the higher-order matrix elements, the resultant cross-sections diverge to a much greater degree and further complicates the problem. If we should reach the 'unitarity limit' ($\sigma = 4\pi\lambda\!\!\!^{-2}$), the weak interaction is obviously as strong as any interaction (including the strong interaction) in the $s$-wave could be. At $E \sim 350$ GeV,

$$\sigma_{\text{weak}} \approx 3{\cdot}5 \times 10^{-32},$$

whereas at 10 GeV, the corresponding cross-section would be $\sim 10^{-41}$ cm$^2$.

Whereas in quantum electromagnetic interactions the divergence is logarithmic, in weak interactions it is quadratic. Although perturbation theory calculations should not be taken too seriously, it, nevertheless, suggests that surprises may await us in the high-energy realm. In this regard, even the intermediate vector boson theory does not help, and is highly divergent. There never has been an accepted explanation of the relationship between strong, electromagnetic, and weak interactions, and the foregoing may possibly indicate that at energies approaching 300 GeV (c.m.s.) may provide the key to understanding the relationships between the different interactions.

## REFERENCES

1. KALLEN, G. (1964). *Elementary particle physics*. Addison-Wesley Publishing Company, New York.
2. GASIOROWICZ, S. (1967). *Elementary particle physics*, John Wiley and Sons, Inc., New York.
3. MARSHAK, R. E., RIAZUDDIN, and RYAN, C. P. (1969). *Theory of weak interactions in particle physics*. Wiley Interscience, New York.
4. FERMI, E. (1934). *Nuovo Cim.* **11,** 1; (1934). *Z. Phys.* **88,** 161.
5. FIERZ, M. (1937). *Z. Phys.* **104,** 553.

6*a*. ALLEN, J. S. (1957). *Rev. mod. Phys.* **31,** 791; *b*. SUDARSHAN, E. C. G. and MARSHAK, R. E. (1957). *Proceedings of the Padua–Venice Conference on Recently Discovered Particles*.

7. PUPPI, G. (1948). *Nuovo Cim.* **5,** 587; KLEIN, O. (1948). *Nature*. **161,** 897; TIOMNO, J., and WHEELER, J. A. (1949). *Rev. mod Phys.* **21,** 144; LEE, T. D., ROSENBLUTH, M., and YANG, C. N. (1949). *Phys. Rev.* **75,** 905.
8. DICAPUA, E., GARLAND, R., PONDROM, L., and STRELZOFF, A. (1964). *Phys. Rev.* **B133,** 1333.
9. BOEHM, F., NOVAY, T. B., BARNES, C. A., and STECH, B. (1957). *Phys. Rev.* **108,** 1497.
10. LEE, T. D. and YANG, C. N. (1956). *Phys. Rev.* **104,** 254.
11. TANNER, N. (1957). *Phys. Rev.* **107,** 1203.
12. WILKINSON, D. H. (1958). *Phys. Rev.* **109,** 1603, 1610, 1614.
13. BOEHM, F. and KANKELEIT, E. (1963). *Phys. Rev. Lett.* **14,** 312.
14. LOBOSHOV, V. M., NAZÓRENKO, V. A., SAENKO, L. F., SMOTRITSKII, L. M., and IOFFE, A. F. (1966). *Soviet Phys. JETP. Lett.* **3,** 76, 268.
15. WU, C. S., AMBLER, E., HAYWARD, R. W., HOPPES, D. D., and HUDSON, R. P. (1957): *Phys. Rev.* **105,** 1413; (1957). **106,** 1361.
16. LEDERMAN, L. M., GARWIN, R. L., and WEINRICH, M. (1957). *Phys. Rev.* **105,** 1415.
17. FRIEDMAN, J. I. and TELEGDI, V. L. (1957). *Phys. Rev.* **105,** 1681.
18. CRAWFORD, JR., F. S., CRESTI, M., GOOD, M. L., GOTTSTEIN, K., LYMAN, E. M., SOLMITZ, F. T., STEVENSON, M. L., and TICHO, H. K. (1957). *Phys. Rev.* **108,** 1102.
19. EISLER, F., PLANO, R., PRODELL, A., SAMIOS, N., SCHWARTZ, M., STEINBERGER, J., BASSI, P., BORELLI, V., PUPPI, G., TANAKA, G., WOLOSCHEK, P., ZOHOLI, V.,

CONVERSI, M., FRANZINI, P., MANNELLI, I., SANTANGELO, R., SILVESTRINI, V., GLASER, D. A., GRAVES, C., and PERL, M. L. (1957). *Phys. Rev.* **108,** 1353.

20. ALIKHANOV, A., GALAKTIONOV, Y., GORODSKOV, Y., ELISEEV, G., and LYUBIMOV, Y. (1960). *Zh. eksp. teor. Fiz. SSSR* **38,** 1918; (1960), *Soviet Phys. JETP* **11,** 1380.
21. BACKENSTOSS, G., HYAMS, B. D., KNOP, G., MARIA, P. C., and STIERLIN, U. (1961). *Phys. Rev. Lett.* **6,** 415.
22. BARDON, M., FRANZINI, P., and LEE, J. (1961). *Phys. Rev. Lett.* **7,** 23.
23. ALLEN, J. S., BROWN, R. L., HEMMSFELD, W. B., STÄHLER, T., and BRAND, T. H. (1959). *Phys. Rev.* **116,** 134.
24. RIDLEY, B. W. (1961). *Nucl. Phys.* **25,** 483.
25. WEYL, H. (1929). *Z. Phys.* **56,** 330.
26. SALAM, A. (1957). *Nuovo Cim.* **5,** 299; LANDAU, L. D. (1957). *Nucl. Phys.* **3,** 127; LEE, T. D. and YANG, C. N. (1957). *Phys. Rev.* **105,** 1671.
27. BOEHM, F. and WAPSTRA, A. (1957). *Phys. Rev.* **107,** 1202.
28. REINES, F. (1960). *A. Rev. nucl. Sci.* **10,** 1; REINES, F. and COWAN, C. L. (1959). *Phys. Rev.* **113,** 273.
29. PRIMAKOFF, H. and ROSEN, S. P. (1959). *Rep. Prog. Phys.* **22,** 121; (1961). *Proc. Phys. Soc.* **78,** 464. See also Marshak *et al.* reference [3].
30. KIRSTEN, T., SCHAEFFER, O. A., NORTON, E., and STOENNER, R. W. (1968). *Phys. Rev. Lett.* **20,** 1300.
31. GOLDHABER, M., GRODZINS, L., and SUNYAR, A. (1958). *Phys. Rev.* **109,** 1015.

32*a*. DANBY, G., GAILLARD, J., GOULIANOS, K., LEDERMAN, L., MISTRY, N., SCHWARTZ, M., and STEINBERGER, J. (1962). *Phys. Rev. Lett.* **9,** 36; *b.* BERNARDINI, G., BIENLEIN, J., VON DARDEL, G., FAISSNER, H., FERRERO, F., GAILLARD, J., GERBER, H., HAHN, B., KAFTANOV, V., FRIENEN, F., MANFREDOTTI, C., REINHARZ, M., and SALMERON, R. (1964). *Phys. Lett.* **13,** 86.

33*a*. BURNS, R., DANBY, G., GOULIANOS, D., HYMAN, E., LEDERMAN, L., LEE, W., MISTRY, N., RETTBERG, J., SCHWARTZ, M., and SUNDERLAND, J. (1965). *Phys. Rev. Lett.* **15,** 42; *b.* BURNS, R., DANBY, G., HYMAN, E., LEDERMAN, L., LEE, W., RETTBERG, J., and SUNDERLAND, J. (1965). *Phys. Rev. Lett.* **15,** 830.

34. BARTLEY, J., BLOCK, M. M., BURMEISTER, H., CUNDY, D. C., EIBEN, B., FRANZINETTI, C., MOLLERUD, R., MYATT, G., NIKOLIC, M., ORKIN-LECOURTIS, A., PATTY, M., PERKINS, D. H., RAMM, C. A., SCHULZE, K., SLETTEN, H., SOOP, K., STUMP, R., VENUS, W., WOTHSMUTH, H. W., and YOSHIKI, H. (1965). Proceedings of the Conference on Weak Interactions, Argonne National Laboratory. p. 257.
35. WILLIS, W. and THOMPSON, J. (1968). *Decays of elementary particles. Advances in particle physics*, Vol. I. Wiley-Interscience, New York. p. 477.
36. PARKER, S., ANDERSON, H., and REY, C. (1964). *Phys. Rev.* **B133,** 768.
37. DUCLOS, J., HEINTZE, J., DE RUJULA, A., and SOERGEL, V. (1964). *Phys. Lett.* **9,** 62.
38. BUHLER, A., CABIBBO, N., FIDECARO, M., MASSAM, T., MULLER, T., SCHNEEGANS, M., and ZICHICHI, A. (1963). *Phys. Lett.* **7,** 368; BLOOM, S., DICK, L., FEUVRAIS, L., HENRY, G., MACQ, P., and SPIGHEL, M. (1964). *Phys. Lett.* **8,** 87.
39. BARDON, M., NORTON, P., PEOPLES, J., SACHS, A. M., and LEE-FRANZINI, J. (1965). *Phys. Rev. Lett.* **14,** 449.

40. SOSNOVSKI, A. N., SPIVAK, P. E., PROKOFIEN, Y. A., KUHKOV, I. E., and DOBININ, Y. P. (1959). *Nucl. Phys.* **10,** 395.
41. BERMAN, S. M. (1958). *Phys. Rev.* **112,** 267.
42. SUDARSHAN, E. C. G. and MARSHAK, R. E. (1958). *Phys. Rev.* **109,** 1860.
43. FEYNMAN, R. P. and GELL-MANN, M. (1958). *Phys. Rev.* **109,** 193.
44. SAKURAI, J. J. (1958). *Nuovo Cim.* **7,** 649.
45. GERSTEIN, S. S. and ZELDOVICH, Y. B. (1956). *Soviet Phys. JETP* **2,** 574; FEYNMAN, R. P. and GELL-MANN, M. (1958). *Phys. Rev.* **109,** 193.
46. FEYNMAN, R. P. and GELL-MANN, M. (1958). *Phys. Rev.* **109,** 193.
47. PARTICLE DATA GROUP. (1968). *Rev. mod. Phys.* **40,** 77; (1970). **42,** 87.
48. RUBBIA, C. (1969). *Weak interaction physics, High energy physics.* Vol. III. Academic Press, New York. p. 283. This review is also useful as a general reference.
49. WILLIS, W. J. (1967). *Proceedings of the Heidelberg Conference on Elementary Particles.* NORTH HOLLAND PUBLISHING COMPANY, AMSTERDAM. pp. 253, 1967.
50. GELL-MANN, M. and PAIS, A. (1955). *Phys. Rev.* **97,** 1387; PAIS, A. and PICCIONI, O. (1955). *Phys. Rev.* **100,** 1487.
51*a*. BELL, J. S. and STEINBERGER, J. (1965). *Proceedings of the Oxford International Conference on Elementary Particles.* P. 195.
51*b*. OKUN, L. B. and RUBBIA, C. (1967). *Proceedings of the Heidelberg Conference on Elementary Particles.* North Holland Publishing Company, Amsterdam. p. 304.
52. CHRISTENSON, J. M., CRONIN, J. W., FITCH, V. R., and TURLAY, R. (1964). *Phys. Rev. Lett.* **13,** 138.
53. ABASHIAN, A., ABRAMS, R. J., CARPENTER, D. W., FISHER, G. P., NEFKENS, B. M. K., and SMITH, J. H. (1964). *Phys. Rev. Lett.* **13,** 243.
54. GALBRAITH, W., MANNING, G., TAYLOR, A. E., JONES, B. D., MALOS, J., ASTBURY, A., LIPMAN, N. M., and WALKER, G. T. (1965). *Phys. Rev. Lett.* **14,** 383.
55. BASILE, P., BINGERT, J., CRONIN, J. W., THEVENER, B., TURLAY, R., ZYLBERACH, S., and ZYLBERSTEIN, A. (1966). *Symposium on weak interactions, Bulaton.*
56. BOTT-BODENHAUSEN, M., DE BOUARD, X., DEKKERS, D., FELST, R., MERMOD, R., SAVIN, I., SCHARFF, P., VIVARGENT, M., WILLITS, T. R., and WINTER, K. (1967). *Phys. Lett.* **B24,** 438.
57. FITCH, V. L., ROTH, R. F., RUSS, J., and VERNON, W. (1967). *Phys. Rev.* **164,** 1711.
58. DE BOUARD, X., DEKKERS, D., JORDAN, B., MERMOD, R., WILLITTS, T. R., WINTER, K., SCHARFF, P., VALENTIN, L., VIVARGENT, M., and BOTT-BODENHAUSEN, M. (1967). *Nuovo Cim.* **52,** 662.
59. WU, T. T. and YANG, C. N. (1964). *Phys. Rev. Lett.* **13,** 380.
60. ABBUD, F., LEE, B. W., and YANG, C. N. (1967). *Phys. Rev. Lett.* **18,** 980.
61. SELOVE, W. (1968). *Proceedings of the XIV International Conference on High Energy Physics, Vienna.* PAPER 801.
62. CRONIN, J. (1968). *Proceedings of the XIV International Conference on High Energy Physics. Vienna.*
63. CAMERINI, U., FRY, W. F., GAIDOS, J. A., HUZITA, H., NATALI, S. V., WILLMAN, R. B., BIRGE, R. W., ELY, R. P., POWELL, W. M., and WHITE, H. (1962). *Phys. Rev.* **128,** 362; CAMERINI, U., CLINE, D., ENGLISH, J. B., FISHBEIN, W., FRY, W. F.,

GAIDOS, J. A., HARTMAN, R., MARCH, R. H., and STARK, R. (1966). *Phys. Rev.* **150,** 1148.

64. MEISNER, G. W., CRAWFORD, B. B., and CRAWFORD, F. S. JR. (1966). *Phys. Rev. Lett.* **16,** 278; (1966). **17,** 492.
65. FITCH, V. L., PIROUE, P. A., and PERKINS, R. B. (1960). *Nuovo Cim.* **22,** 1160.
66. AUBERT, B., BEHR, L., CANAVAN, F. L., CHOUNET, L. M., LOWYS, J. P., MITTNER, P., and PASCAUD, E. (1965). *Phys. Lett.* **17,** 59.
67. BALDO-CEOLIN, M., CALIMANI, E., CIAMPOLILLO, S., FILIPPI-FILOSOFO, C., HUZITA, H., MATTIOLI, F., and MIARI, G. (1965). *Nuovo Cim.* **38,** 684.
68. HILL, D. G., LÜERS, D., ROBINSON, D. K., SAKITT, M., SKJEGGESTADT, M., CANTER, O., CHO, Y., DRALLE, A., ENGLER, A. E., FISK, W. E., KRAEMER, R. W., and MELTZER, C. M. (1967). *Phys. Rev. Lett.* **19,** 668; CAMERINI, U., CLINE, D., ENGLISH, J. B., FISHBEIN, W., FRY, W. F., GARDOS, J. A., HARTMAN, R. D., MARCH, R. H., and STARK, R. (1966). *Phys. Rev.* **150,** 1148.
69. FITCH, V. L., PIROUE, P. A., and PERKINS, R. B. (1961). *Nuovo Cim.* **22,** 1160.
70. GOOD, R. H., MATSEN, R. P., MULLER, F., PICCIONI, O., POWELL, W. M., WHITE, H. S., FOWLER, W. B., and BIRGE, R. W. (1961). *Phys. Rev. Lett.* **124,** 1223.
71. FUJII, T., JOVANICH, J. V., TURKOF, F., and ZORN, G. T. (1964). *Phys. Rev. Lett.* **13,** 253.
72. CHRISTENSON, J. M., CRONIN, W. J., FITCH, V. L., and TURLAY, R. (1964). *Phys. Rev. Lett.* **13,** 138.
73. VISHNEVSKY, M. E., GALANINA, N. D., SEMENOV, Y. A., KRUPCHITSKY, A. P., BEREZIN, V. M., and MURISOV, V. A. (1965). *Phys. Lett.* **18,** 339.
74. ALFF-STEINBERGER, C., HEUER, W., KLEINKNECHT, K., RUBBIA, C., SCRIBANO, A., STEINBERGER, J., TANNENBAUM, M. J., and TITTEL, K. (1966). *Phys. Lett.* **21,** 595.
75. BOTT-BODENHAUSEN, M., DE BOUARD, X., CASSEL, D. G., DEKKERS, D., FELST, R., MERMOD, R., SAVIN, I., SCHARFF, P., VIVARGENT, M., WILLITS, T. R., and WINTER, K. (1966). *Phys. Lett.* **23,** 277.
76. CANTER, J., CHO, T., ENGLER, A., FISK, H. E., KRAEMER, R. W., MELTZER, C., HILL, D. G., ROBINSON, D. K., and SAKITT, M. (1967). *Phys. Rev. Lett.* **17,** 942.
77. MELHOP, W. H. W., GOOD, R. M., PICCIONI, O., SWANSON, R. A., MURTY, S. S., BURNETT, T. M., HOLLAND, C. M., and BOWLES, P. (1967). *Proceedings of the International Conference on Elementary Particles, Heidelberg.*
78. BENNETT, S., NYGREN, D., SAAL, H., STEINBERGER, J., and SUNDERLAND, J. (1967). *Phys. Rev. Lett.* **19,** 993.
79. DORFAN, D., ENSTROM, J., RAYMOND, D., SCHWARTZ, M., WOJCICKI, S., MILLER, D. H., and PACIOTTI, M. (1967). *Phys. Rev. Lett.* **19,** 987.
80. BENNETT, S., NYGREN, D., SAAL, H., SUNDERLAND, J., STEINBERGER, J., and KLEINKNECHT, K. (1968). *Phys. Lett.* **B27,** 244.
81. CABIBBO, N. (1963). *Phys. Rev. Lett.* **10,** 531.
82. ADEMOLLO, M. and GATTO, R. (1964). *Phys. Rev. Lett.* **13,** 264.
83. OKUBO, S. (1959). *Nuovo Cim.* **13,** 292.
84. ONEDA, S. and SUCHER, J. (1965). *Phys. Rev. Lett.* **15,** 927.
85. BARKAS, W. H., BIRNBAUM, W., and SMITH, F. M. (1956). *Phys. Rev.* **101,** 778.

86. Shafer, R. E., Crowe, K. M., and Jenkins, D. A. (1965). *Phys. Rev. Lett.* **14,** 923.
87. Lathrop, J., Lundy, R. A., Telegdi, V. L., Winston, R., and Yovanovitch, D. D. (1960). *Nuovo Cim.* **17,** 109.
88. Charpak, G., Farley, F. J. M., Garwin, R. L., Mueller, T., Sens, J. C., and Zichichi, A. (1965). *Nuovo Cim.* **37,** 1241.
89. Meyer, S. L., Anderson, E. W., Bleser, E., Lederman, L. M., Rosen, J. L., Rothberg, J., and Wang, I. T. (1963). *Phys. Rev.* **132,** 2693.
90*a*. Lobkowicz, F., Melissinos, A. C., Nagashima, Y., Tewksbury, S., Von Briesen, H., and Fox, J. D. (1967). *Phys. Rev. Lett.* **17,** 548.
90*b*. Ayres, D. S., Cormack, A. M., Greenberg, A. J., Kenney, R., Caldwell, D. O., Elings, V. B., Hesse, W. P., Morrison, R. J. (1968). *Phys. Rev. Lett.* **21,** 261.
91*a*. Casella, R. C. (1968). *Phys. Rev. Lett.* **21,** 1128; (1969). **22,** 554; (1969). *Proceedings of the Third Hawaii Topical Conference on Particle Physics.*
91*b*. Steinberger, J. (1969). *Proceedings of the CERN International Conference on Weak Interactions, Geneva.*
92*a*. Lee, T. D. (1966). *Proceedings of the XIII International Conference on High Energy Physics, Berkeley.* (1965). *Proceedings of the Oxford Conference on Elementary Particles.*
92*b*. Treiman, S. B. (1968). *Proceedings of the XIV International Conference on High Energy Physics, Vienna.*
93. Christ, N. and Lee, T. D. (1965). *Phys. Rev. Lett.* **B139,** 1650.
94. Panofsky, W. (1968). *Proceedings of the XIV International Conference on High Energy Physics, Vienna.*
95. Appel, J. A. (1968). *Proceedings of the XIV International Conference on High Energy Physics, Vienna.* paper 483; See also Chen, J. (1968). *Thesis*, Harvard University.
96. Kobzarev, I., Olson, L. B., and Terentyev, M. V. (1965). *Soviet Phys. JETP Lett.* **2,** 289.; see also Dubovik, V. M., and Cheshkov, A. A. (1967). *Soviet Phys. JETP* **24,** 111; Dubovik, V. M., Likhtman, E. P., and Cheshkov, A. A. (1967). *Soviet Phys. JETP* **25,** 464.
97. Prepost, R., Simonds, R. M., and Wick, B. H. (1968). *Proceedings of the XIV International Conference on High Energy Physics, Vienna.* paper 786.
98. Treiman, S. B. (1968). *Proceedings of the XIV International Conference on High Energy Physics, Vienna.*
99. Wolfenstein, L. (1964). *Phys. Rev. Lett.* **13,** 562; Lee, T. D., and Wolfenstein, L. (1965). *Phys. Rev.* B**138,** 1490.
100. Lee, T. D. (1966). *Proceedings of the XIII International Conference on High Energy Physics, Berkeley.* p. 75.
101. Private communication by Miller, P. D., Dreis, W. D., Bond, J. K., and Ramsey, N. F. (Quoted by Cronin. (1968). *Proceedings of the XIV International Conference on High Energy Physics, Vienna.*)
102. Gell-Mann, M. (1958). *Phys. Rev.* **111,** 362.
103. Lee, Y. K., Mo, L. E., and Wu, C. S. (1963). *Phys. Rev. Lett.* **10,** 253.
104. Wu, C. S. (1964). *Rev. mod. Phys.* **36,** 618.

105. Cabibbo, N., Maiani, L., and Preparata, G. (1967). *Phys. Lett.* **25,** 29.
106. Gell-Mann, M., and Levy, M. (1960). *Nuovo Cim.* **16,** 705.
107. Bernstein, J., Fubini, S., Gell-Mann, M., and Thirring, W. (1960). *Nuovo Cim.* **17,** 757.
108. Goldberger, M. and Treiman, S. B. (1958). *Phys. Rev.* **110,** 1178.
109. Gell-Mann, M. (1962). *Phys. Rev.* **125,** 1067; (1964). *Physics* **1,** 63.
110*a*. Weisberger, W. I. (1961). *Phys. Rev. Lett.* **14,** 1047; *b*. Adler, S. L. (1965). *Phys. Rev. Lett.* **14,** 1051.
111. Amati, D., Bouchiat, C., and Nugts, J. (1965). *Phys. Lett.* **19,** 59.
112. Low, F. (1968). *Proceedings of the Trieste International Symposium on Contemporary Physics.*

# 13

# HIGH-ENERGY BEHAVIOUR OF TOTAL CROSS-SECTIONS AND ELASTIC SCATTERING

## 13.1. Introduction

As was mentioned in Chapter 4, the total cross-sections in the few hundred MeV to several GeV region show clearly defined energy structure, or resonance effects, as a function of increasing energy. These effects are quite large below 1 GeV, and become much smaller as the energy increases, and by the time we reach about 10 GeV, there are no known structure effects in total interaction cross-sections. The structure, or lack of it, in total cross-sections is closely related to the question of asymptotic behaviour of high-energy phenomena.

## 13.2. Asymptotic behaviour

For nearly two decades, physicists have been engaged in a search for a high enough energy, beyond which at least the gross features of high energy strong interactions would have exhibited an asymptotic behaviour with increasing energy.

The first theory which contained the elements of an asymptotic theory was the Fermi statistical theory. As discussed in Chapter 6, Fermi proposed that, after reaching a sufficiently high energy, the incident and target particles in a high-energy collision stuck together at rest in the centre of mass system, and formed a high temperature volume in thermodynamic equilibrium, which subsequently boiled off particles according to thermodynamic statistical probabilities. On the basis of analysis of existing cosmic-ray data, Fermi believed that kinetic energies of about 1 GeV were sufficient to establish this thermodynamical equilibrium. It was shown in Chapter 6 that interactions involving several GeV incident particles exhibited striking structure effects, and that isobar and one-particle exchange models were much more appropriate descriptions of the observed phenomena than the Fermi statistical theory. Obviously, asymptotic behaviour could not be expected to begin while new unanticipated structure effects were being observed with increasing energy.

The next round of asymptotic theories dealt with simpler and more quantitatively analysable basic quantities, such as total cross-section and elastic scattering amplitude behaviour.

## 13.3. The Pomeranchuk–Okun rule

The first set of these asymptotic predictions were proposed by Pomeranchuk (1956) [1] and Okun and Pomeranchuk (1956) [2] and provided rules for relating the various elastic scattering amplitudes when particles of an isotopic spin multiplet were incident upon target particles which also belonged to an isotopic spin multiplet. The method was centred on the assumption that, at high enough energies, the charge-exchange amplitude, being one of an arbitrarily large number of inelastic channels, would become arbitrarily small compared to the elastic scattering amplitudes, which represent the coherent effect of many inelastic channels, and by the optical theorem must have a large enough imaginary part to give the proper total cross-section. Okun and Pomeranchuk then employed the assumption of isotopic spin conservation to relate various non-charge-exchange elastic scattering amplitudes to the vanishing charge-exchange amplitude and, thus, to each other.

Let us first consider the following elastic scattering and charge-exchange interactions,

$$\bar{p}+p \to \bar{p}+p, \tag{13.1}$$

$$\bar{p}+n \to \bar{p}+n, \tag{13.2}$$

$$\bar{p}+p \to \bar{n}+n. \tag{13.3}$$

The application of charge symmetry will give us the three corresponding anti-neutron–nucleon reactions, which are equivalent, respectively, to the above three. There are also three similar proton–nucleon reactions,

$$p+p \to p+p \tag{13.4}$$

$$p+n \to p+n \quad \text{(ordinary scattering)}, \tag{13.5}$$

$$p+n \to n+p \quad \text{(charge-exchange scattering)}. \tag{13.6}$$

These latter relations are not in any way related to the first three, since the baryon number has changed from 2 to 0. We prefer to treat the first three, since charge-exchange scattering is clearly defined (i.e., reaction (13.1) is obviously different from reaction (13.2), whereas there is some possibility of confusion in separating reaction (13.5) from reaction (13.6)). However, at very high energy, since reaction (13.6) is supposed to become negligibly small, and be sharply peaked forward, the separation may still, in principle, be possible.

Isotopic spin conservation allows us to express all three reactions (13.1)–(13.3) in terms of $f_{T=1}$ and $f_{T=0}$, the $T = 1$ and $T = 0$ scattering amplitudes, respectively. Thus we can write for these reactions (using the last digit of the equation number as an identification subscript on $\sigma$),

$$\frac{d\sigma_1}{dt} = \frac{1}{4}\frac{\pi}{k^2}|f_{T=0}+f_{T=1}|^2, \tag{13.7}$$

$$\frac{d\sigma_2}{dt} = \frac{\pi}{k^2}|f_{T=1}|^2, \tag{13.8}$$

$$\frac{d\sigma_3}{dt} = \frac{\pi}{4k^2}|f_{T=0}-f_{T=1}|^2. \tag{13.9}$$

Then, if we assume that the charge-exchange scattering cross-section becomes arbitrarily small, compared to the elastic scattering cross-section at high enough energy, we imply that

$$\lim_{s\to\infty} \frac{|f_{T=0}-f_{T=1}|^2}{|f_{T-1}|^2} \to 0, \tag{13.10}$$

$$\lim_{s\to\infty} \frac{|f_{T=0}-f_{T=1}|^2}{|f_{T=0}+f_{T=1}|^2} \to 0. \tag{13.11}$$

Therefore,

$$\lim_{s\to\infty} f_{T=0} \text{ (anti-nucleon–nucleon system)} = $$
$$= f_{T=1} \text{ (anti-nucleon–nucleon system)}. \tag{13.12}$$

Similarly, we obtain

$$\lim_{s\to\infty} f_{T=0} \text{ (nucleon–nucleon system)} = $$
$$= \lim f_{T=1} \text{ (nucleon–nucleon system)}. \tag{13.13}$$

Thus,

$$\lim_{s\to\infty}\frac{d\sigma_1}{dt} = \lim_{s\to\infty}\frac{d\sigma_2}{dt},$$

and

$$\lim_{s\to\infty}\frac{d\sigma_4}{dt} = \lim_{s\to\infty}\frac{d\sigma_5}{dt}.$$

Hence, when $s \to \infty$, the Pomeranchuk–Okun rules yield the following equalities

$$\frac{d\sigma}{dt}(\mathrm{p+p \to p+p}) = \frac{d\sigma}{dt}(\mathrm{p+n \to p+n}) = \frac{d\sigma}{dt}(\mathrm{n+n \to n+n}), \tag{13.14}$$

and

$$\frac{d\sigma}{dt}(\bar{p}+p \to \bar{p}+p) = \frac{d\sigma}{dt}(\bar{p}+n \to \bar{p}+n) = \frac{d\sigma}{dt}(\bar{n}+n \to \bar{n}+n)$$

$$= \frac{d\sigma}{dt}(\bar{n}+p \to \bar{n}+p). \tag{13.15}$$

Similar relations (Pomeranchuk and Okun 1956) were derived for pion–nucleon scattering. Consider the reactions

$$\pi^+ + p \to \pi^+ + p, \tag{13.16}$$
$$\pi^- + p \to \pi^- + p, \tag{13.17}$$
$$\pi^- + p \to \pi^0 + n, \tag{13.18}$$
$$\pi^0 + p \to \pi^0 + p, \tag{13.19}$$
$$\pi^0 + p \to \pi^+ + n, \tag{13.20}$$
$$\pi^+ + n \to \pi^+ + n, \tag{13.21}$$
$$\pi^+ + n \to \pi^0 + p, \tag{13.22}$$
$$\pi^- + n \to \pi^- + n, \tag{13.23}$$
$$\pi^0 + n \to \pi^0 + n, \tag{13.24}$$
$$\pi^0 + n \to \pi^- + p. \tag{13.25}$$

Owing to the isotopic spin conservation, all ten of these reactions are expressible in terms of only two isotopic spin amplitudes $f_{T=\frac{1}{2}}$ and $f_{T=\frac{3}{2}}$. The reactions of practical interest experimentally are (13.16)–(13.18) and their charge symmetric equivalents (13.21)–(13.23). Using isotopic spin conservation, we can write

$$\frac{d\sigma}{dt}(\pi^+ + p \to \pi^+ + p) = \frac{\pi}{k^2}|f_{T=\frac{3}{2}}|^2 = \frac{d\sigma}{dt}(\pi^- + n \to \pi^- + n), \tag{13.26}$$

$$\frac{d\sigma}{dt}(\pi^- + p \to \pi^- + p) = \frac{\pi}{9k^2}|f_{T=\frac{3}{2}} + 2f_{T=\frac{1}{2}}|^2 = \frac{d\sigma}{dt}(\pi^+ + n \to \pi^+ + n), \tag{13.27}$$

$$\frac{d\sigma}{dt}(\pi^- + p \to \pi^0 + n) = \frac{2\pi}{9k^2}|f_{T=\frac{3}{2}} - f_{T=\frac{1}{2}}|^2 = \frac{d\sigma}{dt}(\pi^+ + n \to \pi^0 + p). \tag{13.28}$$

Thus, assuming that as $s \to \infty$ the charge-exchange cross-sections become negligibly small compared to the elastic scattering cross-sections, we obtain

$$\lim_{s\to\infty} f_{T=\frac{3}{2}}(s,\, t) = \lim_{s\to\infty} f_{T=\frac{1}{2}}(s,\, t). \tag{13.29}$$

Hence, we obtain from the Pomeranchuk–Okun rule the following equalities, when $s \to \infty$,

$$\frac{\mathrm{d}\sigma}{\mathrm{d}t}(\pi^+ + \mathrm{p} \to \pi^+ + \mathrm{p}) = \frac{\mathrm{d}\sigma}{\mathrm{d}t}(\pi^- + \mathrm{p} \to \pi^- + \mathrm{p}) = \frac{\mathrm{d}\sigma}{\mathrm{d}t}(\pi^- + \mathrm{n} \to \pi^- + \mathrm{n})$$

$$= \frac{\mathrm{d}\sigma}{\mathrm{d}t}(\pi^+ + \mathrm{n} \to \pi^+ + \mathrm{n}). \tag{13.30}$$

For the pion–anti-nucleon system we, in a similar manner, obtain the following relationships, as $s \to \infty$,

$$\frac{\mathrm{d}\sigma}{\mathrm{d}t}(\pi^+ + \bar{\mathrm{p}} \to \pi^+ + \bar{\mathrm{p}}) = \frac{\mathrm{d}\sigma}{\mathrm{d}t}(\pi^- + \bar{\mathrm{p}} \to \pi^- + \bar{\mathrm{p}}) = \frac{\mathrm{d}\sigma}{\mathrm{d}t}(\pi^- + \bar{\mathrm{n}} \to \pi^- + \bar{\mathrm{n}})$$

$$= \frac{\mathrm{d}\sigma}{\mathrm{d}t}(\pi^+ + \bar{\mathrm{n}} \to \pi^+ + \bar{\mathrm{n}}). \tag{13.31}$$

There is no relationship between the terms in eqn (13.30) and eqn (13.31) because the baryon numbers are +1 and −1, respectively. Eqns (13.29)–(13.31) imply certain relationships between total cross-sections via the optical theorem, which directly relates the imaginary part of the forward scattering amplitude to the total cross-section. Thus, as $s \to \infty$, we can conclude

$$\sigma_{\text{total}}(\pi^- + \mathrm{p}) = \sigma_{\text{total}}(\pi^+ + \mathrm{p}) = \sigma_{\text{total}}(\pi^- + \mathrm{n}) = \sigma_{\text{total}}(\pi^+ + \mathrm{n}), \tag{13.32}$$

and

$$\sigma_{\text{total}}(\pi^+ + \bar{\mathrm{p}}) = \sigma_{\text{total}}(\pi^- + \bar{\mathrm{p}}) = \sigma_{\text{total}}(\pi^- + \bar{\mathrm{n}}) = \sigma_{\text{total}}(\pi^+ + \bar{\mathrm{n}}). \tag{13.33}$$

In the nucleon–nucleon and anti-nucleon–nucleon interactions, the optical theorem can only be applied if we assume spin independence. This certainly appears to be a reasonable assumption as $s \to \infty$. Thus, we obtain the relations, as $s \to \infty$,

$$\sigma_{\text{total}}(\mathrm{p} + \mathrm{p}) = \sigma_{\text{total}}(\mathrm{p} + \mathrm{n}) = \sigma_{\text{total}}(\mathrm{n} + \mathrm{n}), \tag{13.34}$$

$$\sigma_{\text{total}}(\bar{\mathrm{p}} + \mathrm{p}) = \sigma_{\text{total}}(\bar{\mathrm{p}} + \mathrm{n}) = \sigma_{\text{total}}(\bar{\mathrm{p}} + \mathrm{n}) = \sigma_{\text{total}}(\bar{\mathrm{n}} + \mathrm{p}). \tag{13.35}$$

These rules can be applied to other systems such as, for example, the kaon–nucleon systems.

## 13.4. The Pomeranchuk theorem

Pomeranchuk later generalized the above results to include connections between total cross-sections for a target particle when the incident beam particle is replaced by its anti-particle. These relations are known as the Pomeranchuk theorem (1958), which states that

$$\lim_{s\to\infty} \sigma_{\text{total}}(\alpha+\beta) = \lim_{s\to\infty} \sigma_{\text{total}}(\bar{\alpha}+\beta), \tag{13.36}$$

where $\beta$ is the common target particle and $\alpha$ and $\bar{\alpha}$ are a particle–anti-particle pair.

For a nucleon (i.e. p or n) target particle, the Pomeranchuk–Okun rules already gave this result for incident pions, since the anti-particle of a pion belongs to the same isotopic spin multiplet as the original pion. However, in the case of nucleons, the Pomeranchuk–Okun rule did not give these relationships which now state, as $s \to \infty$,

$$\sigma_{\text{total}}(\mathrm{p}+\mathrm{p}) = \sigma_{\text{total}}(\bar{\mathrm{p}}+\mathrm{p}), \tag{13.37}$$

$$\sigma_{\text{total}}(\mathrm{n}+\mathrm{p}) = \sigma_{\text{total}}(\bar{\mathrm{n}}+\mathrm{p}), \tag{13.38}$$

$$\sigma_{\text{total}}(\mathrm{p}+\mathrm{n}) = \sigma_{\text{total}}(\bar{\mathrm{p}}+\mathrm{n}), \tag{13.39}$$

$$\sigma_{\text{total}}(\mathrm{n}+\mathrm{n}) = \sigma_{\text{total}}(\bar{\mathrm{n}}+\mathrm{n}). \tag{13.40}$$

For the kaon–nucleon system the Pomeranchuk theorem states, as $s \to \infty$,

$$\sigma_{\text{total}}(\mathrm{K}^{+}+\mathrm{p}) = \sigma_{\text{total}}(\mathrm{K}^{-}+\mathrm{p}), \tag{13.41}$$

$$\sigma_{\text{total}}(\mathrm{K}^{0}+\mathrm{p}) = \sigma_{\text{total}}(\bar{\mathrm{K}}^{0}+\mathrm{p}), \tag{13.42}$$

$$\sigma_{\text{total}}(\mathrm{K}^{+}+\mathrm{n}) = \sigma_{\text{total}}(\mathrm{K}^{-}+\mathrm{n}). \tag{13.43}$$

Pomeranchuk's original proof made use of some plausible assumptions, and employed the forward dispersion relations. He estimated that the theorem would become applicable at about 10 GeV incident energy, and a check on its validity was one of the interesting early investigations at both the new CERN 28 GeV proton synchrotron (which became available for experiments in 1960) and the 33 GeV Brookhaven AGS (which became available in 1961).

Before discussing the experimental situation, we will summarize the methods of proof of the Pomeranchuk theorem.

## 13.5. Proof of the Pomeranchuk theorem

As we saw in Chapter 5, the basic axioms of quantum field theory allowed us to establish the required analyticity properties of the forward scattering amplitude to obtain the forward dispersion relations. Utilizing these relations and the establishment of the Froissart bound (or even an improved Greenberg–Low bound) allowed us to conclude that a doubly subtracted forward dispersion relation will be convergent.† A general description of the methods of proof of the Pomeranchuk theorem is given in Eden's book (Chap. 5, [20]). We only present the basic elements here. If‡

$$F_s(\alpha+\beta \to \alpha+\beta) \tag{13.44}$$

is the $s$-channel scattering amplitude, then the corresponding $u$-channel scattering amplitude is

$$F_u(\alpha+\bar{\beta} \to \alpha+\bar{\beta}). \tag{13.45}$$

We can define the anti-symmetric forward scattering amplitude as

$$F_A(\omega, 0) = F_s(\omega, 0) - F_u(\omega, 0). \tag{13.46}$$

Thus we can write the doubly subtracted dispersion relation §

$$\operatorname{Re} F_A(\omega, 0) = \omega F_A^{(1)}(0, 0) + \frac{2\omega^3}{\pi} \int_m^\infty \frac{\operatorname{Im} F_A(\omega', 0)}{\omega'^2(\omega'^2 - \omega^2)}\, d\omega', \tag{13.47}$$

where $F_A^{(1)}$ is the first derivative of $F_A$.

Let us now assume that the total cross-sections approach constants (which can be different) as $\omega \to \infty$, denoted by $\sigma_s$ and $\sigma_u$, respectively, for the $s$-channel and the $u$-channel cross-section. Then

$$\sigma_A = \sigma_s - \sigma_u.$$

Employing the optical theorem as $E \to \infty$, we obtain

$$\operatorname{Im} F_A(\omega, 0) \sim \sigma_A \omega \sim (\sigma_s - \sigma_u)\omega. \tag{13.48}$$

Using eqn (13.48) in the dispersion relation, we obtain

$$\operatorname{Re} F_A(\omega, 0) \sim (\sigma_s - \sigma_u)\omega \log \omega. \tag{13.49}$$

For $F_s$, Re $F_s$ is small compared to Im $F_s$. Thus it can be shown that, since

† The singly subtracted $D^+$ in Chapter 5 has the convergence equivalence of a doubly subtracted dispersion relation owing to the crossing symmetry effects discussed in Chapter 5.

‡ $F(s, t) = 8\pi s^{\frac{1}{2}} f(s, t)$, where $\dfrac{d\sigma}{dt} = |f(s, t)|^2$.

§ The even (constant) subtraction term turned out to be zero due to the anti-symmetry, and the reality in the neighbourhood of $\omega = 0$.

the imaginary parts of $F_s$ and $F_u$ behave like $\sigma_s\omega$ and $\sigma_u\omega$ as $\omega \to \infty$, whereas the real parts behave as a constant $\times$ $\sigma_s\omega \log \omega$ and a

$$\text{constant} \times \sigma_u\omega \log \omega$$

then, as $\omega \to \infty$,

$$\frac{\text{Re } F_s}{\text{Im } F_s} \sim \log \omega \tag{13.50}$$

and

$$\frac{\text{Re } F_u}{\text{Im } F_u} \sim \log \omega. \tag{13.51}$$

This is a physically unexpected result, which was generally considered unacceptable and, thus, it was assumed that

$$\sigma_s - \sigma_u \to 0 \tag{13.52}$$

faster than $\dfrac{1}{\log \omega}$, so as to make $\dfrac{\text{Re } F_s}{\text{Im } F_s}$ and $\dfrac{\text{Re } F_u}{\text{Im } F_u} \to 0$ as $\omega \to \infty$, thus giving us the Pomeranchuk theorem

$$\sigma(\alpha+\beta \to \alpha+\beta) = \sigma(\bar{\alpha}+\beta \to \bar{\alpha}+\beta). \tag{13.53}$$

Eden (1966) [4] demonstrated that if we use unitarity, the Pomeranchuk theorem can be proven for the case

$$(\sigma_s - \sigma_u) \sim C_{\text{A}}(\log \omega)^m, \qquad 0 < m \leq 2. \tag{13.54}$$

This would include the case of the cross-section obeying the Froissart bound. In this case, we obtain

$$\text{Re } F_{\text{A}}(\omega, 0) \sim \text{constant} . C_{\text{A}} E(\log E)^{m+1}. \tag{13.55}$$

It was also shown that the Pomeranchuk theorem can be established using unitarity, even if total cross-sections increase as fast as $(\log \omega)^\epsilon$ for any positive $\epsilon$. In this case, the ratio of

$$\frac{\sigma(\alpha+\beta \to \alpha+\beta)}{\sigma(\bar{\alpha}+\beta \to \bar{\alpha}+\beta)} \to 1, \tag{13.56}$$

as $\omega \to \infty$. If we assume that the difference of the cross-sections oscillates [5]–[7] an infinite number of times as $\omega \to \infty$, we cannot establish the Pomeranchuk theorem using crossing symmetry. This problem is discussed by Bessis and Kinoshita (1966) [6] and Eden and Lukasak (1967) [7].

### 13.6. Experimental investigation of high-energy cross-section behaviour

When the CERN 28 GeV proton synchrotron became available for experiments in 1960, among the first investigations [8] was an experimental

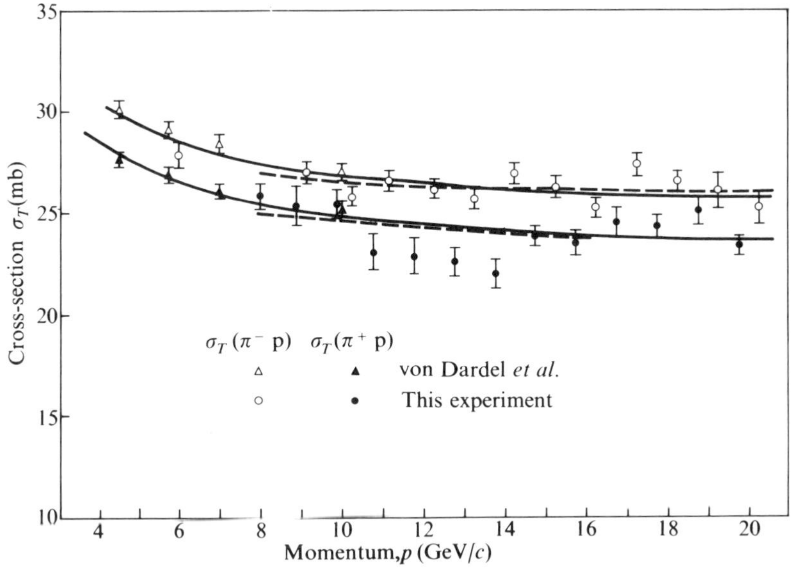


FIG. 13.1. $\pi^-+p$ and $\pi^++p$ total cross-sections. The solid lines and the dashed lines represent, respectively, the fits to the data of references [8] and [9], and the fits to the data of reference [9] of the form $a+(b/p)$. (From [9].) The difference between p–p and $\bar{p}$–p total cross-sections (not shown) was much larger.

investigation of the total cross-sections of incident beam particles, and their corresponding anti-particles. Similar investigations [9] were pursued at the Brookhaven 33 GeV proton synchrotron when it became available one year later. These early results are shown in Fig. 13.1.

It was already clear from the CERN results [8], which reached energies of about 10 GeV, that the Pomeranchuk theorem was not satisfied at this energy, as Pomeranchuk had originally predicted. There were still (see Fig. 13.1) sizeable differences in the particle–proton anti-particle–proton cross-section pairs such as

$$\sigma(\bar{p}+p)-\sigma(p+p), \tag{13.57}$$

$$\sigma(\pi^-+p)-\sigma(\pi^++p), \tag{13.58}$$

$$\sigma(K^-+p)-\sigma(K^++p). \tag{13.59}$$

However, it did appear from these results, especially in the case of

$$\sigma(\bar{p}+p)-\sigma(p+p),$$

that the differences were decreasing with energy sufficiently rapidly that, perhaps at about 20–30 GeV, the Pomeranchuk theorem might be satisfied. The subsequent Brookhaven investigation [9] extended these total cross-section measurements to about 20 GeV. In both investigations, scintillation counter telescopes were used for measuring the total cross-section via the transmission method. Good-geometry Cerenkov counters were employed in a momentum-analysed incident beam to select and identify the pion, kaon, nucleon (or anti-nucleon) desired. Since none of the data showed any statistically significant structure, and appeared to be monotomically decreasing toward a constant with increasing energy, the Brookhaven data was fitted by simple power laws of the form $\sigma = a+\frac{b}{p}$, etc. It was concluded from these fits that the Pomeranchuk theorem would not be satisfied until at least several hundred GeV.

More complete measurements [10], [11] were then made by a number of investigators for beam particles incident on both liquid hydrogen and liquid deuterium. The general behaviour of the data is summarized in Fig. 13.2 (Lindenbaum) [11].

The cross-sections for a particle incident on a neutron were, in general, determined from the relationship

$$\sigma(\alpha+\mathrm{d})-\sigma(\alpha+\mathrm{p}) \approx \sigma(\alpha+\mathrm{n})-\Delta\sigma(\mathrm{G.W.}), \qquad (13.60)$$

where $\Delta(\mathrm{G.W.})$ represents the shadowing (G.W. = Glauber–Wilkin) and other corrections to the sum of $\sigma(\alpha+\mathrm{p})$ and $\sigma(\alpha+\mathrm{n})$, which must be applied to obtain the deuteron cross-section. The $\Delta\sigma(\mathrm{G.W.})$ contains several effects, including

(a) the partial shadowing of the proton by the neutron (and vice versa) in the deuteron
(b) the effects of Fermi momentum
(c) possible multiple scattering processes including double charge exchange
(d) real amplitude effects.

These corrections can be uncertain by as much as 3–5 per cent. Fig 13.2 shows the qualitative features of the behaviour of the data. The typical accuracy of these transmission measurements for the total cross-sections, using a hydrogen target, are about 1–2 per cent, with errors predominantly of a systematic nature. These systematic errors include uncertainties in the extrapolation of the transmission data obtained with the detecting telescope counters after the hydrogen target, which subtend a finite solid angle (partly poor-geometry) to the case of an ideal good-geometry for which the solid angle subtended by the detector is zero. This is required to obtain the total

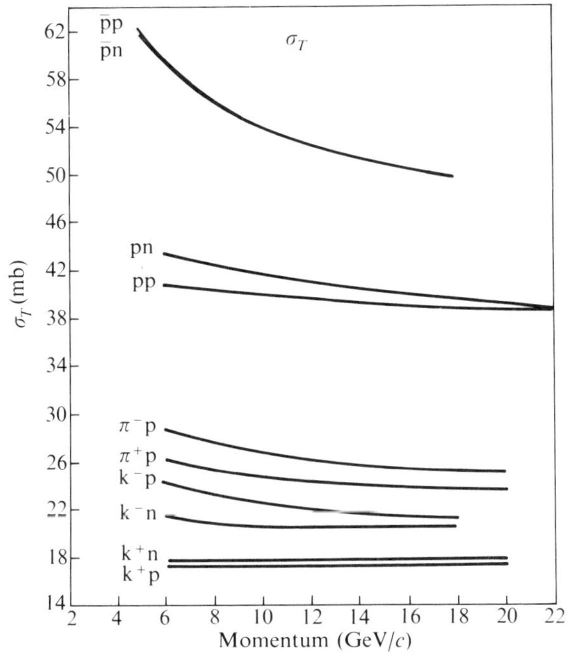


FIG. 13.2. The general behaviour of total cross-section data. (From Lindenbaum [11].)

nuclear cross-section, but Coulomb scattering effects (including multiple scattering), and interference between the real Coulomb amplitude and the real part of the nuclear amplitude, as well as uncertainties in the extrapolation procedure, limit the precision obtainable. Uncertainties in beam contamination, accidental effects, the actual amount of hydrogen absorber in the target, etc. also contribute to the errors. In deducing the cross-section on a neutron target via the difference in cross-section on deuterium and hydrogen, there is an additional uncertainty due to Glauber–Wilkins shadowing, etc. corrections.

Subsequently, very high precision total cross-section measurements for $\pi^{\pm}$–p and p–p of absolute accuracy 0·3 per cent were performed by Foley *et al.* [16], [17], using the counter–hodoscope, and on-line computer techniques, described in Chapter 5. The results of these new measurements, in comparison to the telescope method measurements of much lower accuracy, are shown in Fig. 13.3. The interpretation of these data will be discussed in §§ 13.18–13.19.

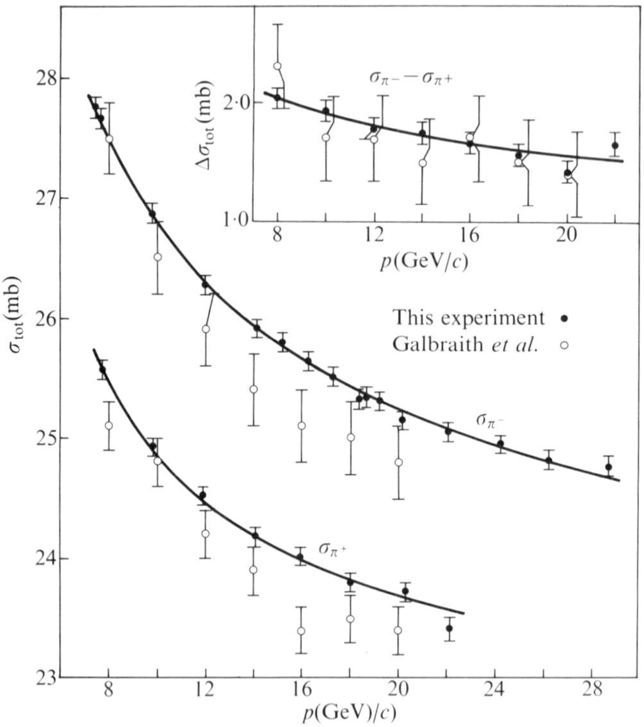


FIG. 13.3.(a) The high precision $\pi^\pm + p$ total cross-sections measurements [16] compared to the previous data. The parametric fits of type I used are given in Fig. 13.30.

The Pomeranchuk theorem is clearly not satisfied for the differences (see Fig. 13.3 and Fig. 13.2)

$$\sigma(\bar{p}-p) - \sigma(p-p), \tag{13.61}$$

$$\sigma(\pi^- + p) - \sigma(\pi^+ + p), \tag{13.62}$$

$$\sigma(K^- + p) - \sigma(K^+ + p), \tag*{(13.62)}$$

$$\sigma(\bar{p}-n) - \sigma(p-n). \tag{13.64}$$

The observed differences are several times the errors, and appear to be decreasing slowly with energy. The total cross-sections for $\pi^+$ and $\pi^-$ on deuterons are in agreement to within 1 per cent, confirming the charge-symmetry prediction. The p–n and p–p cross-sections did now show any significant difference within measurement errors. However, mean values exhibited about 4 per cent difference at 10 GeV/$c$ (with $\sigma_{n-p}$ above $\sigma_{p-p}$) which difference essentially disappeared at 20 GeV/$c$.

Thus we find an anomalous situation. The Pomeranchuk theorem does not appear to be satisfied in any case. The Pomeranchuk–Okun rule appears to be satisfied within the errors, at least at the high-momentum end, in the following cases

$$\sigma(\bar{p}\text{–}p) - \sigma(\bar{p}\text{–}n) \approx 0 \tag{13.65a}$$

$$\sigma(p\text{–}n) - \sigma(p\text{–}p) \approx 0 \tag{13.65b}$$

$$\sigma(K^+\text{–}n) - \sigma(K^+\text{–}p) \approx 0. \tag{13.65c}$$

This last case is particularly marginal, and is barely within the error limits. However, uncertainties in the Glauber–Wilkin corrections make these conclusions quite uncertain. In fact, direct measurements of $\sigma_T$(n–p) by Kreisler *et al.* (1968) [12*a*] indicated that $\sigma_{n-p}$ is systematically below $\sigma_T$(p–p) for the incident momenta in the region 14·6–21·6 GeV/*c*, but may equal $\sigma_{p-p}$ at about 27 GeV/*c*. The preliminary data of Engler *et al* [12*b*] in the 4–15 GeV/*c* region is in agreement with this conclusion.

Recent, more accurate, deuteron difference data still left uncertainties large enough for it not to be clear whether $\sigma_T$(p–n) is below, or above,

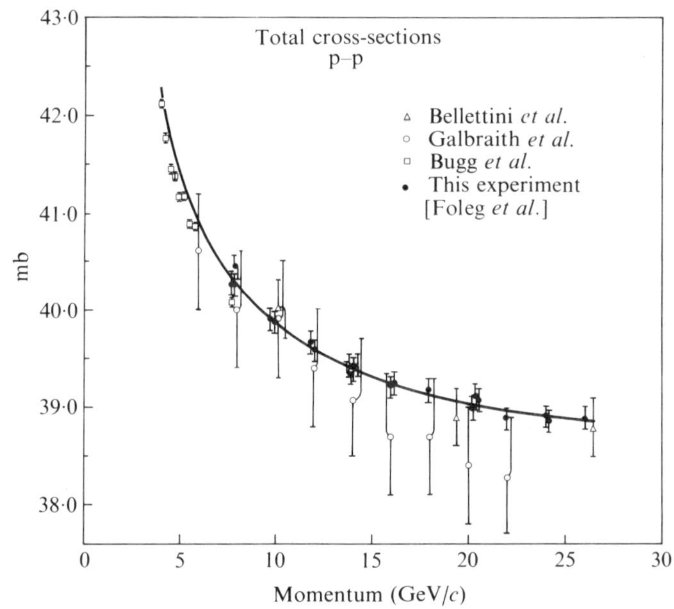


FIG. 13.3.(b). The high precision p–p total cross-section measurements [17] with a power law parametric fit of type I. See Fig. 13.31 for the parameters used.

$\sigma_T$(p–p) in the 4–5 GeV/$c$ region, but it appears more likely that it is below. It is definitely known that in the region of 2–3 GeV/$c$, $\sigma_T$(p–n) is below $\sigma_T$(p–p), and rising toward it with increasing energy. Thus, it appears (see Galbraith review) that conclusions on differences of $\sigma(\alpha+\mathrm{n})-\sigma(\alpha+\mathrm{p})$, based on $\sigma(\alpha+\mathrm{d})-\sigma(\alpha+\mathrm{p})$ experiments, must be considered quite uncertain at present.

In the one case where the Pomeranchuk–Okun rule gives the same result as the Pomeranchuk theorem, namely, $\sigma(\pi^-+\mathrm{p})-\sigma(\pi^++\mathrm{p})$, the difference is clearly not zero considerably outside of error and, thus, both the Pomeranchuk theorem and the Pomeranchuk–Okun rule also are violated. The Pomeranchuk–Okun rule also appears to be violated in the following

$$\sigma(\mathrm{K}^--\mathrm{p})-\sigma(\mathrm{K}^--\mathrm{n}) \neq 0. \tag{13.66}$$

Another striking characteristic observed was that the magnitudes of the $\pi^\pm$–p and $\mathrm{K}^\pm$–p total cross-sections were much smaller than the p–p and $\bar{\mathrm{p}}$–p.

The meson–baryon cross-sections were about 1/2 to 2/3 the baryon–baryon cross-sections. This observation was later to find an explanation in the quark model since, as we have seen in Chapter 10, a meson can be thought of as consisting of two quarks, whereas a baryon would consist of three. Thus, roughly speaking, a meson–baryon collision involves six quark–quark† pair interactions, whereas a baryon–baryon collision involves a nine quark–quark collision. Thus, if we assume that the resultant cross-sections are approximately proportional to the total number of quark–quark or anti-quark–quark pairs, we can, approximately, explain the observed phenomenon.

## 13.7. Elastic scattering at high energies

### 13.7.1. *Introduction*

About two decades ago it was demonstrated, at cyclotron (~several hundred MeV) energies, that high-energy elastic scattering of incident particles by target nuclei, and nucleons, seems to follow the general characteristics of diffraction scattering by a black (or more accurately a grey) sphere, or absorbing potential of well-defined range. This was later confirmed at Cosmotron (~3 GeV) and Bevatron (~6 GeV) energies. We would intuitively expect this if we assume that the nuclear interaction is strong, and of well-defined range. The early applications of this so-called

† Half of these are quark–anti-quark pairs.

optical model [13] were mostly concerned with scattering of incident particles by nuclei, but the same treatment was later applied to proton and deuteron targets.

The early results on high-energy elastic diffraction scattering experiments performed at the CERN PS, and the Brookhaven AGS, in 7–26 GeV/$c$ incident particle energy ranges, and their theoretical interpretation are summarized by Lindenbaum [14]. Before reviewing the experimental situation, let us consider the predictions of various simple theoretical models. The simplest and earliest model was the black sphere (or grey sphere) model, which is essentially the optical model.

### 13.7.2. *The optical model*

If we assume that, at high energies, the elastic scattering amplitude is spin independent, we can represent the high-energy elastic scattering amplitude by one complex function $A(s, t)$. Thus, in our notation,

$$\frac{\mathrm{d}\sigma_{\mathrm{el.}}(s, t)}{\mathrm{d}t} = |A(s, t)|^2. \tag{13.67}$$

The measurements of the total cross-section give us the value of $A(s, 0)$ via the optical theorem. Thus we can write, in our units,

$$\mathrm{Im}\, A(s, 0) = \frac{\sigma_t(s)}{\sqrt{(16\pi)}}. \tag{13.68}$$

As we have previously (see Chapter 5) defined

$$\alpha = \frac{\mathrm{Re}\, A(s, 0)}{\mathrm{Im}\, A(s, 0)}, \tag{13.69}$$

The entire forward scattering amplitude is

$$A(s, 0) = \frac{\sigma_t(s)}{\sqrt{(16\pi)}}(\alpha + \mathrm{i}). \tag{13.70}$$

It was generally assumed (prior to experimental evidence to the contrary) that the effects of the real part of the scattering amplitude would be small, and negligible compared to the imaginary part. This is certainly true for the 'black sphere' model, as we shall see in a moment.

In our units, the partial wave expansion (see Chapter 5) for the invariant scattering amplitude $A(s, t)$, becomes

$$A(s, t) = \frac{\sqrt{\pi}}{k}\left\{\frac{1}{2ik}\sum_l (2l+1)(\mathrm{e}^{2\mathrm{i}\delta_l} - 1)P_l(\cos\theta)\right\}. \tag{13.71}$$

Let

$$\delta_l = (\delta_l)_{\mathrm{Re}} + \mathrm{i}(\delta_l)_{\mathrm{Im}}. \tag{13.72}$$

Then we obtain

$$A(s, t) = \frac{\sqrt{\pi}}{2\mathrm{i}k^2} \sum_l (2l+1)(\eta_l \mathrm{e}^{2\mathrm{i}(\delta_l)_{\mathrm{Re}}} - 1) P_l(\cos\theta), \tag{13.73}$$

where $\eta_l = \mathrm{e}^{-2(\delta_l)_{\mathrm{Im}}}$ and represents the fraction of the incident wave which is not absorbed, whereas $(\delta_l)_{\mathrm{Re}}$ represents the real phase shift. For no absorption of a partial wave, $\eta_l = 1$, and for total absorption, $\eta_l = 0$.

The simplest black sphere model assumes that, at high energies, when many partial waves are involved, total absorption occurs for all values of $l$, up to $l_{\max}$, where $l_{\max}$ is defined in terms of an effective interaction radius, $R$, as follows,

$$l_{\max} \approx kR.$$

Thus, for all $l \lesssim l_{\max}$

$$\eta_l = 0. \tag{13.74}$$

For all $l > l_{\max}$, it is assumed that there is no absorption and thus for $l \gtrsim l_{\max}$

$$\eta_l = 1. \tag{13.75}$$

Hence,

$$A(s, 0) \approx \frac{\mathrm{i}\sqrt{\pi}}{2k^2} \sum_{l=0}^{l=kR} (2l+1) \approx \frac{\mathrm{i}\sqrt{\pi}}{2k^2}(kR)^2 \approx \frac{\mathrm{i}\sqrt{\pi}\, R^2}{2}. \tag{13.76}$$

Thus the amplitude is purely imaginary as expected.

Applying the optical theorem, we obtain

$$\sigma_t(s) = \sqrt{(16\pi)}\,\mathrm{Im}\, A(s, 0) = \frac{\sqrt{(16\pi)}\sqrt{\pi}\, R^2}{2} = 2\pi R^2, \tag{13.77}$$

which is just the expected result for classical black sphere scattering.

From eqn (13.73), if $\eta_l = 1$ within the radius $R$, (i.e. no absorption but scattering, which changes the phase), then, depending on the phase, $\sigma_t$ can vary from 0 to $4\pi R^2$.

In order to derive an expression for the shape of the diffraction scattering from a black sphere at small angles, we can use the small angle approximation

$$P_l(\cos\theta) \approx J_0((l+\tfrac{1}{2})\theta).$$

Thus,

$$A(s, t) = \frac{\mathrm{i}\sqrt{\pi}}{2k^2} \sum_{l=0}^{l=kR} (2l+1) P_l(\cos\theta) \approx \frac{\mathrm{i}\sqrt{\pi}}{2k^2} \int_0^{kR} \{J_0(l\theta)\} 2l\,\mathrm{d}l$$

$$\approx \frac{\mathrm{i}\sqrt{\pi}\, R}{k\theta} J_1(kR\theta). \tag{13.78}$$

Since we are dealing with an invariant amplitude, it is desirable to express the result in a Lorentz invariant form. This can be done easily, since

$$t = -2k^2(1-\cos\theta) \approx -k^2\theta^2.$$

We have employed the small angle approximation for $\cos\theta$, since at high energies the diffraction pattern is concentrated at small angles. Thus

$$A(s, t) \approx \frac{\mathrm{i}\sqrt{\pi}\, R}{\sqrt{|t|}}\, J_1((\sqrt{|t|}R). \tag{13.79}$$

This gives an elastic scattering pattern similar to that for the optical case. A measurement of the shape of the diffraction pattern at small angles would give us a measure of the radius. The sharp diffraction minima (which go to zero) are, as we shall see shortly, not observed. This is entirely understandable, since it is physically unrealistic to expect a sharp boundary between absorption and non-absorption, and it is this sharp boundary which gives the diffraction minima.

A more realistic approach is to use a Gaussian form for $\eta_l$ such as

$$\eta_l = \mathrm{e}^{-l^2/k^2R^2},$$

or other gradually cutting-off forms. This will smooth out, or considerably reduce the depth of, the sharp minima. From eqn (13.79) we obtain

$$\frac{\mathrm{d}\sigma}{\mathrm{d}t} \approx \pi R^2 \left(\frac{J_1(R\sqrt{|t|})}{\sqrt{|t|}}\right)^2. \tag{13.80}$$

For $R \approx 1$ fm, or less, and $|t| \leq 0{\cdot}3$, we can use the Gaussian approximation for $J_1(R\sqrt{|t|})$, and obtain

$$\frac{\mathrm{d}\sigma}{\mathrm{d}t} \approx \frac{\pi R^4 \mathrm{e}^{(-(R/2)^2|t|)}}{4} = \left(\frac{\mathrm{d}\sigma}{\mathrm{d}t}\right)_{\mathrm{opt}} \mathrm{e}^{(-(R/2)^2|t|)}. \tag{13.81}$$

For a Gaussian pure imaginary absorbing potential, a similar† form would correspond to the r.m.s. radius. Therefore, the value of $R$, determined from experimental data using eqn (13.81), is a good way to characterize the effective radius phenomenologically. Having defined the radius, we can then define an effective opacity, as follows,

$$\bar{O} = \frac{(\sigma_{\mathrm{tot}} - \sigma_{\mathrm{el}})}{\pi R^2} = \frac{\sigma_{\mathrm{inel}}}{\pi R^2}. \tag{13.82}$$

If we generalize this treatment, we obtain the so-called impact parameter representation.

† The r.m.s. radius $(\bar{r})$ would be $\dfrac{R}{\sqrt{2}}$, for a pure Gaussian absorbing potential.

Let us return to eqn (13.71), and make the (small angle) approximation

$$P_l(\cos\theta) \approx J_0((l+\tfrac{1}{2})\theta). \tag{13.83}$$

Thus, we obtain

$$A(s,t) = \frac{\sqrt{\pi}}{k}\sum_{l=0}^{l=\infty} \frac{1}{2ik}(2l+1)(e^{2i\delta_l}-1)J_0((l+\tfrac{1}{2})\theta). \tag{13.84}$$

At high energies, where $l_{max}$ is large, we can approximate

$$A(s,t) \approx \frac{\sqrt{\pi}}{k}\int \frac{(2l+1)}{2ik}\, dl(e^{i2\delta_l}-1)J_0((l+\tfrac{1}{2})\theta). \tag{13.85}$$

If we introduce the variable

$$\rho \approx (l+\tfrac{1}{2})/k,$$

which in the classical limit corresponds to the impact parameter, we obtain

$$A(s,t) \approx 2k\sqrt{\pi}\int \rho\, d\rho f(\rho,s)J_0(\rho\sqrt{|t|}), \tag{13.86a}$$

where

$$f(\rho,s) = \left(\frac{e^{2i\delta_l}-1}{2ik}\right). \tag{13.86b}$$

The above is the so-called impact parameter representation. Inverting the Fourier–Bessel transformation gives us the following

$$f(\rho,s) = \frac{1}{2k\sqrt{\pi}}\int_0^\infty (\sqrt{|t|})\, d(\sqrt{|t|})A(s,t)J_0(\sqrt{|t|}\,\rho). \tag{13.87}$$

As we shall discuss in the following section, it was experimentally observed that, at high incident energies ($\gtrsim 10$ GeV), for $\pi^\pm$–p, $K^\pm$–p, p–p, and $\bar{p}$–p, the low $|t|$ diffraction scattering region could be parameterized in the form

$$\frac{d\sigma}{dt} = e^{a+bt}. \tag{13.88a}$$

Thus,

$$A(s,t) = e^{(a/2+b/2)t}. \tag{13.88b}$$

If we assume that the elastic scattering amplitude is, to a good approximation, imaginary, and apply the optical theorem, we obtain

$$e^{a/2} = \frac{i\sigma_{tot}}{4\sqrt{\pi}}. \tag{13.89}$$

Thus,

$$A(s,t) \approx \frac{i\sigma_{tot}}{4\sqrt{\pi}}\, e^{(b)t/2}. \tag{13.90}$$

Substituting eqn (13.90) into eqn (13.87), we obtain

$$f(\rho, s) = \frac{i\sigma_{tot}}{8\pi k}\int_0^\infty Z\,dZ\,J_0(\rho Z)e^{(-b/2)Z^2}, \tag{13.91}$$

where $Z = \sqrt{|t|}$.

$$f(\rho, s) = \frac{i\sigma_t}{8\pi k b}\,e^{-\rho^2/2b}. \tag{13.92}$$

Thus, the 'effective width', or radius, of the Gaussian potential is $\sqrt{2b}$. Comparing eqn (13.81) and eqn (13.88), we obtain

$$R = 2\sqrt{b}. \tag{13.93}$$

Thus the effective radius, or 'width', obtained for a black, or uniformly absorbing, disc is the $\sqrt{2}$ larger than the effective Gaussian 'width', or radius. Owing to the strength of strong nuclear interactions and the accompanying saturation effects, we should probably expect that the uniformly absorbing potential model with a Gaussian (or alternately Yukawa) tail could be expected to fit the data.

### 13.7.3. *Elastic (diffraction) scattering experiments at high energies*

Experiments which determined the p–p elastic scattering from 12·1 GeV to 26·2 GeV, using conventional techniques, were performed at CERN by Diddens *et al.* [15], and the results for $\frac{d\sigma}{dt}$ against $-t$ are presented in Fig. 13.4(a).

The normalization factor

$$\left(\frac{\sigma_{tot}\,(P)}{\sigma_{tot}\,(20\text{ GeV}/c)}\right)^{-2}$$

was chosen for convenience. Assuming no (or a small) real amplitude, and the optical theorem, we have

$$\left(\frac{d\sigma}{dt}\right)_{t=0} = |A(s, 0)|^2 \approx \frac{\sigma_{tot}^2}{16\pi}. \tag{13.94}$$

Thus, by plotting $\sigma_{tot}^{-2}\,\frac{d\sigma}{dt}$, we should find the $y$-axis intercept of this quantity (i.e. its value at $t = 0$) is the same at all momenta, thus facilitating comparison of the shapes of the curves. Since the momentum dependence of the cross-sections is not very large, it was further convenient to include the constant $\sigma_{tot}^2$ (20 GeV/$c$) in the determination of the normalization, so that the

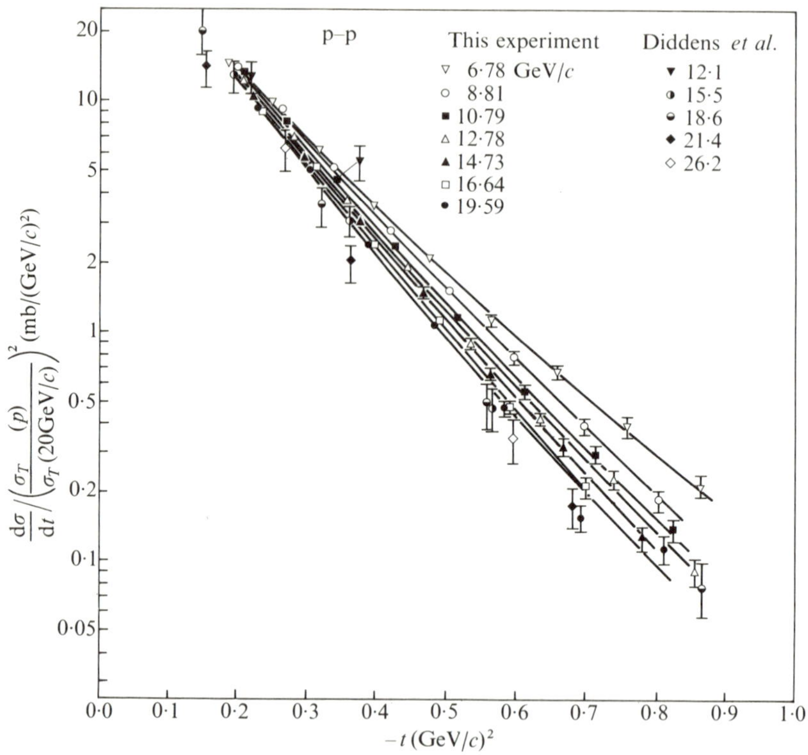


FIG. 13.4.(a) Differential cross-section for elastic p+p scattering. The different symbols represent different incident laboratory momenta. The data of Diddens *et al.* are from [15]. The 12·1 GeV/$c$ point of Diddens *et al.* at $t = -0{\cdot}3\,(\text{GeV}/c)^2$ should be superimposed onto our point at 10·79 GeV as indicated by the arrow. (From Foley *et al.* (1963). *Phys. Rev. Lett.* **10,** 376 [16].)

normalization constant is close to unity and, therefore, the ordinate exhibits the approximate value of $\dfrac{\mathrm{d}\sigma}{\mathrm{d}t}$.

A more complete investigation of the high-energy elastic scattering of p–p, as well as $\pi^{\pm}$–p, $\mathrm{K}^{\pm}$–p, and $\bar{\mathrm{p}}$–p was performed [16] at the Brookhaven AGS, by developing new techniques involving a counter–hodoscope detector system, on-line data handling, and an on-line computer to monitor the experiment, and reduce the data to the final physical form. The part of the apparatus used for the $|t|$ range $0{\cdot}2\,(\text{GeV}/c)^2$ to $1\,(\text{GeV}/c)^2$ is shown in Fig. 13.4(b). The desired incident particle was selected by a gas Cerenkov counter-telescope, and allowed to scatter in a liquid hydrogen target. Both the

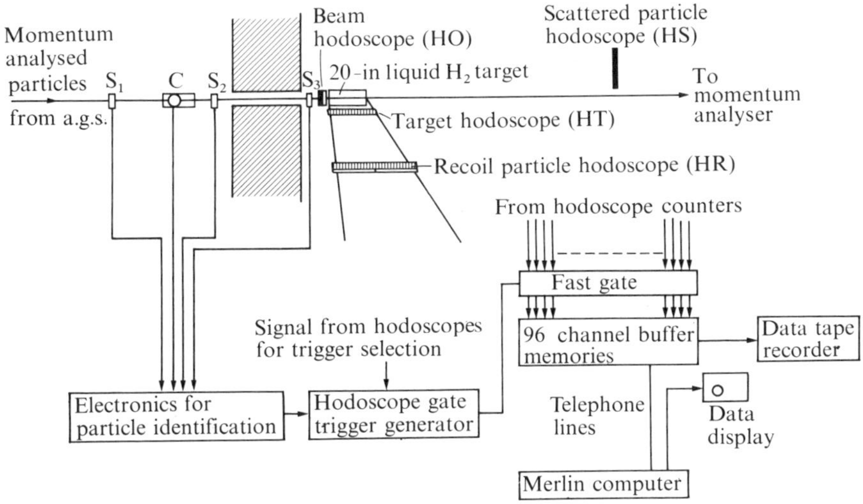


FIG. 13.4.(b) A counter hodoscope (coincidence arrangement) with automatic data handling and on-line computer. Elastic scattering is selected by coplanarity and kinematic angle requirements. (Lindenbaum. (1964). *Nucleon structure*. (Edited by R. Hofstadter). Stanford University Press, Stanford.)

forward scattering particle, and the proton recoil, were detected by a crossed slab hodoscope scintillation counter system. The computer monitored the equipment, provided various oscilloscope displays, and completely analysed the data.

The experimental results obtained for p–p are shown in Fig. 13.4(a), while those for $\pi^{\pm}$–p are shown in Fig. 13.4(c). In all cases ($\pi^{\pm}$–p, $K^{\pm}$–p, $\bar{p}$–p, and p–p) good fits were obtained, using the parametric form

$$\frac{d\sigma}{dt} = e^{a+bt+ct^2}, \tag{13.95}$$

with $c$ significantly positive (except in the case of $\bar{p}$–p) [17].

It is clear from Figs. 13.4(a) and 13.4(c) that the p–p diffraction pattern shrinks (i.e. the effective radius decreases) with increasing energy, whereas the $\pi^{\pm}$–p diffraction pattern does not. These effects will be discussed under the Regge-pole model which will be considered in the next section. In the remainder of this section, we shall discuss the calculated effective particle radii using the simple methods previously discussed.

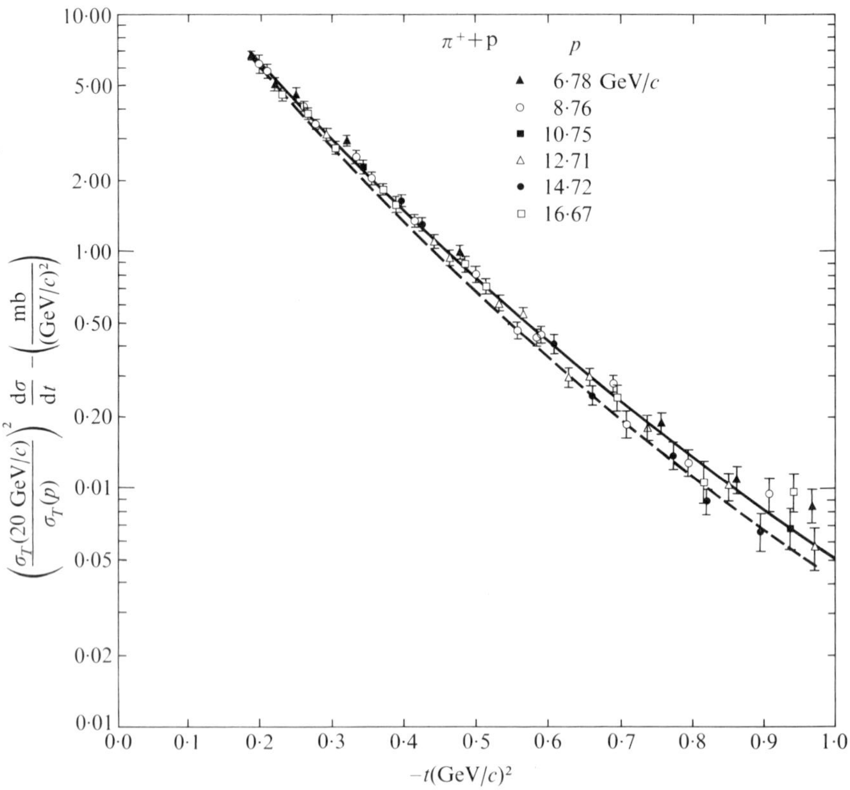


FIG. 13.4.(c) $x(\sigma_{\text{tot}}(\text{p}))^{-2}(\sigma_{\text{tot}}(20\ \text{GeV}/c))^2\ d\sigma/dt$, which is proportional to $(d\sigma/dt)_{\text{opt}}^{-1}\ d\sigma/dt$, against $t$. The solid line is a least-squares fit to all the $\pi^+ + \text{p}$ data; the dashed line is a least-squares fit to all the $\pi^+ + \text{p}$ data from [16]. (From Foley *et al.* (1963). *Phys. Rev. Lett.* **10,** 543.)

### 13.7.4. *Effective particle radii*

Since the data exhibited significant $c$ coefficients, there was no unique definition of an effective radius independent of $t$ and, thus, it seemed appropriate to define $R$ in terms of the slope of the exponential (i.e. derivative of $a+bt+ct^2$ with respect to $t$) at a $t$-value corresponding to a distance which represented a reasonable compromise in averaging over the outer pion cloud and core region, of the interactions. $|t| \approx 0{\cdot}2$ corresponds to a distance of about $\frac{1}{2}$ fm, and, thus, was considered a reasonable compromise. Since a proton target was common in all cases, we may, for convenience, associate the radius of the interaction with the incident particle. The radius was calculated from the relation,

$$\text{slope of exponent} = b + 2ct \equiv (R/2)^2, \tag{13.96}$$

and the results of the Brookhaven [16], and other, experiments are shown in Figs. 13.5, (a)–(c).

Thus, in the case of p–p, we find a trend for the effective radius to increase from $(1{\cdot}06 \pm 0{\cdot}04)$ fm at 7 GeV/$c$ to about $(1{\cdot}18 \pm 0{\cdot}05)$ fm at 20 GeV/$c$. However, in the case of $\pi^- + \text{p}$ and $\pi^+ + \text{p}$, we see no trend for change in the incident momentum. The average effective radius for (7–17) GeV/$c$ $\pi^+$–p is approximately $(1{\cdot}05 \pm 0{\cdot}022)$ fm, whereas for $\pi^- + \text{p}$ the value is $(1{\cdot}11 \pm 0{\cdot}018)$ fm. Thus, except for the small increase with energy in the p–p case, the $\pi^\pm$–p and p–p radii are all on the average near 1·1 fm. However, as can be seen in Fig. 13.5(c) for the $\bar{\text{p}}$–p case, the average effective radius is much larger ($\approx(1{\cdot}35 \pm 0{\cdot}08)$ fm) in the (7–12) GeV/$c$ region and tends to increase at lower momentum. The $\text{K}^- + \text{p}$ ((7–9) GeV/$c$) has an average effective radius of about $(1{\cdot}13 \pm 0{\cdot}04)$ fm, which is comparable to that observed for protons and pions, whereas for $\text{K}^+ + \text{p}$ ((7–15) GeV/$c$), the effective radius was found to be $(0{\cdot}95 \pm 0{\cdot}04)$ fm, which is considerably smaller than any of the others.

### 13.7.5. *Opacity*

We can use eqn (13.82) to calculate the effective average opacity $\bar{\text{O}}$ and from the data [16], [14] it was determined that $\bar{\text{O}} \approx 0{\cdot}74 \pm 0{\cdot}02$ for protons. $\bar{\text{O}}$ is consistent with $\approx 0{\cdot}58 \pm 0{\cdot}003$ for positive or negative pions. $\bar{\text{O}} \approx 0{\cdot}51 \pm 0{\cdot}04$ for positive or negative kaons. For anti-protons (above 7 GeV/$c$), $\bar{\text{O}} \approx 0{\cdot}83 \pm 0{\cdot}07$. Thus there were striking differences observed in the equivalent radii, opacities, and their energy behaviour. If we used the equivalent Gaussian radius, the opacities would be twice as large and thus larger than 1, which would be non-physical, clearly indicating that the absorption, at least over a good deal of the interaction region, is stronger than Gaussian. As remarked previously, a combination of a uniform potential with a Gaussian tail would probably do well in representing the data.

## 13.8. The Regge pole and related models

### 13.8.1. *Non-relativistic Regge theory*

Regge [18] set out to establish the Mandelstam representation for non-relativistic potential scattering.†

† There are reviews of this subject in Eden's book and Gasiorowicz's book.

High-energy elastic scattering of $\pi^{\pm}, p, \bar{p}$, and $K^{\pm}$

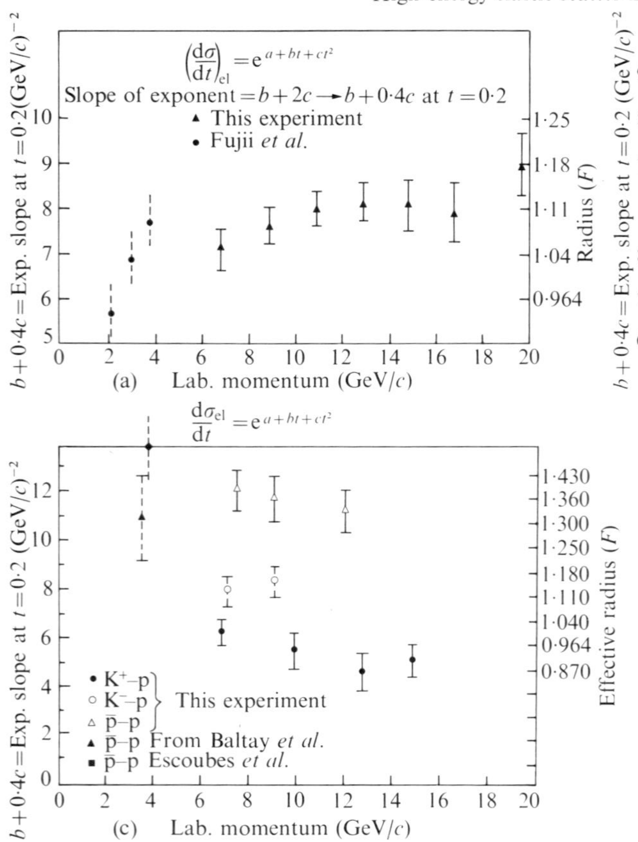


FIG. 13.5.(a) p–p effective radius in fm (right-hand scale) using definition in text. (Lindenbaum. (1964). *Nucleon structure* (Edited by R. Hofstadter). Stanford University Press, Stanford.)

FIG. 13.5.(b) $\pi^{\pm}$–p effective radii. (Lindenbaum. (1964). *Nucleon structure* (Edited by R. Hofstadter). Stanford University Press, Stanford.)

FIG. 13.5.(c) $K^+$–p, $K^-$–p, and $\bar{p}$–p effective radii, this experiment [16]. (From Lindenbaum. (1964). *Nucleon structure* (Edited by R. Hofstadter). Stanford University Press, Stanford.)

Regge started with the partial wave expansion

$$f(k^2, \cos\theta) = \sum_{l=0}^{l=\infty} (2l+1) f_l(k^2) P_l(\cos\theta). \tag{13.97a}$$

The corresponding invariant amplitude is

$$F(s, t) = \frac{8\pi s^{\frac{1}{2}}}{k} \sum_{0}^{\infty} (2l+1) f_l(s) P_l(\cos\theta). \tag{13.97b}$$

The above series only converges inside the Martin–Lehmann ellipse, and is regular inside this region.

In Regge's method, $l$ was allowed to become complex, and the partial wave series was rewritten. The complex function $f(l, s)$ was defined using the Watson–Sommerfeld integral representation, so that

$$f(l, s) = f_l(s),$$

for real integral $l$.

$$F(s, t) = \frac{8\pi s^{\frac{1}{2}}}{k} \int_{C_0} \frac{\mathrm{d}l(2l+1) f(l, s) P_l(-\cos\theta)}{\sin \pi l}, \tag{13.98}$$

where

$$\cos\theta = +1 + \frac{t}{2k^2}.$$

The contour $C_0$ is surrounding the real axis in the complex $l$-plane, as shown in Fig. 13.6(a), and for some physical $s$-values, it is assumed that $f(l, s)$ contains no singularities within the contour $C_0$. The denominator $\sin \pi l$ of the above integral goes through zero at all integral values of $l$ and, thus, selects the residues of the integrals at these zeros, since the residues of

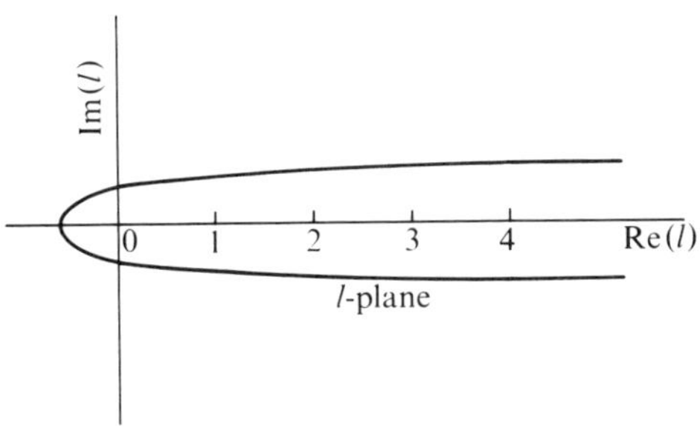


FIG. 13.6.(a)

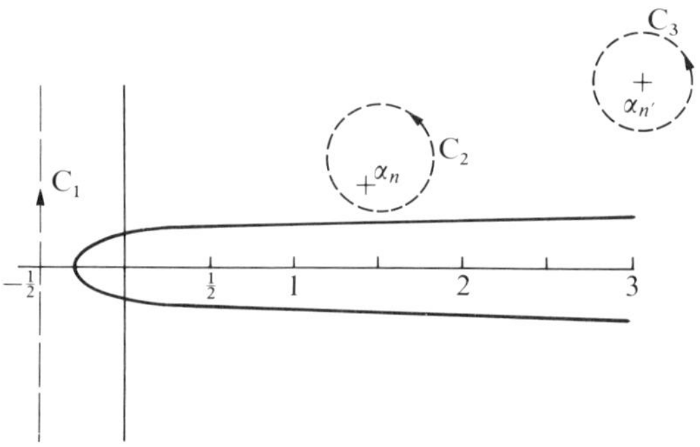


FIG. 13.6.(b)

$\sin \pi l$ are $(-1)^l/\pi$. This then reproduces the partial wave series. For reasonable assumptions on $f(l, s)$, as $\operatorname{Re} l \to \infty$, the integral converges for physical values of $s$ and $t$. The Regge approach then justifies deforming the contour $C_0$ to the contour $C_1$, plus a number of contours around singularities $C_2$, $C_3$ .... In Fig. 13.6(b), the contour $C_1$ lies along $\operatorname{Re} l = -\frac{1}{2}$, and the contours $C_2$, $C_3$, and so on, surround the various singularities of $f(l, s)$ in the domain $\operatorname{Re} l > -\frac{1}{2}$. To justify this step, it is necessary that (a) the analytic partial wave amplitudes have poles as their only singularities for $\operatorname{Re} l > -\frac{1}{2}$, and (b) there is no contribution from the curved contour at $\infty$ that connects $C_0$ with $C_1$.

If we evaluate the integrals around the poles at $C_2$, $C_3$, etc., we obtain the sum

$$-\sum_n \frac{8\pi s^{\frac{1}{2}}}{k} \frac{(2\alpha_{n+1})\beta_n(\alpha_n, s)P_{\alpha_n}(-1-t/2k^2)}{\sin \pi\alpha_n}, \tag{13.99}$$

where $\alpha_n(s) = \alpha_n$ represents a pole of $f(l, s)$ in $\operatorname{Re}(l) > -\frac{1}{2}$ (a Regge pole) and $\beta_n(\alpha_n, s)$ is the residue of $f(l, s)$ at the pole. Thus, we can write

$$F(s, t) = \frac{+4\pi s^{\frac{1}{2}}\mathrm{i}}{k} \int_{-\frac{1}{2}-\mathrm{i}\infty}^{-\frac{1}{2}+\mathrm{i}\infty} (2l+1) \frac{f(l, s)P_l(-1-t/2k^2)}{\sin l\pi}\,\mathrm{d}l - $$
$$-\sum_n \frac{8\pi^2 s^{\frac{1}{2}}}{k} \frac{(2\alpha_n+1)\beta_n(\alpha_n, s)P_{\alpha_n}(-1-t/2k^2)}{\sin \pi\alpha_n}. \tag{13.100}$$

The integral term is generally referred to as the 'background integral'. The sum (second term) represents the sum of the contributions from the individual Regge poles. Using this form, which allows $\cos\theta$ or $t$ to become complex, and still have a valid representation, Regge was able to establish the

Mandelstam representation for a superposition of Yukawa potentials in non-relativistic potential scattering.

An interesting consequence of the above is that, as $z \to \infty$,

$$|P_l(-z)| \sim |z|^{\mathrm{Re}\,l}.$$

We assume that the background integral along $C_1$ varies as $\dfrac{1}{\sqrt{|z|}}$, and can be neglected as $z \to \infty$. Therefore, it follows that the contribution from the Regge pole with the greatest value of $l$ (the pole farthest to the right) exclusively determines the asymptotic behaviour of $F(s, t)$ as $|\cos\theta|$ or $|t|$ become large, and the contributions of all other poles become negligible. For this most right pole, for a particular set of $s$-values,† let us designate $n = 1$. Then, for $|z|$ or $|t| \to \infty$,

$$F_{\alpha_1}(s, t) \approx -\frac{8\pi^2 s^{\frac{1}{2}}(2\alpha_1 + 1)\beta_1(\alpha_1, s)\,|z|^{\mathrm{Re}\,l}}{k \sin \pi\alpha_1}. \tag{13.101}$$

Thus, it follows that, for potential scattering, the Mandelstam representation requires a definitely predictable, and finite, number of subtractions in the $t$-variable.

Regge then demonstrated that for an attractive Yukawa potential these poles behaved in the following way (as illustrated in Fig. 13.7 (Lovelace and Masson [19])). When $k^2 = -\infty$, the Regge poles lie on the real axis at $l \leq -1$. As $k^2$ becomes more positive, the poles move to the right along the real axis and, finally, leave the real axis, as shown, when $k^2 > 0$ ($s = 4m^2$), and follow the trajectories illustrated in the upper-half plane. As $k^2 \to \infty$, the trajectories eventually return to the line $l = -1$.

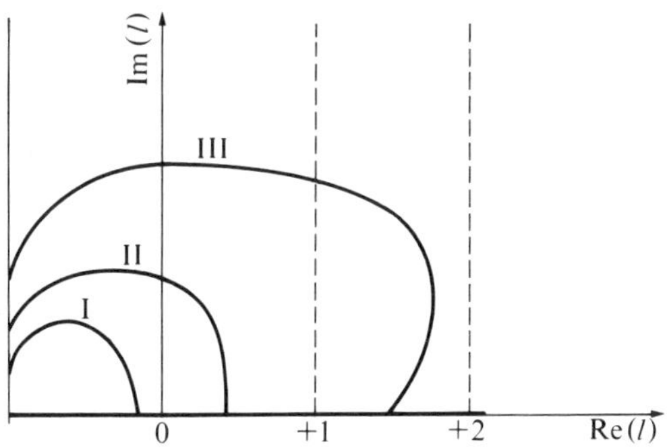


FIG. 13.7.

† The Regge poles move in the complex plane as $s$ changes. Therefore, the most right pole can, in principle, change its identity as $s$ changes.

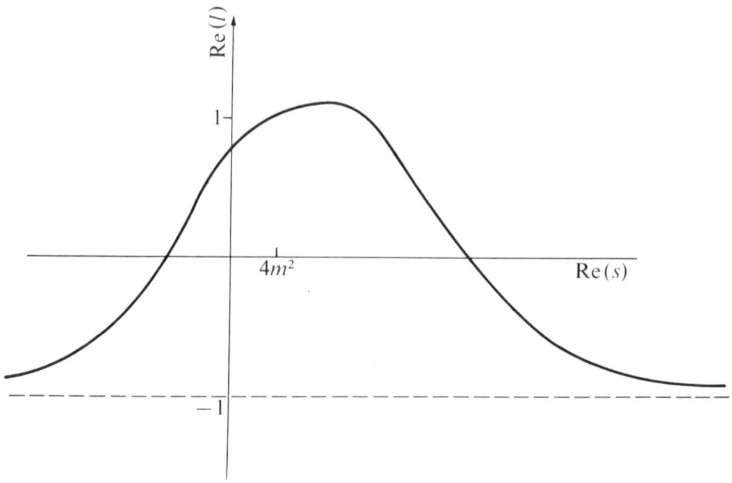


FIG. 13.8. A case of Type II.

Case I represents a weak enough Yukawa potential, so that no bound state exists in the neighbourhood of the pole. If the potential is so strong that, before leaving the real axis, $\alpha(s)$ crosses $l = 0$, while $q^2$ is still negative (Case II), we have a behaviour of the scattering amplitude representing the pole characteristics of an $S$-state resonance, in the vicinity of the pole. Case III corresponds to an even stronger potential, which results in the trajectory crossing $l = 1$, before leaving the real axis, and which corresponds to an $l = 1$ bound state, in the vicinity of the pole. There can, in principle, be a number of bound $l = 0$ states, $l = 1$ states, $l = 2$ states, and so on. Hence, there is a relationship between the bound states of a system, and the moving Regge poles, in the complex $l$-plane. For the case $q^2 > 0$, or $s > 4m^2$, when a trajectory passes close to an integral $l$-value, it can be shown that the particular partial wave scattering amplitude corresponds to a resonance. From eqn (13.100), we have that the partial wave scattering amplitude, due to a particular Regge pole, which is assumed to dominate (in the neighbourhood of a resonance), can be expressed as

$$F(s, t) = \frac{\Gamma(s)P_{\alpha_n}(-\cos\theta)}{\sin\pi\alpha_n(s)}. \tag{13.102}$$

Since

$$\sin\pi\alpha = \frac{\pi}{2}(\alpha - l)(\alpha + l + 1)\int_{-1}^{1} \mathrm{d}(\cos\theta)P_l(\cos\theta)P_l(-\cos\theta),$$

it can be shown that

$$F_l(s) = \frac{\Gamma(s)}{\pi(\alpha_n(s)-l)(\alpha_n(s)+l+1)}. \tag{13.103a}$$

If, at the resonance energy $s = s_r$, Re $\alpha_n(s) = l$, the resultant partial wave amplitude exhibits the following form, appropriate for a resonance,

$$\frac{\Gamma(s)}{(2l+1)\left\{(s-s_r)\left(\frac{\mathrm{d}\,\mathrm{Re}\,\alpha_n(s)}{\mathrm{d}s}\right)_{s_r} + \mathrm{i}\,\mathrm{Im}\,\alpha_n(s)\right\}}. \tag{13.103b}$$

Another plot of interest is Re $l$ against Re $s$. A Case II plot is illustrated in Fig. 13.8.

13.8.2. *The relativistic Regge pole model*

Although there was no way found to derive a relativistic Regge-pole model, it was hypothesized by Chew, Frautschi [20], and others [21] that, in relativistic scattering, the partial wave scattering amplitudes are analytic functions in the complex $J$-plane (i.e. we generalize, so that $J$ represents the total angular momentum) in the domain to the right of a vertical line characterized by Re $J < 0$, except for Regge-type poles. These relativistic Regge poles were then assumed to move on a trajectory, the characteristics of which are determined by stable, and unstable, resonances, in analogy to the non-relativistic case.

There was, however, one striking new element in the relativistic Regge model. Owing to crossing symmetry, the asymptotic expansion in the $s$-channel (which we have just considered), where $t \to \infty$ while $s$ (positive) is small, corresponds in the $t$ channel to a large $s$-variable (i.e. $s \to \infty$) and a small positive $t$-variable. Thus, as $s \to \infty$, the Regge pole farthest to the right would dominate and completely determine the asymptotic behaviour for small $t$. This pole would correspond to small mass resonances in the $t$-channel.

Thus, high-energy low $t$-scattering amplitudes would be determined by the exchange of particles, or resonant states, of low mass in the $t$-channel, thereby relating high-energy scattering amplitudes to the exchange of low-energy resonances or low-mass particles.

From eqn (13.101), we have, as $t \to \infty$,

$$F_{\alpha_1}(s, t) \approx C_1(s)t^{\alpha_1(s)}, \tag{13.104a}$$

where

$$C_1 = -\frac{8\pi^2 s^{\frac{1}{2}}(2\alpha_1+1)\beta_1^{\frac{1}{2}}(\alpha_1, s)}{k \sin \pi\alpha_1}\left(\frac{+2}{4m^2-s}\right)^{\alpha_1(s)}, \tag{13.104b}$$

where we have considered the case for equal mass particles without spin dependence; for example, two scalar bosons interacting. Thus, applying crossing symmetry in the relativistic case, we obtain the following asymptotic form of the scattering amplitude, as $s \to \infty$,

$$F_{\alpha_1}(s, t) = C_1(t)\left(\frac{s}{s_0}\right)^{\alpha_1(t)}. \tag{13.105}$$

The physical region corresponds to $t < 0$, and is related to the non-physical region by analytic continuation. $s_0$ is a parameter with the dimensions of $(\text{mass})^2$ of typical order of magnitude about $(1\ \text{GeV})^2$. At high enough energies to use the approximations, but not so high that the leading pole completely dominates, we will have the following

$$F(s, t) = \sum_n C_n(t)\left(\frac{s}{s_{0n}}\right)^{\alpha_n(t)} \tag{13.106a}$$

or, equivalently,†

$$A(s, t) = \sum_n \beta_n(t, \tau)\left(\frac{s}{s_{0n}}\right)^{\alpha_n(t)-1}. \tag{13.106b}$$

Thus, since $\dfrac{d\sigma}{dt} = |A(s, t)|^2$, we find, from eqns (13.105) and (13.106), that

$$\frac{d\sigma}{dt} = \left|\sum_n \beta_n(t, \tau)\left(\frac{s}{s_{0_n}}\right)^{\alpha_n(t)-1}\right|^2. \tag{13.107}$$

If the leading right-most pole dominates

$$\frac{d\sigma}{dt} = |\beta(t, \tau)|^2 \left(\frac{s}{s_0}\right)^{2\alpha_1(t)-2}. \tag{13.108}$$

Hence,

$$\ln\frac{d\sigma}{dt} = 2\ln|\beta_1(t, \tau)| + (2\alpha_1(t) - 2)\ln\left(\frac{s}{s_0}\right), \tag{13.109a}$$

$$\frac{d\sigma}{dt} = |\beta_1(t, \tau)|^2\, e^{2(\alpha_1(t)-1)\ln\left(\frac{s}{s_0}\right)} \tag{13.109b}$$

In the case of high-energy p–p interactions, the total cross-section experimentally appeared constant (or nearly so) and, in other cases, appeared to be approaching a constant. Therefore, (in the case of p–p) assuming that

† where $\tau$ is the so-called 'signature' (+1 or −1), $\beta_n(t, \tau)$ contains the phase appropriate to the signature, which is $-\{1 + \tau\exp(-i\pi\alpha)\}/2\sin\pi\alpha$. The 'signature' will be discussed in the next sub-section.

the optical theorem applies,† $\dfrac{d\sigma(s, t = 0)}{dt}$ is a constant independent of $s$, and, thus, $\alpha_1(0) = 1$. This leading pole is assumed to have the quantum numbers of the vacuum, and it has been called the vacuum pole, or Pomeranchuk pole, or pomeron, since the Pomeranchuk theorem follows from it.

13.8.3. *Signature requirements*

Even in non-relativistic potential scattering theory, when there are exchange forces, we must perform the analytic extension in $l$ for $f_l(s)$ separately, in the even and odd $l$-terms, in order to satisfy the uniqueness requirements of the analytic extension. In relativistic scattering, it is always necessary to make this distinction. Therefore, we define the so-called 'even and odd signature' partial wave amplitudes

$$f^+(l, s) = f_{l_{\text{even}}}(s), \tag{13.110a}$$

$$f^-(l, s) = f_{l_{\text{odd}}}(s). \tag{13.110b}$$

The signature is often denoted by the symbol $\tau$.

For the even signature partial wave amplitudes, the corresponding set of Regge poles will be referred to as $\alpha_n^+(s)$, whereas for the odd signature partial wave amplitudes, the corresponding set of Regge poles will be referred to as $\alpha_n^-(s)$.

13.8.4. *The Pomeranchuk trajectory and the vacuum pole dominance model*

We have already shown that, based on experimental data, it was assumed that $\alpha_p(t = 0) = 1$ to obtain a constant total cross-section and, also, to satisfy the optical theorem. Let us parameterize $\alpha_p(t)$ near $t = 0$, by the series

$$\alpha_P(t) = \alpha_P(0) + \alpha'_P(0)t + \frac{\alpha''_P(0)}{2}t^2 + \ldots. \tag{13.111a}$$

Near enough to $t = 0$ we need only retain the linear term. Thus,

$$\alpha_P(t) = \alpha_P(0) + \alpha'_P(0)t. \tag{13.111b}$$

In order to fit the experimental data [14], [16], [17] at any one energy, we

† It is also assumed that real amplitude effects are small, and spin independence is assumed in the cases of p–p and $\bar{\text{p}}$–p.

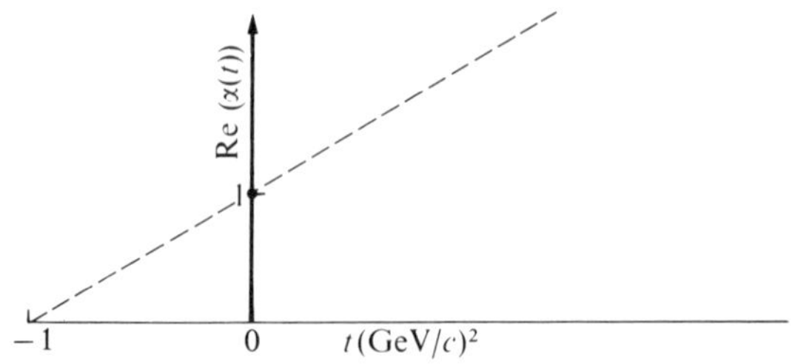


FIG. 13.9.

have already found that

$$\frac{\mathrm{d}\sigma}{\mathrm{d}t} = \mathrm{e}^{a+bt+ct^2} \to \mathrm{e}^{a+bt}, \text{ for small } t.\dagger \tag{13.112}$$

Thus, from eqns (13.109) and (13.112) $|\beta(t)|^2$ is of the form

$$|\beta(t)|^2 = \mathrm{e}^{a'+b't+c't^2}. \tag{13.113}$$

Hence, substituting eqns (13.111) and (13.113) into eqn (13.109*b*), we obtain

$$\begin{aligned}\frac{\mathrm{d}\sigma}{\mathrm{d}t} &= (\mathrm{e}^{a'+b't})(\mathrm{e}^{2\alpha_{\mathrm{P}}'(0)t\ln(s/s_0)}) = \mathrm{e}^{a'+[b'+2\alpha_{\mathrm{P}}'(0)\ln(s/s_0)]t} \\ &= \left(\frac{\mathrm{d}\sigma}{\mathrm{d}t}\right)_{t=0}\mathrm{e}^{[b'+2\alpha_{\mathrm{P}}'(0)\ln(s/s_0)]t} = \left(\frac{\mathrm{d}\sigma}{\mathrm{d}t}\right)_{t=0}\mathrm{e}^{bt},\end{aligned} \tag{13.114}$$

where we have made the linear trajectory approximation (i.e.

$$\alpha_{\mathrm{P}}(t) = \alpha_{\mathrm{P}}(0) + (\alpha_{\mathrm{P}}'(0))t \quad \text{and} \quad b = b' + 2\alpha_{\mathrm{P}}' \ln(s/s_0)$$

but we know experimentally that $b$ is a positive number, which is what is required for a limited range interaction. Also $\alpha_{\mathrm{P}}(0)$ must be $\leq 1$, to satisfy unitarity. Thus $\alpha_{\mathrm{P}}' \geq 0$.

If we remember that we have applied crossing symmetry, we obtain for the Pomeranchuk trajectory the predicted general behaviour in the linear approximation shown in Fig. 13.9. The value of the slope $\alpha_{\mathrm{P}}'(0)(\sim 1\ (\mathrm{GeV}/c)^{-2})$ was roughly estimated by analogy to the non-relativistic Regge theory, where it was shown that, below threshold, the rate at which a non-relativistic $\alpha$ can rise is related to the range of interaction, and the same criteria were

† A study of the elastic scattering, resulting from the inelastic processes in an uncorrelated jet model by Van Hove [92*a*], also resulted in this form. Cottingham and Peierls [92*b*] have modified this method by including a real part of the p–p elastic scattering amplitude represented by a difference of two Gaussians.

applied to relativistic Regge poles. The reasonable assumption that

$$R \approx \frac{\hbar}{2m_\pi c}$$

is the radius of the bound Regge state was made. It then follows that

$$\alpha_p'(0) \approx 1\ (\text{GeV}/c)^{-2},$$

and this is the slope used in Fig. 13.9. In order to satisfy the requirement that $\sigma_t$ is approximately constant for the asymptotic behaviour, the Pomeranchuk pole must occur in the even signature amplitude. This is obvious, since many elastic scattering amplitudes involve a $\pi$–$\pi$ vertex and, thus, $\tau P = +$ and $G = +$ are requirements for the Pomeranchuk pole. To satisfy the Pomeranchuk rule, and ensure that no quantum numbers other than those of the vacuum are exchanged in elastic scattering, it is obvious that $T = 0$, since both $T = 0$ and $G = +$ and we can see that $C = +$.

At $J = 1$ there is no physical particle corresponding to the pomeron. However,† for $s > 4m^2$, if the trajectory should enter the neighbourhood of $J = 2$, a physical particle of spin 2 was predicted as a possibility, and the $J^P = 2^+$, $f^0$, which was later discovered, was considered a likely candidate for this. The value of $b$ is increasing logarithmically with increasing energy, and the radius

$$R = 2\sqrt{b} = 2\sqrt{(b' + 2\alpha_p' \ln(s/s_0))} \approx 2\sqrt{(b' + 2\ln(s/s_0))}, \qquad (13.115)$$

and, thus, we predict the logarithmic shrinkage, with energy, of the diffraction pattern. If $\alpha_p'$ were to become negative, we would obtain the unphysical result that, at high enough energy ($s \to \infty$), $\frac{d\sigma}{dt}$ would increase infinitely rapidly from its value at $t = 0$, with increasing $|t|$. However, with $\alpha_p'(0)$ positive as $s \to \infty$ the diffraction pattern becomes infinitely narrow logarithmically with energy, and the effective radius of interaction (i.e. effective range) becomes infinite logarithmically with energy. If the total interaction cross-section approaches a constant, the effective transparency approaches zero.

It was originally predicted [20], [21] that the vacuum (or Pomeranchuk) pole would dominate at energies $\gtrsim 10$ GeV, and that we would observe universally, in all elastic scattering reactions, a shrinkage of the diffraction peak with increasing energy of the type just discussed. There was a considerable amount of excitement generated at the CERN International Conference

† In Fig. 13.9 the corresponding statement would involve $t > 4m^2$.

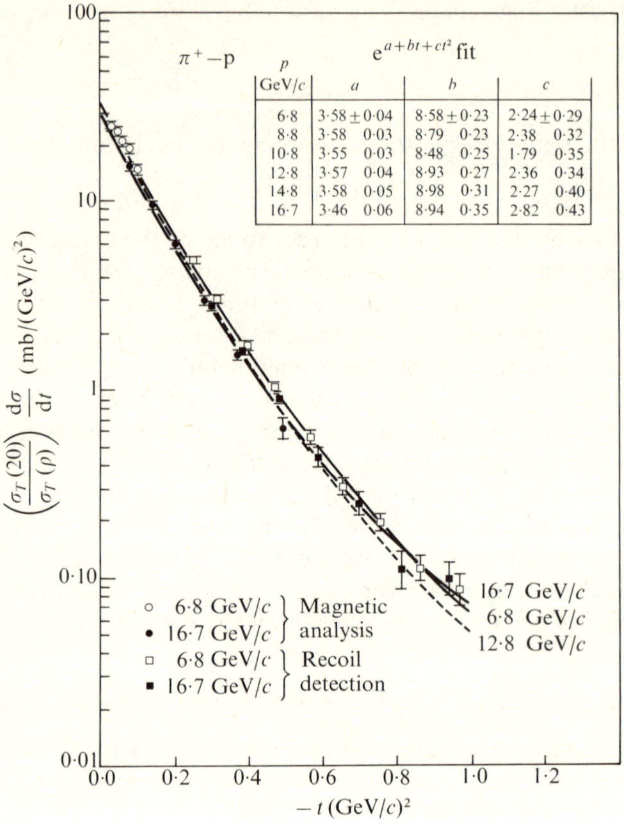


| $p$ GeV/$c$ | $a$ | $b$ | $c$ |
|---|---|---|---|
| 6·8 | 3·58 ± 0·04 | 8·58 ± 0·23 | 2·24 ± 0·29 |
| 8·8 | 3·58 0·03 | 8·79 0·23 | 2·38 0·32 |
| 10·8 | 3·55 0·03 | 8·48 0·25 | 1·79 0·35 |
| 12·8 | 3·57 0·04 | 8·93 0·27 | 2·36 0·34 |
| 14·8 | 3·58 0·05 | 8·98 0·31 | 2·27 0·40 |
| 16·7 | 3·46 0·06 | 8·94 0·35 | 2·82 0·43 |

FIG. 13.10.(a) The differential cross-sections for the highest and lowest momenta measured. In addition to the errors shown, there is a systematic scale uncertainty of ±5 per cent for the p–p case, ±6 per cent for the $\pi$–p cases, and a relative normalization uncertainty of ±3 per cent between the magnetic and correlation data. (a) $\pi^+$–p, (b) $\pi^-$–p, and (c) p–p.

(1962) when a reported study of p–p elastic scattering, using conventional counter telescope techniques, yielded results exhibiting the predicted shrinkage behaviour for p–p (see Fig. 13.4). This was followed closely by the BNL experiments, which utilized counter hodoscopes and on-line computer techniques, to study accurately p–p, $\bar{p}$–p, $\pi^{\pm}$–p, and $K^{\pm}$–p elastic scattering. Typical results are shown in Figs. 13.4 and 13.10.

It is evident from the figures, and established by a one-pole analysis that, in the 7–20 GeV/$c$ region, p–p does show a Regge-type shrinkage, with

$$\alpha_{p-p}(t) = 1{\cdot}07 \pm 0{\cdot}03 + (0{\cdot}83 \pm 0{\cdot}07)t.$$

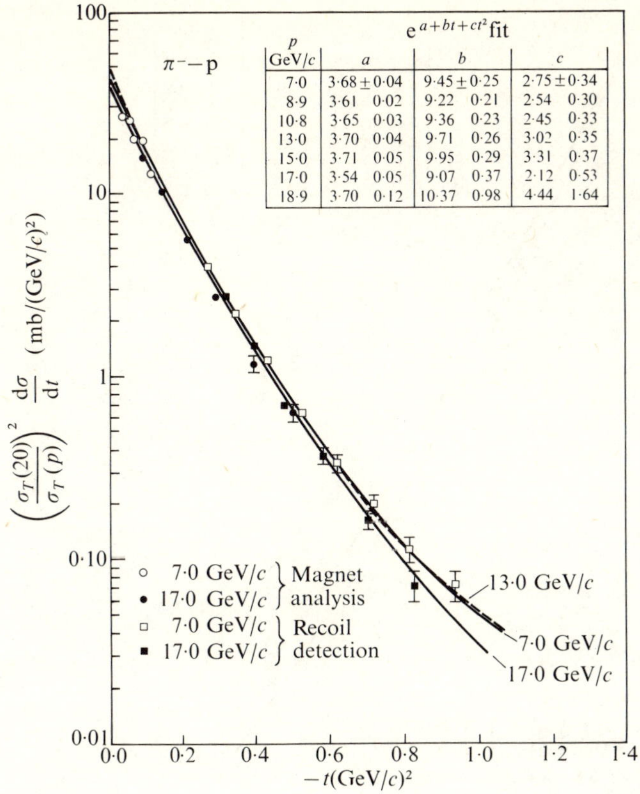


FIG. 13.10.(b)

On the other hand, $\pi^- + p$ and $\pi^+ + p$ did not shrink within the errors.

$$\alpha_{\pi^- + p} = (0{\cdot}96 \pm 0{\cdot}03) + (0{\cdot}008 \pm 0{\cdot}080)t,$$
$$\alpha_{\pi^+ + p} = (0{\cdot}96 \pm 0{\cdot}04) + (0{\cdot}086 \pm 0{\cdot}097)t.$$

$K^+ + p$ elastic scattering also exhibited shrinkage of somewhat less magnitude than p–p, which was further established by later experiments. $K^- + p$ seemed, within error, not to shrink, and this was also further established by later experiments. $\bar{p}$–p, however, exhibited an apparent expansion of the diffraction peak with increasing energy. Thus, it was clear that the vacuum

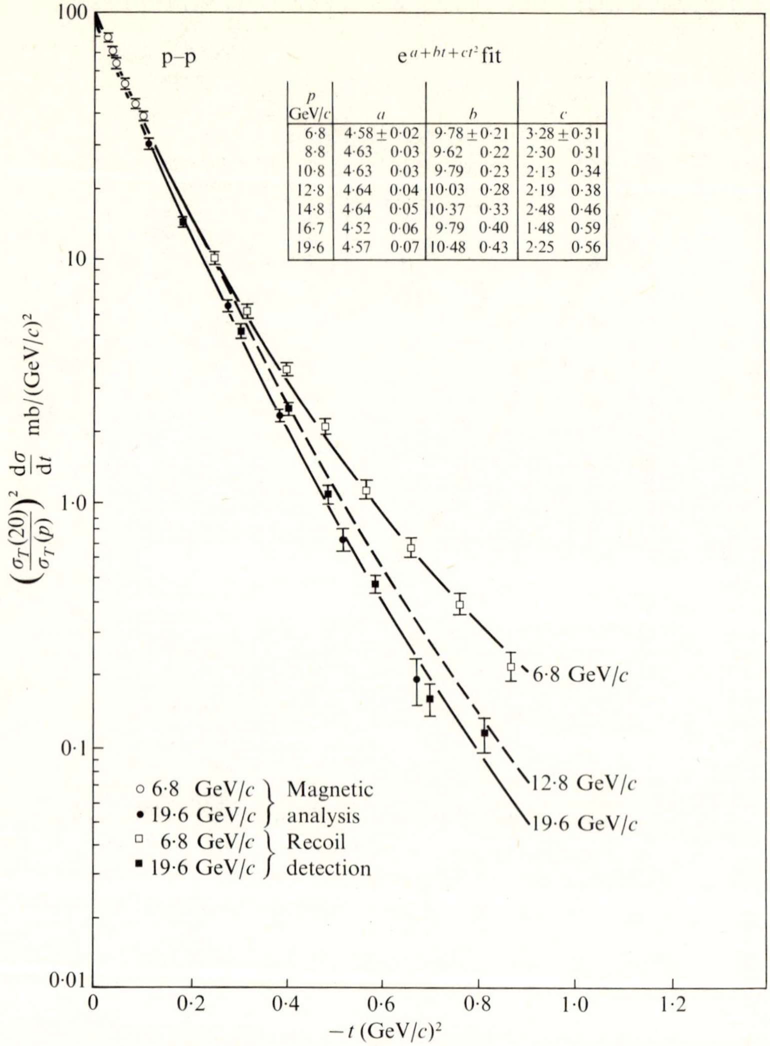
p–p
e^{a+bt+ct²} fit
p GeV/c | a | b | c
6·8 | 4·58 ± 0·02 | 9·78 ± 0·21 | 3·28 ± 0·31
8·8 | 4·63 0·03 | 9·62 0·22 | 2·30 0·31
10·8 | 4·63 0·03 | 9·79 0·23 | 2·13 0·34
12·8 | 4·64 0·04 | 10·03 0·28 | 2·19 0·38
14·8 | 4·64 0·05 | 10·37 0·33 | 2·48 0·46
16·7 | 4·52 0·06 | 9·79 0·40 | 1·48 0·59
19·6 | 4·57 0·07 | 10·48 0·43 | 2·25 0·56
(σ_T(20)/σ_T(p))² dσ/dt mb/(GeV/c)²
100
10
1·0
0·1
0·01
0
0·2
0·4
0·6
0·8
1·0
1·2
−t (GeV/c)²
6·8 GeV/c
12·8 GeV/c
19·6 GeV/c
○ 6·8 GeV/c
• 19·6 GeV/c
Magnetic analysis
□ 6·8 GeV/c
■ 19·6 GeV/c
Recoil detection

FIG. 13.10.(c)

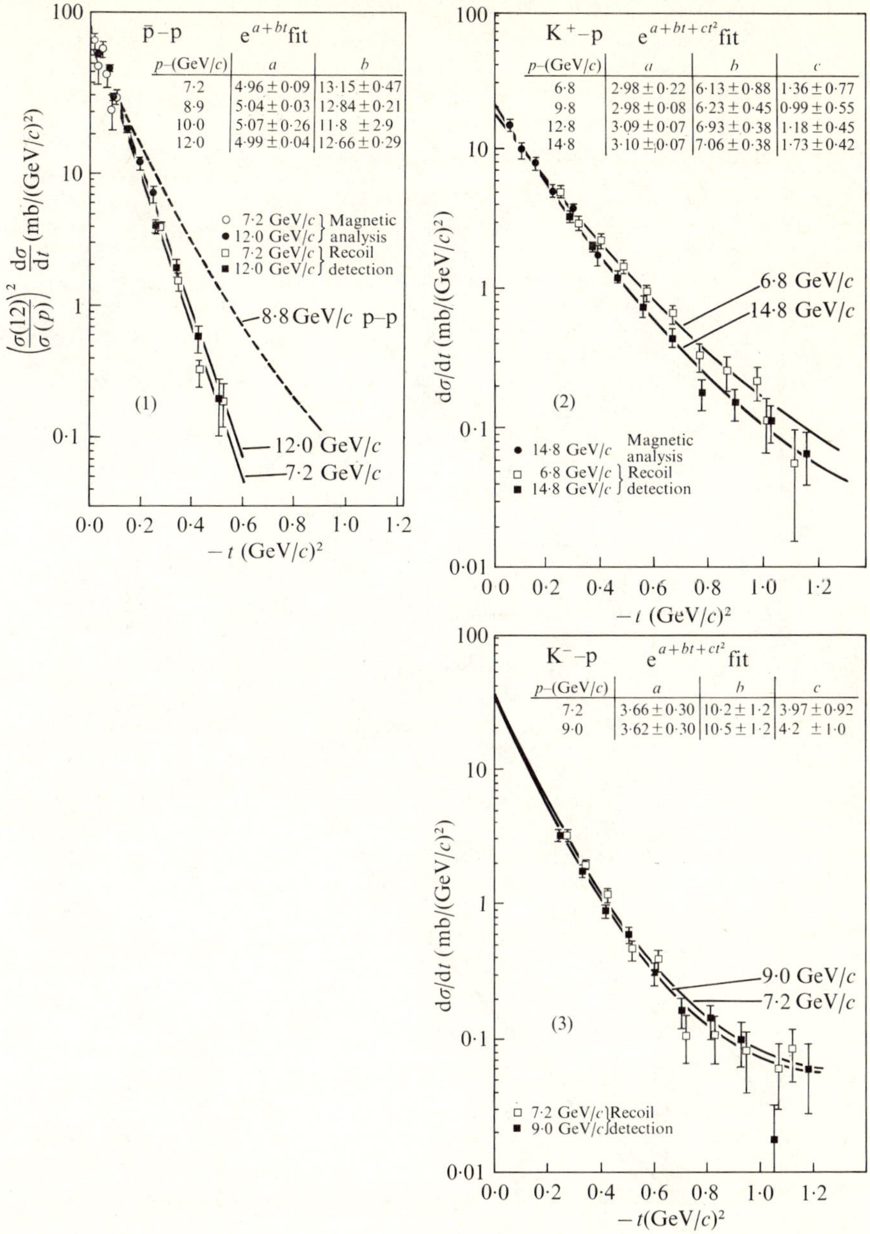


FIG. 13.10.(d) Differential cross-sections for the highest and lowest momenta measured. In addition to the errors shown, there is a systematic scale uncertainty of ±5 per cent for p̄–p, 7 per cent for $K^+$–p, and 3 per cent relative normalization uncertainty between magnetic analysis and recoil detection data. (1) p̄–p, the dashed line is a p–p curve at a comparable momentum. (2) $K^+$–p. (3) $K^-$–p.

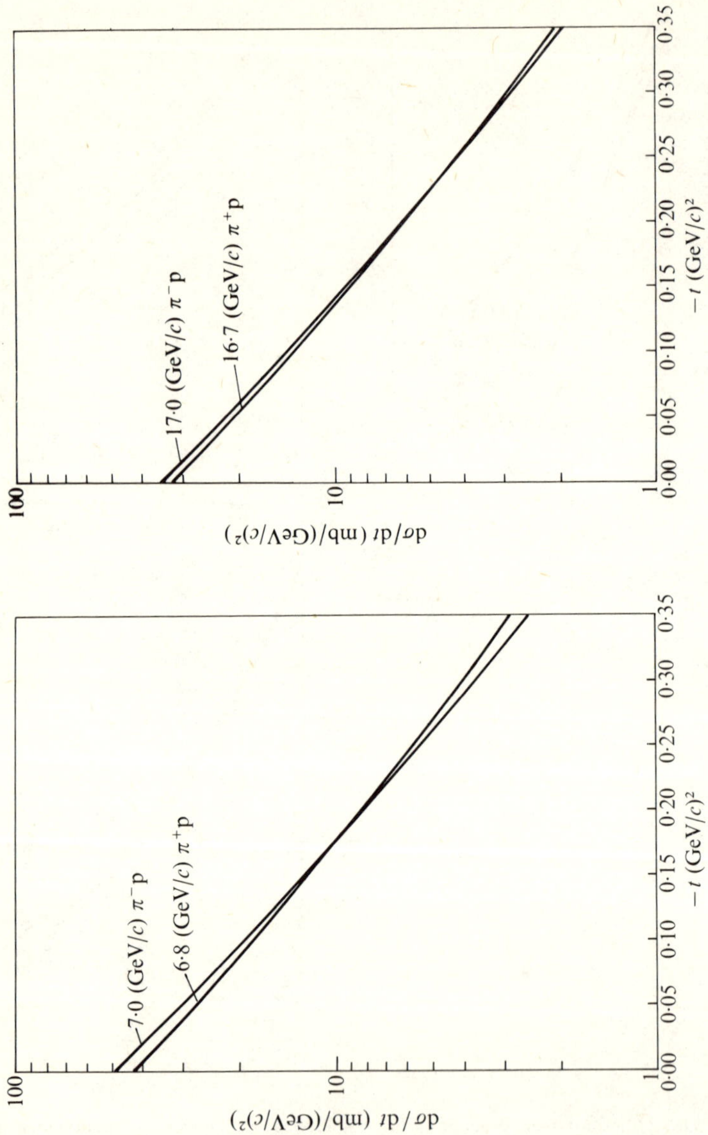


FIG. 13.10.(e) and (f). Crossover effects in $\pi^{\pm}$–p.

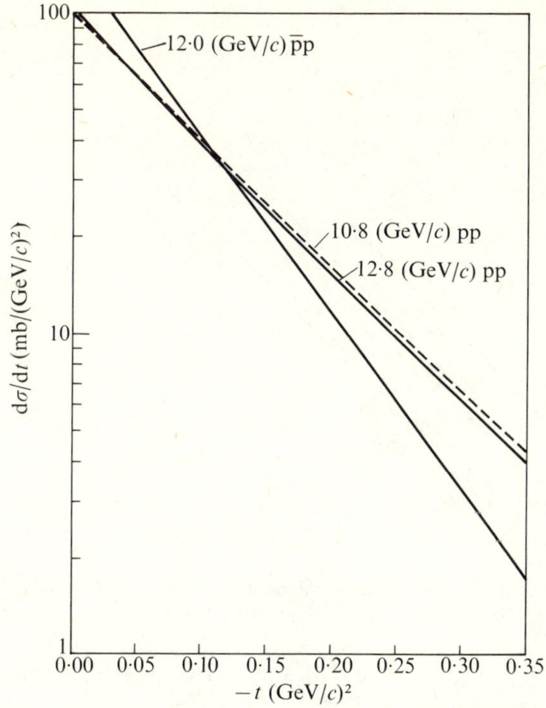


FIG. 13.10.(g) Crossover effects for p̄–p and p–p.

pole dominance model was contradicted in 10–25 GeV energy regions. Instead of universal shrinkage, there was almost uniform distribution among the three possibilities ('shrinkage', 'no shrinkage', or 'expansion'). Thus it was necessary to consider either a sufficient number of additional poles, or the possible effects of neglected cuts. The latter was shown possibly to have a considerable effect by Mandelstam and Amati, Fubini, and Stanghellini and Tonsin [22].

### 13.8.5. *The multi-pole models*

In attempts to fit the energy behaviour of the total cross-sections (Udgaonkar) [23] it became clear that additional poles were necessary. A second vacuum trajectory, P′, (or equivalently a distorted ABC trajectory) ($\alpha_{P'}(0) \approx 0{\cdot}4$) was introduced, as well as two trajectories corresponding to the $\rho$ and $\omega$ vector-mesons. At first, all trajectories were considered parallel

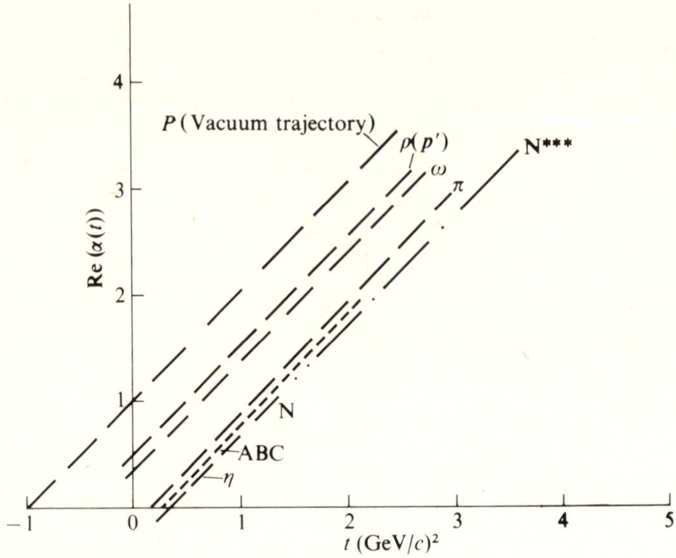


FIG. 13.11. Chew–Frautschi plot. (From Lindenbaum. (1964). *Nucleon structure* (Edited by R. Hofstadter). Stanford University Press, Stanford.)

to the Pomeranchuk trajectories and allowed to pass through the particle with which they were associated.

Figure 13.11 shows a Chew–Frautschi plot of the leading trajectories considered at the time of the Stanford conference (1963). The P, P′, $\rho$, and $\omega$ are the leading trajectories. This led to the four-pole model, which was approximately reduced to the three-pole model (in many cases), since the $\rho$ coupling appeared to be relatively weak.

Spin-averaged amplitudes and parallel, and approximately straight-line, Regge-pole trajectories were assumed, with average slope deduced from the nucleon trajectories. Good fits could be obtained to the total cross-sections and the p+p diffraction shrinkage. However, the behaviour of the $\pi^{\pm}$–p, $K^{\pm}$–p, and $\bar{p}$–p diffraction scattering made it impossible to explain the experimental results with the three- and four-pole models described above. However, Regge-pole fits to these experimental data have always been attainable to date by the simple technique of increasing the number of potentially-inexhaustible theoretical parameters used. For example, Rarita and Phillips [38] used about 50 parameters to explain the elastic scattering, and charge-exchange data. These fits involved five poles (P, P′, $\rho$, $\omega$, and R (or $A_2$)), and allowed arbitrary fits to Regge trajectories, $\alpha(t)$, and the residue functions.

The significance of such multi-parameter fits has been questioned by many people.

We have previously discussed the behaviour of Regge trajectories in non-relativistic potential scattering in the Re $l$, against $s$-planes. Thus, after applying crossing symmetry it is, similarly, assumed by analogy, for Regge poles in relativistic theory, that in the Re $J$ against $t$-plane,

$$\text{Im}\,\alpha(t) = 0, \quad \text{for} \quad t < 4m^2 \text{ (or } k_t^2 < 0),$$

$$\text{Im}\,\alpha(t) > 0, \quad \text{for} \quad t > 4m^2 \text{ (or } k_t^2 > 0).$$

For bosons, $\alpha(t)$ is a regular function in the entire $t$-plane, except for the branch cut on the real axis. Furthermore, as $t \to \infty$,

$$\text{Re}\,\alpha(t) < 0,$$

$$\text{Im}\,\alpha(t) \to 0.$$

Inelastic thresholds would be expected to result in a branch point for the trajectories. However, owing to the fact that the trajectories are usually characterized with $s$ near, but slightly above, the real axis in the $s$-plane, the trajectories are unique, and those deduced empirically are close to straight lines in the Re $J$ against Re $t$ plane.

Regge trajectories have been deduced in two ways.

(a) By placing straight lines through points in the positive Re $J$ against positive $t$ plane quadrant which correspond to observed particles, or resonances, with the same quantum number, but differing in $J$ value by 2 units (in order to preserve the uniqueness of the signature). Various trajectories of Regge recurrences have been assembled this way.

(b) By empirical analysis to fit experimental data. Sometimes particle candidates for these trajectories are later discovered.

### 13.8.6. *Factorization*

In the non-relativistic case we saw the intricate relationship between Regge poles and resonances, and this relationship was also assumed to hold in the relativistic case and is often used to locate Regge trajectories.

As is well known from nuclear reaction theory, the possibility that a resonance can decay via several distinct channels allows us to express the over-all cross-section for its formation from channel $a$, and decay via channel $b$, as the product of the partial width for its formation from channel $a$, and the partial width for its decay into channel $b$. Thus, by treating Regge poles as resonances, we see that, near a pole, the partial wave amplitude should

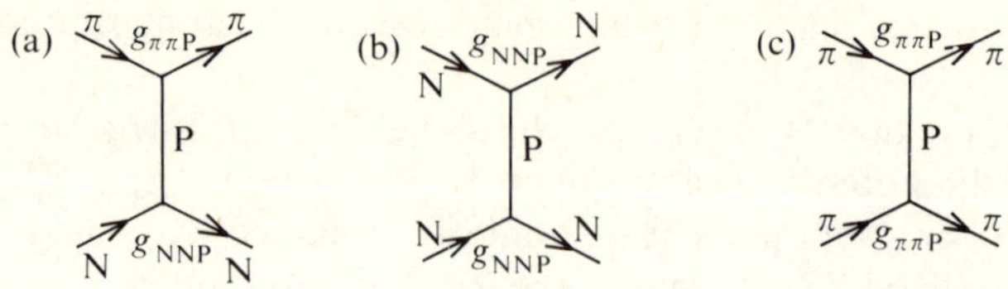


FIG. 13.12. $g_{\pi\pi P}$ represents the pomeron coupling to the $\pi$–$\pi$ vertex, $g_{NNP}$ represents the pomeron coupling to the N–N vertex, and $g_{\pi NP}$ represents the pomeron coupling to the $\pi$–N vertex.

be factorizable as follows,

$$f_{ab}(l, t) = \frac{\Gamma_a \Gamma_b R_{ab}(l, t)}{l - \alpha(t)}, \tag{13.116}$$

where $R_{ab}$ is regular at the poles. Thus, when various reactions in the $s$-channel involve exchange of the same Regge pole in the $t$-channel, we can use the 'factorization hypothesis' to relate their residues, and thus their cross-sections. Consider, for example, pomeron exchanges in the reactions shown in Fig. 13.12 (as $s \to \infty$ so that it dominates the reaction).

Thus, since all three processes involve the exchange of the pomeron, we expect the following. For $\pi$–N scattering

$$\beta_{\pi N} \propto g_{NNP} g_{\pi\pi P}. \tag{13.117a}$$

For N–N scattering

$$\beta_{NN} \propto (g_{NNP})^2. \tag{13.117b}$$

For $\pi$–$\pi$ scattering

$$\beta_{\pi\pi} \propto (g_{\pi\pi P})^2. \tag{13.117c}$$

Thus,

$$(\beta_{\pi\pi})(\beta_{NN}) = (\beta_{\pi N})^2. \tag{13.117d}$$

Applying this to the total cross-section (i.e. let $t = 0$, and take the imaginary part of the scattering amplitudes), we obtain

$$(\sigma_t)_{\pi\pi}(\sigma_t)_{NN} = ((\sigma_t)_{\pi N})^2. \tag{13.118}$$

Gribov and Pomeranchuk (1962) [24] obtained a similar generalized result, based on generalized unitarity, which led to

$$(\sigma_t)_{\pi\pi}(\sigma_t)_{KK} = ((\sigma_t)_{\pi K})^2. \tag{13.119}$$

All residues are assumed to be analytic in $t$, hence, the factorization theorem holds at all values of $t$, even though we have only illustrated the cases for $t = 0$.

In general, when there are many possibilities for the initial state ($a$) and the final state ($b$), the factorization theorem results demonstrated in potential theory are assumed to apply relativistically also. In the $t$-channel reaction

$$\alpha+\beta\underset{\longrightarrow}{\left(+\text{Regge Pole}\right)}\gamma+\delta \quad \text{and,} \tag{13.120a}$$

$$\beta_{\gamma\delta,\alpha\beta} \propto g_{\alpha\beta \mathrm{P}_n} g_{\gamma\delta \mathrm{P}_n}. \tag{13.120b}$$

Here $g_{\alpha\beta\mathrm{P}_n}$ and $g_{\gamma\delta\mathrm{P}_n}$ are universal for all reactions involving particles ($\alpha$ and $\beta$) and ($\gamma$ and $\delta$), respectively, when the Regge pole $\mathrm{P}_n$ is exchanged.

13.8.7. *Fermion Regge poles*

So far we have explicitly treated only boson Regge poles, since this case is relatively straightforward. In the case of $\pi$–N scattering shown in Fig. 13.13, the $t$-channel Regge poles which control high-energy forward scattering

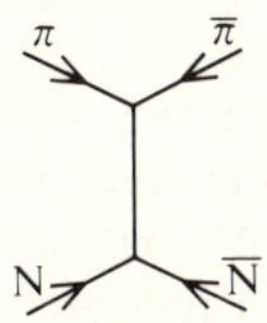


FIG. 13.13.

are bosons, but the $u$-channel Regge poles must be fermions. The $\pi$–N scattering amplitude in the $s$-channel can be expressed in terms of the helicity partial wave expansion. We need to express the partial waves in definite parity combinations for Regge-pole applications. General crossing relations for helicity amplitudes have been obtained (Trueman and Wick 1964) [25] which allow continuation of the $s$-channel amplitudes into the crossed channels for $t$ or $u$. The treatment of fermion Regge poles [26] leads to inconvenient kinematic singularities in the partial wave amplitudes† at $W = \sqrt{s} = 0$. For example, in $\pi$–N scattering in the $u$-channel these singularities are caused by kinematic factors introduced by the nucleon spin, when $k^2 = 0$. Additional branch points occur for non-physical $J$. Thus, for Regge trajectories, we must specify the signature $\tau$, or the $J$-parity, the space parity, and the isotopic spin. Hence, if we assume all particles (fermion as well as boson) are Regge-ized, the nucleon will have, associated with it, a trajectory specified by $T = \frac{1}{2}$, $\tau = +$, or $J = \frac{1}{2}, \frac{5}{2}, \frac{9}{2}, \ldots$ .

The N*(1238) will have, associated with it, a trajectory specified by $T = \frac{1}{2}$, $\tau = +$, and $J = \frac{3}{2}, \frac{7}{2}, \ldots$ .

The exchange of baryon Regge poles will be particularly relevant in our subsequent treatment of backward $\pi^{\pm}$–p scattering which will, within the

† For fermions, it is more convenient to employ the $W = \sqrt{s}$-plane, instead of the $s$-plane.

framework of the Regge pole model, obviously involve baryon pole exchange in the $u$-channel. Backward $\pi^- + p \rightarrow \pi^- + p$ in the $s$-channel will be dominated by a Regge exchange in the corresponding $u$-channel. For backward $\pi^- + p$ scattering, the pole which is exchanged in the $u$-channel must be doubly charged so that the forward-going $\pi^-$ in the $s$-channel becomes a forward-going proton in the $s$-channel. Thus the nucleon pole cannot enter this process, but the doubly charged N*(1236), with $T = \frac{3}{2}$, can. However, in the backward scattering of $\pi^+ + p$, the neutron can be exchanged.

We can gain insight into the source of the kinematic singularities which occur in the forward direction by the following. The most general scattering amplitude for nucleon–nucleon scattering is restricted by the usual symmetry principles, and involves five complex functions. If we consider the partial wave expansion, we then need five partial wave amplitudes that are functions of $J$ and $W = \sqrt{s}$. However, we have previously remarked that, in the forward direction, two of the five functions vanish, leaving only three, thus implying that relations exist between the various partial wave amplitudes when $t = 0$. Hence, it is implied that the locations of the Regge poles of the various states are not independent at $t = 0$, and are related via constraint equations. A similar situation exists for $\pi$–p where one (i.e. the spin, or helicity, flip) of the two amplitudes vanish at $t = 0$. Various ways of satisfying these constraints have been proposed, and have resulted in 'daughter trajectories', 'conspiracy', or 'evasion'.

### 13.8.8. *Daughter trajectories*

As a result of kinematic singularities and constraint equations, in order to retain the asymptotic form $s^{\alpha(t)}$ near $t = 0$, Regge trajectories must occur in families, with intercepts at $t = 0$, equally spaced by integers [27]. The original Regge trajectory is referred to as the parent, and the accompanying trajectories as the 'daughters'. It was concluded that the tensor fields representing elementary particles contain extra spin components, when the intermediate particle is off its mass-shell. These extra spin components then give rise to the daughter trajectories which, thus, should be parallel for all values of $t$. When extra spin components were eliminated by subsidiary conditions, it was concluded that the daughter trajectories should not make any contribution at physical values of angular momentum and, thus, would not contain physical particles.

Toller [28] proposed that scattering amplitudes (for equal mass scattering), at $t = 0$, obey four-dimensional symmetry, and Regge-ized the amplitudes in terms of the homogeneous Lorentz group. This led to an infinite family of Regge poles, when expanded in the three-dimensional representation.

These poles are referred to as Lorentz, or Toller, poles. In essence, daughter trajectories seem to follow from the requirements of 0(4) symmetry.

13.8.9. *Various additional requirements on Regge pole amplitudes and solutions to the constraint equations*

Various types of solutions to the constraint equations, at $t = 0$, were proposed [29]. These included the following, [30]†

(a) *Evasion.* In this case the residue functions must satisfy certain relations. However, the trajectories are not restricted in any way. The so-called 'trivial evasive solution' has all residues vanishing simultaneously for $t = 0$.

(b) *Conspiracy.* In this case relations between the residues and the intercepts of different Regge (with different internal quantum numbers) poles are required such that the constraint equations are satisfied.

(c) *Daughter-like solutions.* A parent Regge trajectory conspires with a sequence of Regge poles having the same internal quantum numbers as the parent. The families correspond to those that were discussed under Toller poles.

There have been numerous attempts to incorporate the predictions of SU(3), and higher symmetries, into Regge pole fits, thus restricting the number of free parameters available.

Loosely speaking, if the exchange potential's effects were small compared to the direct potential, the even and odd signature amplitudes would become equal‡ (i.e. $f^+ = f^-$) and, thus, $\alpha^+ = \alpha^-$, $\beta^+ = \beta^-$. In the Mandelstam representation, '*exchange degeneracy*' follows from the assumption that the contribution of the spectral functions $\rho_{su}$ and $\rho_{tu}$ are negligible compared to the contribution from $\rho_{st}$.

For even signature trajectories, if $\alpha^+(t) = 0$ for $t < 0$, we have a so-called '*ghost*' *state* which represents a particle of imaginary mass in the $t$-channel. Thus, it is assumed that the coefficient in the amplitude multiplying the pole term vanishes at this value of $t$, which kills the ghost. The suggestion was made [29*b*] both for $\alpha^+$ and $\alpha^-$, that whenever $n = 0, -1, -2, -3 \ldots$, the coefficient multiplying the particular pole should vanish. However, these ghost-killing mechanisms in the $t$-channel lead to predicted dips in the $s$-channel differential cross-sections. In some cases these dips fit the behaviour of the data, and in many others they do not.

If the helicity of a channel is such that $h \leq \alpha$, the channel is referred to as a 'sense' channel, and the subscript 's' is often used to denote this. If $h > \alpha$,

† A general review of Regge pole models is given in reference [30].

‡ This leads to so called 'exchange degeneracy'.

the channel is referred to as a non-sense channel and the subscript 'n' is used. Gell-Mann, Chew, and others have considered various methods of defining Regge residues in terms of couplings to s- and n-channels, to avoid undesirable effects (kill ghosts, etc.). These techniques lead to experimental effects (dips, etc.) which we shall consider later.

It was shown [31], via a study of Feynman graphs, that the simultaneous exchange of two Regge poles generated a cut. Mandelstam [32] observed that (assuming the absence of the third double spectral function) such cuts are partially cancelled by other multi-particle diagrams, but their persistence (to some extent) would be caused by other more complicated diagrams. He related these cuts to the essential singularity of Gribov and Pomeranchuk [33]. It was later shown [34] that the Mandelstam cut combined with the Gribov–Pomeranchuk essential singularity results in a simple fixed pole which has a signature which does not affect the physical amplitude. The Mandelstam cut mechanism, and the assumption $\alpha_p(0) = 1$, results in an infinite series of Regge cuts. This leads to the conclusion of vanishing total cross-sections, or constant diffraction peaks [35].

If there were a branch point in $\text{Re}\, l > -\frac{1}{2}$, located at $l = \alpha_c(t)$, for large $s$, the result of integrating over this branch point would be of the form

$$F(s, t) \sim \beta(t) s^{\alpha_c(t)} (\log s)^{-|\gamma_c(s)|}. \tag{13.121a}$$

If we fix the cut at $\alpha_c(0)$, when $t < 0$, then for $t < 0$, as $s \to \infty$,

$$F(s, t) \sim \beta(t) s^{\alpha_c(0)} (\log s)^{-|\gamma_c(t)|}. \tag{13.121b}$$

Considering the case where exchange of a pomeron is allowed, we have

$$\alpha_c(0) = \alpha_p(0) = 1. \tag{13.121c}$$

Thus, when $s \to \infty$, only when $t = 0$ can the pole dominate over the branch cut. When $t < 0$, the branch cut would dominate.

If we apply the principle of factorization to a set of Regge poles with definite quantum numbers, we obtain a symmetry generally referred to as '*line-reversal*'. This symmetry refers to two states, the second of which is related to the first by a particular operation such as $C$, $G$, etc. In that case, for a particular Regge pole the two vertex functions will be related by the eigenvalue of the Regge pole for the operator. For example, consider the case

$$\mathrm{N} + \overline{\mathrm{N}} \to \overline{\mathrm{N}} + \mathrm{N}. \tag{13.122a}$$

The two vertices are related by $C$ which transforms one into the other, thus the phase of each vertex will be equal to the $C$-eigenvalue for the Regge pole involved. A general statement of line-reversal [36] is that if

$$\alpha+\beta \rightarrow \gamma+\delta, \tag{13.122b}$$

$$\alpha+\bar{\delta} \rightarrow \gamma+\bar{\beta}, \tag{13.122c}$$

are related, so that they exchange the same Regge poles in the crossed ($t$) channel, the contribution of the poles to either process has the same magnitude, but the sign (phase) changes according to the eigenvalue of the operator relating the two processes.

### 13.8.10. *Regge pole models applied to high-energy scattering at low and moderate t*

Let us now consider, in more detail, the status of Regge pole models in relation to their explanation of the high-energy diffraction scattering region.

We shall consider $\pi$–p elastic scattering first. So far we have tacitly ignored spin dependence, and discussed only the spinless particles case (i.e. $\pi$–$\pi$, $\pi$–K, or K–$\pi$ scattering), or have assumed that we can represent the process by a spin averaged amplitude. If we now generalize eqn (13.106b) to the $\pi$–N case, where the nucleon has a spin, it will become of the form [36]

$$A(s,t) \sim \sum_n \beta_n(\tau,t)\left(\frac{s}{s_{0_n}}\right)^{\alpha_n(t)-1}((\eta_\pi)_n(\eta_N)_n+(\eta_\pi)_n(\phi_N)_n), \tag{13.123}$$

where $\beta_n(\tau, t)$ contains the signature factor, and $t$ dependence of the residue. $(\eta_\pi)_n$ and $(\eta_N)_n$ represent the non-helicity flip coupling (for the $n$-pole exchange) of the pion and the nucleon. $(\phi_N)_n$ represents the helicity flip coupling for the nucleon. Thus, applying the optical theorem to obtain total cross-sections, and remembering that the helicity flip terms vanish in the forward direction, we obtain

$$(\sigma_{\pi N})_t \sim \sum_n \beta_n(t,\tau)\left(\frac{s}{s_{0_n}}\right)^{\alpha_n(0)-1}(\eta_\pi)_n(\eta_N)_{n'}. \tag{13.124}$$

For the elastic differential scattering cross-section, we obtain

$$\frac{d\sigma_{\pi N}}{dt} = |A(s,t)|^2 = \sum_{n,n'} \text{Re}\ \beta_n^*(t,\tau)\beta_{n'}(t,\tau)\left(\frac{s}{s_{0_n}}\right)^{\alpha_{n'}-1}\left(\frac{s}{s_{0_{n'}}}\right)^{\alpha_n(0)-1}\{(\eta_\pi)_n(\eta_N)_{n'}\}\times$$
$$\times\{(\eta_N)_n(\eta_N)_{n'}+(\phi_N)_n(\phi_N)_{n'}\}. \tag{13.125}$$

The results for a similar treatment of the nucleon–nucleon case can be obtained from previous references [36].

13.8.11. *Polarization, crossover, etc.*

It is characteristic of the Regge-pole expressions that the phase between the helicity flip and helicity non-flip amplitudes is such that the resulting polarization in any reaction, caused by the exchange of a single pole, is zero, assuming that the couplings $\eta$ and $\phi$ are entirely real. This is expected to be true, provided $t$ is less than the lowest $t$-channel threshold. Therefore, for example, a test for single-pole exchange is to observe zero polarization. As $s \to \infty$, the Regge pole expression, for the polarization in elastic scattering, becomes

$$P\left(\frac{d\sigma}{dt}\right)_{\pi N} \sim \sum_{nn'} \operatorname{Im} \beta_n^*(t,\tau)\beta_{n'}(t,\tau)\left(\frac{s}{s_{0_n}}\right)^{\alpha_n-1}\left(\frac{s}{s_{0_{n'}}}\right)^{\alpha_{n'}-1}(\eta_\pi)_n(\eta_\pi)_{n'}(\phi_N)_n(\phi_N)_{n'}. \tag{13.126}$$

As previously remarked, it was possible to fit the total cross-section data with the P, P′, $\rho$, and $\omega$ poles, and also fit the p–p shrinkage of the diffraction peak data. However, the experimental results on elastic scattering for $\pi^{\pm}$–p, which showed no shrinkage, and $\bar{\text{p}}$–p which showed expansion instead of shrinkage, as well as $K^+ + p$ which showed shrinkage and $K^- + p$ (consistent with no shrinkage) led to the proposal of the five-pole model P, P′, $\rho$, $\omega$, and R (or $A_2$). The residues and trajectories and fits for the parameters were selected to fit the experimental data.

As we can see in Fig. 13.10, comparing the differential elastic scattering cross-sections of $\pi^-$ and $\pi^+$ as a function of $t$, we can see that

$$\frac{d\sigma}{dt}(\pi^- + p)_{t=0} > \frac{d\sigma}{dt}(\pi^+ + p)_{t=0},$$

but that the exponential slope, $b$, of $\frac{d\sigma}{dt}(\pi^- + p)$ is larger so that the $\pi^- + p$ curves (originally above) *cross over* the $\pi^+ + p$ curves in the low $|t|$ region. This also occurs for the $\bar{\text{p}}$–p curve which (originally higher) crosses the p–p curve. The $K^- + p$ and $K^+ + p$ curves exhibit the same phenomena. The value of $t_c$ where the *crossover* occurs seems to be contained in the region $-0{\cdot}1$ to $-0{\cdot}25$. On the simple absorbing potential model this effect could be explained qualitatively in a simple way. We need only recognize that in $\bar{\text{p}}$–p, $\pi^- + p$, and $K^- + p$, there are a larger number of channels open than in the corresponding p–p, $\pi^+ + p$, and $K^+ + p$ reactions. $\bar{\text{p}}$–p has the annihilation channels which are absent in p+p. $K^- + p$ has the possibility of transforming the proton into a strangeness $-1$ hyperon and also of charge-exchange (both absent in $K^+ + p$). $\pi^- + p$ has the possibility of charge-exchange (not possible in $\pi^+ + p$). Thus we could imagine that, owing to these additional channels, both the total cross-section and effective radii are larger for the $\bar{p}$, $K^-$, $\pi^-$ incident on p. Therefore, with a purely complex potential model, we would expect the crossover effect. However, for Regge-pole theory, explaining this effect causes severe problems.

From line-reversal symmetry (via the $C$ operator), it is clear that the same Regge trajectories contribute to both processes (since $C = +$ for P, P′, $A_2$, $\pi$, while $C = -$ for $\rho$, $\omega$, $\phi$). Thus,

$$\Delta\left(\frac{d\sigma_{\bar{\alpha}+\beta}}{dt} - \frac{d\sigma_{\alpha+\beta}}{dt}\right) \sim \sum \mathrm{Re}(T^*_{c=+} T_{c=-}),$$

where $T_+$ = the sum of contributions from $C = +$ trajectories (i.e. P, P′, $A_2$, $\pi$), and $T_-$ = the sum of contributions from $C = -$ trajectories (i.e. $\rho$, $\omega$, $\phi$). To make this interference term disappear at the crossover, for the observed reactions, has been a great problem for Regge-pole fits. A number of attempts [37] have been made to solve the problem, none of which are entirely devoid of difficulties,† and some of which use the technique of introducing new trajectories. The status of the crossover effect explanations to date are doubtful enough for some doubt to be shed on the factorization principle.

The explanation of the crossover effect seems to require zeros in residues. These are in addition to those naturally required, at special values of $\alpha$, by ghost-killing mechanisms, etc. The results of these zeros are to predict dips in differential cross-section, some of which are in agreement with the data and some of which are not. We shall discuss these phenomena in a later section.

*In order to explain the absence of shrinkage in* $\pi$–p, it has been found in various Regge fits [38], [39] that the slope of $\alpha_p$ is constrained to be small, whereas values more or less close to $\alpha'_p = 1{\cdot}0$ are appropriate for all the other trajectories. It was found [38] in a study of $\pi^\pm$–p, p–p, and $\bar{p}$–p that

$$\alpha_p = 1{\cdot}0 + (0{\cdot}0 \text{ to } 0{\cdot}3)t,$$

whereas

$$\alpha'_\rho = 0{\cdot}75 + 1{\cdot}5t,$$

or

$$\alpha_\rho = 0{\cdot}57 + 2{\cdot}17t.$$

† For example, the normal explanation of the $\bar{p}$–p and p–p crossover effect [37*a*] requires the $\omega$ trajectory residue to vanish at $t \approx -0{\cdot}18$ (GeV/$c$)$^2$. Then from the factorization theorem, it follows that all reactions for which the dominant mechanism is $\omega$-exchange should exhibit a dip at $t \approx -0{\cdot}15$ (GeV/$c$)$^2$. However, both the reactions $\pi + p \rightarrow \rho + n$ and $\gamma + p \rightarrow \pi^0 + p$ are thought to be dominated by the $\omega$-exchange, and neither show the predicted dip [37*b*, *c*].

Cabibbo, Kokkedee, Horwitz, and Ne'eman [40] have used fits to the data with $\alpha_{\mathrm{p}}(0)$ slightly less than 1. This obviously causes all cross-sections to approach zero as $s \to \infty$. However, the residues then satisfy the $\mathrm{U}(3) \times \mathrm{U}(3)$ algebra. The consensus of the various studies of the P and P′ trajectories, based on AGS and CERN data, was that the P is flat, or almost flat, and, therefore, has no known Regge recurrence. Although originally the $f^0$, with spin 2 and positive parity, was considered to be a Regge recurrence of the P, it was later considered to be a Regge recurrence of the P′.

The characteristics of the $\rho$, $\omega$, $\mathrm{A}_2$, etc. are better studied in charge-exchange, and other channels. The changes in the analysis proposed as a result of the recent experiments at Serpukhov will be considered later in this Chapter.

### 13.8.12. *Pion–nucleon charge-exchange scattering*

There have been a number of investigations of pion–nucleon charge exchange at high energies. [41] The characteristics of charge-exchange scattering are illustrated in Fig. 13.14(a).

The charge-exchange curve exhibits a peculiar shape in that starting from very low $|t|$, $\dfrac{\mathrm{d}\sigma}{\mathrm{d}t}$ seems to (probably) increase for a while and then round off and decrease with a typical pion–nucleon scattering slope from $|t| = 0{\cdot}2$ to $|t| = 0{\cdot}5$. It then flattens off and is followed by a second peak in the neighbourhood of $t \approx -1\ (\mathrm{GeV}/c)^2$. The small $|t|$ behaviour of the pion charge-exchange suggests that there are comparable components of spin flip and non-spin flip amplitudes. Both components of the charge-exchange amplitudes have squares (i.e. contributions to the cross-section) small ($\lesssim \frac{1}{100}$) compared to elastic scattering amplitudes squared which, according to experimental evidence is mainly of the non-spin flip type in the small $|t|$ diffraction region. Since

$$f_{\mathrm{ch-ex}} = \frac{(f_{\pi^+-\mathrm{p}} - f_{\pi^--\mathrm{p}})}{\sqrt{2}} \tag{13.127a}$$

and

$$f_{\pi^++\mathrm{p}} \approx f_{\pi^--\mathrm{p}}, \tag{13.127b}$$

it is not too surprising that in taking differences, we find comparable contributions from ordinary and spin flip amplitudes. The ratio

$$\frac{\mathrm{Re}\, f_{\mathrm{ch-ex}}}{\mathrm{Im}\, f_{\mathrm{ch-ex}}} \sim 1.$$

This can be understood, since we are taking differences.

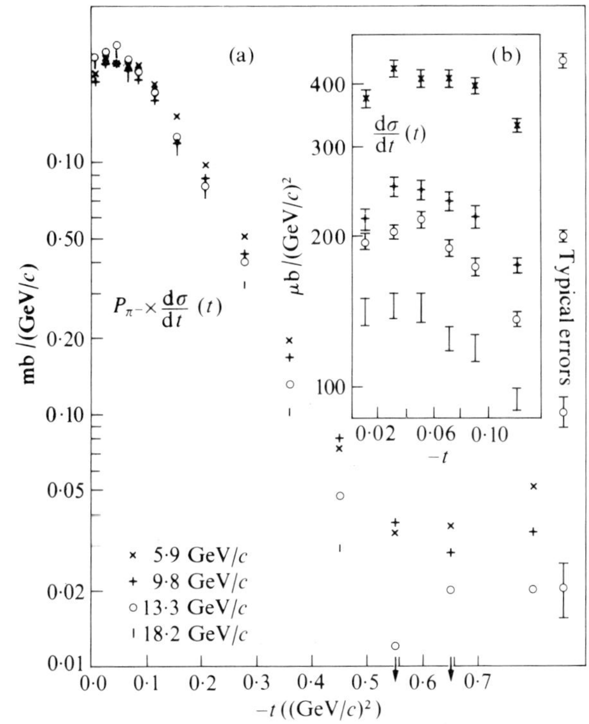


FIG. 13.14.(a) (1) $(P_{\pi^-})\dfrac{d\sigma}{dt}(t)$ for the reaction $\pi^- + p \rightarrow \pi^\circ + n$ at 5·9, 9·8, 13·3, and 18·2 GeV/$c$. Typical errors are shown on the right. (2) $\dfrac{d\sigma}{dt}(t)$ in the region of the forward peak. (From [41*b*], Stirling *et al.*)

The energy dependence of the charge-exchange cross-section, $\dfrac{d\sigma}{dt}(t = 0)$, and the estimated integrated total charge-exchange cross-section, both show $\dfrac{1}{P_\pi}$ dependence, since when $P\dfrac{d\sigma}{dt}$ is plotted (Fig. 13.14(a)) it appears constant. A most significant use for these measurements is that we can, by extrapolation, determine $\left(\dfrac{d\sigma}{dt}\right)_{t=0}$. There is, of course, some uncertainty in this extrapolation due to the peculiar shape of $\dfrac{d\sigma}{dt}$ near $t = 0$. However, this has been done using a power series fit. At $t = 0$, the spin flip amplitude must vanish, and we are left with just the contribution of the ordinary charge-exchange amplitude.

From the analysis of the previously-discussed measurements of $\pi^{\pm}$–p small-angle scattering and total cross-section measurements, we have determined both the real and imaginary parts of $f_{\pi^- + p}$ and $f_{\pi^+ - p}$, at $t = 0$. Thus we can, assuming charge independence, exactly predict the charge-exchange amplitude at $t = 0$. Unfortunately, there is no way of directly obtaining the charge-exchange amplitude itself, experimentally. However, we can square the predicted charge-exchange amplitude and, thus, predict $\left(\frac{d\sigma}{dt}\right)_{t=0}$, and compare this with the experimentally-determined results for charge-exchange as shown in Fig. 13.14(b).

All sources of systematic, as well as relative, errors have been included in the predicted results for $\pi^{\pm}$–p forward scattering and total cross-section measurements. Considering the errors, the agreement is satisfactory, particularly when we note that no error has been included for the uncertainty in functional form to be used in extrapolating the charge-exchange data to $t = 0$, and absolute normalization calibrations in the charge-exchange data

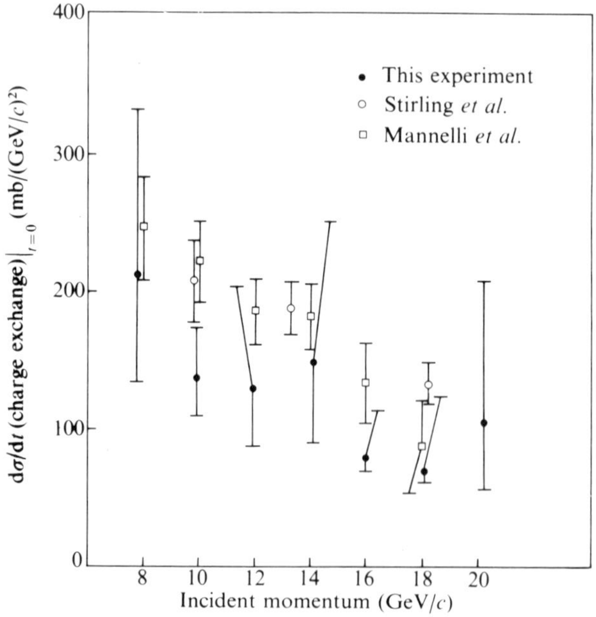


FIG. 13.14.(b) Comparison of the charge-exchange cross-section at $t = 0$ and the values deduced from this experiment, with the assumption of charge independence. (From Foley *et al.* (1969). *Phys. Rev.* **181,** 1775.)

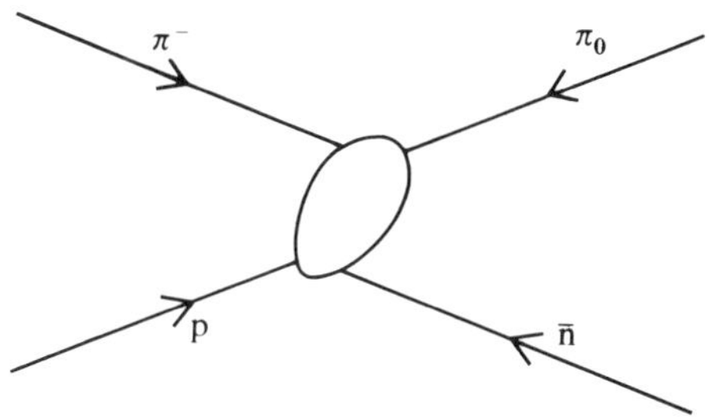


FIG. 13.15.

are difficult to perform. Even though the check is only to an accuracy of 10 to 20 per cent of the charge-exchange amplitude, this amplitude is itself less than about 10 per cent of the $T = \frac{3}{2}$, or $T = \frac{1}{2}$, amplitudes. Thus, if there are charge-dependent terms in either of the $T = \frac{1}{2}$, and $T = \frac{3}{2}$, amplitudes, they are limited to 1–2 per cent of the amplitudes. Hence we have a sensitive and definitive check of charge independence for the pion–nucleon system up to 20 GeV/$c$.

The presence of a sizeable spin flip term would fit the picture of a vector meson exchange. Let us look at the requirements on single particle exchange in Fig. 13.15, which represents pion–nucleon charge exchange. It is clear that the exchanged particle must have $I = 1, G = +$ (or negative signature) and $s = 0$. The only known particle with these properties was the $\rho$-meson.† Calculation of ordinary vector-meson exchange gave a constant cross-section, independent of energy, corresponding to a real scattering amplitude with the usual associated unitarity difficulties. The proper incorporation of unitarity requirements in these calculations would lead to a decreasing cross-section with energy.

The pion–nucleon charge-exchange reaction was considered to have great significance for the Regge pole model since it is the first reaction studied where only one known pole (i.e. the $\rho$) could be exchanged. Furthermore, Regge-ization, because of the damping due to the decrease of $\alpha(t)$ with increasing $t$, removes the violations of unitarity, which always troubled fixed-pole models for vector exchange. Thus we would expect the predicted Regge shrinkage, since this is a single-pole exchange reaction. From an analysis of the charge-exchange data, it was determined that [42]

$$\alpha_\rho(t) \simeq 0{\cdot}55 + 0{\cdot}9t.$$

This trajectory (within the error) passes through the $\rho$-meson and, thus,

† It subsequently appeared that the B-meson has the necessary quantum numbers also, but, in any case, it would be expected to be much less important than the $\rho$.

this was interpreted as a considerable triumph for the Regge pole model, since it explained [42], [38*a*], [43] both the energy dependence, and the shrinkage, and the second dip.

In any vector exchange model, including the Regge model, the dip near $t = 0$ can be attributed to the required vanishing of the spin (or helicity) flip amplitude. However, the Regge model also appeared to explain the second dip at $t \approx -0{\cdot}6\ (\mathrm{GeV}/c)^2$, since here $\alpha_P = 0$ and, thus, there should be a zero in the helicity flip amplitude at this 'non-sense' point. However, fixed poles are supposed to occur at wrong-signature non-sense points and could, in principle, cancel the dip, unless we further assume that they are coupled weakly. If the third spectral function vanishes, or is small, fixed poles and Regge cuts are both not necessary.

The same dip mechanism for the $\rho$ should lead to dips in $\pi^+ + \mathrm{n} \rightarrow \omega + \mathrm{p}$ and $\pi^+ + \mathrm{p} \rightarrow \omega + \mathrm{N}^{*++}$ due to $\rho$ exchange dominating. However, these dips do not appear. In some other cases, predicted Regge dips do occur, such as in $\gamma + \mathrm{p} \rightarrow \pi^0 + \mathrm{p}$. This reaction is dominated by $\omega$ (and also B) exchange, and a dip similar to the $\rho$ case is expected, and observed [44].

Byers and Yang [45] point out that the similarity of the exponential slope of $\dfrac{\mathrm{d}\sigma}{\mathrm{d}t}$, in the $|t|$ range 0·2–0·5, to the diffraction-like pattern that is observed for $\pi^\pm$–p scattering was not necessarily what we would expect, since the charge-exchange amplitude is dealing with the difference between two amplitudes. Therefore, they concluded that the results imply that high-energy elastic charge-exchange scattering may be the coherent result of a distribution of local interactions occurring in a small region, which are effectively localized by the lack of time for single particle exchange taking place over the whole volume. Thus, it is argued that the single-particle exchange model is not justified. The second peak observed in the charge-exchange scattering in the neighbourhood of $t \approx -1\ (\mathrm{GeV}/c)^2$, in the incident momentum range of 3 to 18 GeV/$c$, further suggests diffraction-like mechanisms.

### 13.8.13. *Charge-exchange polarization*

In any single Regge-pole exchange model the polarization must be zero, since the helicity flip and non-flip amplitudes have the same phase which, therefore, requires that the polarization be zero. The polarization in charge-exchange has been measured [46] and been found to be comparable with the polarization previously observed in $\pi^\pm$–p scattering. The magnitude, and the energy dependence, of the polarization in pion–nucleon charge-exchange made the single $\rho$-pole hypothesis untenable. Several explanations for a $P$, more or less independent of energy, have been put forth, as follows.

(a) That there is a second trajectory [47*a*] with the $\rho$ quantum numbers, which is located somewhat below the $\rho$ trajectory, and this is referred to as the $\rho'$.

(b) A Regge cut probably due to $(P+\rho)$ exchange [47*b*]. This cut would lie as high as the $\rho$ and explain the energy dependence.

(c) Contributions from direct channel resonances [47*c*].

In any event, it is clear that once again the eagerly embraced simplicity of a Regge-pole fit has, with the passage of time, given way to ever increasing complexity in the analysis.

### 13.8.14. *Forward and backward peaks and Regge pole models*

It has been observed throughout many high-energy experiments that for the simple pole exchange diagram for two body collisions shown below,

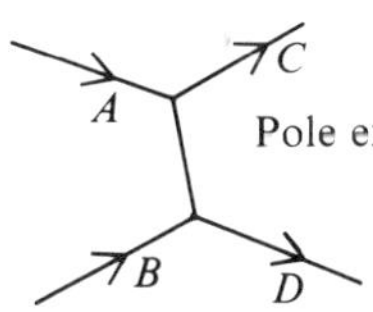


the $\frac{d\sigma}{dt}$ (A → C) will always exhibit peaks at small $t$, whenever there exists a known particle (or resonance) in the $t$-channel with the right quantum numbers for the exchange.† Conversely, if there is no such known particle, there is no peak observed at small $t$. If we assume a power law dependence for the cross-section then $\sigma = \sigma_0 P_L^{-n}$, where $P_L$ is the incoming particle laboratory momentum, we obtain [48], [57], for vacuum exchange, $n \sim 0$, and for charge or isospin exchange, $n \sim 2$, strangeness exchange $n \sim 2{\cdot}5$, and baryon number exchange $n \sim 3$–$4$. This observation is intuitively expected for any (single) pole exchange model including the Regge pole model.

When we examine the problem in detail, a simple Regge pole model cannot explain forward peaks due to the kinematic constraint conditions and conspiracy, or some other equally complicated schemes, must be invoked. There is no unique solution which withstands the 'test of time' for any appreciable period.

The OME model, with absorption corrections, has a simpler task than the Regge model in explaining the forward peaks, when it is relevant.

† If A is a meson and B a baryon, and the exchange particle is a baryon, so that C becomes a baryon, this leads to the observation of backward peaks in the meson–baryon interaction.

13.8.15. *The multi-Regge model*

The Regge model for two body reactions is illustrated by Fig. 13.16(a) which considers the two body reaction $a+b \rightarrow c+d$. However, since either, or both, $c$ and $d$ could be strong interaction resonances with $c$ decaying in the final state to $c_1+c_2$, and $d$ decaying to $d_1+d_2$, this diagram could also represent the processes of Figs. 13.16(b), 13.16(c), or 13.16(d). Thus Fig. 13.16(b) and Fig. 13.16(c) represent three particle final states, whereas Fig. 13.16(d) represents four particle final states. However, in all these cases the

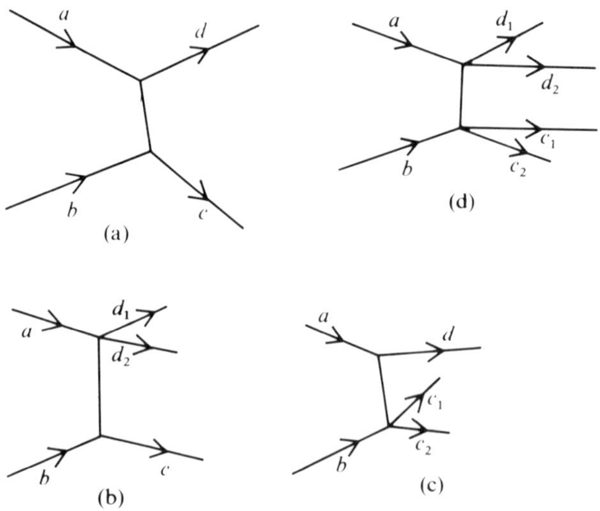


FIG. 13.16.

intermediate state was characterized always as consisting of two identifiable resonances, or particles, with definite quantum numbers. In order to treat final states with 1, 2, 3, ... $N$ final state particles which do not form two resonance (or particle) combinations in the intermediate state, a more generalized approach is necessary. The multi-peripheral model [49] treated this problem (it has subsequently been Regge-ized and it is referred to as the multi-Regge model) [50]. The simplest multi-Regge graph is shown in Fig. 13.17(a), where $\alpha_1$ and $\alpha_2$ are two distinct Regge poles. The generalization of this diagram is shown in Fig. 13.17(b). The generalized amplitude [51] for Fig. 13.17(b) is of the form

$$A \rightarrow \Phi_1(t_{\alpha_1}) s_{1,2}^{\alpha_1(t_{\alpha_1})} \Phi_2(t_{\alpha_1}, \phi_2, t_{\alpha_2}) s_{2,3}^{\alpha_2(t_{\alpha_2})} \times \Phi_3(t_{\alpha_2}, \phi_3, t_{\alpha_3}) \ldots$$
$$\times \Phi_{N-1}(t_{\alpha_{N-1}}, \phi_{N-1}, t_{\alpha_N}) s_{N-1,N}^{\alpha_N(t_{\alpha_N})} \Phi_N(t_{\alpha_N}),$$

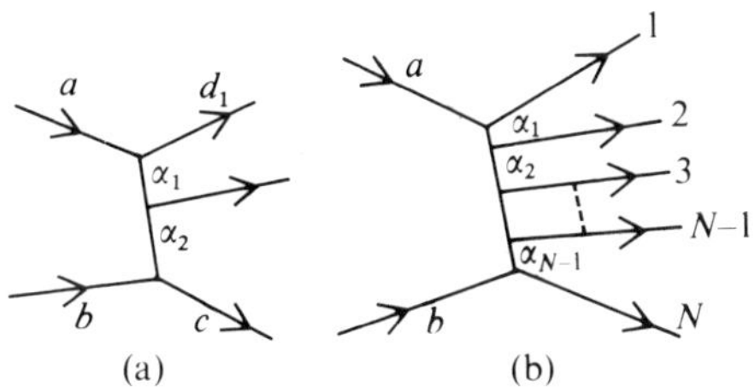


FIG. 13.17.

where $\Phi_i$ contains the vertex functions, signature factors, and trajectory parameters of the Regge poles. $\phi$ is the Toller variable which, for example, for Fig. 13.17 represents the azimuthal angle between the planes of $(axd_1)$ and (bxc), measured in the rest frame of particle $d_2$. All the graphs, corresponding to all the possible permutations of the final state particles, contribute with strengths which are strongly influenced by the quantum numbers, and other characteristics of the Regge poles exchanged. For the multi-Regge model to apply, the incident energy and all of the individual $s_{ij}$ must be high enough to expect the Regge model to work and also to avoid low-energy resonance effects. These conditions are difficult to achieve, in practice, at CERN PS and Brookhaven AGS energies, therefore a modified multi-Regge model has been developed along the following lines.

(a) For those cases where all $s_{ij}$ are large enough, the scattering amplitude is considered to become fully Regge-ized.

(b) For those cases where a group of final state particles, taken together in a cluster, exhibit the mass and quantum numbers of a known resonance, a single quasi-particle representing the resonance is used to replace them.

(c) For those cases where a non-resonant low-mass cluster occurs, it is assumed that the part of the amplitude representing interactions inside the cluster is approximately representable by a constant.

So far, the multi-Regge model has had some success in explaining multiparticle production in $\pi$–p collisions from 5·5 to 16 GeV/$c$ and other reactions, by using Regge characteristics deduced from experimental reactions, but without re-introducing the enormous number of additional parameters which were characteristic of the elastic scattering fits. Even though this is encouraging, we must remember the narrow range of applicability of the model at present energies.

The multi-Regge model involves exchange of Regge poles in series (i.e. one after another). If we exchange Regge poles simultaneously (i.e. in parallel) we obtain cuts.

### 13.8.16. *Duality*

It was pointed out [52] that there is an inherent ambiguity in separating experimental resonances from background. This ambiguity is a fundamental difficulty. We have previously discussed how a circle, or a spiral (of the proper type), in the Argand diagram of a partial wave accompanies a resonance in that partial wave. Therefore, it was generally assumed that the observation of a circle, in the Argand diagram of a partial wave, corresponds to a resonance. However, Schmid has now shown that the exchange of a Regge pole in the $t$-channel also results in circles in the Argand diagram, which are in many respects similar to the circles corresponding to an $s$-channel resonance.

It appears clear that any physical scattering amplitude can, in principle, be parametrized by either a series of $s$-channel resonances plus a general background or, equivalently, by exchange of a series of $t$-channel poles and other $t$-channel singularities. When one description is very simple, and the other more complicated, it is fruitful to favour the simpler solution. However, the test of whether a physical $s$-channel resonance really exists has to depend on an identifiable particle with the quantum numbers, and mass, of the resonance observable in many different reaction channels, when there are no selection rules forbidding its formation, and that these observations include production (including multi-particle production) as well as formation experiments (when possible), performed over a wide range of $s$ and $t$. The very fact that we have the possibility of the *duality* of the representation in either $s$ or $t$ channels leads to (possibly) severe problems of double counting, which we must avoid when using combinations of $s$-channel resonances and $t$-channel exchanges in a calculation.

There have been some successes in explaining $s$-channel resonances in terms of simple $t$-channel Regge-pole exchanges, such as Schmid's explanation of some high-mass N* states, but Kaps and Logan observed quite different and contradictory results [53]. In some cases it appears that either a simple $t$-channel exchange, or a simple $s$-channel resonance, can explain some observed resonance-like phenomena. In these cases, it may well be that (crudely speaking) the former ($t$-channel exchange) represents the mechanism by which the $s$-channel resonance is formed.

### 13.8.17. *The Veneziano model*

Veneziano [54] has recently proposed a model for the amplitude (Veneziano model) which attempts properly to incorporate the combined effects of low-energy resonances and Regge asymptotic behaviour (in all channels), crossing symmetry, analyticity, and duality. Veneziano imposed additional conditions on the trajectories, and this led to values of slopes and intercepts for both the $\rho$ and $A_2$ trajectories which explain the related experimental results. He was able to derive the scattering lengths for $\pi$–$\pi$ scattering and to demonstrate that the finite energy sum rules (FESR) are satisfied. A prescription for three particle decays, and for a continuation of the masses of external pions to the value of zero, as is required by the Adler self-consistency requirement and various current algebra constraints, were provided. Some preliminary tests of this model have led to success [35].

The model has been generalized to multi-particle processes, including reactions involving five particles (the five-point function). There have been a number of recent reports on the attempts to generalize the model to the $N$-multi-particle amplitude case, and these indicate encouraging progress [56].

13.8.18. *Evasion versus conspiracy*

Studying the photoproduction of positive pions from hydrogen [57],

$$\gamma + p \rightarrow \pi^+ + n, \tag{13.128}$$

is a good way of resolving the competition between conspiracy and evasion solutions in the constraint problem. The pion is the only known particle, with the right quantum numbers, to be exchanged in this reaction.

For an amplitude which is predominantly formed by a single $t$-channel trajectory exchange with a definite parity, evasion can be the only solution; thus there must be a zero (or dip) in $d\sigma/dt$, at $t = 0$. Experimentally, a peak has been observed, in accurate measurements down to $t \sim 10^{-3}$ $(\text{GeV}/c)^2$, thereby clearly contradicting 'evasion' as the only possibility, and implying [58] conspiracy as the mechanism which should be employed. The existence of a peak near $t = 0$, in the charge-exchange reaction,

$$p + n \rightarrow n + p, \tag{13.129a}$$

also implies that the conspiracy mechanism should be employed. However, if we assume that the $\pi$-trajectory, and the $\pi'$, its opposite parity partner, with all other quantum numbers the same, conspire in the photoproduction of $\pi^+$ then, assuming factorization, the reaction

$$\pi^+ + p \rightarrow \rho^0 + N^{*++}(1238) \tag{13.129b}$$

should exhibit a dip near $t = 0$. The experimental results, on the contrary, show a peak [57], [59]. Therefore the situation is rather self-contradictory, but we can (following past precedents) consider the combined effects of other poles, cuts, and parameter-increasing mechanisms to alleviate these difficulties—or, perhaps, it is better to say emphasize how serious the difficulties really are.

### 13.8.19. *The ω trajectory*

As a result of many fits to the experimental data, a somewhat confusing picture has resulted for the $\omega$-trajectory and, in a number of cases, an $\omega'$ has also been thought to be necessary (especially for explaining crossover effects) [60]. However, the $\omega$-trajectory appears to behave approximately as if it were exchange degenerate with the P′. This conclusion was based on the nearly-constant total cross-section for p+p scattering at high energy, and the absence of dips in the elastic differential cross-section, which correspond to the observed dips in the $\bar{p}$–p case.

The exchange degeneracy of P′ and $\omega$ also explains the near-constancy with energy, and near-equality of the $K^+ + p$ and $K^+ + n$ total cross-sections. The anti-shrinkage of $\bar{p}$–p with increasing $s$, in contrast to the shrinkage of p–p has been explained as an interference effect of the $\omega$ with the P and P′.

### 13.8.20. *The $A_2$ trajectory*

Older Regge fits had employed the R-trajectory, which subsequently was associated with the $A_2$ (~1300 MeV) with $J^P = 2^+$, $G = -$, $C = +$, and $\tau = +$. The characteristics of the reaction are such that exchange of an

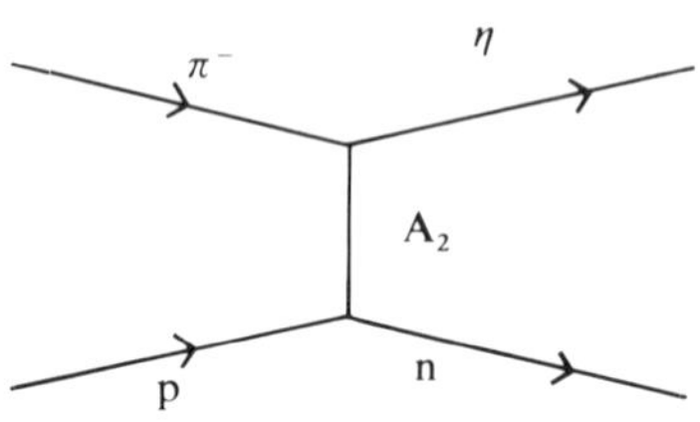


$A_2$ in the $t$-channel is expected to be the dominant mechanism.

In analysing the kaon–nucleon exchange reactions

$$K^- + p \rightarrow \bar{K}^0 + n \quad \text{and} \tag{13.130a}$$

$$K^+ + n \rightarrow K^0 + p, \tag{13.130b}$$

the $A_2$ and $\rho$ are usually both assumed to be dominant and, approximately, exchange-degenerate [61]. However, phenomenological determinations of the $A_2$ trajectory are still inaccurate, and its slope appears uncertain by a factor of as much as two. The $A_1$ and B trajectories (with $P = +$, $T = 1$,

$C = +, \tau = -$) are not used very often, and then usually when fits with the previously-mentioned poles alone do not work, or to fill in dips in charge-exchange reactions (where their $T = 1$ should, presumably, make them relatively more important).

13.8.21. *The pion trajectory*

Before the Regge model, the OPE model assigned the dominant role in inelastic interactions to the pion exchange. The pion was thought to dominate because of its low mass, and nearness of the corresponding pole to the physical region.

In the Regge model, on the other hand, the dominance is mostly determined by $\alpha(0)$ and, thus, the pion became unimportant. Furthermore, an accompanying conspirator ($\pi'$), or cuts, or other poles are required, thus further confusing the status of the pion. Nevertheless, it should again be remarked that the OPE model with absorption, where it is applicable, more easily explains forward diffraction peaks than the Regge model.

13.8.22. *Sum rules and superconvergence relations*

Let $\nu = \dfrac{s-u}{4m}$. If the analytic amplitude $F(\nu, t)$ is odd under crossing ($\nu \to -\nu$), and is such that

$$|F(\nu, t)| < \nu^{\gamma}, \quad \text{as} \quad \nu \to \infty, \tag{13.131a}$$

where $\gamma < -1$ and $\nu$ is the energy variable, we can write a convergent fixed-$t$ dispersion relation† as follows,

$$F(\nu) = \frac{1}{\pi}\int_0^{\infty} \frac{\mathrm{d}\nu' \ \mathrm{Im}\ F(\nu')}{\nu' - \nu}, \tag{13.131b}$$

but, from our bound condition,

$$\nu F(\nu) \to 0, \quad \text{as} \quad \nu \to \infty, \tag{13.131c}$$

and the dispersion relation, it also follows that [62]

$$\int_0^{\infty} \mathrm{Im}\ F(\nu')\,\mathrm{d}\nu' = 0. \tag{13.131d}$$

The above is described as a superconvergence relation (SCR), or a superconvergence sum rule. The $t$-channel reaction amplitudes, which do not

† Where we omit the $t$ for convenience.

possess kinematic singularities, and have a leading trajectory for which

$$\alpha(t) < m-1,$$

where $m$ is the maximum $t$-channel helicity flip, also satisfy eqn (13.131d).

If we further require that, if the amplitudes possess fixed poles, then their residues vanish, similar relations have been derived for $\nu^n$ Im $f(\nu)$,

$$0 > n \text{ (an integer)} > m-\alpha(t)-1.$$

Suppose we cannot write an SCR for $F$. We can, by subtracting a sufficient sum of Regge poles from $F$, write a superconvergence sum rule for

$$F' = F - F_{\mathrm{R}}. \tag{13.132a}$$

These latter relationships are often referred to as generalized superconvergence relations (GSCR). An example of the application of a GSCR [63] (sometimes referred to as a sum rule) is

$$\int_{\mu}^{\infty} (\nu^2-\mu^2)^{\frac{1}{2}}(\sigma_{\pi^-+\mathrm{p}}(\nu)-\sigma_{\pi^++\mathrm{p}}(\nu))\,\mathrm{d}\nu = -4\pi\sum_i \beta_i P_{\alpha_i}(\nu/\mu)+4\pi f^2. \tag{13.132b}$$

The optical theorem is employed and the pole term $f^2$ is explicitly inserted. When the above relation was used, it was found that the $\rho$ trajectory alone was enough, and there was no need for additional contributions from a cut or $\rho'$ pole.

If we further assume that, beyond some sufficiently high energy, the $\nu\,(>\nu_0)$ contribution to the infinite integral of $\mathrm{Im}(F-F_{\mathrm{R}})$ becomes negligible, we can conclude

$$\int_0^{\nu_0} \mathrm{Im}(F-F_{\mathrm{R}})\,\mathrm{d}\nu = 0. \tag{13.133}$$

Hence,

$$S_0 \equiv \frac{1}{\nu_0}\int_0^{\nu_0} \mathrm{Im}\,F\,\mathrm{d}\nu = \sum \frac{\beta\nu_0^{\alpha}}{\alpha+1}.$$

Furthermore, if $\nu^{|n|+1}(F-F_{\mathrm{R}}) \to 0$ as $\nu \to \infty$, where $n$ is an integer, we can deduce the finite energy sum rules (FESR),

$$S_n \equiv \frac{1}{\nu_0^{|n|+1}}\int_0^{\nu_0} \nu^{|n|}\,\mathrm{Im}\,F\,\mathrm{d}\nu = \sum \frac{\beta\nu_0^{\alpha}}{\alpha+n+1}.$$

An enormous number of such sum rules could be generated by considering various moments, $\nu^{\gamma}\exp(-\mathrm{i}\pi\gamma/2)F'(\nu)$, in place of $F'(\nu)$. The values of $\gamma$

can then be selected (it does not have to be integral) so as to emphasize the desired region in $\nu$, and will involve (in general) the real, as well as the imaginary, part of the amplitude. This gives rise to the continuous moment sum rules (CMSR). Enough Regge poles have to be used to make the integral convergent. The integrals can be evaluated using phase-shift analysis results, when available, or the technique of saturating the integral with known resonances. Typically, $\nu_0$ is selected around 1–2 GeV, since experimental data are best known up to this region. It is, of course, questionable whether Regge behaviour should be assumed to be applicable at these low energies. For treating the special case, $t = 0$, total cross-section data can be used, and these are generally available to much better precision.

Some of the applications of FESR have been the use of low energy $\pi$–N experimental data to more accurately determine the P, P$'$, $\rho$, and $\rho'$ contributions at high energy and the studying of $\pi$ and $A_2$-trajectories and low-energy pion photoproduction data. FESR calculations have been used as a bootstrap mechanism [64]. The known Born terms, and resonances, are used to saturate the FESR integral, and the FESR relation is then employed to bootstrap trajectories in the cross channel. This approach is referred to as the 'Dolen–Horn–Schmid duality'.

The fact that the combination of dispersion relations with analytic Regge amplitude representations gives consistent results is, of course, interesting. However, we must be somewhat careful as to the interpretation of these procedures. It certainly appears that, from all experimental information to date, physical amplitudes are analytic except for certain types of singularities, such as poles and cuts. Therefore, when both high- and low-energy data are employed in dispersion relations determining Regge pole fits (which, after all, represent a parameterization for an analytic amplitude) it is not clear that the resultant parameterization is necessarily physically significant.

## 13.9. Higher $|t|$ pion–nucleon scattering

### 13.9.1. *General characteristics*

We have previously discussed the forward diffraction peak region in $\pi^{\pm}$–p scattering.

Fig. 13.18(a) shows the $\pi^- + p \rightarrow \pi^- + p$ angular distribution over the complete $t$-range of [66] observations. At lower (2·5–4·0 GeV/$c$) energies, it was observed, as shown in Fig. 13.18(b), that there was a clearly-defined dip at $t \approx -0{\cdot}6$ (GeV/$c$)$^2$, followed by a peak, or shoulder, in the neighbourhood of $\approx -1{\cdot}5$ (GeV/$c$)$^2$. As the incident pion energy increased, this structure gradually disappeared, but a dip in the neighbourhood of

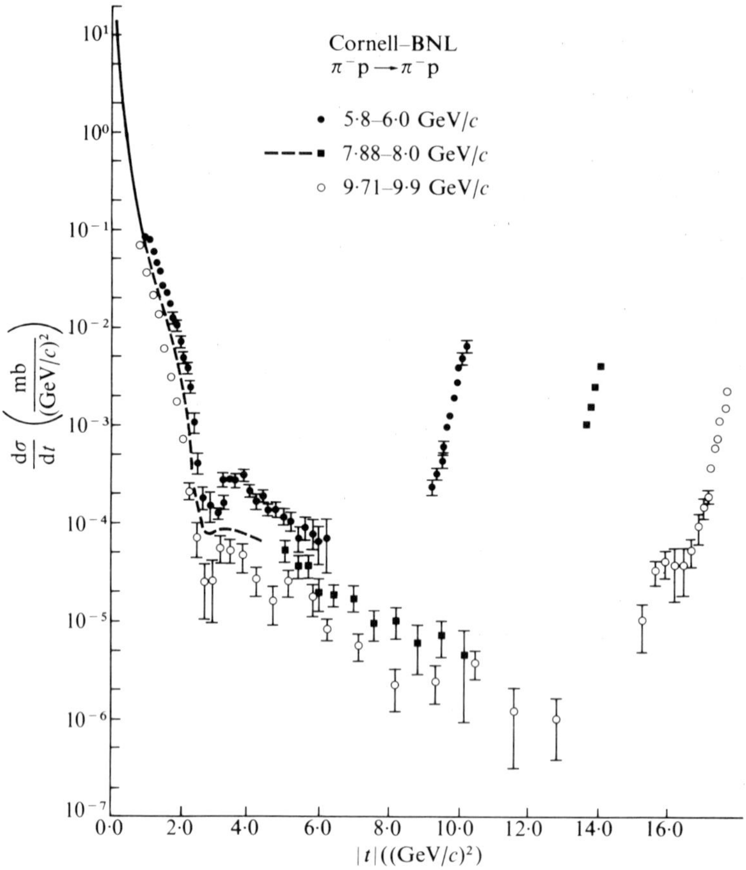


FIG. 13.18.(a) Full $\pi^-$p angular distributions at ~6, ~8, and ~10 GeV/$c$ incident momentum. The rough shape of the cross-sections at $|t| \lesssim 0{\cdot}8$ (GeV/$c$)$^2$ is shown as a full line. To allow a clear presentation, the data at ~8 GeV/$c$ for $|t| \lesssim 5$ (GeV/$c$)$^2$ are not shown and are replaced by a broken line. (From [66] and [70].)

$-3$ (GeV/$c$)$^2$, followed by a peak, and a broad maximum, which decreased more slowly until the backward direction in the c.m.s. was approached, where sharp backward peaks were observed with about four orders of magnitude lower cross-section than the forward peaks. Figs. 13.18(b) and (c) compare the $\pi^+ + p$ and $\pi^- + p$ angular distributions, as presented at the Berkeley conference [65].

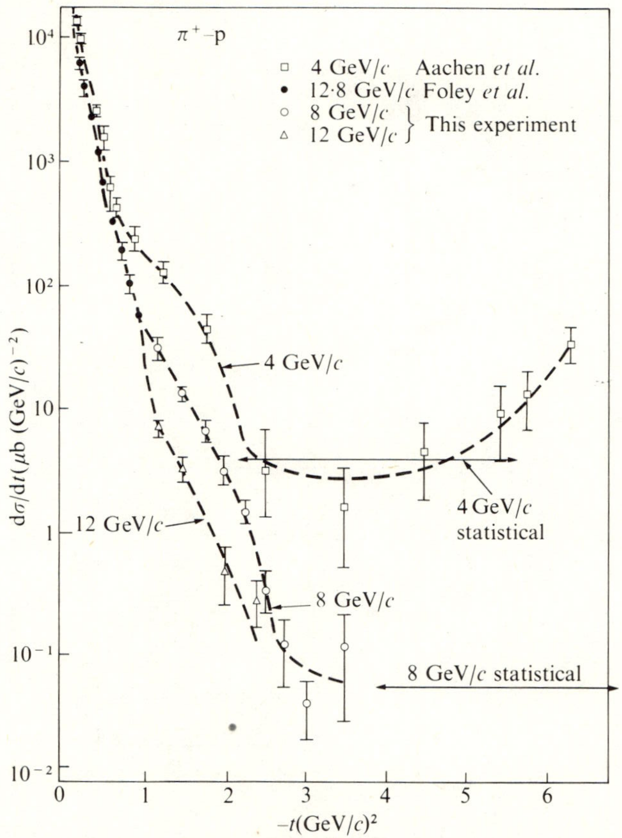


FIG. 13.18.(b) $\pi^+$–p elastic scattering distributions between 4 and 12 GeV/$c$. The data was compiled by Orear *et al.* [24] who gave the points labelled 'this experiment'. (From [65*b*], *Proceedings of the XIII International Conference on high energy physics.* Berkeley. (1966). p. 274.)

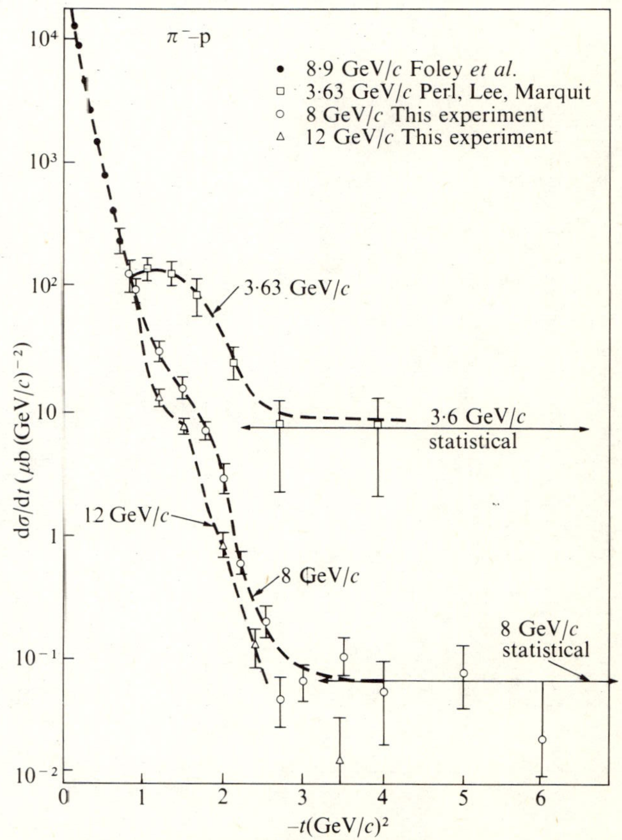


FIG. 13.18.(c) $\pi^-$–p elastic scattering distributions between 3·6 and 12 GeV/$c$. The data are compiled by Orear *et al.* [24], who gave the points labelled 'this experiment'. (Reference as for Fig. 13.18(b)).

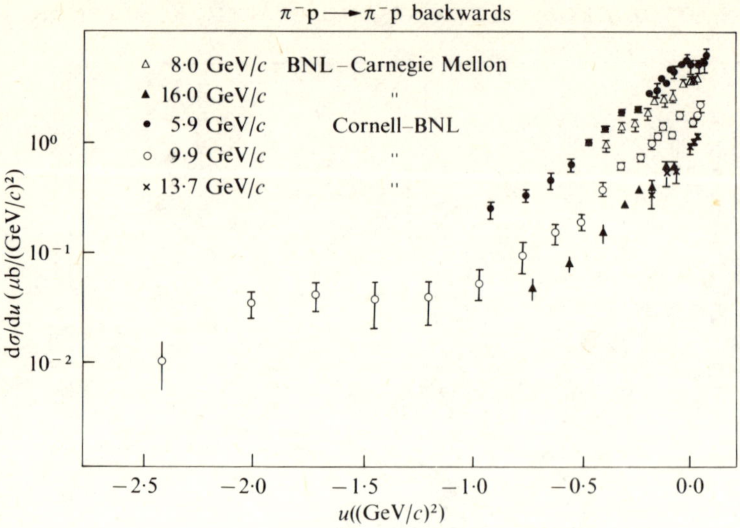


FIG. 13.19.(a) Backward $\pi^- p$ scattering data.

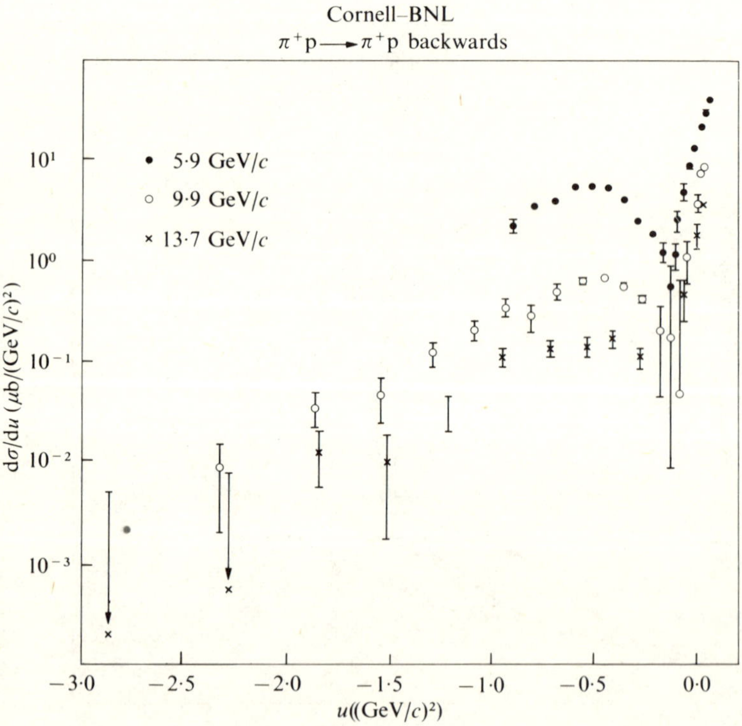


FIG. 13.19.(b) Backward $\pi^+ p$ differential cross-section between ~6 and ~14 GeV/$c$.

### 13.9.2. *Backward peaks*

The backward peaks were studied in detail for both $\pi^+ + p$ and $\pi^- + p$, and the results are shown in Fig. 13.19(a) and (b) plotted against $u$, which is the most appropriate variable for displaying the results. The backward $\pi^+ + p$ scattering results are particularly striking in that they exhibit a sharp dip at $u \approx -0{\cdot}15\ (\text{GeV}/c)^2$. The backward peak can be fitted by an exponential in $u$ of slope approximately 13 $(\text{GeV}/c)^2$ at about 6 GeV/$c$. As the energy increases to about 14 GeV/$c$, the slope increases (i.e. the peak shrinks) to approximately 18 $(\text{GeV}/c)^2$. A backward peak in $\pi^\pm$–p scattering immediately suggests that a baryon exchange mechanism may possibly be responsible. From the Regge point of view, exchange of a baryon in the

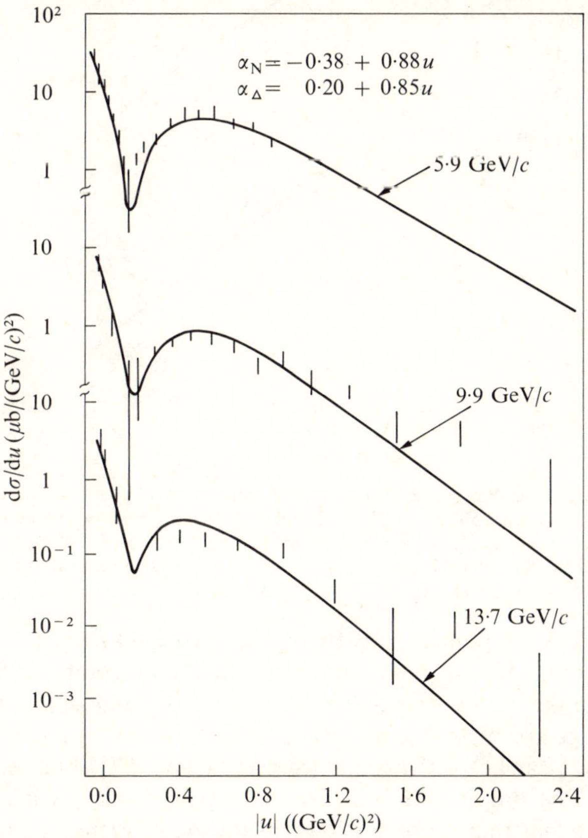


FIG. 13.19.(c) Regge pole model fits to backward $\pi^+$p scattering data. The indicated $\alpha_N$ and $\alpha_\Delta$ trajectories are as obtained in the fit to both $\pi^-$p and $\pi^+$p data. (All figures are from p. 337 of [66].)

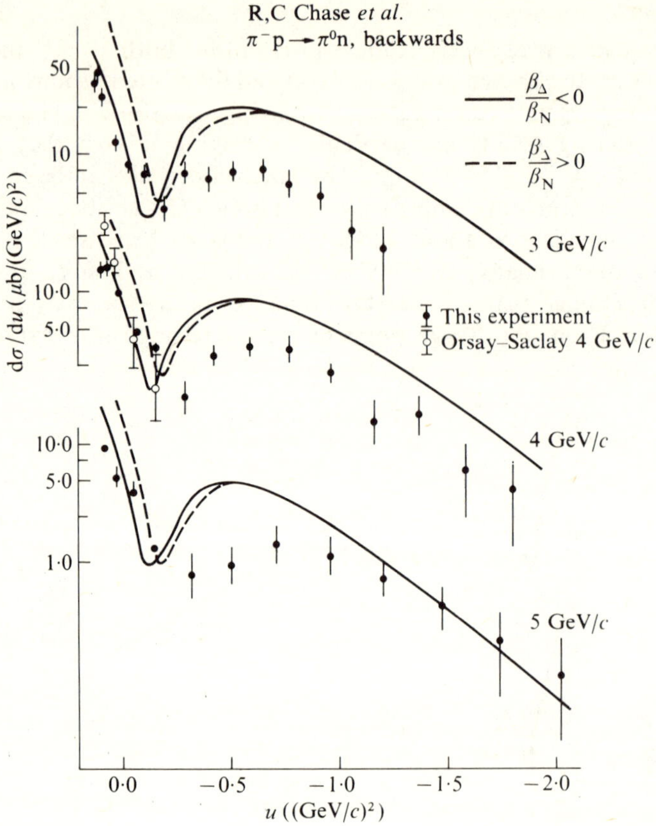


FIG. 13.20. Preliminary results by the Minnesota experiment on $\pi^-$p charge-exchange in the backward region at 3, 4, and 5 GeV/$c$. The full and broken curves show predictions by the Barger and Cline Regge pole model for the two indicated signs for the ratios of the pole residues of the Δ and $N_\alpha$ trajectories. Preliminary data at 4 GeV/$c$ by the Orsay–Saclay group are also shown. (From p. 344 of [66].)

$u$-channel would be responsible for these observed backward peaks. Considering the neutron and $N^*_{\frac{3}{2},\frac{3}{2}}$ states to be the most important, we obviously conclude that, in the $\pi^-+$p case only, $N^*_{\frac{3}{2},\frac{3}{2}}$ can be exchanged, (i.e. exchange of a doubly charged particle is required), whereas, in the $\pi^++$p case, either a neutron or an $N^*_{\frac{3}{2},\frac{3}{2}}$ can be exchanged. Barger and Cline [67] obtained a Regge fit to the data. The dip at $u \approx -0{\cdot}15$ in $\pi^++$p backward scattering was attributed to the fact that there is a zero in the $N_\alpha$ exchange amplitude at $N_\alpha = -\frac{1}{2}$, and this occurs, naturally, at the observed value of $u$ for the dip. Furthermore, for a pure $N^*_{\frac{3}{2},\frac{3}{2}}$ exchange the expected ratio of the differential cross-section of $\pi^++$p to $\pi^-+$p is about 9, which is consistent with the observations. However, we would expect similar dips in $\pi^- + p$ scattering when $(N^*_{\frac{3}{2},\frac{3}{2}})_\alpha = -\frac{3}{2}$, and these dips did not occur, thus raising the possibility of the dip being a coincidence in the $\pi^+ + p$ case.

Diffraction model calculations have also been able to explain the major features of backward pion–proton scattering [68].

13.9.3. *Backward charge exchange*

Backward $\pi^- + p$ charge-exchange reactions have been studied [69]. The experimental results (Fig. 13.20) at these lower energies (3–5 GeV/*c*) are only qualitatively, but certainly not quantitatively, in agreement with the Regge pole predictions.

## 13.10. Higher |*t*| $\bar{p}$–p and $\bar{p}$–n scattering

In Fig. 13.21, we observe [70] that, after the forward diffraction peak in $\bar{p}$–p scattering, a shoulder develops in the neighbourhood of *t* approximately

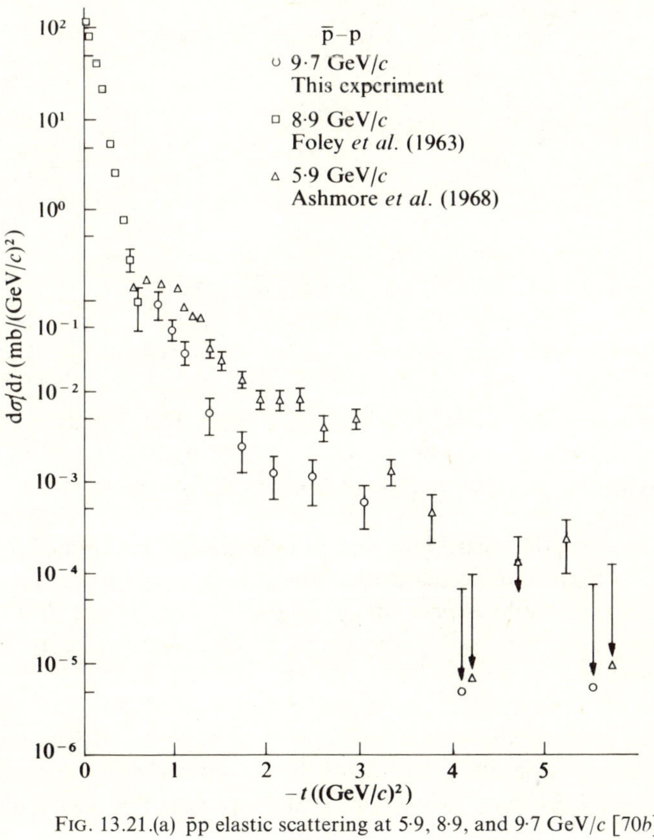


FIG. 13.21.(a) $\bar{p}p$ elastic scattering at 5·9, 8·9, and 9·7 GeV/*c* [70*b*].

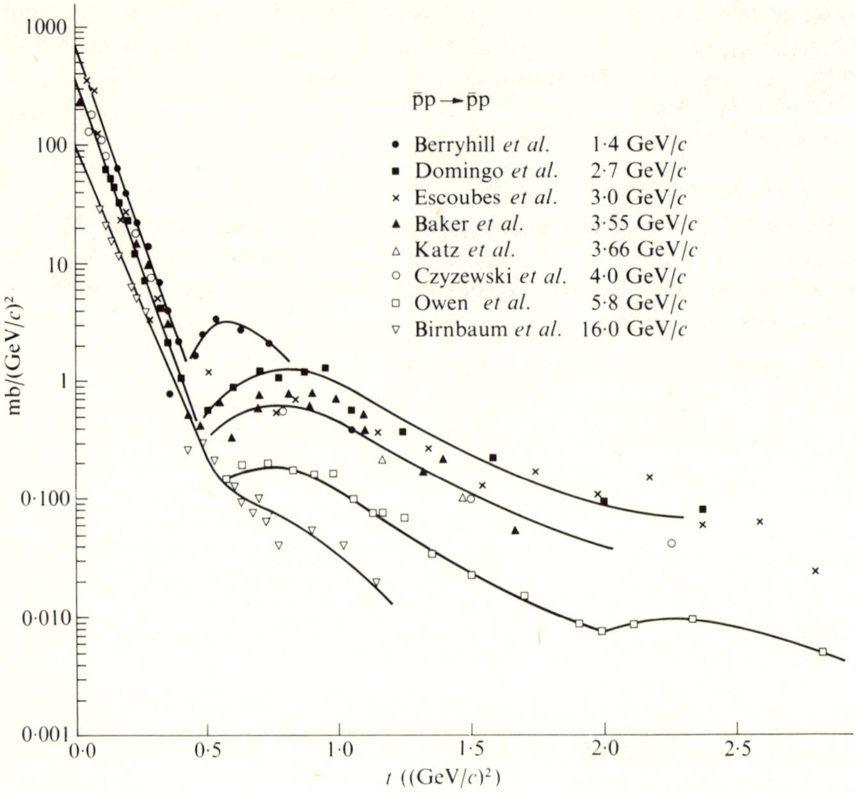


FIG. 13.21.(b) Behaviour of p̄–p elastic scattering as a function of the total energy ($\sim\sqrt{s}$). (From Montanet. (1969). *Proceedings of the International Conference on Elementary Particles*.)

equal to $-0{\cdot}6\ (\text{GeV}/c)^2$ in the 5·7–5·9 GeV/$c$ data. This shoulder tends to become less marked, and moves outward, toward higher $|t|$, as the energy increases. At lower incident energies ($\lesssim$4 GeV), this shoulder becomes a well-defined peak, as shown in Fig. 13.21(b).

At 1·4 GeV/$c$ [71] the elastic p̄–p and p̄–n were both determined from the same run of a deuterium bubble chamber, on which anti-protons were incident. Thus, we would expect that Glauber–Wilkin effects should tend to affect both deuterium cross-sections similarly, and enhance the significance of the comparison.

As can be seen in Fig. 13.22, the shoulder at $t \approx -0{\cdot}6\ (\text{GeV}/c)^2$ becomes a well-defined peak at these energies, and p̄–n appears, at low $|t|$, to be larger (or at least comparable), and then passes below p̄–p in the neighbourhood of $t \approx -0{\cdot}2\ (\text{GeV}/c)^2$. However, it then crosses p̄–p again at about

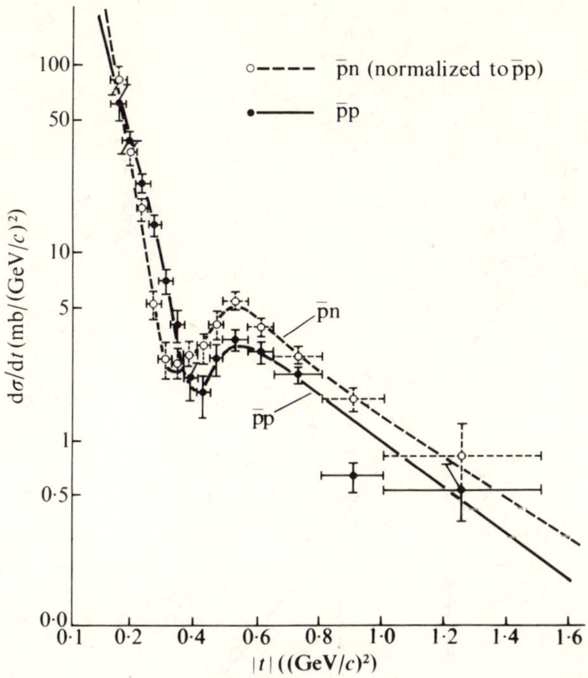


FIG. 13.22. The p̄p (full circles) and p̄n (open circles) differential cross-sections at 1·4 GeV/$c$. Hand-drawn curves through the points are also shown. (From J. Berryhill and D. Cline. (1968). *Proceedings of the XIV International Conference on High Energy Physics, Vienna.* paper 280 and *Phys. Rev. Lett.* **21,** 770.

$t = -0{\cdot}4\,(\mathrm{GeV}/c)^2$, and appears to stay above it thereafter. There appears to be a second shoulder in the neighbourhood of $-(2\text{–}3\,(\mathrm{GeV}/c)^2)$.

## 13.11. Higher $|t|$ K–p elastic scattering

Fig. 13.23(a)–(b) show the behaviour of $K^- + p \rightarrow K^- + p$ scattering at higher $|t|$, in the 5·49 to 13·57 GeV/$c$ incident momentum region. The data have been displaced by one or two decades on the ordinate, to facilitate observation of the individual curves.

For $|t| \lesssim 1$ there is an apparent expansion of the forward peak with increasing energy. Thus the $K^+ + p$ and $K^- + p$ pair behave similarly to the $p + p$ and $\bar{p} + p$ pair, respectively. For $|t| > 1$ there is a decrease in slope, which is more marked for lower energy and, in fact, shows some structure in the neighbourhood of $-(5{\cdot}5\text{–}5{\cdot}8\,(\mathrm{GeV}/c)^2)$. In the backward direction there is no peak in the $K^- + p$ scattering, such as that observed in $\pi^- + p$ scattering and, instead, the backward cross-section drops very

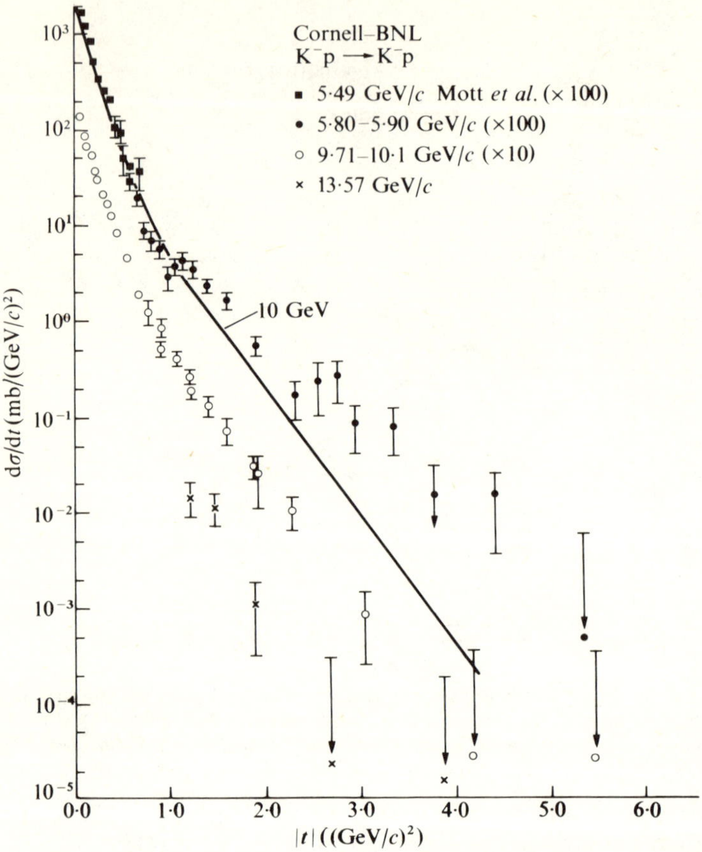


FIG. 13.23.(a) Cornell–BNL data on elastic $K^-p$ scattering. The 5·49 GeV/*c* data by Mott *et al.* are also included. The ~6 GeV/*c* data are plotted after multiplying by 100, and the ~10 GeV/*c* data after multiplying by 10. The full line shows the qualitative behaviour of the 10 GeV/*c* data, after multiplying by 100. (From [66].)

sharply with both increasing $s$ and $t$. This observation lends support to the hypothesis that backward peaks are due to baryon exchange. Obviously, a strangeness $S = +1$ baryon would have to be exchanged in the $u$-channel to yield a backward peak in $K^- + p$ scattering. Since an $S = +1$ baryon has never been discovered, it either does not exist, or is probably, at most, weakly coupled and, thus, from the baryon exchange mechanism, we would not expect a backward peak in $K^- + p$. However, since there are many $S = -1$ known baryons, we would expect a backward peak in $K^+ + p$ scattering, and as shown in Fig. 13.23(d)–(e), this is observed [72].

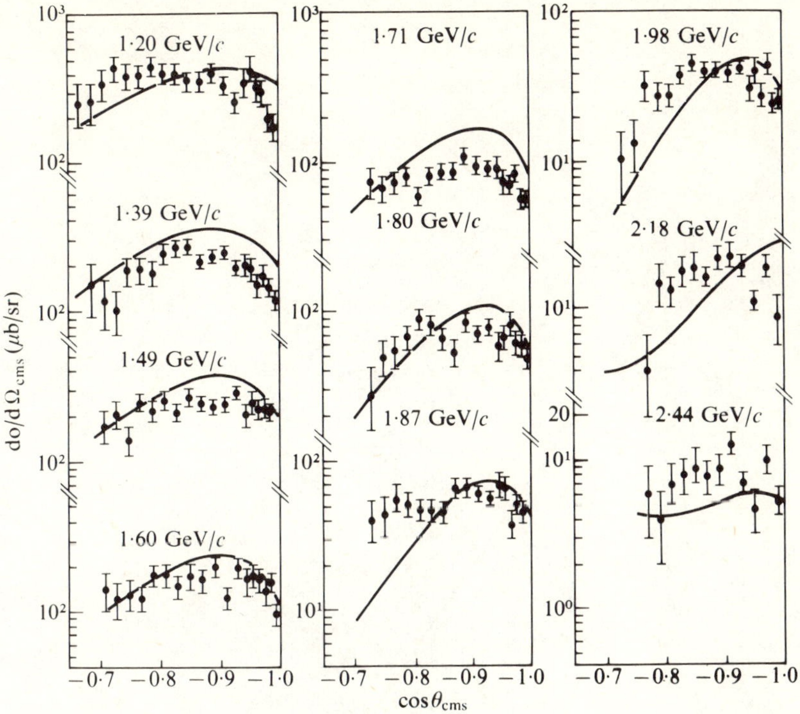


FIG. 13.23.(b) $K^-p$ elastic scattering differential cross-sections in the backward region between 1·2 and 2·44 GeV/$c$ [50]. The full curves are best fit curves using a superposition of the known $s$-channel resonances. (From [66].)

## 13.12. Higher $|t|$ p–p elastic scattering

Higher $|t|$ p–p elastic scattering has been studied by a number of groups. Summaries of the progressive development of this field are given in references [11] and [73].

A number of attempts at obtaining universal parameterizations were made (Orear, Kirch, etc.). However, as more recent data became available, their universality was no longer applicable, and it now appears that use of either of the Lorentz invariant plots, $\frac{d\sigma}{dt}$ against $p_\perp$, or $\frac{d\sigma}{dt}$ against $t$, provide the most elucidating display of the data.

A summary of the results of all groups [74] for $\frac{d\sigma}{dt}$ against $|t|$ is given in Fig. 13.24(a). The curves, of course, end at $\theta = 90°$ owing to the indistinguishability of the two protons.

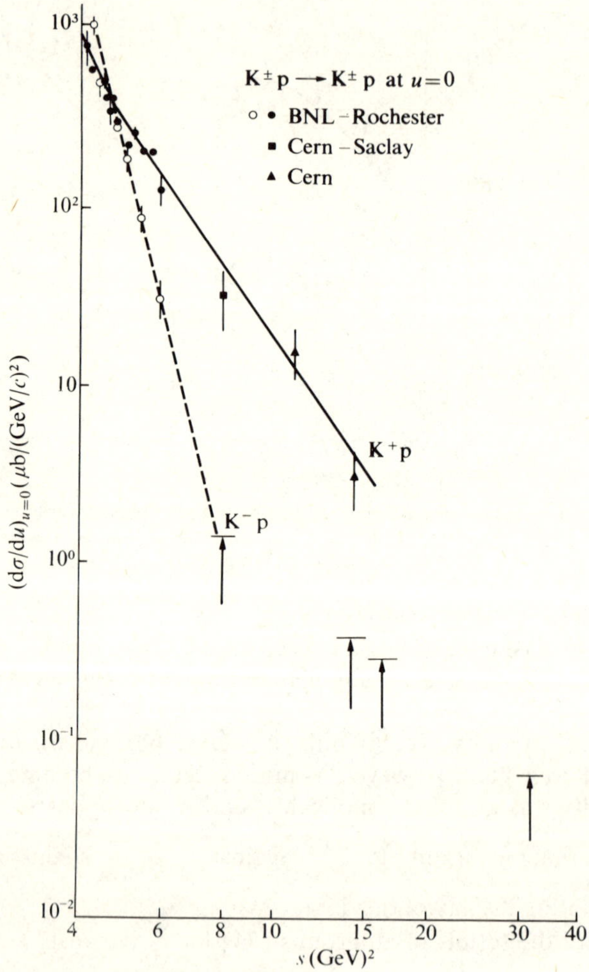


FIG. 13.23.(c) Energy behaviour of the $u = 0$, $K^-p$ and $K^+p$ elastic scattering cross-section. (From [66].)

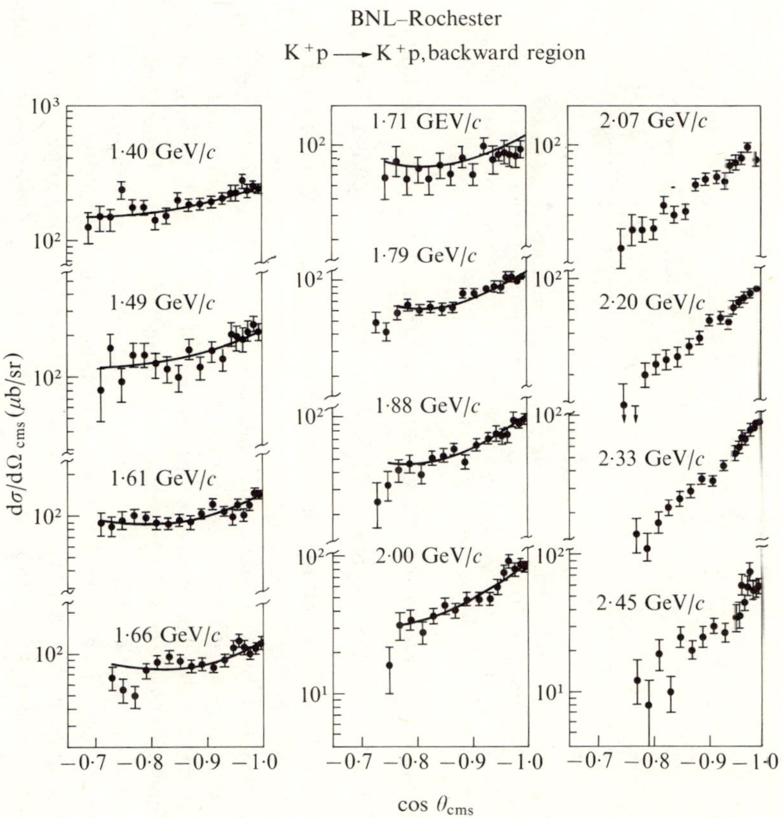


FIG. 13.23.(d) $K^+p$ elastic scattering differential cross-sections in the backward region, between 1·40 and 2·45 GeV/$c$. The full lines are hand-drawn curves through the data [66].

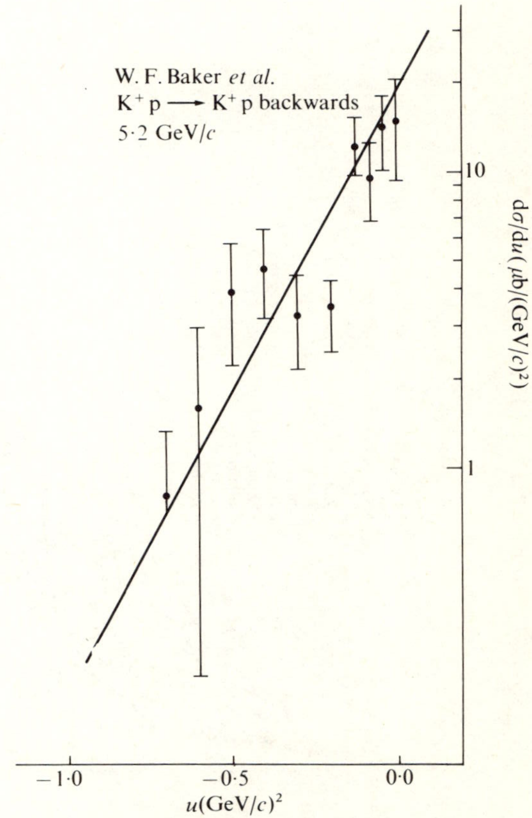


FIG. 13.23.(e) Backward $K^+p$ data. The full line shows the forward differential cross-section (slope $\sim 5$ $(\text{GeV}/c)^{-2}$) divided by 1000 and plotted at $|u| = |t|$ [66]. (From Bellettini. (1968). *Proceedings of the XIV International Conference on High Energy Physics, Vienna.*)

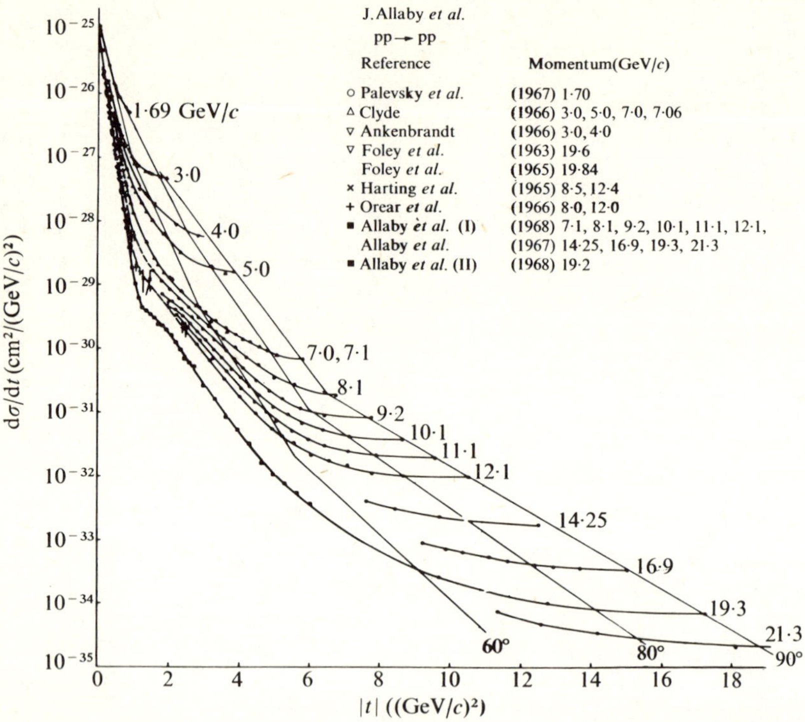


FIG. 13.24.(a) Proton–proton elastic scattering differential cross-sections as functions of $|t|$ (Allaby *et al*. Reference [6] of [66].) The curves joining the experimental points are hand-drawn to guide the eye. The loci of cross-sections for fixed c.m.s. angle are indicated for 60°, 80°, and 90°. (From Bellettini. (1969). *Proceedings of the XIV International Conference on High Energy Physics, Vienna*.)

It is apparent that, as the energy increases, there develops a substantial shoulder effect at $|t| \sim 1\ (\mathrm{GeV}/c)^2$. It also appears that $\dfrac{\mathrm{d}\sigma}{\mathrm{d}t}$ tends to level off (as expected) at about 90° in c.m.s., at ever decreasing values as the incident energy increases.

The various dips, shoulders, and peaks exhibited in elastic scattering data are suggestive of diffraction scattering minima and maxima effects. It is clear from our discussion of the optical model that such diffraction peak and minima structure are very sensitive to the exact form of the effective tail of the effective complex potential for the interaction. It appears that, whenever there is a sizeable Regge shrinkage, the diffraction structure appears to be minimized and, conversely, to become more apparent as the shrinkage either disappears as in $\pi$–p, or as it changes sign to anti-shrinkage such as in $\bar{\mathrm{p}}$–p.

For a review of p–p, and other nucleon–nucleon scattering, especially of the lower energies, the student should see an article by Breit [75].

### 13.13. n–p elastic scattering

Measurements of n–p elastic scattering were performed [76] by allowing a collimated neutron beam at the LRL Bevatron (and also later at the AGS) to strike a hydrogen target, and then measuring the momentum and angle of the recoil proton with a pair of thin plate spark chambers before and after a magnet. The scattered neutrons were detected by requiring that a neutral particle (selected by an anti-counter) interact, and create one, or more, charged particles in an array of seven steel plate spark chambers. The higher energy results are shown in Fig. 13.24(b).

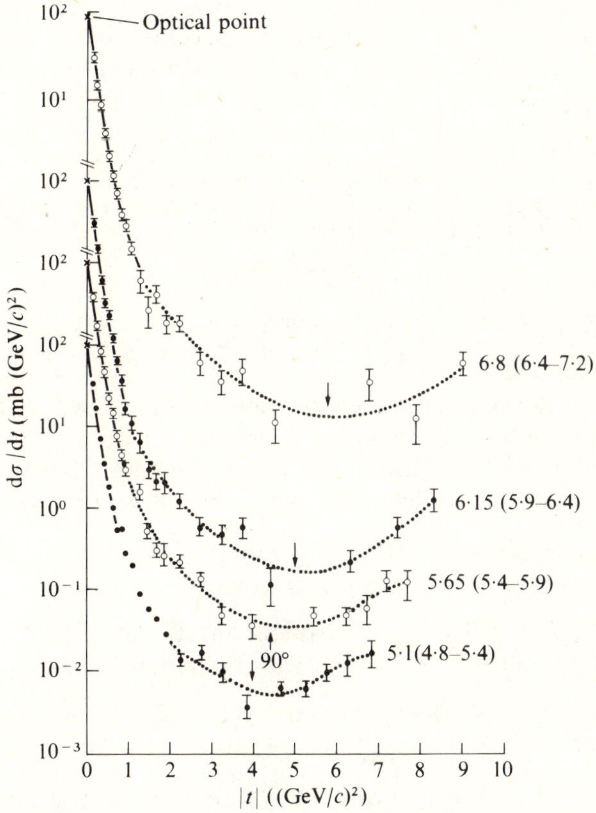


FIG. 13.24.(b) $d\sigma/dt$ for n–p against $|t|$. The numbers in parenthesis are the range of incident neutron energies. The number to the left is the central value. (From Perl *et al. SLAC Publication*, 622.)

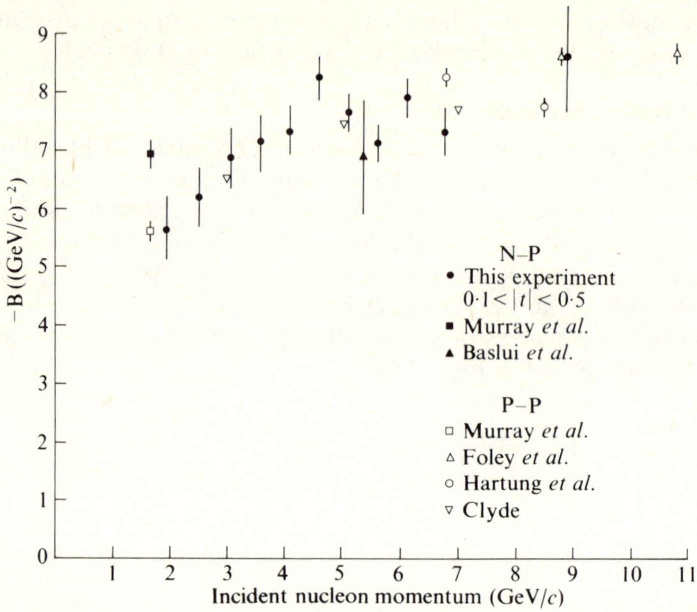


FIG. 13.24.(c) The slope $(-B\,(\mathrm{GeV}/c)^2)$ of the diffraction peak for n–p (solid points) and p–p (open points) against incident momentum. (From Perl *et al. SLAC Publication*, 622.)

The slope ($B$), of the diffraction peak, defined as

$$\left(\frac{\mathrm{d}\sigma}{\mathrm{d}t}\right) = \left(\frac{\mathrm{d}\sigma}{\mathrm{d}t}\right)_{t=0} \mathrm{e}^{-B|t|},$$

determined in this, and other investigations, is shown in Fig. 13.24(c), and compared to the various values of $B$ determined from p–p experiments. n–p diffraction behaves, within the error, the same as p–p diffraction. However, it was deduced that, although in the diffraction region,

$$\left(\frac{\mathrm{d}\sigma}{\mathrm{d}t}\right)_{\mathrm{n+p}} \approx \left(\frac{\mathrm{d}\sigma}{\mathrm{d}t}\right)_{\mathrm{p-p}},$$

at high $|t|$, the $T = 0$ differential cross-section falls below the $T = 1$ differential cross-section. This is consistent with the requirements of charge independence at $\theta = \frac{\pi}{2}$, which predict that

$$\frac{\left(\frac{\mathrm{d}\sigma}{\mathrm{d}t}\right)^{\mathrm{n+p}}\left(\frac{\pi}{2}\right)}{\left(\frac{\mathrm{d}\sigma}{\mathrm{d}t}\right)^{\mathrm{p-p}}\left(\frac{\pi}{2}\right)} \geq \frac{1}{4},$$

while the observed average value is

$$\frac{\left(\frac{d\sigma}{dt}\right)^{n+p}\left(\frac{\pi}{2}\right)}{\left(\frac{d\sigma}{dt}\right)^{p-p}\left(\frac{\pi}{2}\right)} = 0{\cdot}63 \pm 0{\cdot}09 \geq \tfrac{1}{4}.$$

Wu and Yang [77] attributed the rapid decrease of $\frac{d\sigma}{dt}$, with increasing $|t|$, for hadron–hadron elastic collisions, as due to the fact that the hadrons, considered as extended objects, have an effective internal rigidity equivalent to a few hundred MeV. Thus, in a hadron–hadron collision, there is an associated difficulty, which increases with increasing $|t|$, for transferring transverse momentum without 'breaking it up', which is required for an elastic collision. Exchanging charge is considered to occur easily in this model and, thus, elastic differential cross-sections are approximately equal in all $T$-spin states. The above leads to the prediction, for the region near $\pi/2$, in

$$\frac{\left(\frac{d\sigma}{d\Omega}\right)_{n+p}}{\left(\frac{d\sigma}{d\Omega}\right)^{p-p}} = \frac{1}{2},$$

which is consistent with the experimental observations considering the errors.

### 13.14. n–p charge exchange

Fig. 13.25 summarizes results on n–p (forward) charge-exchange scattering. Fig. 13.25(a) reveals that the n–p charge-exchange cross-sections obtained at 8·15 GeV/$c$ are similar, as a function of transverse momentum, to those observed at lower energies.

Fig. 13.25(b) shows that the results obtained at 3 GeV were not fitted even by the combination of single $\pi$ and $\rho$ exchange, with absorption. The fit was, at best, qualitative for small angles ($|t|^{\frac{1}{2}} \lesssim 100$ MeV/$c$) but did not fit at all for larger $|t|$.

A two-exponential fit to the n–p charge-exchange angular distribution gave the following results.

(a) The equivalent radius of the steeper small $|t|$ exponential was greater than 2 fm (i.e. greater than for even the case of $\bar{p}$–p scattering).

(b) The data for $|t| \gtrsim 0{\cdot}1$ to $0{\cdot}2$ shows much flatter experimental results which are equivalent to a radius of less than 1 fm. The shape of the angular distribution remains similar down to 1–2 GeV/$c$ [78]. This

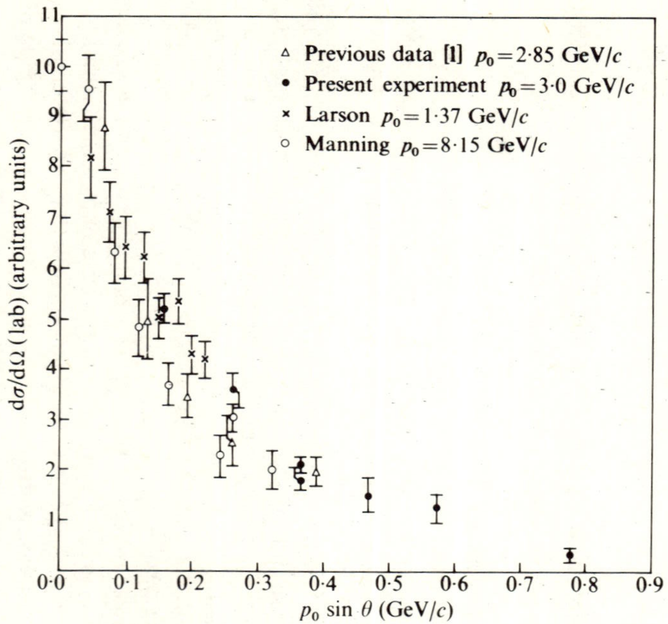


FIG. 13.25.(a) n–p elastic charge-exchange angular distribution as a function of transverse momentum $p_\perp = p \sin\theta$. The zero-degree cross-sections, for each incoming momentum, have been normalized to the same value. The errors are statistical.

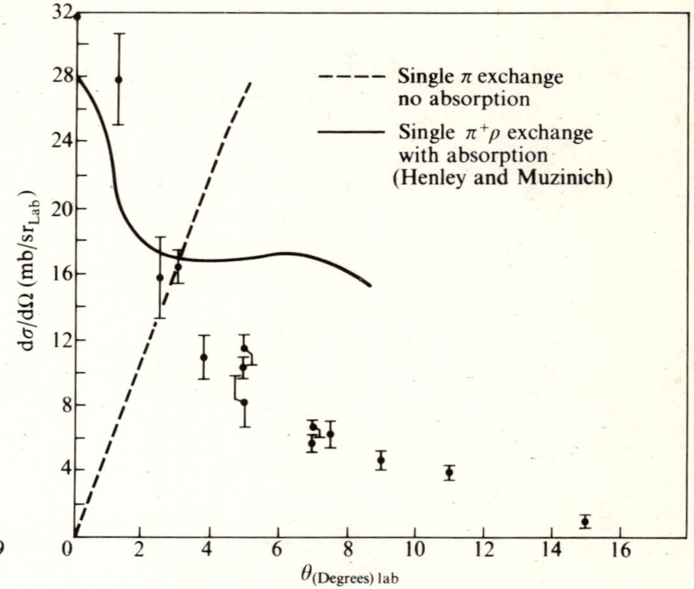


FIG. 13.25.(b) Measured angular dependence of n–p elastic charge-exchange at 3 GeV/$c$ compared to the calculations of Henley and Muzinich based on simple $\pi$ and $\rho$ exchange with absorption. The zero-degree cross-section is $(31{\cdot}6 \pm 5)$ mb/str. (Lindenbaum. (1965). *Proceeding of Oxford International Conference on Elementary Particles.*)

behaviour of the n–p charge-exchange is drastically different from the behaviour of $\pi$–p charge-exchange where, at small $|t|$, a flat, or rounded, plateau is observed. The momentum dependence of $\frac{d\sigma}{dt}$, at $t = 0$, is estimated to be $\sim \frac{1}{p^2}$.

## 13.15. K–p charge exchange

Fig. 13.26(a) shows the results obtained [79] for

$$\frac{d\sigma}{dt}(K^- + p \rightarrow \bar{K}^0 + n).$$

The rounded plateau near $t = 0$ is reminiscent of the $\pi$–p charge-exchange, and suggests vector meson exchange. The solid curves are based on Regge-pole fits to the 9·5 GeV/$c$ data using $\rho$ or R (or $A_2$) poles. In contrast to the $\pi$–p case, no dip is seen at t $\approx$ 0·6(GeV/$c$)$^2$.

We could anticipate that reactions involving the same particle exchanged in the $t$-channel should be similar. Fig. 13.26(b) compares [80] various reactions which involve either $\rho$ or $A_2$, or both, as follows.

$$\pi^- + p \rightarrow \pi^0 + n, \qquad \rho \text{ exchange.}$$

$$K^- + p \rightarrow \bar{K}^0 + n, \qquad (\rho + A_2) \text{ exchange.}$$

$$\pi^- + p \rightarrow \eta^0 + p, \qquad A_2 \text{ exchange.}$$

$$\pi^+ + p \rightarrow \eta^0 + N^{*++}, \qquad A_2 \text{ exchange.}$$

$$K^+ + p \rightarrow K^0 + N^{*++}, \qquad (\rho + A_2) \text{ exchange.}$$

In general, $\rho$ exchange leads to a flattening (i.e. plateau) of the $\frac{d\sigma}{dt}$ curve near $t = 0$, and a dip near $t \approx$ 0·5–0·6. On the other hand, $A_2$ exchange produces a flat region from $t \approx 0$ to $-0{\cdot}25$. SU(3) symmetry allows us to predict what would happen in a reaction which involved the sum of the two exchanges. The prediction obtained is that some flattening should be observed near $t = 0$, but no dip of any appreciable size at $t \approx$ 0·5–0·6. The experimental results are obviously in agreement with this prediction.

## 13.16. $\bar{p} + p \rightarrow \bar{n} + n$

Astbury *et al.* [81] have observed the charge-exchange reaction

$$\bar{p} + p \rightarrow \bar{n} + n,$$

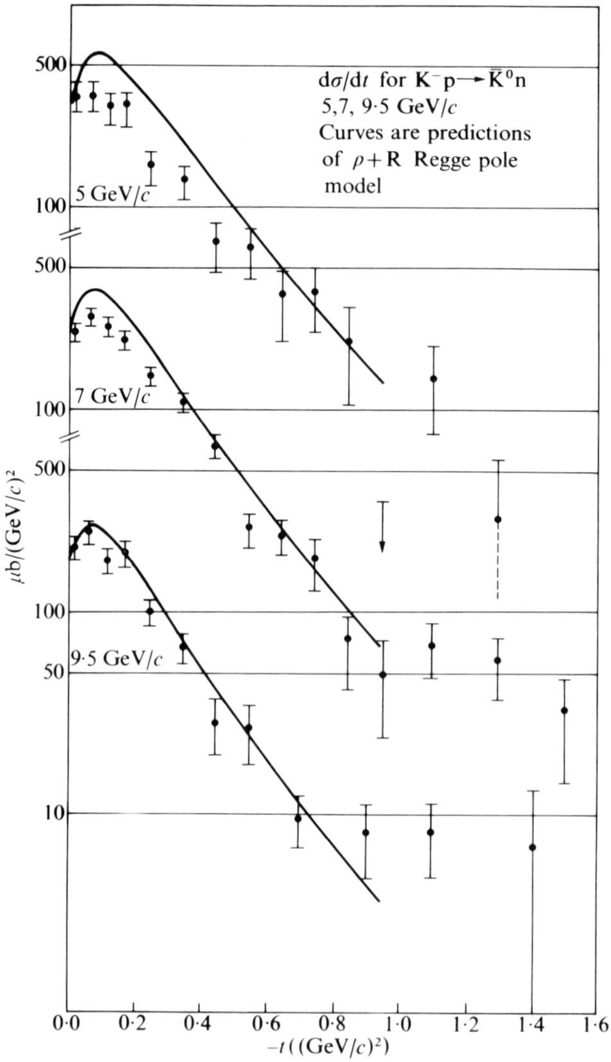


FIG. 13.26.(a) The differential cross-section for the reaction $K^-p \rightarrow \bar{K}^\circ n$ at laboratory system momenta of 5, 7, and 9·5 GeV/$c$, reported by Astbury *et al*. The theoretical curves are for a Regge pole model fit by Phillips and Rarita. In this fit, only the 9·5 GeV/$c$ data were used and, therefore, the 5 and 7 GeV/$c$ theoretical curves are predictions according to the model. (From Van Hove. (1966). *Proceedings of the XIII International Conference on High Energy Physics, Berkeley*. p. 259).

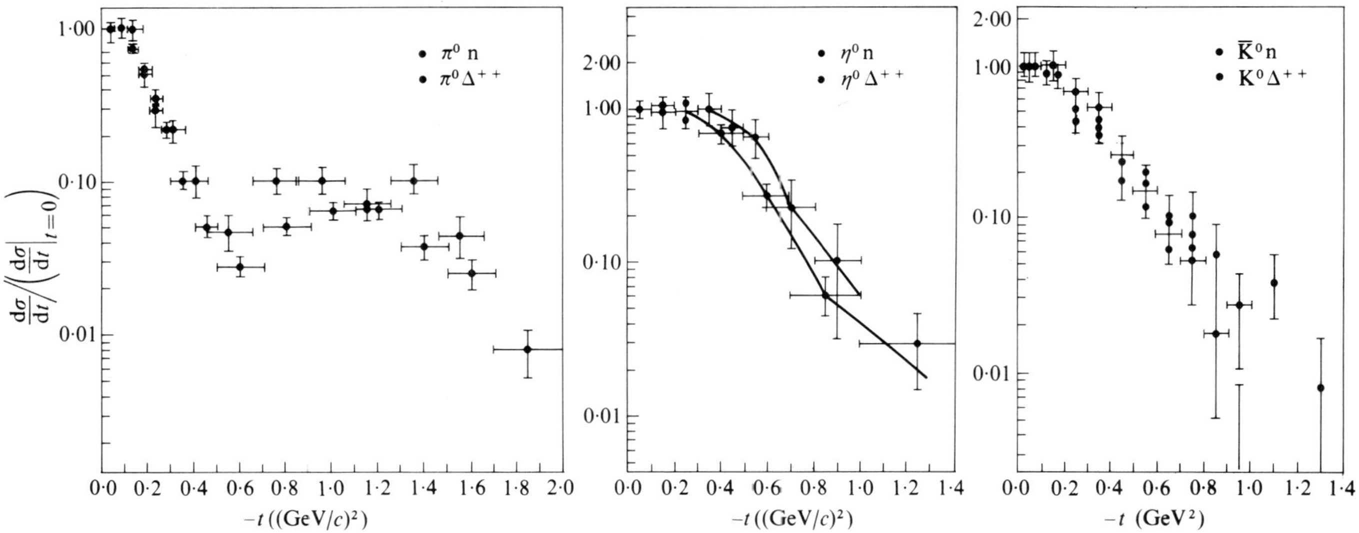


FIG. 13.26.(b) Differential cross-section ($d\sigma/dt$) distributions normalized to their value at $t = 0$, against $-t$. The first graph is for the reactions $\pi^-p \to \pi^\circ n$ and $\pi^+p \to \pi^\circ N^{*+-}$. The second is for $\pi^-p \to \eta^\circ n$ and $\pi^+p \to \eta^\circ N^{*++}$. The third is for $K^-p \to K^\circ n$ and $K^+p \to K^\circ N^{*++}$. (From Mathews. (1969). *Nucl. Phys.* **B111,** 339 and Morrison. (1969). *Proceedings of the International Conference on High Energy Physics, Lund.*)

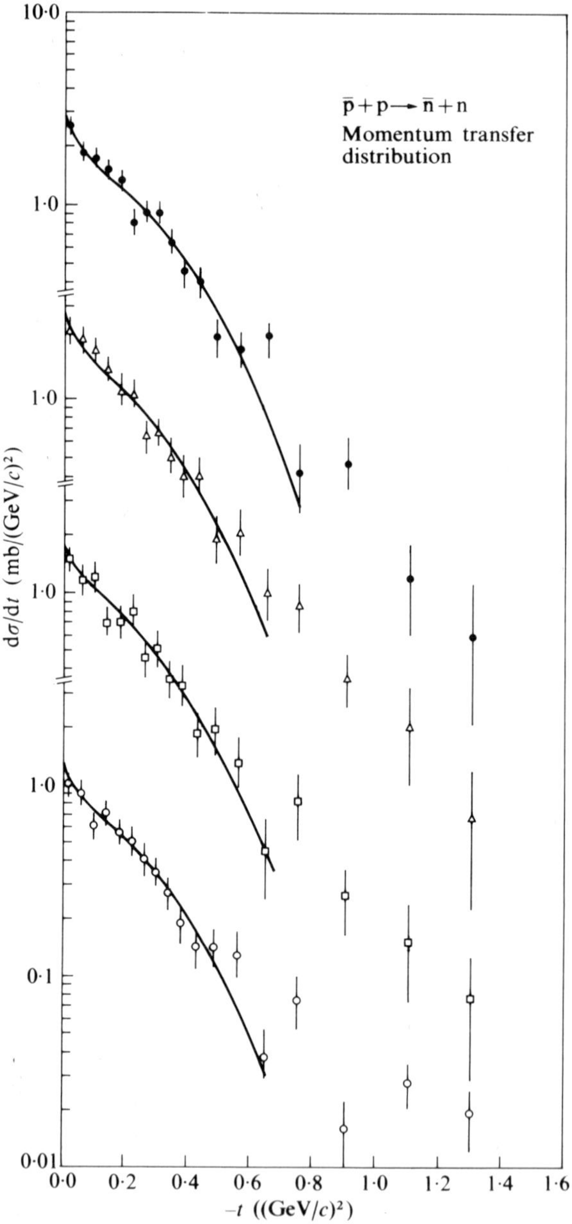


FIG. 13.27. The differential cross-section for the reaction $\bar{p}p \rightarrow \bar{n}n$ for the momenta 5, 6, 7, and 9 GeV/$c$, as measured by Astbury *et al*. The theoretical predictions are for the coherent droplet model. Errors indicated are statistical only. $P_{in}$ (in GeV/$c$): ●5; △6; □7; ○9. (From [65*a*].)

in the 5–9 GeV/$c$ region. The results are shown in Fig. 13.27. The solid curves are calculations with the coherent drop model. However, in order to explain the sharp peak at small $|t|$ in n–p, Byers had to introduce a one-pion exchange term, in addition to the coherent drop model. The $\rho$ and R (or $A_2$) Regge fits did not work, since they give a different $s$ dependence to that observed.

## 13.17. Polarization in high-energy scattering

### 13.17.1. *Introduction*

(General reviews of this subject are given by Van Rossum [82] and Breit [75].)

The development and use of polarized targets has made it possible to obtain a considerable amount of information on the polarization parameter, $P$, in elastic scattering and charge-exchange reactions.

### 13.17.2. *Pion–nucleon scattering*

For pion–nucleon elastic scattering, the polarization parameter $P(t)$ is given by

$$P(t) = \frac{2\,\mathrm{Im}(f^*g)}{|f|^2+|g|^2}, \tag{13.134}$$

where $f$ is the ordinary (non-spin flip) amplitude, and $g$ is the spin flip amplitude.

The behaviour of the $\pi^{\pm}$–p polarization [83] in the range 5–14 GeV/$c$ is shown in Fig. 13.28(a)–(b). The characteristic features are:

(a) All polarizations approach zero as $t \to 0$, which is entirely expected from eqn (13.134), since the spin flip amplitude vanishes as $t \to 0$.

(b) The maximum of the polarizations occur in the neighbourhood of $t \approx -0{\cdot}1$ to $-0{\cdot}2$ (GeV/$c$)$^2$.

(c) $P$ passes through zero at $t \approx -0{\cdot}6$.

(d) The $\pi^+ + \mathrm{p}$ and $\pi^- + \mathrm{p}$ polarizations seem to be of similar magnitude, but of opposite sign.

(e) The polarizations seem to be independent, or vary only slowly with energy.

The Regge-pole fits originally explained these effects [83*d*] as predominantly due to interference between the spin flip part of the $\rho$-trajectory, and the non-spin flip part of the P and P′. This explains the change of sign between $\pi^- + \mathrm{p}$ and $\pi^+ + \mathrm{p}$ polarizations. Also, since the $\rho$ spin flip part vanishes at $t = -0{\cdot}6$ (GeV/$c$)$^2$, this was considered a triumph for the Regge pole model. The only difficulty with the explanation at this time was the

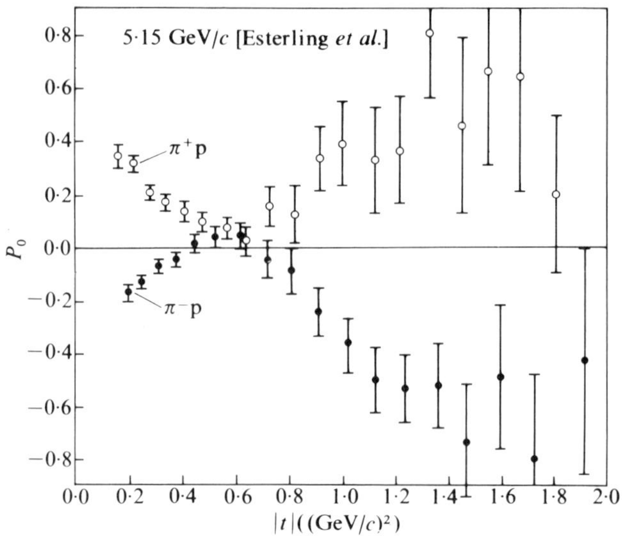


FIG. 13.28.(a) Polarization parameter in $\pi^{\pm}$p scattering at 5·15 GeV/$c$. [82].

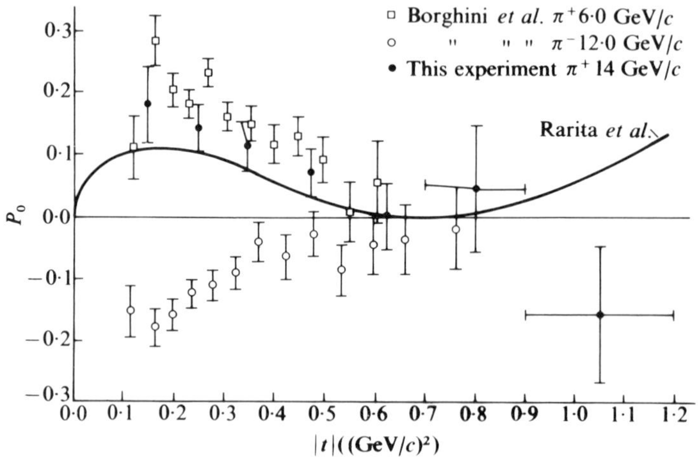


FIG. 13.28.(b) Preliminary results for the polarization parameter in $\pi^+$p scattering at 14 GeV/$c$ (full points). $\pi^+$p and $\pi^-$p data by Borghini *et al.*, at lower energies, are shown for comparison. The curve represents the theoretical prediction of Rarita *et al.* for $P_0$ in $\pi^+$p scattering at 14 GeV/$c$. (From Bellettini. (1968). *Proceedings of the International Conference on High Energy Physics, Vienna*. p. 347.)

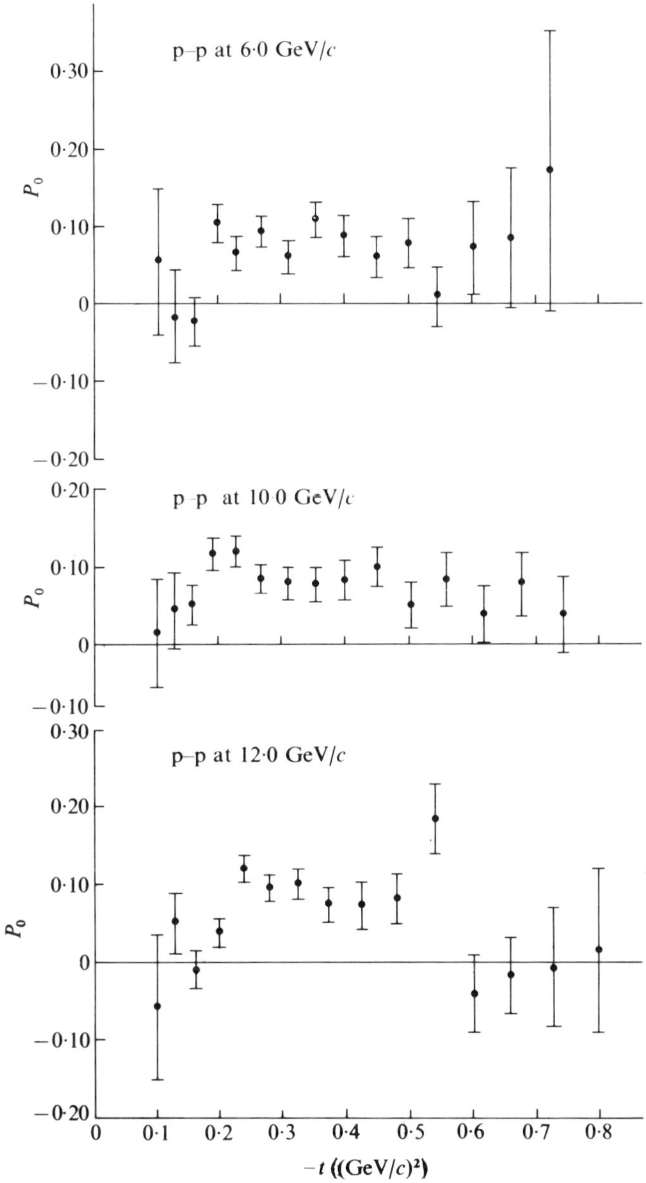


FIG. 13.29.(a) The polarization parameter $P_0$ plotted against the invariant four-momentum transfer $t$, in pp scattering, measured by Borghini *et al.* (from Wetherell p. 275 of [65*b*].)

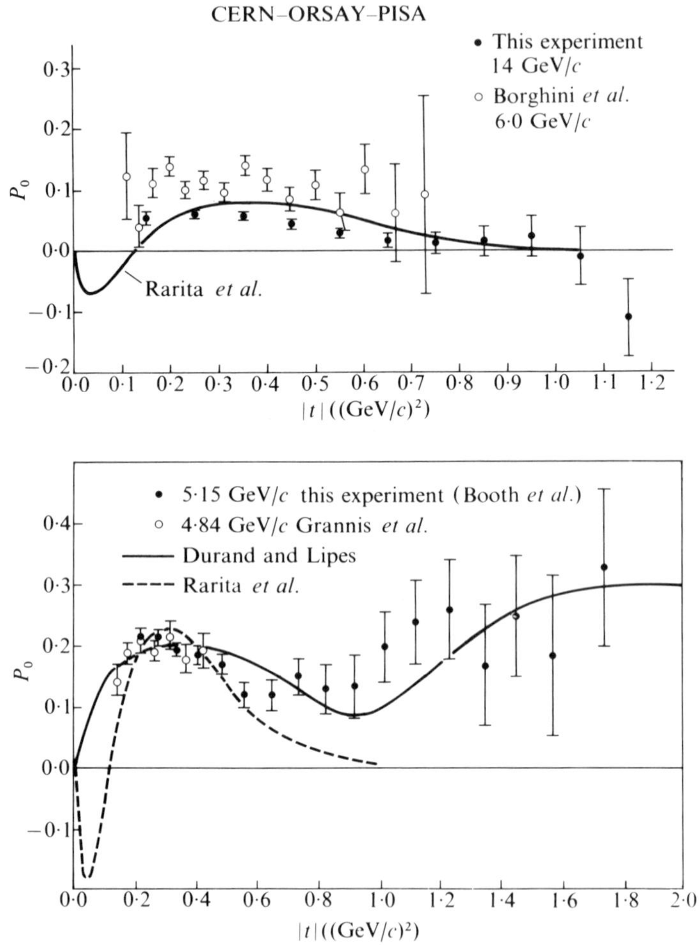


FIG. 13.29.(b) *Top graph*. The pp polarization data at 5·15 GeV/$c$ by the Chicago–Argonne group. Data by Grannis *et al.* at 4·84 GeV/$c$ are in good agreement with these data. However, some normalization problem seems to exist on comparing with the 6 GeV/$c$ data by Borghini *et al.* The theoretical curves are predicted by the models of Durand and Lipes and Rarita *et al.* *Bottom graph*. Preliminary results for the polarization parameter in pp scattering at 14 GeV/$c$. Data by Borghini *et al.* at 6·0 GeV/$c$ is shown for comparison. (Graphs from [66].)

near-constancy of the observed polarization with increasing $s$. The crucial test which was awaited was the experimental confirmation of the Regge model prediction that, in $\pi^- + p \rightarrow \pi^0 + n$, the polarization should be zero, since only the $\rho$ pole was involved and, for exchange of any single Regge pole, $P$ is necessarily zero. However, it was soon observed (Saclay–Orsay–Pisa collaboration) that the polarization in charge exchange at 5·9 GeV/$c$ and 11·2 GeV/$c$, was of similar magnitude to that of $\pi^+ + p$ [81]. This led to the usual proliferation of parameters (including the introduction of a $\rho'$) or cuts, or $s$-channel resonances (which may represent double counting).

### 13.17.3. p–p *polarization*

Fig. 13.29(a) shows the behaviour of p–p polarization as a function of $t$, for incident momentum in the 6–12 GeV/$c$ range, and Fig. 13.29(b) compares the results at 6 GeV/$c$ and new results at 14 GeV/$c$.

At lower energies the p–p polarization, at its maxima, reaches a peak of about 50 per cent in the neighbourhood of 0·8 GeV incident energy. The peak value then drops to about $(10 \pm 5)$ per cent in the neighbourhood of 5 GeV, and seems to maintain this level until 14 GeV/$c$ incident momentum.

The Booth *et al.* data [85] (Fig. 13.29(b)) show that $P$ increases when $|t| \gtrsim 0{\cdot}7\ (\mathrm{GeV}/c)^2$ and may reach a higher peak. The Regge-pole fits with $P + P' + \omega$ could not reproduce this effect. However, optical model calculations by Durand and Lipes could easily account for it.

## 13.18. Predictions of asymptotic behaviour based on experiments up to 30 GeV incident energy

The ever-increasing complexity of Regge-pole fits in this energy range, and the peculiar behaviour of the almost-flat Pomeranchuk trajectory, make it manifestly clear that asymptotic behaviour of even the forward scattering amplitudes could not be predicted with any degree of reliability via Regge-pole fits. However, it had been demonstrated experimentally,† by measurement of the complete forward scattering amplitudes (i.e. real and imaginary parts), that a doubly subtracted dispersion relation was valid for $D^+$ until 20 GeV/$c$, regardless of any conceivable change in asymptotic behaviour. Furthermore, a singly subtracted dispersion relation was satisfied for the symmetric scattering amplitude, and an unsubtracted dispersion relation was satisfied for the anti-symmetric amplitudes, thus strongly implying that the Pomeranchuk theorem was valid.

† See Chapter 5, § 5.8, and Foley *et al.* reference [29].

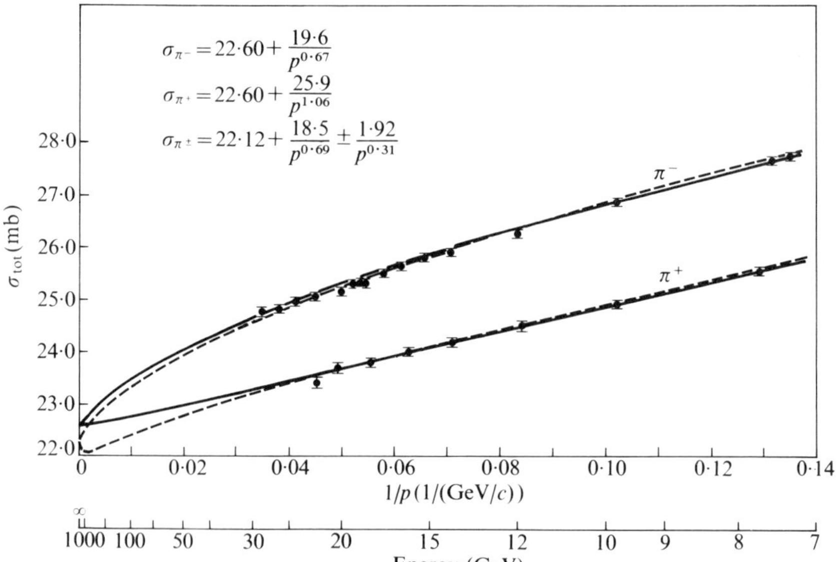


FIG. 13.30. The predicted values of $\sigma_{\text{total}}(\pi^{\pm}\text{–p})$ above 8 GeV/$c$ plotted against a linear plot of $1/P$ in the laboratory system, with a non-linear energy scale (GeV) also shown. The solid lines represent a fit of type I. The dashed lines represent a fit of Type II. (From Lindenbaum, Trieste Symposium [86*d*].)

The high precision total cross-section measurements in the 8–29 GeV/$c$ region were discussed in Chapter 5. It was found that the so-called Fit I gave good agreement between the predicted $D^-$ and the experimentally determined $D^-$. The so-called Fit II, which is the equivalent of a three-pole Regge fit‡ (P, P′ and $\rho$) gave poor agreement with $D^-$. Both of these fits are shown plotted (Fig. 13.30) against $\frac{1}{P}$. Thus the asymptotic region lies at the origin of this plot.

It is interesting to estimate where the Pomeranchuk theorem is expected to become true to within the best precision we can reasonably hope to attain. When we consider the various sources of systematic errors and electromagnetic correction uncertainties, it appears that the best precision we can reasonably expect to attain in measuring the differences of $\pi^+ + \text{p}$ and $\pi^- + \text{p}$, even in the forseeable future, is about 0·1 per cent.

Since Fit I gives good agreement between the predicted $D^-$ and the experimentally determined $D^-$, it is the one selected as an appropriate vehicle for predicting, within the errors, the point where the Pomeranchuk

‡ Fit I has the equivalent effect of two Regge poles in determining the difference, whereas Fit II has only one; both have only five parameters total for the $\pi^{\pm}$–p system.

theorem will come true. For this fit we find that at $\gtrsim$25 000 GeV, the Pomeranchuk theorem will be satisfied, with the best expected experimental precision of 0·1 per cent. For Fit II (the three-Regge-pole type), which gives a rather poor fit to $D^-$, the Pomeranchuk theorem would not be satisfied until about 1 600 000 GeV.

The author [86] has referred to 'Asymptopia'—the theoretically-promised land, where all asymptotic theorems come true. Therefore, an operational definition of where 'Asymptopia' lies might be at those energies for which the Pomeranchuk theorem becomes true. In this case, it is predicted that 'Asymptopia' lies at least beyond 25 000 GeV and, perhaps, beyond 1 000 000 GeV.

As seen in Fig. 13.30, the constant cross-section at infinity is predicted to be in the range 22·1–22·6 mb. Assuming spin independence, fits similar to Fit I were made for the p–p and p̄–p elastic scattering data and were consistent with dispersion relation calculations, considering the many uncertainties involved in the p–p case. This fit is shown in Fig. 13.31. The result is that 'Asymptopia' is predicted to lie beyond 35 000 GeV, which is good confirmation of the general magnitude of the number predicted in the $\pi^{\pm}$–p analysis.

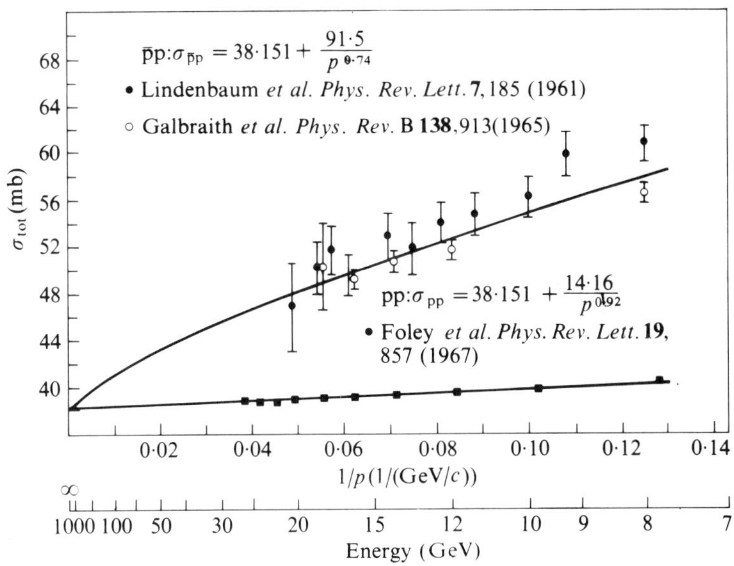


FIG. 13.31. The predicted values of $\sigma_{total}$(p–p) and $\sigma_{total}$(p̄–p) by a fit of type I to the experimental data, against a linear plot of $1/P$ in the laboratory system with a non-linear energy scale (GeV) also shown. (From Lindenbaum, Trieste Symposium [86*d*].)

As far as predictions of the behaviour of the diffraction peak are concerned it should be abundantly clear that these should not be attempted in view of the confusing experimental situations and phonomenological analyses.

### 13.19. Recent higher energy results for total and elastic scattering cross-section measurements at Serpukhov

The successful operation of the 70 GeV Serpukhov accelerator, and the resulting early experimental results have been of great general interest and, in particular, provide a new check on asymptotic predictions made on the basis of the up to 30 GeV/$c$ region.

The first results [87] (Allaby *et al.*) on total cross-sections of $\pi^-$, $K^-$, and $\bar{p}$, incident on protons and deuterons in the 20–65 GeV/$c$ momentum range, were obtained by the Institute for High Energy Physics, Serpukhov and a CERN collaboration. Secondary particles were produced in an internal target of the synchrotron, and only negative beams could be extracted at near-zero production angles, thus providing high flux at high energies. The incident beam particles were identified by a combination of a threshold, and a good-geometry, differential Cerenkov counter. Transmission measurements were performed with a counter telescope, which was located downstream of the hydrogen or deuterium target, and contained twelve circular transmission counters, with diameters ranging from 6 to 40 cm, mounted on a trolley positioned so as to allow extrapolation to zero solid angle, with minimal energy-dependent systematic errors. Owing to requirements connected with the early phases of accelerator operation, a liquid hydrogen and deuterium target was not available. Thus, instead, target cells (which were cylinders 3 m long and 12 cm in diameter, one of which was filled with hydrogen while the other was filled with deuterium, at about 70 atm. pressure, at liquid nitrogen temperatures) were employed. The equivalent of 7 g cm$^{-2}$ of hydrogen and 14 g cm$^{-2}$ of deuterium were placed in the beam path, which also passed through 1·3 mm thick stainless steel end-windows. Temperature and pressure monitoring were employed, and the uncertainty on the hydrogen density was estimated to be $\pm 0{\cdot}7$ per cent, and the deuterium density uncertainty to be $\pm 0{\cdot}9$ per cent.

The extrapolation to zero solid angle was carried out using the relation

$$\sigma_i = \sigma_{\text{tot}} e^{at_i + bt_i^2},$$

where $\sigma_i$ is the partial cross-section measured by the channel. Fits, in the $t$-range

$$0{\cdot}014 < |t_i| < 0{\cdot}065\ (\text{GeV}/c)^2,$$

were employed, and the magnitude of the $t$-range was used to determine limits on $t$-range effects on the final results. Corrections for muon contamination in the pion beam were made, but were negligible in the kaon case.

The results are presented in Fig. 13.32. The errors were compounded results of statistical errors, extrapolation errors, and point-to-point uncertainties on the muon corrections. There is a scale error, originating primarily from uncertainties in the gas density, and in the correction for muon contamination. The error is typically $\pm 1$ per cent for hydrogen data and $\pm 1{\cdot}2$ per cent for deuterium data. In addition, the scale error for the neutron data contains the extra uncertainty of the Glauber correction, resulting in a total error of 2·5 per cent for $\pi^- + \mathrm{n}$ and $\mathrm{K}^- + \mathrm{n}$, and 3·0 per cent for $\bar{\mathrm{p}}$–n.

Fig. 13.32 shows the new total cross-section data, compared to previous lower-energy data. In general, the agreement of the lower-energy data with earlier results is good.† The data appears to imply some striking changes in the behaviour of total cross-sections in this energy range. Beyond 30 GeV/$c$ it appears that the pion total cross-sections on hydrogen have become rather energy independent (within the error of $\pm 0{\cdot}2$ mb on each point) averaging about $(24{\cdot}6 \pm 0{\cdot}3)$ mb. In this error estimate the scale error has been included. The $\mathrm{K}^- + \mathrm{p}$ cross-section also appears to be energy independent, averaging $(20{\cdot}9 \pm 0{\cdot}3)$ mb. The anti-proton cross-section continues to decrease, reaching a value of about 44 mb at 50 GeV/$c$, which still exceeds the estimated p–p cross-section by about 10–15 per cent.

Using the Glauber–Wilkin correction,‡ the cross-sections on neutrons were deduced, from the difference of the deuteron and hydrogen cross-sections. The following deduced differences,

$$\sigma(\pi^- + \mathrm{p}) - \sigma(\pi^- + \mathrm{n}) = (1{\cdot}3 \pm 0{\cdot}7)\ \mathrm{mb},$$
$$\sigma(\mathrm{K}^- + \mathrm{p}) - \sigma(\mathrm{K}^- + \mathrm{n}) = (1{\cdot}0 \pm 0{\cdot}6)\ \mathrm{mb},$$
$$\sigma(\bar{\mathrm{p}} - \mathrm{p}) - \sigma(\bar{\mathrm{p}} - \mathrm{n}) \approx 0,$$

suggest that, considering uncertainties in the error estimates above 30 GeV/$c$, the Okun–Pomeranchuk rule is satisfied. It is clear that there is an apparent change in the observed behaviour of total cross-sections of $\pi^- + \mathrm{p}$, $\pi^- + \mathrm{n} \approx \pi^+ + \mathrm{p}$ and $\mathrm{K}^- + \mathrm{p}$, and $\mathrm{K}^- + \mathrm{n}$ at these higher energies, namely, the uniform decrease with increasing energy observed at lower energies seems to disappear, and the new total cross-section data seem to suggest

† With the possible exception of the $\pi^- + \mathrm{p}$ and $\pi^- + \mathrm{d}$ cross-sections of Galbraith *et al.* This apparent discrepancy does not seem significant, in view of the systematic error in both experiments.

‡ The additional effects, pointed out by Wilkin, become unimportant at these energies.

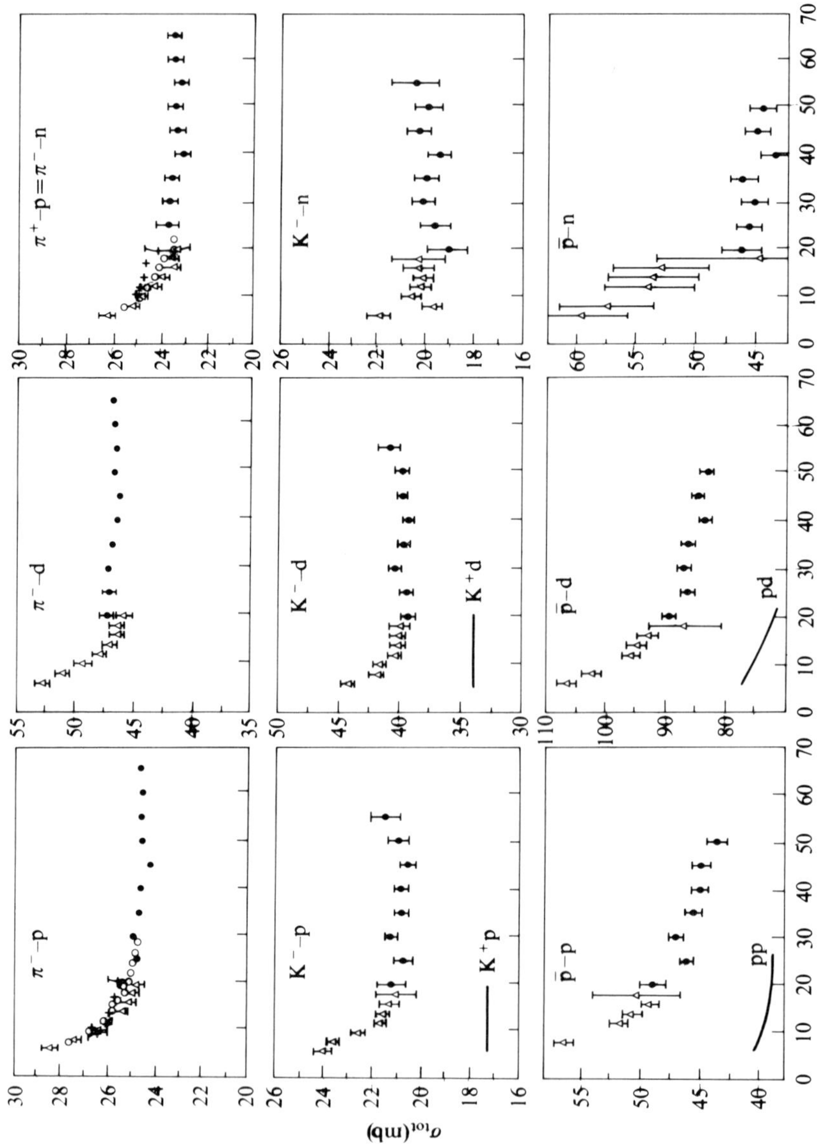


FIG. 13.32. The Serpukhov total cross-section measurements (solid black points) compared to the lower energy data. ([86c], original source [87]).

either energy independence, or perhaps even a minimum, and slight increase with increasing energy. There is, furthermore, the suggestion that, for the $K^- + p$ and $K^+ + p$ pair, the convergence toward a common Pomeranchuk limit may have ceased, at least at these energies. This apparent trend is in contradiction to the extrapolation of lower-energy data fits of either the phenomenological power law, or three Regge pole, type which gave monotonically decreasing total cross-sections in the Serpukhov energy range. Thus the new data imply that either a new effect, or additional parameters, have to be included in the lower-energy fits. To first determine whether these qualitative conclusions were statistically significant, the previous lower-energy power law fits were redone with the inclusion of the new Serpukhov data.

Fig. 13.33(a) shows a fit of the Fit II type (equivalent to a three pole P, P′, $\rho$ Regge fit) for $\pi^{\pm}$–p which is attainable with a reasonable $\chi^2$. This fit gives predictions for $D^+$ which agree reasonably with the experimental observations at lower energies but, just as in the case of the lower-energy data alone,

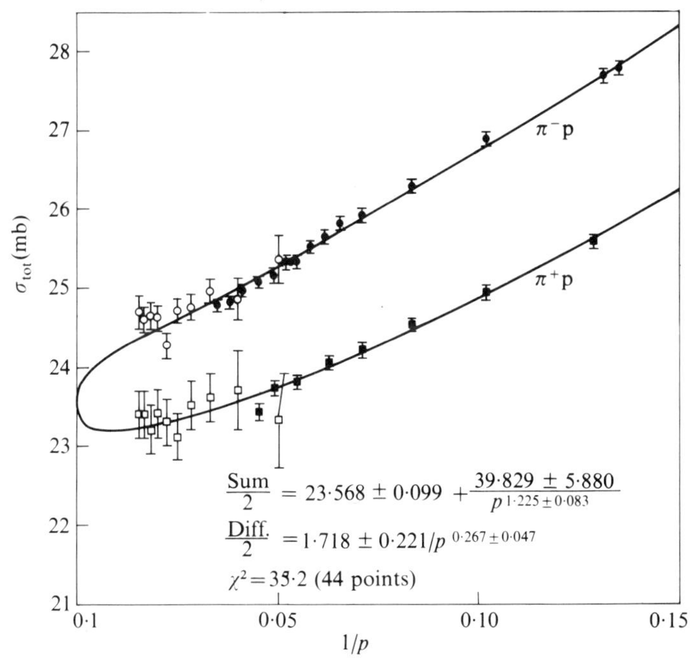


FIG. 13.33.(a) A fit of type II (three pole fit) to the Serpukhov data and the lower energy data. (From Lindenbaum, Austin Symposium [86c].)

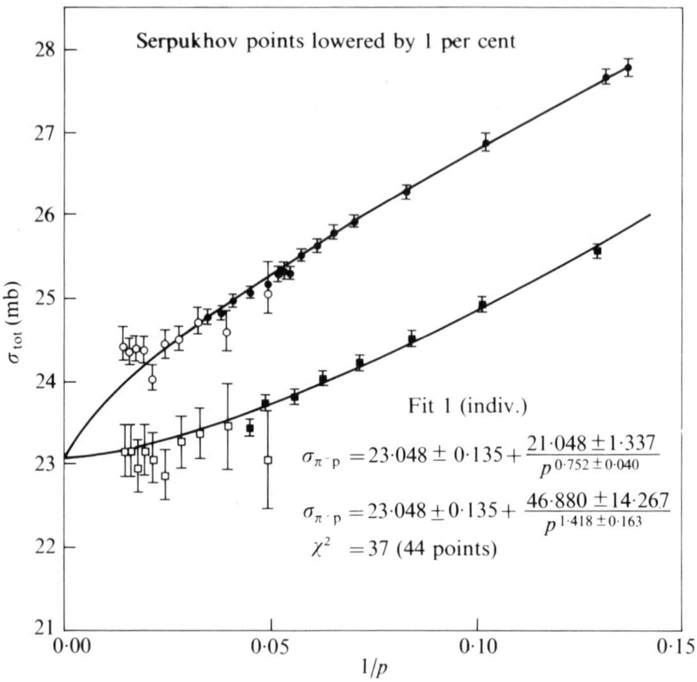


FIG. 13.33.(b) A fit of type I (individual power law fits) to the Serpukhov data and the lower energy data. (From Lindenbaum, Austin Symposium, [86*c*].)

this fit converges far too slowly to give predicted $D^-$ values from the unsubtracted $D^-$, which agree with the experimental measurements up to 20 GeV/*c*. Thus, this fit should probably be rejected as unsatisfactory. Furthermore, it is clear that the new data for $\pi^- + p$, at the highest momenta, are systematically above the fit and behave qualitatively differently.

In attempting to obtain an acceptable $\chi^2$ for a fit of the Fit I type, it was necessary to take account of the approximately 1 per cent scale-uncertainty in the recent Serpukhov data. Fig. 13.33(b) shows the resultant fit to both the lower-energy and higher-energy data, with the Serpukhov data lowered by 1 per cent (i.e. the best possible accommodation of the two sets of data to the assumed monotonically decreasing power law forms). Although we are able to obtain an acceptable $\chi^2(= 37)$ for 44 points, it is clear that the new Serpukhov high-momentum points lie systematically above it and contribute most of the $\chi^2$ and, thus, by themselves, fit this parametric form poorly. To clarify this point, we should remember that, since all errors are predominantly systematic, therefore, do not necessarily exhibit point-to-point fluctuations of a statistical nature, it is possible for the lower-momentum points to give a much lower $\chi^2$ than the expected value for statistical errors. From the above it can be concluded [86*c*] that we cannot yet assume that future, more accurate, experiments will not lead to consistency with the previously-used lower-energy parametric forms, but the odds at present favour a change in parametric form at the Serpukhov energies.

The new Fit I gives $D^-$, as well as the $D^+$ predictions, which fit the experimental data at the lower energies [86*c*]. A good fit of the Fit II type could also be obtained, but the fit to $D^-$ is poor. It was straightforward to include the Serpukhov data in the p–p and $\bar{p}$–p fits and obtain a fit of the Fit I type with a good $\chi^2$, and which gave predicted $D^+$ and $D^-$ values in agreement with the experimental measurements, assuming spin independence. Thus it appears that, in order to get good fits to the total cross-section and real forward amplitude data, new parameters, which increase both the $\pi^{\pm}$–p cross-section, but essentially do not change the convergence of the difference, are implied.

Barger and Phillips [88] proposed that the observed levelling out of $\pi^-+N$ and $K^-+N$ total cross-sections could be provided by Regge cuts [89]. Since the leading vacuum Regge cuts dominate over secondary Regge poles at high energy, if the cuts then contribute negatively, we could expect $\sigma_T(\pi^-+N)$ and $\sigma_T(K^-+n)$ to fall initially, and then to level out, finally to rise toward their limits at infinite energy.

Their fits to the data are shown in Fig. 13.34. The cut contributions rise logarithmically to the constant limit at infinite energy. It is clear that the choice of a Regge vacuum cut, as a source of additional parameters, is certainly not unique. For example, the addition of a term $\dfrac{-|\text{constant}|}{p^n}$ to both $\pi^++p$ and $\pi^-+p$ cross-sections (or, equivalently to the sum of the cross-sections) could be used to fit the data. Furthermore, we could probably obtain a fit with, for example, the form

$$\sigma_{\pi^\pm+p} = a\left\{\ln\left(\frac{s}{s_0}\right)\right\}^n + \frac{b}{p^{n'}} \pm \frac{c}{p^m},$$

$$\sigma_{K^\pm+p} = a'\left\{\ln\left(\frac{s}{s_0}\right)\right\}^n + \frac{b'}{p^{n'}} \pm \frac{c'}{p^{m'}},$$

where $n \leq 2$ to satisfy the Froissart bound.† Therefore, once again, new data has required new parameters to be introduced into the theory.

From the point of view of the Regge-pole theory, it has become clear that, in addition to an ever-increasing number of poles, we must acknowledge

† The $n$ in $\sigma_{\pi\pm+p}$, and that in $\sigma_{K\pm+p}$, could be different.

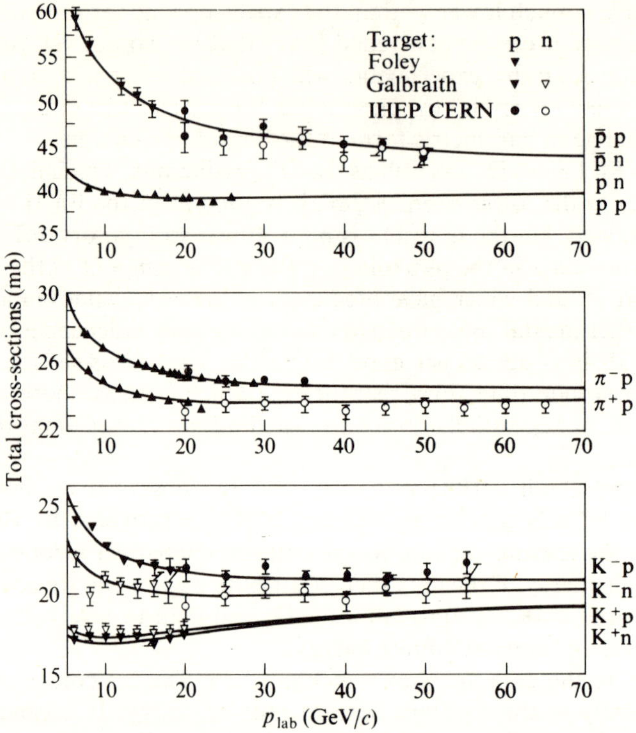


FIG. 13.34. Fit to Serpukhov and lower-energy data using Regge cuts, etc. (From [87], [88]).

the importance of cuts. The complexity of the theory has clearly achieved the point where we must seriously re-evaluate its usefulness.

It is also relevant to the above discussion that a Serpukhov group [90] has determined the slope parameter of the differential cross-section for p–p elastic scattering, in the incident energy range 12–70 GeV, and in the $|t|$ range $0{\cdot}008 < |t| < 0{\cdot}12\ (\mathrm{GeV}/c)^2$.

From a one-pole fit,

$$\alpha_{p'} = 0{\cdot}40 \pm 0{\cdot}09$$

was obtained, which agrees with the experimental values obtained for $\alpha_{p'}$ at the high-energy end of the AGS experiments. In the light of this, all lower-energy Regge-pole fits, with a flat Pomeranchuk trajectory, would be in doubt. However, a combination of the Regge vacuum cuts, a suitably sloping Pomeranchuk trajectory, and secondary poles, could probably be found to fit all the data.

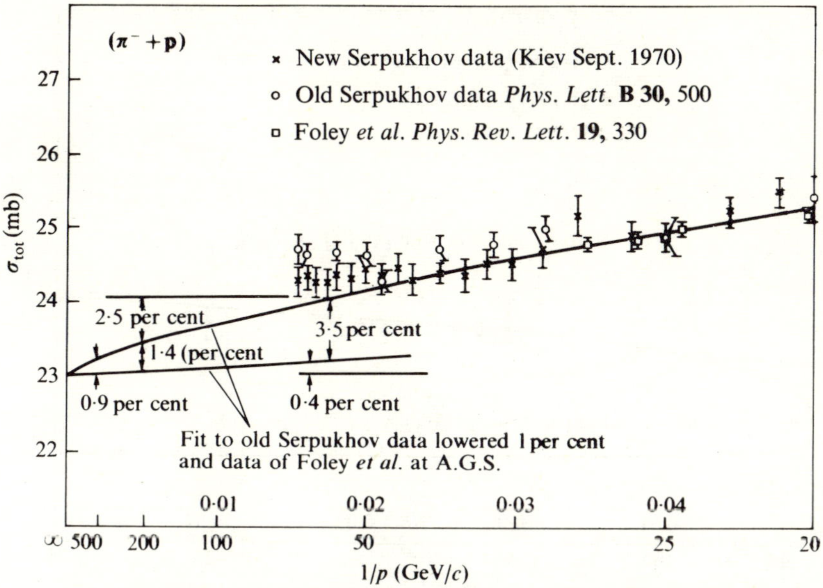


FIG. 13.35.(a) Same as Fig. 13.30 with the new Serpukhov data added.

At the Kiev Conference (15th International Conference on High Energy Physics (1970)) a new set of more accurate total cross-section measurements, using a liquid hydrogen target, but still employing counter telescope techniques, were reported. The results for $\pi^{\pm}$–p are shown in Fig. 13.35(a), superimposed on the previous Fit I. It is clear that about half of the departure of the higher-energy data from the old fit has now disappeared, and the new data are practically consistent with the old fit. The original Serpukhov results for $K^-+p$ and $K^++p$ ($\approx K^-+n$) were also originally fitted by the same method (Fit I), which is shown in Fig. 13.35(b). The new Serpukhov data, taken with a liquid hydrogen target, has been added, and agrees reasonably well with the old fit. Refitting, including the new higher-energy data, has been done in both cases, and the results are good $\chi^2$ fits, which are essentially the same as those shown.

*Finally, we shall consider the connection between the Pomeranchuk theorem and asymptotic behaviour.* The fits to the combination of the lower- (10–30 GeV) energy data, and the new Serpukhov energy data, have not yet (within the uncertainties involved) substantially changed the convergence of the differences $\Delta(\sigma(\pi^-+p)-\sigma(\pi^++p))$ and $\Delta(\sigma(\bar{p}+p)-\sigma(p+p))$ and, thus, the prediction of where the Pomeranchuk theorem will apply has not been

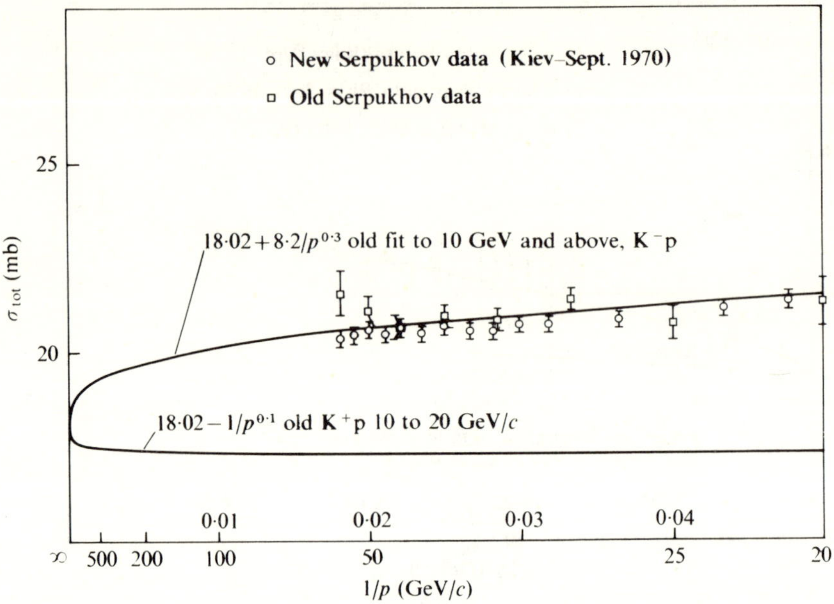


FIG. 13.35.(b) A fit of type I for $\sigma_{\text{tot}}(K^{\pm}-p)$ using the old Serpukhov data, with the new Serpukhov data added.

substantially altered. However, what has been brought into question is the asymptotic form of the behaviour of each individual cross-section, and their sum. The Serpukhov data still, to some degree, implies that all cross-sections may either approach a constant at infinite energy from below, or perhaps increase indefinitely logarithmically as the energy goes to infinity (but probably slower than the Froissart bound), although this has not yet been demonstrated. This eventuality would be consistent within present measurements of $D^+$, for the $\pi^{\pm}$–p, p–p, and $\bar{p}$–p cases. In the case of $K^{\pm}$–p, the situation is considerably less clearcut, since the available accuracy of the measurements is lower and there are no high-energy measurements available for the real part of the forward amplitude.

Although fits consistent with the Pomeranchuk theorem are attainable, the lack of real amplitude measurements prevents realistic estimates of the point where the Pomeranchuk theorem is expected to come true.

*In conclusion* it is clear that considerably higher energies, and many high precision measurements, will be required before we can hope to determine all the features of asymptotic behaviour. Taking into account all the relevant information obtained to date, it appears that a number of reliable pointers indicate that we should try to attain the equivalent of $\gtrsim$100 000 (to perhaps 1 000 000) GeV laboratory energies in order to either reach the asymptotic region or, alternatively, to enter a new domain in which the basic characteristics of high-energy phenomena will change drastically.

## REFERENCES

1. Pomeranchuk, I., Ya. (1956). *Soviet Phys. JETP* **3,** 306.
2. Okun, L. B. and Pomeranchuk, I. Ya. (1956). *Soviet Phys. JETP* **3,** 307.
3. Pomeranchuk, I. Ya. (1958). *Soviet Phys. JETP* **7,** 499.
4. Eden, R. J. (1966). *Phys. Rev. Lett.* **16,** 39.
5. Weinberg, S. (1961). *Phys. Rev.* **124,** 2049.
6. Bessis, J. D. and Kinoshita, T., *CERN Preprint 661/482/5, TH-729.*
7. Eden, R. J. and Lukasak, L. (1967). *Nuovo Cim.* **47,** 817.
8. Von Dardel, G., Mermod, R., Piroue, P. A., Vivargent, M., Weber, G., and Winter, K. (1961). *Phys. Rev. Lett.* **7,** 127.
9. Lindenbaum, S. J., Love, W. A., Niederer, J. A., Ozaki, S., Russell, J. J., and Yuan, L. C. L. (1961). *Phys. Rev. Lett.* **7,** 352.

10*a*. Galbraith, W., Jenkins, E. W., Kycia, T. F., Leontic, P. A., Phillips, R. H., Read, A. L., and Rubinstein, R. (1965). *Phys. Rev.* B**138,** 913.

10*b*. Galbraith, W. (1969). *Rep. Prog. Phys.* **32,** No. 5. pp. 547–606.

11. Lindenbaum, S. J. (1965). *Proceedings of the Oxford International Conference on Elementary Particles.*

12*a*. Kriesler, M. N., Jones, L. W., Longo, M. J., and O'Fallon, J. R. (1968). *Phys. Rev. Lett.* **20,** 468.

12*b*. Engler, J., Horn, K., Konig, J., Monnig, F., Schludlecker, P., Schopper, H., Sievers, P., Ullrich, H., and Runge, K. (1968). *Proceedings of the XIV International Conference on High Energy Physics, Vienna* and *Phys. Lett.* B**27,** 599.

13. Fernbach, S., Serber, R., and Taylor, T. B. (1949). *Phys. Rev.* 1382; Matthews, P. T., and Salam, A. (1961). *Nuovo Cim.* **21,** 823.
14. Lindenbaum, S. J. (1964). *Nucleon Structure.* (edited by Hofstader, R. and Schiff, L.). Stanford University Press, Stanford.
15. Diddens, A., Lillethun, E., Manning, G., Taylor, A., Walker, T. G., and Wetherell, A. M. (1962). *Phys. Rev. Lett.* **9,** 108.
16. Foley, K. J., Lindenbaum, S. J., Love, W. A., Ozaki, S., Russell, J. J., and Yuan, L. C. L. (1963). *Phys. Rev. Lett.* **10,** 376, 543, **11,** 425, 503.
17. Foley, K. J., Gilmore, R. S., Jones, R. S., Lindenbaum, S. J., Love, W. A., Ozaki, S., Willen, E. H., Yamada, R., and Yuan, L. C. L. (1965). *Phys. Rev. Lett.* **15,** 45.
18. Regge, T. (1959). *Nuovo Cim.* **14,** 951.
19. Lovelace, C. and Masson, D. (1962). *Proceedings of the CERN Conference on High Energy Physics.* (1963). *Nuovo Cim.* **26,** 472.
20. Chew, G. F. and Frautschi, S. C. (1961). *Phys. Rev. Lett.* **7,** 394.
21. Blankenbecler, R. and Goldberger, M. L. (1962). *Phys. Rev.* **126,** 766; Chew,

G. F., Frautschi, S. C., and Mandelstam, S. (1962). *Phys. Rev.* **126,** 1202; Frautschi, S. C., Gell-Mann, M., and Zachariasen, F. (1962). *Phys. Rev.* **126,** 2204.

22. Amati, D., Fubini, S., Stanghellini, A., and Tonsin, M. (1961). *Nuovo Cim.* **22,** 569.
23. Udgaonkar, B. M. (1962). *Phys. Rev. Lett.* **8,** 142.
24. Gribov, V. N. and Pomeranchuk, I. Ya. (1962). *Phys. Rev. Lett.* **8,** 346; (1962). **8,** 412.
25. Trueman, T. L. and Wick, G. C. (1964). *Ann. Phys.* 1926, **26,** 322.
26. Singh, V. (1963). *Phys. Rev.* **129,** 1889.
27. Durand, L. (1967). *Phys. Rev.* **154,** 1537; (1968). **166,** 1680; Oakes, R. J. (1967). *Phys. Lett.* B**24,** 154.
28. Toller, M. (1965). *Nuovo Cim.* **37,** 631.

29*a*. Frautschi, S. and Jones, L. (1968). *Phys. Rev.* **167,** 1335; Fox, G. C., Leader, E., and Rogers, W. T. (1967). *Proceedings of the Heidelberg International Conference on High Energy Physics.* paper 107.

29*b*. Arbab, F. and Chiu, C. B. (1966). *Phys. Rev.* **147,** 1045.

30. Hite, G. E. (1969). *Rev. mod. Phys.* **41,** 669.
31. Amati, D., Fubini, S., and Stanghellini, A. (1962). *Phys. Lett.* **1,** 29.
32. Mandelstam, S. (1962). *Nuovo Cim.* **30,** 1127.
33. Pomeranchuk, I., Ya. and Gribov, V. N. (1962). *Phys. Lett.* **2,** 239.
34. Mandelstam, S. and Wang, L. L. (1967). *Phys. Rev.* **160,** 1490; Schwarz, J. (1967). *Phys. Rev.* **162,** 1671.
35. Gervais, J. and Yndurain, F. J. (1968). *Phys. Rev. Lett.* **20,** 27.

36*a*. Wagner, G. and Sharp, D. H. (1962). *Phys. Rev.* **128,** 2899.

36*b*. Wagner, W. G. (1963). *Phys. Rev. Lett.* **10,** 202.

37*a*. Barger, V. and Durand, L. (1967). *Phys. Rev. Lett.* **19,** 1295.

37*b*. Contogouris, A. P., Tran-Thanh-Van, J., and Lubatti, H. J. (1967). *Phys. Rev. Lett.* **19,** 1352.

38*a*. Phillips, R. J. N. and Rarita, W. (1965). *Phys. Rev.* B**139,** 1336.

38*b*. Rarita, W., Riddell, R. J., Chiu, C. B., and Phillips, R. J. N. (1968). *Phys. Rev.* **165,** 1615.

39. Binford, T. O. and Desai, B. R. (1965). *Phys. Rev.* B**138,** 1167.
40. Cabibbo, N., Kokkedee, J. J. J., Horwitz, L., and Ne'eman, Y. (1966). *Nuovo Cim.* A**45,** 275.

41*a*. Manelli, I., Bigi, A., Carrara, R., Wahlig, M., and Sodickson, L. (1965). *Phys. Rev. Lett.* **14,** 408.

41*b*. Stirling, A. V., Sonderegger, P., Kirz, J., Falk-Vairant, P., Guisan, O., Bruneton, C., Borgeaud, P., Yvert, M., Guillaud, J. P., Caverzasio, C., and Amblard, B. (1965). *Phys. Rev. Lett.* **14,** 763. Also (1966). *Phys. Lett.* **20,** 75.

42. Logan, R. (1965). *Phys. Rev. Lett.* **14,** 414.
43. Hohler, G., Baacke, J., Schlaile, H., and Sonderegger, P. (1966). *Phys. Lett.* **20,** 79.
44. Locher, M. P. and Rollnik, H. (1966). *Phys. Lett.* **22,** 696.

45. BYERS, N., and YANG, C. N. (1966). *Phys. Rev.* **142,** 976.
46. BONAMY, P., BORGEAUD, P., BREHIN, S., BRUNETON, C., FALK-VAIRANT, P., GUISAN, O., SONDEREGGER, P., CAVERZASIO, C., GUILLAUD, J. P., SCHNEIDER, J., YVERT, M., MANNELLI, I., SERGIAMPIETRI, F., and VINCELLI, M. L. (1966). *Proceedings of the XIII International Conference on High Energy Physics, Berkeley.* Paper 12a, 23.
47*a*. HOGAASEN, H. and FISCHER, W. (1966). *Phys. Lett.* **22,** 516; LOGAN, R. K., BEAUPRE, J., and SERTORIO, L. (1966). *Phys. Rev. Lett.* **18,** 259; SERTORIO, L., and TOLLER, M. (1967). *Phys. Rev. Lett.* **19,** 1146.
47*b*. DE LANY, V. M., GROSS, D. J., MUZINICH, I. J., and TEPLITZ, V. L. (1967). *Phys. Rev. Lett.* **18,** 149.
47*c*. LOGAN, R. and SERTORIO, L. (1966). *Phys. Rev. Lett.* **17,** 834; DESAI, B., GREGORICH, D., and RAMACHANDRAN, R. (1967). *Phys. Rev. Lett.* **18,** 565.
48. MORRISON, D. R. O. (1966). *Proceedings of the Conference on Two Body Reactions, Stony Brook, New York.*
49. AMATI, D., FUBINI, S., and STANGHELLINI, A. (1962). *Nuovo Cim.* **26,** 896.
50. CHAN, HONG-MO. (1969). *Rep. Prog. Phys.* **32,** p.p. 607–631.
51. BALI, N., CHEW, G. F., and PIGNOTTI, A. (1967). *Phys. Rev. Lett.* **19,** 614; (1967). *Phys. Rev.* **163,** 1572.
52. SCHMID, C. (1968). *Phys. Rev. Lett.* **20,** 689; DOLEN, R., HORN, D., and SCHMID, C. (1968). *Phys. Rev.* **166,** 1768.
53. HARARI, H. (1968). *Proceedings of the XIV International Conference on High Energy Physics, Vienna.*
54. VENEZIANO, G. (1968). *Nuovo Cim.* **57,** 190.
55. BARDECKI, K. and RUEGG, H. (1969). *Phys. Lett.* **28,** 342, 671.
56. CHAN, HONG-MO. (1969). *CERN Th. 1057.*
57. CHAN, HONG-MO. (1968). *Proceedings of the XIV International Conference on High Energy Physics, Vienna.*
58. BALL, J. S., FRAZER, W. R., and JACOB, M. (1968). *Phys. Rev. Lett.* **20,** 518.
59. AACHEN-BERLIN-CERN COLLABORATION. (1968). *Proceedings of the XIV International Conference on High Energy Physics, Vienna.* p. 452; CHAN, HONG-MO, *CERN Preprint, Ref. D. Ph. II, (DROM), M. M.*
60. BARGER, V. and DURAND II., L. (1967). *Phys. Rev. Lett.* **19,** 1298.
61. ARBAB, F., BALL, N. F., and DASH, J. W. (1967). *Phys. Rev.* **158,** 1515.
62. ALFARO, V., FUBINI, S., ROSSETTI, G., and FURLON, G. (1966). *Phys. Lett.* **21,** 576.
63. LOGUNOV, A. A., SOLOVIEV, L. D., and TAVKHELDIGE, A. N. (1967). *Phys. Lett.* B**24,** 181; IGI, K., and MATSUDA, S. (1967). *Phys. Rev. Lett.* **18,** 625.
64. DOLEN, R., HORN, D., and SCHMID, C. (1967). *Phys. Rev. Lett.* **19,** 402.
65*a*. VAN HOVE, L. (p. 253), *b*. WETHERELL, A. (p. 272). (1969). *Proceedings of the XIII International Conference on High Energy Physics, Berkeley.*
66. BELLETTINI, G. (1968). *Proceedings of the XIV International Conference on High Energy Physics, Vienna.*

67. Barger, V. and Cline, D. (1968). *Phys. Rev. Lett.* **21,** 392.
68. Chiu, C. B. and Finkelstein, J. (1968). *CERN Preprint Th. 914.*
69. Chase, R. C., Coleman, E., Courant, H. W. J., Marquit, E., Petraske, E. W., Romer, H., Ruddick, K. (paper 115); Schneider, J., Lepeltier, V., Bonamy, P., Borgeaud, P., Guisan, O., Sonderegger, P. (paper 942); Crouch, H. R. Jr., Hargraves, R., Lanou, R. E. Jr., Massimo, J. T., Pifer, A. E., Shapiro, A. M., Widgoff, M., Brenner, A. E., Ioffredo, M., Rudnick, F. D., Calvelli, G., Gasparini, F., Guerriero, L., Salandin, G. A., Tomasin, A., Voci, C., Waldner, F., Eisenberg, Y., Ronat, E. E., Toaf, S., Bastien, P., Brabson, B., Feld, B. T., Kistiakowsky, V., Goldschmidt-Clermont, Y., Miller, D., Pless, I. A., Rogers, A., Rosenson, L., Ventura, L., Watts, T. L., Yamamoto, R. K. (paper 731). (1968). *Proceedings of the XIV International Conference on High Energy Physics, Vienna.*
70*a*. Ashmore, A., Dammerell, C. J. S., Frisken, W. R., Rubinstein, H., Orear, J., Owen, D. P., Petersen, F. C., Read, A. L., Ryan, D. G., and White, D. H. (1968). *Phys. Rev. Lett.* **21,** 389.
70*b*. Orear, J., Owen, D. P., Peterson, F. C., Read, A. L., Ryan, D. G., White, D. H., Ashmore, A., Dammerell, C. J. S., Frisken, W. R., and Rubinstein, H. (1968). *Proceedings of the XIV International Conference on High Energy Physics, Vienna.* paper 513.
70*c*. Bockmann, K., Nellen, B., Paul, E., Wagini, B., Borecka, I., Diaz, J., Heeren, U., Liebermeister, U., Lohrmann, E., Raubold, E., Soding, P., Wolff, S., Kidd, J., Mandelli, L., Mosca, L., Pelosi, V., Ratti, S., and Tallone, L. (1966). *Nuovo Cim.* A**42,** 954.
70*d*. Foley, K. J., Lindenbaum, S. J., Love, W. A., Ozaki, S., Russell, J. J., and Yuan, L. C. L. (1963). *Phys. Rev. Lett.* **11,** 503.
71. Berryhill, J. and Cline, D. (1968). *Phys. Rev. Lett.* **21**, 770.
72. Baker, W. F., Berkelman, K., Carlson, P. J., Fisher, G. P., Fleury, P., Harthill, D., Kalbach, R., Lundby, A., Mukhin, S., Nierhaus, R., Pretzl, K. P., and Woulds, J. (1968). *Proceedings of the XIV International Conference on High Energy Physics, Vienna.* paper 229.
73. Di Lella, L. (1967). *Proceedings of the Heidelberg International Conference on Elementary Particles.* North Holland Publishing Company, Amsterdam.
74. Allaby, J. V., Binon, F., Diddens, A. N., Duteil, P., Klovning, A., Meunier, R., Peigneux, J. P., Sacharidis, E. J., Schlupmann, K., Spighel, M., Stroot, J. P., Thorndike, A. M., and Wetherell, A. M. (1968). *Proceedings of XIV International Conference on High Energy Physics, Vienna.* paper 605(a); (1968). *Phys. Lett.* B**27,** 49.
75. Breit, G., and Haracz, R. D. (1967). *High energy physics*, Vol. I. Academic Press, New York.
76. Perl, M. L., Cox, J., Longo, M. J., and Kriesler, M. N. (1969). *SLAC PUB 622.*
77. Wu, T. T. and Yang, C. N. (1965). *Phys. Rev.* B**137,** 708.
78. Manning, G., Parham, A. G., Jafar, J. D., Vander Raay, H. B., Reading, D. H., Ryan, D. G., Jones, B. D., Malos, J., and Lipman, N. H. (1966). *Nuovo Cim.* A**41,** 167.

79. ASTBURY, P., FINNOCCHIARO, G., MICHELINI, A., VERKERK, C., WEBSDALE, D., WEST, C. H., BEUSCH, W., GOBBI, B., PEPPIN, M., POUCHON, M. A., and POLGOR, E. (1965). *Phys. Lett.* **16,** 328.

80. MORRISON, D. R. O. (1969). *Proceedings of the Lund International Conference on Elementary Particles;* MATHEWS, R. D. (1969). *Nucl. Phys.* B**11,** 339.

81. ASTBURY, P., BRAUTTI, G., FINOCCHIARO, G., MICHELINI, A., WEBSDALE, D., WEST, C. H., POLGOR, E., BEUSCH, W., FISCHER, W. E., GOBBI, B., and PEPPIN, M. (1967). *Proceedings of the XIV International Conference on High Energy Physics, Berkeley.* paper 12(a), 10.

82. VON ROSSUM, L. (1968). *Proceedings of the Topical Conference on High Energy Collisions of Hadrons, CERN.*

83*a*. BORGHINI, M., COIGNET, C., DICK, L., KURODA, K., DI LELLA, L., MACQ, P. C., MICHAILOWICZ, A., and OLIVIER, J. C. (1967). *Phys. Lett.* B**24,** 77.

83*b*. ESTERLING, R. J., BOOTH, N. E., CONFORTO, G., PARRY, J., SCHEID, J., SHERDEN, D., and YOKASOWA, A. (1968). *Proceedings of the XIV International Conference on High Energy Physics, Vienna,* paper 311.

83*c*. BELL, R. T., BORGHINI, M., DICK, L., GREGOIRE, G., DI LELLA, L., OLIVIER, J. C., POULET, M., SCHARFF-HANSEN, P., CRONENBERGER, D., KURODA, K., MICHAILOWICZ, A., BELLETTINI, G., BRACCINI, P. L., DE PRETE, T., FOA, L., SANGUINETTI, G., and VALDATA, M. (1968). *Proceedings of the XIV International Conference on High Energy Physics, Vienna,* paper 239.

83*d*. CHIU *et al.* (1966). *UCRL 16940.*

84. BONAMY, P., BORGEAUD, P., BRUNETON, C., FALK-VARIANT, P., GUISAN, O., SONDEREGGER, P., CAVERZASIO, C., GUILLAUD, J. P., SCHNEIDER, J., YVERT, M., MANNELLI, I., SERGIAMPIETRI, L., and VINCELLI, M. L. (1966). *Phys. Lett.* **23,** 501.

85. BOOTH, N. E., (1968). *Phys. Rev. Lett.* **21,** 465; (1968). *Proceedings of the XIV International Conference on High Energy Physics, Vienna.* Paper 239.

86*a*. LINDENBAUM, S. J. (1967). *Symmetry principles at high energy,* W. H. Freeman and Co., San Francisco. pp. 122–164; *b*. (1969). *Pion-nucleon scattering.* (Edited by G. L. Shaw and D. Y. Wong), John Wiley and Sons, New York. pp. 81–113; *c*. (1970). Experimental verification of dispersion theory and high energy scattering, *Proceedings of the Symposium on the Past Decade in Particle Theory, University of Texas, Austin; d.* (1969). *Contemporary Physics Trieste Symposium.* Vol. II. International Atomic Energy Agency, Vienna. pp. 123–143.

87. ALLABY, J. V., BUSHNIN, YU., B., DENISOV, S. P., DIDDENS, A. N., DOBINSON, R. W., DONSKOV, S. V., GIACOMELLI, G., GORIN, YU. P., KLOVNING, A., PETRUKHIN, A. I., PROKOSHKIN, YU. D., SHUVALOV, R. S., STAHLBRANDT, C. A., and STOYANOVA, D. A. (1969). *Phys. Lett.* B**30,** 500.

88. BARGER, V. and PHILLIPS, R. J. N. (1970). *Phys. Rev. Lett.* **24,** 291.

89. GRIBOV, V. N. (1968). *Soviet Phys. JETP* **26,** 44; GRIBOV, V. N., *et al.* (1965). *Physica* **2,** 361; FRAUTSCHI, S. and MARGOLIS, B. (1968). *Nuovo Cim.* A**56,** 1155.

90. BEZNOGIKH, G. G., BUJAK, A., IOVCHEV, K. I., KIRILLOVA, L. F., MARKOV, P. K., MOROZOV, B. A., NIKITIN, V. A., NOMOKONOV, P. V., SHAFRANOVA, M. G., SVIRIDOV, V. A., TRUONG, B., ZAYACHKI, V. I., ZHIDDKOV, N. K., ZOLIN, L. S.

(Joint institute for nuclear research, Dubna) and NURUCHEV, S. B., and SOLOVIANOV, V. P. (IHEP Serpukhov). (1969). *Proceedings of the Lund International Conference on Elementary Particles.*

91. VASELEV, L. M., GORIN, YU. P., DENISOV, S. P., DIMITREVSKI, Y. P., DONSKOV, S. V., MELNIK, YU, M., PETRUKHIN, A. I., PROKOSHKIN, YU, D., SELZENEV, V. S., STROYANOVA, D., and SHUVAVLOV, R. (1970). *Proceedings of the XV International Conference on High Energy Physics, Kiev, USSR.*

92*a*. VAN HOVE, L. (1964). *Rev. mod. Phys.* **36,** 655; *b*. COTTINGHAM, W. N. and PEIERLS, R. F. (1965). *Phys. Rev.* **B137,** 147.

# APPENDIX

Excerpts from Review of Particle Properties by the Particle Data Group, *Physics Letters*, Vol. 39B, (1972),* published by North-Holland Publishing Co.

1

# TABLES OF PARTICLE PROPERTIES

April 1972

"Now go, write it before them in a table, and note it in a book,
that it may be for the time to come for ever and ever."
Isaiah 30:7-8

". . . or at least until the next edition."
Particle Data Group

N. Barash-Schmidt, A. Barbaro-Galtieri, J. Bartels, C. Bricman, V. Chaloupka, J. E. Enstrom, T. A. Lasinski, A. Rittenberg, M. Roos, A. H. Rosenfeld, P. Söding, and T. G. Trippe

(Closing date for data: Feb. 1, 1972)

## Stable Particle Table

For additional parameters, see Addendum to this table.

*Quantities in italics have changed by more than one (old) standard deviation since April 1971.*

| Particle | $I^G(J^P)C_n$ | Mass (MeV) Mass² (GeV)² | Mean life (sec) $c\tau$ (cm) | Partial decay mode: Mode | Fraction[a] | p or $p_{max}$[b] (MeV/c) |
|---|---|---|---|---|---|---|
| $\gamma$ | $0,1(1^-)^-$ | $0(<2)10^{-21}$ | stable | stable | | |
| $\nu$ $\nu_e$ | $J=\frac{1}{2}$ | 0(<60 eV) | stable | stable | | |
| $\nu_\mu$ | | 0(<1.2) | | | | |
| e | $J=\frac{1}{2}$ | 0.5110041 | stable | stable | | |
| | | ±.0000016 | $(>2\times10^{21}$y) | | | |
| $\mu$ | $J=\frac{1}{2}$ | 105.6594 | *$2.1994\times10^{-6}$* | $e\nu\bar{\nu}$ | 100 | 53 |
| | | ±.0004 | *±.0006* S=1.1* | $e\gamma\gamma$ | $(<1.6)10^{-5}$ | 53 |
| | | $m^2=0.0112$ | $c\tau=6.593\times10^4$ | 3e | $(<6)10^{-9}$ | 53 |
| | | $m_\mu-m_{\pi^\pm}=-33.917$ | | $e\gamma$ | $(<2.2)10^{-8}$ | 53 |
| | | ±.011 | | | | |
| $\pi^\pm$ | $1^-(0^-)$ | 139.576 | $2.6024\times10^{-8}$ | $\mu\nu$ | 100 % | 30 |
| | | ±.011 | ±.0024 | $e\nu$ | $(1.24\pm0.03)10^{-4}$ | 70 |
| | | $m^2=0.0195$ | $c\tau=780.2$ | $\mu\nu\gamma$ | $^c(1.24\pm0.25)10^{-4}$ | 30 |
| | | | $(\tau^+-\tau^-)/\bar{\tau}=$ | $\pi^0e\nu$ | $(1.02\pm0.07)10^{-8}$ | 5 |
| | | | (0.05±0.07)% | $e\nu\gamma$ | $^c(3.0\pm0.5)10^{-8}$ | 70 |
| | | | (test of CPT) | $e\nu e^+e^-$ | $(<3.4)10^{-8}$ | 70 |
| $\pi^0$ | $1^-(0^-)^+$ | 134.972 | $0.84\times10^{-16}$ | $\gamma\gamma$ | (98.84±0.04)% | 67 |
| | | ±.012 | ±.10 S=2.1* | $\gamma e^+e^-$ | (1.16±0.04)% | 67 |
| | | $m^2=0.0182$ | $c\tau=2.5\times10^{-6}$ | $\gamma\gamma\gamma$ | $(<5)10^{-6}$ | 67 |
| | | $m_{\pi^\pm}-m_{\pi^0}=4.6043$ | | $e^+e^-e^+e^-$ | $^d(3.47)10^{-5}$ | 67 |
| | | ±.0037 | | | | |

* *This updated version of the P.D.G. Tables was inserted while the book was in production,* for the readers' convenience for use as a reference source. However, the reader should note that in the book many references to the P.D.G. Tables refer to the earlier versions (January 1970 and August 1970) which differ somewhat from this version.

# Stable Particle Table *(cont'd)*

| Particle | $I^G(J^P)C_n$ | Mass (MeV) Mass$^2$ (GeV)$^2$ | Mean life (sec) $c\tau$ (cm) | Partial decay mode: Mode | Fraction[a] | | p or $p_{max}$[b] (MeV/c) |
|---|---|---|---|---|---|---|---|
| $K^{\pm}$ | $\frac{1}{2}(0^-)$ | 493.84 | $1.2371\times10^{-8}$ | $\mu\nu$ | (63.77±0.28)% | S=1.1* | 236 |
| | | ±0.10 | ±.0026 S=1.9* | $\pi\pi^0$ | (20.92±0.29)% | S=1.2* | 205 |
| | | | $c\tau$=370.8 | $\pi\pi^-\pi^+$ | (5.58±0.03)% | S=1.1* | 126 |
| | | $m^2$=0.244 | $(\tau^+-\tau^-)/\bar{\tau}$= | $\pi\pi^0\pi^0$ | (1.68±0.04)% | | 133 |
| | | | (.11±.09)% | $\mu\pi^0\nu$ | (3.20±0.11)% | S=1.8* | 215 |
| | | | (test of CPT) | $e\pi^0\nu$ | (4.86±0.07)% | S=1.1* | 228 |
| | | | S=1.2* | $e\pi^0\pi^0\nu$ | $(1.8^{+2.4}_{-0.6})\ 10^{-5}$ | | 207 |
| | | $m_{K^\pm}-m_{K^0}$=−3.95 | | $\pi\pi^{\mp}e^{\pm}\nu$ | *(3.7 ±0.2) $10^{-5}$* | | 204 |
| | | ±0.13 | | $\pi\pi^{\pm}e^{\mp}\nu$ | (<5) $10^{-7}$ | | 204 |
| | | S=1.1* | | $\pi\pi^{\mp}\mu^{\pm}\nu$ | (0.9 ±0.4) $10^{-5}$ | | 151 |
| | | | | $\pi\pi^{\pm}\mu^{\mp}\nu$ | (<3) $10^{-6}$ | | 151 |
| | | | | $e\nu$ | (1.30±0.18) $10^{-5}$ | | 247 |
| | | | | $e\nu\gamma$ | c (<7) $10^{-5}$ | | 247 |
| | | | | $\pi\pi^0\gamma$ | c (2.2 ±0.7) $10^{-4}$ | | 205 |
| | | | | $\pi\pi^+\pi^-\gamma$ | c (10 ±4) $10^{-5}$ | | 126 |
| | | | | $\pi e\nu\gamma$ | c (3.7 ±1.4) $10^{-4}$ | | 227 |
| | | | | $\pi e^+e^-$ | (< 0.4) $10^{-6}$ | | 227 |
| | | | | $\pi^{\mp}e^{\pm}e^{\pm}$ | (< 1.5) $10^{-5}$ | | 227 |
| | | | | $\pi\mu^+\mu^-$ | (< 2.4) $10^{-6}$ | | 172 |
| | | | | $\pi\gamma\gamma$ | c (< 3.5) $10^{-5}$ | | 227 |
| | | | | $\pi\gamma\gamma\gamma$ | c (< 3) $10^{-4}$ | | 227 |
| | | | | $\pi\nu\bar{\nu}$ | (< 1.2) $10^{-6}$ | | 227 |
| | | | | $\pi\gamma$ | (< 4) $10^{-6}$ | | 227 |
| $K^0$ | $\frac{1}{2}(0^-)$ | 497.79 | 50% $K_{Short}$, 50% $K_{Long}$ | | | | |
| | | ±0.15 | | | | | |
| $K^0_S$ | $\frac{1}{2}(0^-)$ | S=1.1* | $0.862\times10^{-10}$ | $\pi^+\pi^-$ | (68.85 ±0.31)% | S=1.1* | 206 |
| | | $m^2$=0.248 | ±.006 S=1.2* | $\pi^0\pi^0$ | (31.15 ±0.31)% | S=1.1* | 209 |
| | | | $c\tau$=2.58 | $\mu^+\mu^-$ | (< .7) $10^{-5}$ | | 225 |
| | | | | $e^+e^-$ | (<35) $10^{-5}$ | | 249 |
| | | | | $\pi^+\pi^-\gamma$ | c (2.3 ±0.8) $10^{-3}$ | | 206 |
| | | | | $\gamma\gamma$ | (< 2) $10^{-3}$ | | 249 |
| $K^0_L$ | $\frac{1}{2}(0^-)$ | | $5.172\times10^{-8}$ | $\pi^0\pi^0\pi^0$ | (21.4 ±0.7)% | S=1.1* | 139 |
| | | | ±0.042 | $\pi^+\pi^-\pi^0$ | (12.6 ±0.3)% | | 133 |
| | | | $c\tau$=1550 | $\pi\mu\nu$ | (26.8 ±0.6)% | | 216 |
| | | | | $\pi e\nu$ | (39.0 ±0.6)% | | 229 |
| | | | | $\pi e\nu\gamma$ | c (1.3 ±0.8)% | | 229 |
| | | $m_{K_L}-m_{K_S}$ = $0.5402\times10^{10}\,\hbar\,sec^{-1}$ | | $\pi^+\pi^-$ | (0.157±0.005)% | | 206 |
| | | ± 0.0035 | | $\pi^0\pi^0$ | (0.094±0.019)% | S=1.5* | 209 |
| | | | | $\pi^+\pi^-\gamma$ | c (< 0.4) $10^{-3}$ | | 206 |
| | | | $\frac{\Gamma(K_S\to\pi^+\pi^-\pi^0)}{\Gamma(K_L\to\pi^+\pi^-\pi^0)}$ <0.45 | $\pi^0\gamma\gamma$ | (< 2.4) $10^{-4}$ | | 231 |
| | | | (test of CP) | $\gamma\gamma$ | *(4.9 ±0.4) $10^{-4}$* | | 249 |
| | | | | $e\mu$ | (< 1.6) $10^{-9}$ | | 238 |
| | | | | $\mu^+\mu^-$ | (< 1.9) $10^{-9}$ | | 225 |
| | | | | $e^+e^-$ | (< 1.6) $10^{-9}$ | | 249 |
| $\eta$ | $0^+(0^-)^+$ | 548.8 | $\Gamma$=(2.63±0.58)keV | $\gamma\gamma$ | (38.0 ±1.0)% | S=1.2* | 274 |
| | | ±0.6 | Neutral decays 71.1% | $\pi^0\gamma\gamma$ | e (3.1 ±1.1)% | S=1.2* | 258 |
| | | S=1.4* | | $3\pi^0$ | (30.0 ±1.1)% | S=1.1* | 180 |
| | | $m^2$=0.301 | Charged decays 28.9% | $\pi^+\pi^-\pi^0$ | (24.0 ±0.6)% | S=1.1* | 175 |
| | | | | $\pi^+\pi^-\gamma$ | (4.9 ±0.2)% | S=1.1* | 236 |
| | | | | $\pi^0e^+e^-$ | (< 0.04)% | | 258 |
| | | | | $\pi^+\pi^-e^+e^-$ | (0.1 ±0.1)% | | 236 |
| | | | | $\pi^+\pi^-\pi^0\gamma$ | (< 0.2)% | | 175 |
| | | | | $\pi^+\pi^-\gamma\gamma$ | (< 0.2)% | | 236 |
| | | | | $\mu^+\mu^-$ | (2.2 ±0.8) $10^{-5}$ | | 253 |
| | | | | $\mu^+\mu^-\pi^0$ | (< 5) $10^{-4}$ | | 211 |
| p | $\frac{1}{2}(\frac{1}{2}^+)$ | 938.2592 | stable | | | | |
| | | ±0.0052 | (> $2\times10^{28}$y) | | | | |
| | | $m^2$=0.8803 | | | | | |
| n | $\frac{1}{2}(\frac{1}{2}^+)$ | 939.5527 | e $(0.935\pm0.014)10^3$ | $pe^-\nu$ | 100 % | | 1 |
| | | ±0.0052 | $c\tau$ = $2.80\times10^{13}$ | | | | |
| | | $m^2$=0.8828 | | | | | |
| | | $m_p-m_n$=−1.29344 | | | | | |
| | | ±0.00007 | | | | | |

# Stable Particle Table *(cont'd)*

| Particle | $I^G(J^P)C_n$ | Mass (MeV) / Mass² (GeV)² | Mean life (sec) / cτ (cm) | Partial decay mode: Mode | Fraction[a] | p or $p_{max}$[b] (MeV/c) |
|---|---|---|---|---|---|---|
| Λ | $0(\frac{1}{2}^+)$ | 1115.59 | $2.521\times10^{-10}$ | $p\pi^-$ | $(64.2\pm0.5)\%$ | 100 |
| | | ±0.05 | ±.021 S=1.2* | $n\pi^0$ | $(35.8\pm0.5)\%$ | 104 |
| | | S=1.2* | cτ = 7.56 | $pe\nu$ | $(8.13\pm0.29)10^{-4}$ | 163 |
| | | $m^2$ = 1.245 | | $p\mu\nu$ | $(1.62\pm0.35)10^{-4}$ | 131 |
| $\Sigma^+$ | $1(\frac{1}{2}^+)$ | 1189.42 | $0.800\times10^{-10}$ | $p\pi^0$ | $(51.6\pm0.7)\%$ | 189 |
| | | ±0.11 | ±.006 | $n\pi^+$ | $(48.4\pm0.7)\%$ | 185 |
| | | S= 1.7* | cτ = 2.40 | $p\gamma$ | c$(1.24\pm0.18)10^{-3}$ S=1.4* | 225 |
| | | $m^2$ = 1.415 | | $n\pi^+\gamma$ | c$(1.31\pm0.24)10^{-4}$ | 185 |
| | | $m_{\Sigma^+}-m_{\Sigma^-}$ = -7.92 | | $\Lambda e^+\nu$ | $(2.02\pm0.47)10^{-5}$ | 72 |
| | | ±.13 | $\frac{\Gamma(\Sigma^+\to\ell^+n\nu)}{\Gamma(\Sigma^-\to\ell^-n\nu)} < .035 \leftarrow$ | $n\mu^+\nu$ | $(<2.4)10^{-5}$ | 202 |
| | | S= 1.6* | | $ne^+\nu$ | $(<1.0)10^{-5}$ | 224 |
| | | | | $pe^+e^-$ | $(<7)10^{-6}$ | 225 |
| $\Sigma^0$ | $1(\frac{1}{2}^+)$ | 1192.48 | $<1.0\times10^{-14}$ | $\Lambda\gamma$ | 100 % | 74 |
| | | ±0.11 | $c\tau<3\times10^{-4}$ | $\Lambda e^+e^-$ | d$(5.45)10^{-3}$ | 74 |
| | | S= 1.3* | | | | |
| | | $m^2$=1.422 | | | | |
| $\Sigma^-$ | $1(\frac{1}{2}^+)$ | 1197.34 | e$1.484\times10^{-10}$ | $n\pi^-$ | 100 % | 193 |
| | | ±0.10 | ±.019 S=1.6* | $ne^-\nu$ | $(1.10\pm0.05)10^{-3}$ | 230 |
| | | S= 1.5* | cτ = 4.45 | $n\mu^-\nu$ | $(0.45\pm0.04)10^{-3}$ | 210 |
| | | $m^2$ = 1.434 | | $\Lambda e^-\nu$ | $(0.60\pm0.06)10^{-4}$ | 79 |
| | | $m_{\Sigma^0}-m_{\Sigma^-}$=-4.86 | | $n\pi^-\gamma$ | c$(1.0\pm0.2)10^{-4}$ | 193 |
| | | ±.06 | | | | |
| $\Xi^0$ | $\frac{1}{2}(\frac{1}{2}^+)^f$ | 1314.7 | $3.03\times10^{-10}$ | $\Lambda\pi^0$ | 100 % | 135 |
| | | ±0.7 | ±.18 | $p\pi^-$ | $(<0.9)10^{-3}$ | 299 |
| | | $m^2$= 1.729 | cτ = 9.08 | $pe^-\nu$ | $(<1.3)10^{-3}$ | 323 |
| | | $m_{\Xi^0}-m_{\Xi^-}$ = -6.6 | | $\Sigma^+e^-\nu$ | $(<1.5)10^{-3}$ | 119 |
| | | ±.7 | | $\Sigma^-e^+\nu$ | $(<1.5)10^{-3}$ | 112 |
| | | | | $\Sigma^+\mu^-\nu$ | $(<1.5)10^{-3}$ | 64 |
| | | | | $\Sigma^-\mu^+\nu$ | $(<1.5)10^{-3}$ | 49 |
| | | | | $p\mu^-\nu$ | $(<1.3)10^{-3}$ | 309 |
| $\Xi^-$ | $\frac{1}{2}(\frac{1}{2}^+)^f$ | 1321.30 | $1.660\times10^{-10}$ | $\Lambda\pi^-$ | 100 % | 139 |
| | | ±0.15 | ±.037 S=1.1* | $\Lambda e^-\nu$ | g$(0.70\pm0.21)10^{-3}$ | 190 |
| | | $m^2$= 1.746 | cτ = 4.98 | $\Sigma^0e^-\nu$ | $(<0.5)10^{-3}$ | 123 |
| | | | | $\Lambda\mu^-\nu$ | $(<1.3)10^{-3}$ | 163 |
| | | | | $\Sigma^0\mu^-\nu$ | $(<0.5)\%$ | 70 |
| | | | | $n\pi^-$ | $(<1.1)10^{-3}$ | 303 |
| | | | | $ne^-\nu$ | $(<1.0)\%$ | 327 |
| $\Omega^-$ | $0(\frac{3}{2}^+)^f$ | 1672.5±.5 | $1.3^{+0.4}_{-0.3}\times10^{-10}$ | $\Xi^0\pi^-$ | Total of | 294 |
| | | $m^2$= 2.797 | cτ = 3.9 | $\Xi^-\pi^0$ | 28 events | 290 |
| | | | | $\Lambda K^-$ | seen | 21[illegible] |

*S = Scale factor = $\sqrt{\chi^2/(N-1)}$, where N ≈ number of experiments. S should be ≈ 1. If S > 1, we have enlarged the error of the mean, δx, i.e., δx→Sδx. This convention is still inadequate, since if S >> 1, the experiments are probably inconsistent, and therefore the real uncertainty is probably even greater than Sδx. See text and ideogram in Stable Particle Data Card Listings.

a. Quoted upper limits correspond to a 90% confidence level.

b. In decays with more than two bodies, $P_{max}$ is the maximum momentum that any particle can have.

c. See Stable Particle Data Card Listings for energy limits used in this measurement.

d. Theoretical value; see also Stable Particle Data Card Listings.

e. See note in Stable Particle Data Card Listings.

f. P for Ξ and $J^P$ for $\Omega^-$ not yet measured. Values reported are SU(3) predictions.

g. Assumes rate for $\Xi^-\to\Sigma^0e^-\nu$ small compared with $\Xi^-\to\Lambda e^-\nu$.

ADDENDUM TO

# Stable Particle Table

**Magnetic moment**

**e**: $1.001\ 159\ 6577 \pm .000\ 000\ 0035$ $\dfrac{e\hbar}{2m_e c}$

**μ**: $1.001\ 166\ 16 \pm .000\ 000\ 31$ $\dfrac{e\hbar}{2m_\mu c}$

**μ Decay parameters** [a]

$\rho = 0.752 \pm 0.003$  $\eta = -0.12 \pm 0.21$

$\xi = 0.972 \pm 0.013$  $\delta = 0.755 \pm 0.009$  $h = 1.00 \pm 0.13$

$|g_A/g_V| = 0.86^{+0.33}_{-0.11}$  $\phi = 180° \pm 15°$

## $K^{\pm}$

| Mode | Partial rate $(sec^{-1})$ | |
|---|---|---|
| $\mu\nu$ | $(51.55 \pm 0.25)10^6$ | $S=1.2^*$ |
| $\pi\pi^0$ | $(16.91 \pm 0.24)10^6$ | $S=1.2^*$ |
| $\pi\pi^+\pi^-$ | $(4.51 \pm 0.02)10^6$ | $S=1.1^*$ |
| $\pi\pi^0\pi^0$ | $(1.36 \pm 0.04)10^6$ | |
| $\mu\pi^0\nu$ | $(2.59 \pm 0.09)10^6$ | $S=1.8^*$ |
| $e\pi^0\nu$ | $(3.92 \pm 0.06)10^6$ | $S=1.2^*$ |

**$\Delta I = \frac{1}{2}$ rule for $K^{\pm} \to 3\pi$**

$\pi^+\pi^+\pi^-$ $c_g = -.206 \pm .007$

$\pi^-\pi^+\pi^-$ $c_g = -.194 \pm .007$

$\pi^+\pi^0\pi^0$ $c_g = .527 \pm .017$

See also Stable Particle Data Card Listings and Appendix I

**Form factors for leptonic decays**

See Stable Particle Data Card Listings for λ and ξ

## $K^0_S$

| Mode | Partial rate $(sec^{-1})$ | |
|---|---|---|
| $\pi^+\pi^-$ | $(0.799 \pm .006)10^{10}$ | $S=1.2^*$ |
| $\pi^0\pi^0$ | $(0.361 \pm .004)10^{10}$ | $S=1.2^*$ |

## $K^0_L$

| Mode | Partial rate $(sec^{-1})$ | |
|---|---|---|
| $\pi^0\pi^0\pi^0$ | $(4.13 \pm 0.13)10^6$ | $S=1.1^*$ |
| $\pi^+\pi^-\pi^0$ | $(2.43 \pm 0.06)10^6$ | |
| $\pi\mu\nu$ | $(5.18 \pm 0.12)10^6$ | |
| $\pi e\nu$ | $(7.54 \pm 0.13)10^6$ | |
| $\pi^+\pi^-$ | $(3.03 \pm 0.10)10^4$ | |
| $\pi^0\pi^0$ | $(1.81 \pm 0.36)10^4$ | $S=1.5^*$ |

**CP violation parameters**

$|\eta_{+-}| = (1.96 \pm 0.03)10^{-3}$, $\phi_{+-} = (43 \pm 3)°$

$|\eta_{00}| = (2.09 \pm 0.12)10^{-3}$, $\phi_{00} = (43 \pm 19)°$ $S=1.2^*$

Charge asymmetry:

$$\delta = \frac{\Gamma(K^0_L \to \ell^+) - \Gamma(K^0_L \to \ell^-)}{\Gamma(K^0_L \to \ell^+) + \Gamma(K^0_L \to \ell^-)} = .32 \pm .03 \times 10^{-2}$$

**$I = \frac{1}{2}$ rule for $K^0_L \to 3\pi$**

$\pi^+\pi^-\pi^0$ $c_g = .60 \pm .03$ $S=3.1^*$

See Data Cards & App. I

**$\Delta S = -\Delta Q$**

Re $x = -.003 \pm .026$  $S=1.5^*$

Im $x = -.007 \pm .039$  $S=1.2$

**Form factors for leptonic decays**

See Stable Particle Data Card Listings for λ, ξ

## η

| Mode | Asymmetry parameter |
|---|---|
| $\pi^+\pi^-\pi^0$ | $(1.2 \pm 0.5)\%$ $S=1.3^*$ |
| $\pi^+\pi^-\gamma$ | $(1.1 \pm 1.3)\%$ |

| | Magnetic moment $(e\hbar/2m_p c)$ | Decay parameters [b] | Measured α | Measured φ(degree) | Derived γ | Derived Δ(degree) | $g_A/g_V$ [b] | $g_V/g_A$ [b] |
|---|---|---|---|---|---|---|---|---|
| p | 2.792782 ±.000017 | | | | | | | |
| n | -1.913148 ±.000066 | $pe^-\nu$ | | | | | -1.242±0.008 $S=1.2^*$ [δ=(178.6±0.9)°] | |
| Λ | -0.67 ±.06 | $p\pi^-$ | 0.645±0.016 | (-6.3±3.5)° | 0.76 | $(7.4^{+4.0}_{-4.1})°$ | | |
| | | $n\pi^0$ | 0.649±0.046 | | | | | |
| | | $pe\nu$ | | | | | -0.66±0.06 $S=1.2^*$ | |
| $\Sigma^+$ | 2.59 ±.46 | $p\pi^0$ | -0.991±0.019 | (22±90)° | 0.12 | $(183^{+11}_{-12})°$ | | |
| | | $n\pi^+$ | +0.066±0.016 | (167±20)° $S=1.1^*$ | -0.97 | $(-73^{+136}_{-10})°$ | | |
| | | $p\gamma$ | $-1.03^{+.52}_{-.42}$ | | | | | |
| $\Sigma^-$ | | $n\pi^-$ | -0.069±0.008 | (10±15)° | 0.98 | $(249^{+12}_{-115})°$ | | |
| | | $ne^-\nu$ | | | | | See Data Cds. | |
| | | $\Lambda e^-\nu$ | | | | | | 0.35±0.18 |
| $\Xi^0$ | | $\Lambda\pi^0$ | -0.35±0.08 | (25±21)° $S=1.3^*$ | 0.85 | $(228^{+16}_{-37})°$ | | |
| $\Xi^-$ | | $\Lambda\pi^-$ | -0.40±0.03 | (-4±8)° $S=1.1^*$ | 0.91 | $(170^{+18}_{-17})°$ | | |

*S = scale factor. Quoted error includes scale factor; see footnote to main Stable Particle Table for definition.

a. $|g_A/g_V|$ defined by $g_A^2 = |C_A|^2 + |C'_A|^2$, $g_V^2 = |C_V|^2 + |C'_V|^2$, and $\Sigma\langle \bar{e}|\Gamma_i|\mu\rangle\langle\bar{\nu}|\Gamma_i(C_i + C'_i\gamma_5)|\nu\rangle$;

$\phi$ defined by $\cos\phi = -R_e(C^*_A C'_V + C'_A C^*_V)/g_A g_V$ [for more details, see text Section IV E]

b. The definition of these quantities is as follows [for more details on sign convention, see text Section IV H]:

$$\alpha = \frac{2|s||p|\cos\Delta}{|s|^2+|p|^2}; \quad \beta = \sqrt{1-\alpha^2}\sin\phi; \quad g_A/g_V \text{ defined by } \langle B_f|\gamma_\lambda(g_V - g_A\gamma_5)|B_i\rangle;$$

$$\beta = \frac{-2|s||p|\sin\Delta}{|s|^2+|p|^2}. \quad \gamma = \sqrt{1-\alpha^2}\cos\phi. \quad \delta \text{ defined by } g_A/g_V = |g_A/g_V|e^{i\delta}.$$

c. The definition of the slope parameter of the Dalitz plot is as follows: $|M|^2 = 1 + g\left(\frac{s_3 - s_0}{m^2_{\pi^+}}\right)$.

# Meson Table

April 1972

*Quantities in italics have changed by more than one (old) standard deviation since April 1971*

| Name | $I^G(J^P)C_n$ estab. | Mass M (MeV) | Full Width Γ (MeV) | $M^2$ $\pm\Gamma M^{(a)}$ $(GeV)^2$ | Partial decay mode: Mode | Fraction % | p or $p_{max}^{(b)}$ (MeV/c) |
|---|---|---|---|---|---|---|---|
| $\pi^{\pm}(140)$ | $1^-(0^-)+$ | 139.58 | 0.0 | 0.019483 | See Stable Particle Table | | |
| $\pi^0(135)$ | | 134.97 | 7.8 eV ±.9 eV | 0.018217 | | | |
| η(549) | $0^+(0^-)+$ | 548.8 ±0.6 | 2.63 keV ±.58 keV | 0.301 ±.000 | All neutral | 71 | See Stable Particle Table |
| | | | | | $\pi^+\pi^-\pi^0 + \pi^+\pi^-\gamma$ | 29 | |
| ε | $0^+(0^+)+$ | | | | $\delta_0^0$ is near 80°-90° in mass region 800-1000 MeV, with probably only slow variation below and cusp at $K\bar{K}$ threshold. Inelasticity ≈0 below 2 $m_K$. | | |
| See note on ππ S wave¶. | | | | | | | |
| ρ(765) | $1^+(1^-)-$ | 765 ±10 (c) | 135 ±20 (c) | 0.585 ±.103 | ππ | ≈100 | 356 |
| | | | | | $e^+e^-$ | *0.0042±.0004* (d) | 382 |
| | | | | | $\mu^+\mu^-$ | 0.0067±.0012 (d) | 368 |
| | | | | | For upper limits, see footnote (e) | | |
| ω(784) | $0^-(1^-)-$ | 783.9 ±0.3 S=1.3* | *10.0 ±0.6* | 0.614 ±.008 | $\pi^+\pi^-\pi^0$ | 89.7±4.0 S=1.1* | 328 |
| | | | | | $\pi^+\pi^-$ | *1.2±0.3 S=1.4** | 366 |
| | | | | | $\pi^0\gamma$ | 9.0±1.0 | 380 |
| | | | | | $e^+e^-$ | 0.0075±.0016 S=1.8* | 392 |
| | | | | | For upper limits, see footnote (f) | | |
| η′(958) or $X^0$ | $0^+(\ ^-)+$ $J^P = 0^-$ or $2^-$ | 957.1 ±0.6 | < 4 | 0.916 <.004 | ηππ | 68.1±2.2 | 230 |
| | | | | | $\pi^+\pi^-\gamma$ (mainly $\rho^0\gamma$) | 30.1±2.3 S=1.1* | 458 |
| | | | | | γγ | *1.8±0.3* | 479 |
| | | | | | For upper limits, see footnote (h) | | |
| $\pi_N(975)$ | $1^-(0^+)+$ | ~ 975 | ~ 60 | 0.950 | ηπ (g) | | 304 |
| Possibly related to the I=1 $K\bar{K}$ threshold enhancement | | | | | | | |
| $S^*$ | $0^+(0^+)+$ | ~ 1000 | | 1.000 | Seen as I=0 $K\bar{K}$ threshold enhancement; appears coupled to the ππ channel. | | |
| See notes on ππ and $K\bar{K}$ S wave¶. | | | | | | | |
| Φ(1019) | $0^-(1^-)-$ | 1019.1 ±0.5 S=1.8* | *4.4 ±.3* | 1.039 ±.004 | $K^+K^-$ | 49.1±2.0 | 126 |
| | | | | | $K_LK_S$ | *30.7±2.4 S=1.4** | 109 |
| | | | | | $\pi^+\pi^-\pi^0$ (incl. ρπ) | 17.5±2.5 S=1.1* | 461 |
| | | | | | ηγ | *2.6±1.2 S=1.9** | 362 |
| | | | | | $e^+e^-$ | .032±.003 S=1.9* | 510 |
| | | | | | $\mu^+\mu^-$ | .025±.003 | 498 |
| | | | | | For upper limits, see footnote (i) | | |
| $A_1(1070)$ | $1^-(1^+)+$ | ~ 1070 | | 1.14 | ρπ | ~ 100 | 232 |
| Broad enhancement in the $J^P=1^+$ ρπ partial wave; not clear if resonant¶. | | | | | | | |
| B(1235) | $1^+(1^+)-$ | 1233 ±10 § | 100 ±20 § | 1.52 ±.12 | ωπ | ≈ 100 | 348 |
| | | | | | ππ | < 30 (Absence suggests $J^P$ = Abnormal) | 600 |
| | | | | | $K\bar{K}$ | < 2 (Absence suggests $J^P$ = Abnormal) | 369 |
| | | | | | For other upper limits, see footnote (j) | | |
| f(1260) | $0^+(2^+)+$ | 1269 ±10 § | 156 ±25 § | 1.60 ±.20 | ππ | ≈ 80 | 619 |
| | | | | | $2\pi^+2\pi^-$ | 6±2 | 556 |
| | | | | | $K\bar{K}$ | ≈ 6 | 393 |

# Meson Table *(cont'd)*

| Name | $I^G(J^P)C_n$ estab. | Mass M (MeV) | Full Width Γ (MeV) | $M^2$ $\pm\Gamma M$[a] $(GeV)^2$ | Partial decay mode: Mode | Fraction % | | p or $p_{max}$[b] (MeV/c) |
|---|---|---|---|---|---|---|---|---|
| D(1285) | $0^+$(A )+ | 1286 | *21* | 1.65 | $\eta\pi\pi$ | Probably seen | | 484 |
| | | ±4 | *±10* | ±.03 | $K\bar{K}\pi$ | Seen | | 305 |
| | | S=1.3* | *S=1.3** | | †[$\pi_N(975)\pi$ | Seen] | | 246 |
| | $J^P = 0^-, 1^+, 2^-$, with $1^+$ favoured | | | | $2\pi^+2\pi^-$ (prob. $\rho^0\pi^+\pi^-$) | Seen | | 565 |
| $A_2(1310)$ | $1^-(2^+)+$ | 1310§ | 100§ | 1.72 | $\rho\pi$ | 76.8±1.8 | | 416 |
| | | ±10 | ±20 | ±.13 | $\eta\pi$ | 16.3±1.5 | | 529 |
| | | | | | $K\bar{K}$ | 5.8±0.8 | | 428 |
| Controversy whether unsplit or split¶. | | | | | $\eta'(958)\pi$ | 1.1±1.1 | S=1.3* | 280 |
| E(1422) | $0^+($ $)+$ | 1422 | 69 | 2.02 | $K^*\bar{K} + \bar{K}^*K$ | 50±10 | | 154 |
| | | ±4 | ±8 | ±.10 | $\pi_N(975)\pi$ | 50±10 | | 357 |
| ‡ | $J^P = 0^-$ or $1^+$ | | | | $\pi\pi\eta$ | < 60 | | 568 |
| f'(1514) | $0^+(2^+)+$ | 1514 | 73 | 2.29 | $K\bar{K}$ | 72±12 | (k) | 570 |
| | | ±5 | ±23 | ±.11 | $K^*\bar{K} + \bar{K}^*K$ | 10±10 | | 294 |
| | | | S=1.8* | | $\pi\pi$ | < 14 | | 744 |
| | | | | | $\eta\pi\pi$ | 18±10 | | 624 |
| | | | | | $\eta\eta$ | < 40 | | 521 |
| $F_1(1540)$ | $\frac{1}{2}$ (A ) | 1540 | 40 | 2.37 | $K^*\bar{K} + \bar{K}^*K$ | Only mode seen | | 321 |
| | | ±5 | ±15 | ±.06 | | | | |
| Evidence based on only one experiment | | | | | | | | |
| $A_3(1640)$ | $1^-(2^-)+$ | ~ 1640 | | 2.69 | $f\pi$ | Dominant | | 306 |
| | | | | | $\omega\pi\pi$ (m) | | | 597 |
| Broad enhancement in the $J^P = 2^-$ $f\pi$ partial wave; not clear if resonant¶. | | | | | | | | |
| Φ(1675) | $0^-(N\ )-$ | 1664 | 141 | 2.77 | $\rho\pi$ | Dominant | | 647 |
| | | ±13 | ±17 | ±.23 | $3\pi$ | Possibly observed | | 804 |
| | | S=1.2* | | | $5\pi$ | 10±10 | | |
| g(1680) | $1^+(3^-)-$ | 1680§ | 160§ | 2.82 | $2\pi$ | ~ 40 | (ℓ) | 828 |
| | | ±20 | ±30 | ±.27 | $4\pi$ (incl. $\pi\pi\rho,\rho\rho,A_2\pi,\omega\pi$) | ~ 50 | | 781 |
| | | | | | $K\bar{K}$ | ~ 3 | | 677 |
| $J^P$, M and Γ from the $2\pi$ mode(ℓ). | | | | | $K\bar{K}\pi$ (incl. $K^*\bar{K}$) | ~ 3 | | 617 |
| ‡ S(1930) | $1^+($ $)-$ | ~ 1930 | 30§ | 3.72 | $\bar{p}p$ | Possibly seen | | 226 |
| | | | ±20 | ±.06 | $\pi\pi$ | Possibly seen | | 955 |
| Seen in $\pi^-p \to (MM)^-p$ ; may be related to the structure seen in 460 MeV/c $\bar{p}p$ backwards scattering¶. | | | | | | | | |

→ See note (p) for other possible heavy states.

| Name | $I(J^P)$ | Mass M (MeV) | Full Width Γ (MeV) | $M^2$ $\pm\Gamma M$ $(GeV)^2$ | Mode | Fraction % | p or $p_{max}$ (MeV/c) |
|---|---|---|---|---|---|---|---|
| $K^+(494)$ | $1/2(0^-)$ | 493.84 | | 0.244 | See Stable Particle Table | | |
| $K^0(498)$ | | 497.79 | | 0.248 | | | |
| $K^*(892)$ | $1/2(1^-)$ | *891.7* | 50.1 | 0.795 | $K\pi$ | ≈ 100 | 288 |
| | | *±0.5* | ±1.1 | ±.045 | $K\pi\pi$ | < 0.2 | 216 |
| → | | (Charged mode; $m^0 - m^\pm = 6.1\pm1.5$ MeV) | | | | | |
| κ | $1/2(0^+)$ | $\delta_0^{1/2}$ is near 90°, with slow variation, in mass region 1200-1400 MeV. In addition, $\delta_0^{1/2}$ may be resonant at M ~ 890 MeV, Γ ~ 30 MeV. | | | | | |
| See note on $K\pi$ S wave¶. | | | | | | | |

# Meson Table *(cont'd)*

| Name (G\I: 0 1; − φ π; + η ρ) | $I^G(J^P)C_n$ ⊢⊣ estab. | Mass M (MeV) | Full Width Γ (MeV) | $M^2 \pm \Gamma M$ (a) $(GeV)^2$ | Partial decay mode: Mode | Fraction % | p or $p_{max}$ (b) (MeV/c) |
|---|---|---|---|---|---|---|---|
| Q: $K_A(1240)$ or C | 1/2(1$^+$) | 1242 ±10 | 127 ±25 | 1.54 ±.16 | Kππ | Only mode seen | |
| seen in $\bar{p}p$ at rest | | | | | †[K$^*$π | Large] | |
| Q: $K_A$(1280 to 1400) | 1/2(1$^+$) | 1280 to 1400 | | | †[Kρ | Seen ] | |
| → → Resonance interpretation unclear (n). | | | | | | | |
| $K_N(1420)$ → → | 1/2(2$^+$) | *1421* ±5 § | 100 ±10 § | 2.02 ±.14 | Kπ | Inequalities explained in note (o). < 56.3±3.0 | 616 |
| | | See note (o). | | | K$^*$π | > 27.8±2.7 | 415 |
| | | | | | Kρ | > 9.5±2.5 | 324 |
| | | | | | Kω | > 4.5±1.7 | 304 |
| | | | | | Kη | > 2.0±2.0 | 482 |
| L(1770) ‡ | 1/2(A) | 1763 ±10 § | $100^{+100}_{-50}$ § | 3.11 ±.18 | Kππ | Dominant | 787 |
| | $J^P=2^-$ favoured, $1^+$ and $3^+$ not excluded. | | | | Kπππ | Seen | 756 |
| | | | | | †[$K_N(1420)\pi$ and other subreactions¶] | | |

→ Data on the following candidates, excluded above, are listed among the data cards¶ :

M(953), H(990), $\eta_N(1080)$, $A_{1.5}(1170)$, $X_{I=0}(1430)$, $X_{I=1}(1440)$, $X^-(1795)$, η/ρ(1830), φ/π(1830), ρ(2100), T(2200), ρ(2275), $N\bar{N}(2350)$, U(2375), $N\bar{N}(2375)$, $X^-(2500)$, $X^-(2620)$, X(2800), $X^-(2880)$, X(3030), $X^-(3075)$, $X^-(3145)$, $X^-(3475)$, $X^-(3535)$; $K_A^{I=3/2}(1175)$, $K_A^{I=3/2}(1265)$, $K_N(1370)$, $K_N(1660)$, $K_N(1760)$, $K_N(1850)$, $K^*(2200)$, $K^*(2800)$.

¶ See Meson Data Card Listings.

* Quoted error includes scale factor $S = \sqrt{\chi^2/(N-1)}$. See footnote to Stable Particle Table.

† Square brackets indicate a subreaction of the previous (unbracketed) decay mode(s).

§ This is only an educated guess; the error given is larger than the error of the average of the published values. (See Meson Data Card Listings for the latter.)

(a) ΓM is approximately the half-width of the resonance when plotted against $M^2$.

(b) For decay modes into ≥ 3 particles, $p_{max}$ is the maximum momentum that any of the particles in the final state can have. The momenta have been calculated by using the averaged central mass values, without taking into account the widths of the resonances.

(c) The values given for M(ρ) and Γ(ρ) and their errors are not average values from various experiments, but rather are intended to give the range where we believe the actual values are most likely to fall. Contrast the results tabulated in this note (references in the Meson Data Card Listings).

| | M(MeV) | Γ(MeV) | |
|---|---|---|---|
| $\rho^0$ | 775±7 | 149±23 | From $e^+e^- \to \pi^+\pi^-$, fitted to Gounaris-Sakurai formula. |
| $\rho^0$ | 768±10 | 140±14 | |
| $\rho^-$ | 764±2 | 147±4 | From physical region fits to $\pi N \to \pi\pi N$, using energy-dep. width. |
| $\rho^0$ | 775±3 | 145±9 | |
| $\rho^0$ | 768±2 | 132±13 | |
| $\rho^0_0$ | 759±7 | 119±20 | From pole extrapol. in → ππN |
| $\rho^-$ | 760 | 131 | |

(d) The $e^+e^-$ branching ratio is from $e^+e^- \to \pi^+\pi^-$ experiments only. The ωρ interference is then due to ωρ mixing only, and is expected to be small. See note in Meson Data Card Listings. The $\mu^+\mu^-$ branching ratio is compiled from 3 experiments; each possibly with substantial ωρ interference. The error reflects this uncertainty; see notes in Meson Data Card Listings. If eμ universality holds, $\Gamma(\rho^0 \to \mu^+\mu^-) = \Gamma(\rho^0 \to e^+e^-) \times$ phase space correction.

(e) Empirical limits on fractions for other decay modes of ρ(765) are $\pi^\pm\gamma$ < 0.5%, $\pi^\pm\eta$ < 0.8%, $\pi^+\pi^+\pi^-\pi^-$ < 0.15%, $\pi^\pm\pi^+\pi^-\pi^0$ < 0.2%.

(f) Empirical limits on fractions for other decay modes of ω(784) are $\pi^+\pi^-\gamma$ < 5%, $\pi^0\pi^0\gamma$ < 1%, η + neutral(s) < 1.5%, $\mu^+\mu^-$ < 0.02%, $\pi^0\mu^+\mu^-$ < 0.2%.

(g) See Meson Data Card Listings for a typed note and an entry "π(950-1020)", which contains the data referred to as δ(962), $\pi_N(975)$, and $\pi_N(1016)$ in our April 1971 edition.

(h) Empirical limits on fractions for other decay modes of η'(958): $\pi^+\pi^-$ < 2%, $\pi^+\pi^-\pi^0$ < 5%, $\pi^+\pi^+\pi^-\pi^-$ < 1%, $\pi^+\pi^+\pi^-\pi^-\pi^0$ < 1%, 6π < 1%, $\pi^+\pi^-e^+e^-$ < 0.6%, $\pi^0e^+e^-$ < 1.3%, $\eta e^+e^-$ < 1.1%, $\pi^0\rho^0$ < 4%, $\pi^0\omega$ < 8%.

# Meson Table *(cont'd)*

(i) Empirical limits on fractions for other decay modes of $\phi(1019)$ are $\pi^+\pi^- < 0.03\%$, $\pi^+\pi^-\gamma < 4\%$, $\omega\gamma < 5\%$, $\rho\gamma < 2\%$, $\pi^0\gamma < 0.35\%$.

(j) Empirical limits on fractions for decay modes of B(1235): $\pi\pi < 30\%$, $K\bar{K} < 2\%$, $4\pi < 50\%$, $\phi\pi < 1.5\%$, $\eta\pi < 25\%$, $(K\bar{K})^{\pm}\pi^0 < 8\%$, $K_S K_S \pi^{\pm} < 2\%$, $K_S K_L \pi^{\pm} < 6\%$.

(k) There is only a weak indication for a $K^*\bar{K} + \bar{K}^*K$ mode of the f'(1514). If this mode does not exist, the $K\bar{K}$ branching fraction will have to be reported as 80 ± 13% (rather than 72 ± 12% as given in the table), and $\eta\pi\pi$ as 20 ± 13%.

(ℓ) We assume as a working hypothesis that peaks with $I^G = 1^+$ observed around 1.7 GeV all come from g(1680). For indications to the contrary see Meson Data Card Listings.

(m) A possible $\omega\pi\pi$ decay mode of the A3 has mass 1690 MeV and width 80 MeV.

(n) See Q-region note in Meson Data Card Listings. Some investigators see a broad enhancement in mass $(K\pi\pi)$ from 1250-1400 MeV (the Q region), and others see structure. Only the $K_A(1240)$ or C seems well established, whereas possible structures from 1280 to 1400 MeV cannot be disentangled. For the whole Q region the decay rate into $K^*(892)\pi$ is large, and a $K\rho$ decay is seen. The $K\eta$, $K\omega$, and $K\pi$ are less than a few percent.

(o) $K_N(1420)$ properties are uncertain because both principal modes have energy-dependent backgrounds:- $K\pi$ mode: Firestone et al. (LBL 516, subm. Phys. Rev. 1972) find a large S-wave phase shift with $\sin^2\delta_0$ peaking at ≈1350 MeV, which probably caused older experiments to overestimate both Γ and the $K\pi$ branching fraction. Instead of our average of 56%, $K\pi$ fraction could be 40-50%, with other fractions raised accordingly. $K\pi\pi$ mode is contaminated with diffractively produced $Q^{\pm}$.
The tabulated mass of 1421 MeV comes only from charged $K_N(1420) \to K\pi$ measurements (to avoid $Q^{\pm}$ contamination); the average of the neutral $K\bar{N}(1420)$ mass is also 1420 MeV (i.e., $m^0 - m^{\pm} \approx 0$) but see typed note "K*(892) Mass" in Meson Data Card Listings.

(p) We tabulate here Y = 0 bumps with M ≥ 1700 MeV, for which no satisfactory grouping into particles is yet possible. See Meson Data Card Listings.

| Name | $I^G$ | $J^P$ | M (MeV) | Γ (MeV) | Decay modes observed | Tentative grouping |
|---|---|---|---|---|---|---|
| $K\bar{K}$(1740) | 1 | | 1740 | ≈ 120 | $K^0K^{\pm}$ | R(1750) |
| R3(1750) | 1,2 | | 1748 ± 15 | ≤ 38 | $(MM)^-$ | |
| ππ(1764) | | | 1764 ± 15 | $87^{+14}_{-20}$ | $\pi^+\pi^-$ | |
| $K\bar{K}\pi$(1820) | 0,1,2 | | 1820 ± 12 | 50 ± 23 | $K_SK^0\pi^0$ | 1830 region |
| R4(1830) | 1,2 | | 1830 ± 15 | ≤ 30 | $(MM)^-$ | |
| η/ρ(1830) | + | | 1832 ± 6 | 42 ± 11 | $\pi^+\pi^-\pi^+\pi^-$ | |
| φ/π(1830) | - | | 1848 ± 11 | 67 ± 27 | $\omega\pi^+\pi^-$, possibly $\omega\rho^0$ | |
| $X^-$(2086) | 1,2 | | 2086 ± 38 | ≈ 150 | $(MM)^-$ backward | ρ(2100) |
| ρ (2120) | $1^+$ | $3^-$(?) | 2120 | < 249 | $\pi^+\pi^-$, $\bar{p}p$ | |
| ππ(2157) | $1^+$ | (odd)- | 2157 ± 10 | 68 ± 22 | $\pi^+\pi^0$ | T region Seems to require >1 resonance |
| $K\bar{K}\omega$(2176) | $0^-,1^+$ | | 2176 ± 5 | $20^{+16}_{-2}$ | $K_SK_S\omega$ | |
| $N\bar{N}$(2190) | $1^-$ | | 2190 | 20-80 | $\rho^0\rho^0\pi^0$, $\bar{p}p$ | |
| | 1 | | 2190 ± 10 | ≈ 85 | Structure in $N\bar{N}$ total σ | |
| T (2195) | 1,2 | | 2195 ± 15 | ≤ 13 | $(MM)^-$ | |
| 3π(2207) | ≤ $3^-$ | | 2207 ± 13 | 62 ± 52 | $\pi^+\pi^-\pi^0$ | |
| 4π(2207) | $1^+,2^+,3^+$ | | 2207 ± 22 | ≈ 130 | $\rho^-\pi^+\pi^-$ | |
| $X^-$(2260) | 1,2 | | 2260 ± 18 | ≤ 25 | $(MM)^-$ backward | ρ(2275) |
| ρ (2290) | $1^+$ | $5^-$(?) | 2290 | < 165 | $\pi^+\pi^-$, $\bar{p}p$ | |
| $N\bar{N}$(2350) | 1 | | 2350 ± 10 | ≈ 140 | Structure in $N\bar{N}$ total σ | U region |
| $X^-$(2370) | 1,2 | | 2370 ± 17 | ≈ 57 | $(MM)^-$ backward | |
| $N\bar{N}$(2375) | 0 | | 2375 ± 10 | ≈ 190 | Structure in $N\bar{N}$ total σ | |
| U (2380) | 1,2 | | 2382 ± 24 | < 30 | $(MM)^-$ | |
| $X^-$(2500) | 1,2 | | 2500 ± 32 | ≈ 87 | $(MM)^-$ backward | |
| $X^-$(2620) | 1,2 | | 2620 ± 20 | 85 ± 30 | $(MM)^-$ | 2650 region |
| 4π(2676) | (1,2,3)+ | | 2676 ± 27 | ≈ 150 | $\rho^-\pi^+\pi^-$ | |
| $X^-$(2800) | 1,2 | | 2800 ± 20 | 46 ± 10 | $(MM)^-$ | 2850 region |
| $X^+$(2820) | 1,2 | | 2820 ± 10 | 50 ± 10 | $(K\bar{K}\pi\pi)+$ | |
| $X^-$(2880) | 1,2 | | 2880 ± 20 | ≤ 15 | $(MM)^-$ | |
| $X^+$(3013) | $1^-$ | | 3013 ± 5 | < 40 | 7π | 3020 region |
| $X^-$(3025) | 1,2 | | 3025 ± 20 | ≈ 25 | $(MM)^-$ | |
| NN(3035) | + | | 3035 ± 25 | 200 ± 60 | 4π, 6π | |
| $X^-$(3075) | 1,2 | | 3075 ± 20 | ≈ 25 | $(MM)^-$ | |
| $X^-$(3145) | 1,2 | | 3145 ± 20 | ≤ 10 | $(MM)^-$ | |
| $X^-$(3475) | 1,2 | | 3475 ± 20 | ≈ 30 | $(MM)^-$ | |
| $X^-$(3535) | 1,2 | | 3535 ± 20 | ≈ 30 | $(MM)^-$ | |

# Baryon Table

April 1972

[See notes on N's and Δ's, on possible Z*'s, and on Y*'s at the beginning of those sections in the Baryon Data Card Listings; also see notes on individual resonances in the Baryon Data Card Listings.]

| Particle[a] | I $(J^P)$ ⊢⊣ estab. | π or K Beam T(GeV) p(GeV/c) $\sigma = 4\pi\lambda\!\!\!^{-2}$ (mb) | Mass $M^b$ (MeV) | Full Width $\Gamma^b$ (MeV) | $M^2$ $\pm\Gamma M^c$ $(GeV^2)$ | Partial decay mode: Mode | Fraction % | p or $p_{max}$[d] (MeV/c) |
|---|---|---|---|---|---|---|---|---|
| p<br>n | 1/2(1/2$^+$) | | 938.3<br>939.6 | | 0.880<br>0.883 | See Stable Particle Table | | |
| N'(1470) | 1/2(1/2$^+$) $P'_{11}$ | T=0.53πp<br>p=0.66<br>σ=27.8 | 1435 to<br>1505 | 165 to<br>400 | 2.16<br>±0.34 | Nπ<br>Nππ | 60<br>40 | 420<br>368 |
| N'(1520) | 1/2(3/2$^-$) $D'_{13}$ | T=0.61<br>p=0.74<br>σ=23.5 | 1510 to<br>1540 | 105 to<br>150 | 2.31<br>±0.18 | Nπ<br>Nππ<br>[Δ(1236)π][e]<br>Nη | 50<br>50<br>[dominant][e]<br>~0.6 | 456<br>410<br>224<br> |
| N'(1535) | 1/2(1/2$^-$) $S'_{11}$ | T=0.64<br>p=0.76<br>σ=22.5 | 1500 to<br>1600 | 50 to<br>160 | 2.36<br>±0.18 | Nπ<br>Nη[n]<br>Nππ[n] | 35<br>55<br>~10 | 467<br>182<br>422 |
| N'(1670)[i] | 1/2(5/2$^-$) $D'_{15}$ | T=0.87<br>p=1.00<br>σ=15.6 | 1655 to<br>1680 | 105 to<br>175 | 2.79<br>±0.24 | Nπ<br>Nππ<br>[Δ(1236)π][e]<br>ΛK<br>Nη | 40<br>60<br>[44][e]<br>< .3<br>< 1[j] | 560<br>525<br>357<br>200<br>368 |
| N'(1688)[i] | 1/2(5/2$^+$) $F'_{15}$ | T=0.90<br>p=1.03<br>σ=14.9 | 1680 to<br>1692 | 105 to<br>180 | 2.85<br>±0.21 | Nπ<br>Nππ<br>[Δ(1236)+π][e]<br>ΛK<br>Nη | 60<br>40<br>[26][e]<br>< .2<br>< .5[j] | 572<br>538<br>371<br>231<br>388 |
| N''(1700)[i] | 1/2(1/2$^-$) $S''_{11}$ | T=0.92<br>p=1.05<br>σ=14.3 | 1665 to<br>1765 | 100 to<br>400 | 2.89<br>±0.42 | Nπ<br>ΛK<br>Nη | ~65<br>5<br> | 580<br>250<br>340 |
| N''(1780)[i] | 1/2(1/2$^+$) $P''_{11}$ | T=1.07<br>p=1.20<br>σ=12.2 | 1650 to<br>1860 | 50 to<br>450 | 3.17<br>±0.51 | Nπ<br>ΛK<br>Nη | 30<br>~ 7<br>~10[j] | 633<br>353<br>476 |
| N(1860) | 1/2(3/2$^+$) $P_{13}$ | T=1.22<br>p=1.36<br>σ=10.4 | 1770 to<br>1900 | 180 to<br>330 | 3.46<br>±0.57 | Nπ<br>Nππ<br>ΛK<br>Nη | 25<br><br>~ 5<br>~ 4[j] | 685<br>657<br>437<br>545 |
| N(2190) | 1/2(7/2$^-$) $G_{17}$ | T=1.94<br>p=2.07<br>σ=6.21 | 2000 to<br>2260 | 270 to<br>325 | 4.80<br>±0.67 | Nπ<br>Nππ | 25 | 888<br>868 |
| N(2220) | 1/2(9/2$^+$) $H_{19}$ | T=2.00<br>p=2.14<br>σ=5.97 | 2200 to<br>2245 | 260 to<br>330 | 4.93<br>±0.65 | Nπ<br>Nππ | 15 | 905<br>887 |
| N(2650) | 1/2( ?$^-$) | T=3.12<br>p=3.26<br>σ=3.67 | 2650 | 360 | 7.02<br>±0.95 | Nπ<br>Nππ | (J+1/2)x<br>=0.45[f] | 1154<br>1140 |
| N(3030) | 1/2( ? ) | T=4.27<br>p=4.41<br>σ=2.62 | 3030 | 400 | 9.18<br>±1.21 | Nπ<br>Nππ | (J+1/2)x<br>=0.05[f] | 1366<br>1354 |
| Δ'(1236)[m] | 3/2(3/2$^+$) $P'_{33}$ | T=0.195 (++)<br>p=0.304<br>σ=91.8 | 1230 to<br>1236 | 110 to<br>122 | 1.53<br>±0.14 | Nπ<br>Nπ$^+$π$^-$<br>Nγ | 99.4<br>0<br>~0.6 | 231<br>90<br>262 |
| | | Pole Position[m]: | 1211 | ± i50 | | | | |
| Δ(1650) | 3/2(1/2$^-$) $S_{31}$ | T=0.83<br>p=0.96<br>σ=16.4 | 1615 to<br>1695 | 130 to<br>200 | 2.72<br>±0.28 | Nπ<br>Nππ | 28<br>72 | 547<br>511 |

# Baryon Table *(cont'd)*

| Particle[a] | I ($J^P$) estab. | π or K Beam T(GeV) p(GeV/c) $\sigma = 4\pi\lambda\!\!\!^{-2}$ (mb) | Mass $M^b$ (MeV) | Full Width $\Gamma^b$ (MeV) | $M^2$ $\pm\Gamma M^c$ ($GeV^2$) | Partial decay mode: Mode | Fraction % | p or $p_{max}$[d] (MeV/c) |
|---|---|---|---|---|---|---|---|---|
| Δ(1670) | 3/2(3/2⁻) $D_{33}$ | T=0.87<br>p=1.00<br>σ=15.6 | 1650 to 1720 | 175 to 300 | 2.79<br>±0.40 | Nπ<br>Nππ | 15 | 560<br>525 |
| Δ(1890) | 3/2(5/2⁺) $F_{35}$ | T=1.28<br>p=1.42<br>σ=9.88 | 1840 to 1920 | 135 to 350 | 3.57<br>±0.49 | Nπ<br>Nππ | 17 | 704<br>677 |
| Δ(1910) | 3/2(1/2⁺) $P_{31}$ | T=1.33<br>p=1.46<br>σ=9.54 | 1780 to 1935 | 230 to 420 | 3.65<br>±0.62 | Nπ<br>Nππ | 25 | 716<br>691 |
| Δ(1950) | 3/2(7/2⁺) $F_{37}$ | T=1.41<br>p=1.54<br>σ=8.90 | 1930 to 1980 | 140 to 220 | 3.80<br>±0.39 | Nπ<br>Δ(1236)π<br>ΣK<br>Σ(1385)K | 45<br>≈50<br>~ 2<br>1.4 | 741<br>571<br>460<br>232 |
| Δ(2420) | 3/2(11/2⁺) | T=2.50<br>p=2.64<br>σ=4.68 | 2320 to 2450 | 270 to 350 | 5.86<br>±0.75 | Nπ<br>Nππ | 11<br>>20 | 1023<br>1006 |
| Δ(2850) | 3/2( ?⁺ ) | T=3.71<br>p=3.85<br>σ=3.05 | 2850 | 400 | 8.12<br>±1.14 | Nπ<br>Nππ | (J+1/2)x<br>=0.25[f] | 1266<br>1254 |
| Δ(3230) | 3/2( ? ) | T=4.94<br>p=5.08<br>σ=2.25 | 3230 | 440 | 10.4<br>±1.4 | Nπ<br>Nππ | (J+1/2)x<br>=0.05[f] | 1475<br>1464 |
| Z* | Evidence for states with hypercharge 2 is controversial. See the Baryon Data Card Listings for discussion and display of data. | | | | | | | |
| Λ | 0(1/2⁺) | | 1115.6 | | 1.24 | See Stable Particle Table | | |
| Λ'(1405) | 0(1/2⁻) $S'_{01}$ | p<0 K⁻p | 1405<br>±5[g] | 40<br>±10[g] | 1.97<br>±0.06 | Σπ | 100 | 142 |
| Λ'(1520) | 0(3/2⁻) $D'_{03}$ | p=0.389<br>σ=84.5 | 1518<br>±2[g] | 16<br>±2[g] | 2.30<br>±0.02 | $N\bar{K}$<br>Σπ<br>Λππ<br>Σππ | 45±1<br>42±1<br>9.6±.7<br>1.0±.1 | 234<br>258<br>250<br>140 |
| Λ''(1670) | 0(1/2⁻) $S''_{01}$ | p=0.74<br>σ=28.5 | 1670 | 15 to 38 | 2.79<br>±0.04 | $N\bar{K}$<br>Λη<br>Σπ | ~20<br>~35<br>~45 | 410<br>64<br>393 |
| Λ''(1690) | 0(3/2⁻) $D''_{03}$ | p=0.78<br>σ=26.1 | 1690 | 27 to 85 | 2.86<br>0.09 | $N\bar{K}$<br>Σπ<br>Λππ<br>Σππ | ~20[h]<br>~60<br>~ 2<br>≤18 | 429<br>409<br>415<br>352 |
| Λ'(1815) | 0(5/2⁺) $F'_{05}$ | p=1.05<br>σ=16.7 | 1820<br>±5[g] | 64 to 104 | 3.30<br>±0.15 | $N\bar{K}$<br>Σπ<br>Σ(1385)π | 62<br>11<br>4 | 542<br>508<br>362 |
| Λ'(1830) | 0(5/2⁻) $D'_{05}$ | p=1.09<br>σ=15.8 | 1835 | 74 to 150 | 3.37<br>±0.20 | $N\bar{K}$<br>Σπ<br>Λππ | ~10<br>~30<br>~11 | 554<br>519<br>536 |
| Λ(2100) | 0(7/2⁻) $G_{07}$ | p=1.68<br>σ=8.68 | 2100 | 60 to 140 | 4.41<br>±0.22 | $N\bar{K}$<br>Σπ<br>Λη<br>ΞK<br>Λω | 25<br>~ 5<br>< 3<br><br><10 | 748<br>699<br>617<br>483<br>443 |
| Λ(2350) | 0( ? ) | p=2.29<br>σ=5.85 | 2350 | 140 to 324 | 5.52<br>±0.55 | $N\bar{K}$ | (J+1/2)x<br>=0.7[f] | 913 |

# Baryon Table *(cont'd)*

| Particle[a] | I ($J^P$) ⊢⊣ estab. | $\pi$ or K Beam<br>T(GeV)<br>p(GeV/c)<br>$\sigma = 4\pi\lambda\!\!\!^{-2}$ (mb) | Mass<br>$M^b$<br>(MeV) | Full Width<br>$\Gamma^b$<br>(MeV) | $M^2$<br>$\pm\Gamma M^c$<br>($GeV^2$) | Partial decay mode: Mode | Fraction % | p or $p_{max}$[d] (MeV/c) |
|---|---|---|---|---|---|---|---|---|
| $\Sigma$ | $1(1/2^+)$ | | (+)1189.4<br>(0)1192.5<br>(-)1197.3 | | 1.41<br>1.42<br>1.43 | See Stable Particle Table | | |
| $\Sigma'(1385)$ | $1(3/2^+)P'_{13}$ | p< 0 $K^-p$ | (+)1383±1<br>S=1.3*<br>(-)1386±2<br>S=2.2* | (+)36±3<br>S=1.9*<br>(-)36±6<br>S=3.5*! | 1.92<br>±0.05 | $\Lambda\pi$<br>$\Sigma\pi$ | 89±5<br>11±5<br>S=1.9* | 208<br>117 |
| $\Sigma'(1670)$[k] | $1(3/2^-)$ $D'_{13}$ | p=0.74<br>$\sigma$ -28.5 | 1670 | 50 | 2.79<br>±0.08 | $N\bar{K}$<br>$\Sigma\pi$<br>$\Lambda\pi$<br>$\Sigma\pi\pi$<br>$[\Lambda(1405)\pi]^e$<br>$\Lambda\pi\pi$ | ~8 | 410<br>387<br>447<br>326<br>207<br>397 |
| | Mass, width and elasticity are the values obtained in partial wave analyses for a $D_{13}$ resonance. For more results see the Baryon Data Card Listings and footnote k. | | | | | | | |
| $\Sigma''(1750)$ | $1(1/2^-)$ $S''_{11}$ | p=0.91<br>$\sigma$=20.7 | 1750 | 50 to<br>80 | 3.06<br>±0.11 | $N\bar{K}$<br>$\Lambda\pi$<br>$\Sigma\eta$ | ~15<br>seen<br>seen | 483<br>507<br>54 |
| $\Sigma(1765)$ | $1(5/2^-)$ $D_{15}$ | p=0.94<br>$\sigma$=19.6 | 1765<br>±5[g] | ~120 | 3.12<br>±0.21 | $N\bar{K}$<br>$\Lambda\pi$<br>$\Lambda(1520)\pi$<br>$\Sigma(1385)\pi$<br>$\Sigma\pi$ | ~42<br>~15<br>~14<br>~ 4<br>~1 | 496<br>518<br>187<br>315<br>461 |
| $\Sigma'(1915)$[i] | $1(5/2^+)$ $F'_{15}$ | p=1.25<br>$\sigma$=13.0 | 1910 | 70 | 3.65<br>±0.13 | $N\bar{K}$<br>$\Lambda\pi$<br>$\Sigma\pi$ | ~11 | 612<br>619<br>568 |
| Formation and production experiments do not agree on $\Sigma\pi/\Lambda\pi$ ratio. | | | | | | | | |
| $\Sigma(2030)$ | $1(7/2^+)$ $F_{17}$ | p=1.52<br>$\sigma$=9.93 | 2030 | 100 to<br>170 | 4.12<br>±0.27 | $N\bar{K}$<br>$\Lambda\pi$<br>$\Sigma\pi$<br>$\Xi K$ | ~ 20<br>~ 20<br>~ 3<br>< 2 | 700<br>700<br>652<br>412 |
| $\Sigma(2250)$ | 1( ? ) | p=2.04<br>$\sigma$=6.76 | 2250 | 100 to<br>230 | 5.06<br>±0.37 | $N\bar{K}$<br>$\Sigma\pi$<br>$\Lambda\pi$ | (J+1/2)x<br>=0.3[f] | 849<br>842<br>799 |
| $\Sigma(2455)$ | 1( ? ) | p=2.57<br>$\sigma$=5.09 | 2455 | ~120 | 6.03<br>±0.29 | $N\bar{K}$ | (J+1/2)x<br>=0.2[f] | 979 |
| $\Sigma(2620)$ | 1( ? ) | p=2.95<br>$\sigma$=4.30 | 2620 | ~175 | 6.86<br>±0.46 | $N\bar{K}$ | (J+1/2)x<br>=0.3[f] | 1064 |
| $\Xi$[l] | $1/2(1/2^+)$ | | (0)1314.7<br>(-)1321.3 | | 1.73<br>1.75 | See Stable Particle Table | | |
| $\Xi(1530)$[l] | $1/2(3/2^+)$<br>p-wave | | (0) 1531.3±0.5<br>S=1.4*<br>(-) 1535.8±1.0 | (0) 9.2±0.8<br>(-) 16.2±4.6 | 2.34<br>±0.01 | $\Xi\pi$ | 100 | 144 |
| $\Xi(1820)$[l] | 1/2( ? ) | | 1795 to<br>1870 | 12 to<br>99 | 3.31<br>±0.10 | $\Lambda\bar{K}$<br>$\Xi\pi$<br>$\Xi(1530)\pi$<br>$\Sigma\bar{K}$ | | 396<br>413<br>234<br>306 |
| | All four decay modes have been seen. Branching ratios not quoted because there may be more than one state here. | | | | | | | |
| $\Xi(1940)$[l] | 1/2( ? ) | | 1894 to<br>1961 | 42 to<br>140 | 3.72<br>±0.18 | $\Xi\pi$<br>$\Xi(1530)\pi$ | | 499<br>336 |
| | Seen in both final states; not clear if one, or more, states present. | | | | | | | |
| $\Omega^-$ | $0(3/2^+)$ | | 1672.5 | | 2.80 | See Stable Particle Table | | |

# Baryon Table *(cont'd)*

* Quoted error includes an S(scale)factor. See footnote to Stable Particle Table.

→ An arrow at the left of the Table indicates a candidate that has been omitted because the evidence for the existence of the effect and (or) for its interpretation as a resonance is open to considerable question. See the Baryon Data Card Listings for information on the following: N(1700) $D''_{13}$, N(1990) $F_{17}$, N(2040) $D'''_{13}$, N(2100) $S'''_{11}$, N(2100) $D''_{15}$, N(2175) $F''_{15}$, N(3245), N(3690), N(3755), Δ(1690) $P''_{33}$, Δ(1960) $D_{35}$, Δ(2160) $P'''_{33}$, $Z_0$(1780), $Z_0$(1865), $Z_1$(1900), Λ(1330), Λ(1750) $P_{01}$, Λ(1860) $P_{03}$, Λ(1870) $S'''_{01}$, Λ(2010) $D'''_{03}$, Λ(2020) $F_{07}$, Λ(2110) $F''_{05}$ or $D''_{05}$, Λ(2585), Σ(1440), Σ(1480), Σ(1620) $S'_{11}$, Σ(1620) $P'_{11}$ , $\Sigma(1670)^k$, Σ(1690), Σ(1880) $P''_{11}$, Σ(1940) $D''_{13}$, Σ(2070) $F''_{15}$, Σ(2080) $P''_{13}$, Σ(2100) $G_{17}$, Σ(3000), Ξ(1630), Ξ(2030), Ξ(2250), Ξ(2500).

a. For the baryon states, the name [such as N'(1470)] contains the mass, which may be different for each new analysis. The value chosen is the rounded average from Table I of the note on N's and Δ's in the Baryon Data Card Listings. For $Y^*$'s and $\Xi^*$'s, the mass is an educated guess obtained by looking at the reported values. The convention for using primes in the names is as follows: when there is more than one resonance on a given Argand diagram, the first has been designated with a prime, the second with a double prime, etc. The name (col. 1) is the same as can be found in large print in the Baryon Data Card Listings.

b. See note on N's and Δ's in the Baryon Data Card Listings. For M and Γ of most baryons we report here an interval instead of an average. Averages are appropriate if each result is based on independent measurements, but inappropriate here where the spread in parameters arises because different models or procedures have been applied to a common set of data. Where only one value is given it is either because only one experiment reports that state or because the various experiments agree. An error is quoted only when the various experiments averaged have taken into account the systematic errors.

c. For this column M is the rounded average which also appears in the name column. For the N's and Δ's, Γ is the average quoted on Table II of the N's and Δ's note in the Baryon Data Card Listings; for the $Y^*$'s and $\Xi^*$'s, Γ is taken as the center of the interval given in the column labeled "Γ".

d. For decay modes into ≥3 particles $p_{max}$ is the maximum momentum that any of the particles in the final state can have. The momenta have been calculated using the averaged central mass values, without taking into account the widths of the resonances.

e. Square brackets indicate a sub-reaction of the previous unbracketed decay mode.

f. This state has been seen only in total cross sections. J is not known; x is $\Gamma_{el}/\Gamma$.

g. This is only an educated guess; the error given is larger than the error of the average of the published values (see the Baryon Data Card Listings for the latter).

h. In previous editions we quoted a larger elasticity. It was required by unitarity because of the large value of $x \cdot x_e$ for the $\Lambda\pi\pi$ decay. A partial wave analysis of new data of the $\Lambda\pi\pi$ channel yields smaller values of $x \cdot x_e$ allowing a smaller elasticity, which is more consistent with partial-wave analysis in the elastic channel.

i. Only information coming from partial-wave analyses has been used here. For the production experiments results see the Baryon Data Card Listings.

j. Value obtained in an energy-dependent partial-wave analysis which uses a t-channel-poles-plus-resonance parametrization. The values of the couplings obtained for the resonances may be affected by double counting.

k. In this energy region the situation is still confused. Formation experiments suggest two states: $P_{11}$(1620) decaying mainly into Σπ, and $D_{13}$(1670) with branching fractions Σπ(40%), Λπ(10%), Σππ(< 14%). Production experiments report four states: Σ(1620) seen only in the Λπ mode, $\Sigma_1$(1660) with appreciable Λπ and Σπ modes. $\Sigma_2$(1660) with main decay mode Λ(1405) + π (that is, Σππ), and Σ(1690) seen in the Λπ mode. Of these four, $\Sigma_1$ and $\Sigma_2$ seem to be on firmer ground than the other two and both seem to have $J^P = 3/2^-$ like the $D_{13}$(1670) seen in formation experiments. Two resonances of the same spin and parity have been hypothesized as the origin of much of the complexity observed in production experiments. With the addition of the $P_{11}$(1620), there are three candidates that eventually might be required to clarify the situation.

ℓ. Only Ξ(1530) is firmly established; information on the other states comes from experiments that have poor statistics due to the fact that the cross sections for S = -2 states are very low. For Ξ states, because of the meager statistics, we lower our standards and tabulate resonant effects if they have at least a four-standard-deviation statistical significance and if they are seen by more than one group. So Ξ(2030), with main decay mode $\Sigma\bar{K}$, reported as a 3.5-standard-deviation effect, is not tabulated. See the Baryon Data Card Listings for the other states.

m. See note on Δ(1236) in the Baryon Data Card Listings. Values of mass and width are dependent upon resonance shape used to fit the data. The pole position appears to be much less dependent upon the parametrization used.

n. The preliminary results of DIEM 70 quoted in the Baryon Data Card Listings have been revised so that they are now in agreement with the values quoted in the present table (G. Smadja, private communication).

13

# PHYSICAL AND NUMERICAL CONSTANTS*

## PHYSICAL CONSTANTS

| | |
|---|---|
| $N$ | $= 6.022169(40)\times10^{23}$ mole$^{-1}$ (based on $A_{C^{12}} = 12$) |
| $c$ | $= 2.9979250(10)\times10^{10}$ cm sec$^{-1}$ |
| $e$ | $= 4.803250(21)\times10^{-10}$ esu $= 1.6021917(70)\times10^{-19}$ coulomb |
| 1 MeV | $= 1.6021917(70)\times10^{-6}$ erg |
| $\hbar$ | $= 6.582183(22)\times10^{-22}$ MeV sec |
| | $= 1.0545919(80)\times10^{-27}$ erg sec |
| $\hbar c$ | $= 1.9732891(66)\times10^{-11}$ MeV cm $= 197.32891(66)$ MeV fermi |
| | $= 0.6240088(21)$ GeV mb$^{1/2}$ |
| $\alpha$ | $= e^2/\hbar c = 1/137.03602(21)$ |
| $k_{\text{Boltzmann}}$ | $= 1.380622(59)\times10^{-16}$ erg K$^{-1}$ |
| | $= 8.61708(37)\times10^{-11}$ MeV K$^{-1}$ $= 1$ eV$/11604.85(49)$K |
| $m_e$ | $= 0.5110041(16)$ MeV $= 9.109558(54)\times10^{-31}$ kg |
| $m_p$ | $= 938.2592(52)$ MeV $= 1836.109(11)\, m_e = 6.72211(63)\, m_{\pi^\pm}$ |
| | $= 1.00727661(8)\, m_1$ (where $m_1 = 1$ amu $= \frac{1}{12} m_{C^{12}} = 931.4812(52)$ MeV) |
| $r_e$ | $= e^2/m_e c^2 = 2.817939(13)$ fermi (1 fermi $= 10^{-13}$ cm) |
| $\lambda\!\!\!^{-}_e$ | $= \hbar/m_e c = r_e \alpha^{-1} = 3.861592(12)\times10^{-11}$ cm |
| $a_{\infty\,\text{Bohr}}$ | $= \hbar^2/m_e e^2 = r_e \alpha^{-2} = 0.52917715(81)$ A (1A $= 10^{-8}$ cm) |
| $\sigma_{\text{Thomson}}$ | $= \frac{8}{3}\pi r_e^2 = 0.6652453(61)\times10^{-24}$ cm$^2$ $= 0.6652453(61)$ barns |
| $\mu_{\text{Bohr}}$ | $= e\hbar/2m_e c = 0.5788381(18)\times10^{-14}$ MeV gauss$^{-1}$ |
| $\mu_{\text{nucleon}}$ | $= e\hbar/2m_p c = 3.152526(21)\times10^{-18}$ MeV gauss$^{-1}$ |
| $\frac{1}{2}\omega^e_{\text{cyclotron}}$ | $= e/2m_e c = 8.794014(27)\times10^{6}$ rad sec$^{-1}$ gauss$^{-1}$ |
| $\frac{1}{2}\omega^p_{\text{cyclotron}}$ | $= e/2m_p c = 4.789484(27)\times10^{3}$ rad sec$^{-1}$ gauss$^{-1}$ |

Hydrogen-like atom (nonrelativistic, $\mu$ = reduced mass):

$$\frac{v}{c}\Big)_{\text{rms}} = \frac{ze^2}{n\hbar c};\; E_n = \frac{\mu}{2}v^2 = \frac{\mu z^2 e^4}{2(n\hbar)^2};\; a_n = \frac{n^2\hbar^2}{\mu z e^2}$$

$R_\infty = m_e e^4/2\hbar^2 = m_e c^2 \alpha^2/2 = 13.605826(45)$ eV (Rydberg)

$pc = 0.3\, H\rho$(MeV, kilogauss, cm); 0.3 (which is $10^{-11}c$) enters because there are $\approx$ 300 "volts"/esu volt.

| | |
|---|---|
| 1 year (sidereal) | $= 365.256$ days $= 3.1558\times10^7$ sec ($\approx \pi\times10^7$ sec) |
| density of dry air | $= 1.205$ mg cm$^{-3}$ (at 20 °C, 760 mm) |
| acceleration by gravity | $= 980.62$ cm sec$^{-2}$ (sea level, 45°) |
| gravitational constant | $= 6.6732(31)\times10^{-8}$ cm$^3$ g$^{-1}$ sec$^{-2}$ |
| 1 calorie (thermochemical) | $= 4.184$ joules |
| 1 atmosphere | $= 1033.2275$ g cm$^{-2}$ |
| 1 eV per particle | $= 11604.85(49)$ °K (from $E = kT$) |

## NUMERICAL CONSTANTS

| | | | | | |
|---|---|---|---|---|---|
| $\pi$ | $= 3.1415927$ | 1 rad | $= 57.2957795$ deg | $\sqrt{\pi}$ | $= 1.7724539$ |
| $e$ | $= 2.7182818$ | $1/e$ | $= 0.3678794$ | $\sqrt{2}$ | $= 1.4142136$ |
| $\ln 2$ | $= 0.6931472$ | $\ln 10$ | $= 2.3025851$ | $\sqrt{3}$ | $= 1.7320508$ |
| $\log_{10} 2$ | $= 0.3010300$ | $\log_{10} e$ | $= 0.4342945$ | $\sqrt{10}$ | $= 3.1622777$ |

*Compiled by Stanley J. Brodsky, based mainly on the adjustment of the fundamental physical constants by B. N. Taylor, W. H. Parker, and D. N. Langenberg, Rev. Mod. Phys. 41, 375 (1969). The figures in parentheses correspond to the 1 standard deviation uncertainty in the last digits of the main number.

# INDEX